Hydrodynamics and Transport Processes of Inverse Bubbly Flow

Hydrodynamics and Transport Processes of Inverse Bubbly Flow

Subrata Kumar Majumder

Department of Chemical Engineering,
Indian Institute of Technology Guwahati,
Assam, India

Amsterdam • Boston • Heidelberg • London • New York • Oxford
Paris • San Diego • San Francisco • Singapore • Sydney • Tokyo

Elsevier
Radarweg 29, PO Box 211, 1000 AE Amsterdam, Netherlands
The Boulevard, Langford Lane, Kidlington, Oxford OX5 1GB, UK
50 Hampshire Street, 5th Floor, Cambridge, MA 02139, USA

British Library Cataloguing-in-Publication Data
A catalogue record for this book is available from the British Library

Library of Congress Cataloging-in-Publication Data
A catalog record for this book is available from the Library of Congress

ISBN: 978-0-12-803287-9

For information on all Elsevier publications
visit our website at https://www.elsevier.com/

Typeset by Thomson Digital

Transferred to Digital Printing in 2016

Dedication

This book is dedicated to my parents, Suniti Majumder (late) and Sudhamay Majumder.

Table of Contents

Preface

Recently, inverse bubble flow systems have been gaining interest for chemical processes, particularly when the interfacial mass transfer area is the rate-controlling step. Today many industries are facing interfacial challenges caused by complex phenomena of gas–liquid or gas–liquid–solid flow in multiphase reactors for different chemical and biochemical processes. A strong need exists for new and innovative concepts to achieve hydrodynamics and transport phenomena in multiphase processes. The inverse bubbly flow column in which the choice of suitable liquid and gas throughput rates permits gas residence times to be adjusted up to maximum gas content within certain limits presents a neat solution. The inverse flow bubble seems to be particularly advantageous when considerably higher gas residence time, intense mixing, better heat, and mass transfer are needed. However, the studies regarding the hydrodynamics, mixing characteristics, bubble size distribution, and specific interfacial area in the inverse bubble flow condition are scanty. Therefore, a precise knowledge of the hydrodynamics and transport phenomena in the inverse bubble flow would be of considerable interest in the industrial, scientific, and academic communities. This book describes the science and fundamentals behind hydrodynamic characteristics, including flow regimes, gas entrainment, pressure drop, hold-up characteristics mixing, bubble size distribution, and interfacial area of inverse bubble flow regimes. The book also describes heat and mass transfer processes in the inverse bubbly flow regime compared with the conventional bubbly flow regime.

This book aims to be useful for researchers in academia and industry working in chemical and biochemical engineering and intends to help facilitate a better understanding of the phenomena of multiphase flow systems as used in the chemical and biochemical industries.

With the continuous increase of archives of research articles on the multiphase flow system, it is difficult to present a treatise that includes all of the important research work to compare the results with the present inverse bubbly flow system. Although every effort has been made to include the most relevant available literature, I had to limit myself to journal publications as authentic research works. If there are any omissions, it is simply ignorance of the work on the part of the author, which will be corrected in the future.

Subrata Kumar Majumder

Acknowledgment

I would first like to acknowledge the editorial and production departments of Elsevier who were extremely cooperative in the endeavor. The Indian Institute of Technology Guwahati and Indian Institute of Technology Kharagpur are highly acknowledged for providing a sound infrastructure for research and for writing the book. I would like to thank Professor Gautam Kundu and Professor Dibyendu Mukherjee for their continuous support in different ways for developing my career and the need to be persistent to accomplish any goal. Without their encouragement and continuous support, I could not have produced this work. I would like to express my gratitude to the reviewers for their constructive suggestions and comments on writing the book. I am grateful to my former student Mr Mekala Sivaiah for his contribution to some parts of the book. I am also grateful to other friends or colleagues for their direct and indirect support whenever sought for.

The most appreciation goes to my wife, Aditi Majumder, who endured many missing weekends while I worked alone in the office.

Subrata Kumar Majumder

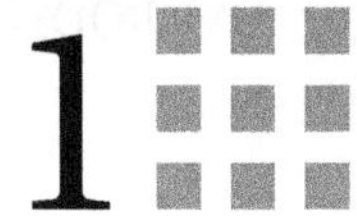

Introduction

Bubbly Flow

Many of the processes in the chemical industry involve the flow and contact of multiple phases. Fine dispersion of one phase into another phase is desirable to promote intense mixing between immiscible phases, to increase the interfacial area of contact, and to create high turbulence in the continuous phase in order to increase mass, momentum, and energy transfer. An example is gas-particle flow in a fluidized bed or the gas–liquid flow in a bubble column. According to Tatterson (1991), (25)% of all chemical reactions occur between gas and liquid. A major class of gas–liquid flows is the one in which the liquid phase is continuous and the gas phase is dispersed in the form of bubbles. The term *bubbly flow* is used to refer to a flow of a gas or a mixture of gas in a continuous liquid as a dispersed phase of bubbles. The bubbly flow behavior is affected by the interfacial tension forces, the wetting characteristics of the liquid on the channel wall, the contact angle and the exchange of mass, momentum, and the energy between the bubbles and liquid.

The bubbly flow patterns in a conduit depend on several factors:

- Dynamic variables (e.g., flow rate of phases, phase fractions)
- Geometric variables (e.g., diameter, length, shape, inclination of conduit, particle size, hole size of the phase distributor, bends, valves)
- Thermodynamic variables (e.g., pressure, temperature, adiabatic or diabatic condition).
- Physical properties of the phases (e.g., density, surface tension, viscosity)

A number of different efficient methods have been developed to contact the gas and liquid phases in the process industry. These contacting systems can be classified according to the flow of phases as cocurrent, countercurrent, and cross current system. Among these three systems, the countercurrent and crosscurrent systems have been used extensively for gas–liquid contacting compared with the cocurrent system. But the co-current system has some inherent advantages as reported by King (1974). These are relative simplicity, low cost of operation, high dispersed phase holdup, intense agitation, continuous operation over a wide range of flow rates of both the phases without any flooding, low pressure drop, and higher interfacial area and transfer coefficients. Hence, in recent years, there has been a growing interest in cocurrent contacting systems. The cocurrent system of gas–liquid generally refers to the flow of gas as a dispersed phase of bubble in a liquid moving in the same direction of the gas bubble. The inverse bubbly flow is referred to as the flow of gas as a dispersed phase of bubble against its buoyancy in a liquid that may be still or moving.

Hydrodynamics and Transport Processes of Inverse Bubbly Flow. http://dx.doi.org/10.1016/B978-0-12-803287-9.00001-1

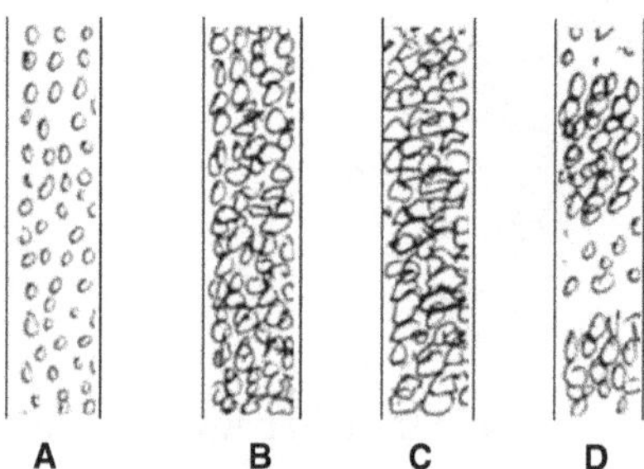

FIGURE 1.1 Different types of bubbly flow. **A,** Separated bubbly flow. **B,** Interacting bubbly flow. **C,** Churn-turbulent bubbly flow. **D,** Clustered bubbly flow.

Typical Features of Bubbly Flow

The bubbly flow unites the characteristics of deformable interfaces, channel shapes, flow direction, and the compressibility of the gas phase. Typical features of bubbly flow are (1) movement and deformation of interfaces of bubbles in time and space domains and (2) complex interactions among the interfaces, the bubbles, and the liquid flow. According to the magnitude of these interactions, bubbly flow is classified into four different types as shown in Figure 1.1:

1. Ideally separated bubbly flow
2. Interacting bubbly flow
3. Churn turbulent bubbly flow
4. Clustered bubbly flow
5. Slug bubbly flow

In ideally separated bubbly flow, all bubbles individually as a single bubble flow without interacting to each other directly or indirectly.

In this case, the bubble number density is low. When the bubble number density increases, the bubbles begin to interact with each other directly or indirectly because of collisions or the effects of wakes caused by other bubbles. With a further increase in bubble number density, the bubbles tend to coalesce to form bigger bubbles, and the shape of the bubbles changes because of more interactions and the momentum exchange to so-called cap bubbles. This causes the flow changes to churn turbulent bubbly flow. At this churn turbulent flow, the formed cap bubbles and other remaining smaller bubbles are highly agitated because of the interactions between bubble motions and turbulent flow. The level of bubble motion and turbulence depends on the local turbulence energy production and dissipation. Relative motions between the bubbles and the liquid flow produce both additional turbulence and dissipation caused by a viscous effect, depending on the scale of interfaces and turbulence eddies (Serizawa and Kataoka, 1992). The accepted physical explanations for the mechanisms have not yet been developed. However, turbulence development may be predicted to some extent by numerical methods (Michiyoshi and Serizawa, 1986; Lahey, 1988; Serizawa and Kataoka, 1992). The large bubbles occasionally form clustering of bubbles, and they behave like a single gas slug. After a certain travel, they sometimes coalesce to form a gas slug, and sometimes they separate into individual bubbles. This flow regime is thus a transition from bubbly flow to slug or churn flow. The various bubbly flows are characterized by phase distribution phenomena, which exhibit different lateral void fraction profiles, depending on the volumetric flow rate of the gas and liquid phases. Wallis (1969) reported that quantitative criteria for the transitions from ideally separated to interacting bubbly flow and from interacting bubbly flow to churn turbulent bubbly flow are roughly 0.01 and 0.06 in void fraction, respectively.

Types of Gas–Liquid Contacting Devices

Gas–liquid contactors may be classified into three main groups depending on the type of contacting system or distribution of the gas and liquid phases. The different groups are:

1. Liquid is present as discrete drops that are distributed within a continuous gas phase. Examples of this system are spray towers and atomizer units.
2. Both the liquid and gas are present as separate continuous phases with a single contact boundary surface. An example of this system is a falling film reactor.
3. The gas is dispersed as bubbles in a continuous liquid phase. Examples of these systems are mechanically agitated reactor vessels such as flotation cells, air-lift reactors, and bubble columns.

For selecting an appropriate gas–liquid contactor, the following considerations should be taken into account:

- Maximum conversion of reactants
- Greatest selectivity for desired products
- Minimum environmental impacts
- Ease of automation and process control
- Simplicity of scale-up
- Low capital and operating cost

Each of these gas–liquid contactors is suitable for specific services, each with some advantages and disadvantages. However, recently bubble columns as bubbly flow devices are widely being used in many chemical process industries (e.g., oxidation, hydrogenation, halogenations, fermentation, coal liquefaction) because of its unique advantages). Contactors or reactors belonging to the jet-mixing category with cocurrent or countercurrent contacting of phases such as ejectors, venturis, and other similar devices for gas–liquid or liquid–liquid contacting are gaining in importance nowadays because of high interfacial areas and mass transfer coefficients obtained in such systems. These are all cocurrent flow devices in which the kinetic energy of a fluid is used to achieve fine dispersion and mixing between the phases.

Bubbly Flow Device

Bubble column reactors belong to the general class of multiphase reactors, which are classified as shown in Figure 1.2. In its most simple form, the device is a vertical cylinder that is called a simple bubble column reactor (Figure 1.3). The gas enters at the bottom through a gas distributor, which may vary in design. The liquid phase may be supplied in batch form, or it may move with or against the flow of the gas. In contrast to the physical mass transfer operations, counterflow offers no significant advantages because the reaction itself ensures a sufficient concentration drop during material exchange. The top of this type of the reactor is often widened to facilitate gas separation (Gerstenberg, 1979). The bubble

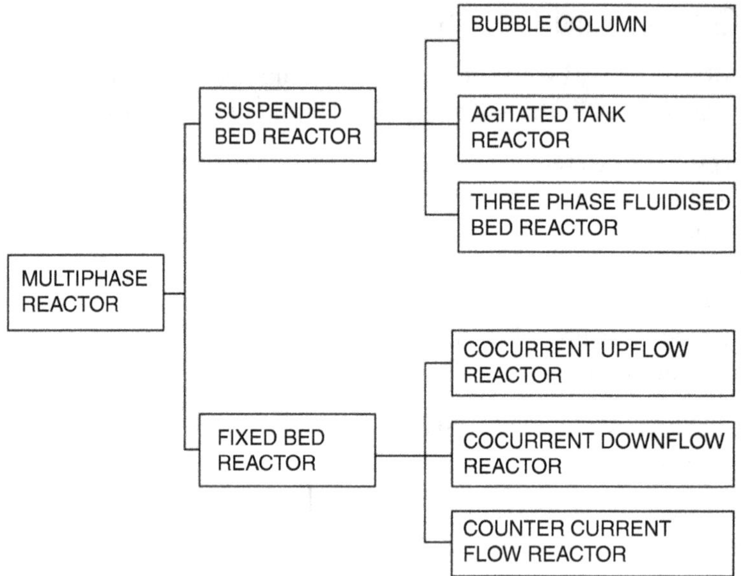

FIGURE 1.2 Classification of multiphase reactors (Coulson et al., 2003).

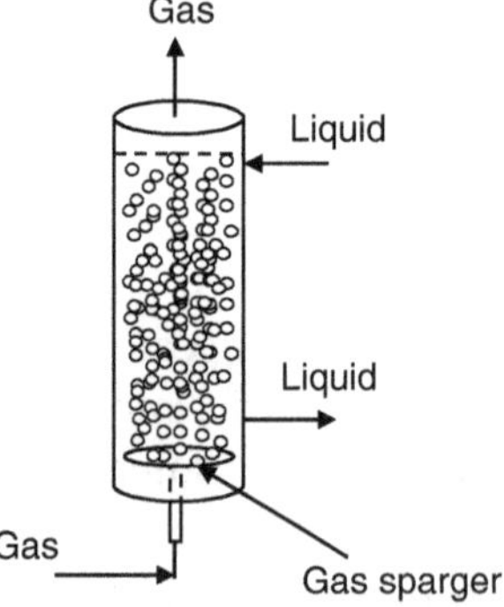

FIGURE 1.3 Simple bubble column.

column reactor is characterized by the lack of any mechanical means of agitation; hence, gas is distributed more evenly in the liquid phase. Short residence time of gas in the reactor is a further disadvantage when conversions are expected to be high, although this period can be varied if a suitable modified bubble column reactor is used.

Importance of Bubbly Flow Devices

The bubbly flow devices are widely used for carrying out reactions and mass transfer operations in which a gas or a mixture of gases is distributed in the form of dispersed phase of bubbles in a continuous liquid phase. In the liquid there can be suspended or fluidized, reactive or catalytic solids. Accordingly, the device is termed a two- or three-phase (slurry) bubble column. Bubble column reactors have been used in chemical, pharmaceutical, biochemical, and other processes for many years.

Some examples of industrial applications of two-phase bubble column reactors are listed in Table 1.1. In fact, a few decades ago, no great scientific interest was shown in bubble column reactors. Very little literature was forthcoming from either industries or universities (Chisti, 1989; Deckwer, 1992). However, since the mid-1970s, the research and

Table 1.1   Examples of Industrial-Scale Processes in Two-Phase Bubble Columns

Processes	With System	Main Products
Absorption of	CO_2 in ammoniated brine	Soda ash
	Buffer solutions and amines	—
	Isobutylene, butenes in aqueous solutions of H_2SO_4	—
Alkylation of	Phenols with Iso- butylene diluted with inert gas	—
	Methanol	Acetic acid
	Benzene	Ethylbenzene, cumene
Carbonylation of	Methanol	Acetic acid
Chlorination of	Aliphatic hydrocarbon	Chloroparaffin
	Aromatic hydrocarbons	Chlorinated aromatics
Hydration of	Isobutene	tert-Butanol
Hydroformylation of	Olefins	Aldehydes and alcohols
Oxidation of	Ethylene (partially)	Acetaldehyde
	Ethylene in acetic acid solutions	Vinyl acetate
	Acetaldehyde, sec-Butanol	Acetic acid
	Acetaldehyde	Acetic anhydride
	Butanes	Acetic acid and methyl ethyl ketone
	Toluene	Benzoic acid
	Xylene	Phthalic acid
	Cumene	Phenol and acetone
	Ethylbenzene	Acetophenone
	Waste water (wet oxidation)	—
Oxychlorination of	Ethylene	Dichloroethane
Oxysulphonation of	Paraffin	Paraffin sulphonate
Separation of	Oily water	

development interest in various types of bubble columns has dramatically increased. This is mainly because of

- A general recognition of the advantages such as
  - Simple construction and low capital cost
  - No moving parts and minimum maintenance
  - Ability to handle solids
  - Ease of temperature control
  - Reasonable interfacial mass transfer area
- The opening up of new fields of applications, especially in biotechnological areas such as effluent treatment, single-cell protein production, animal cell culture, and antibiotic fermentation (Chisti, 1989).
- A revival of interest in coal liquefaction and slurry phase Fisher-Tropsch synthesis, both relying greatly on bubble column technology, resulting from the oil crisis of 1973 and the subsequent search for alternative raw materials and synthesis fuels. Bubble column also played an important role in the development of C_1 chemistry (Deckwer, 1992).

Apart from large number of advantages bubble column reactors have few drawbacks. This can be minimized by making appropriate modifications. Disadvantages of bubble column reactors are (Deckwer, 1992):

- Considerable degree of back mixing in the liquid and gas
- Higher pressure drop with to packed bed columns
- Rapid decreasing of interfacial area above values of the aspect ratio (>12) because of the increased rate of coalescence
- Shorter residence time

Many of the processes making oil substitutes from natural gas such as the Fischer-Tropsch synthesis step in the production of middle distillates from synthesis gas ($CO + H_2$), the methanol synthesis from synthesis gas, and the coal liquefaction or coal hydrogenation processes all rely greatly on bubble column reactors. This is because the reactors have such advantages as mentioned earlier and because CO-enriched synthesis gas (H_2: $CO = 0.6$–0.7) can directly be used in the reactors. Using this composition, low cost and high thermal efficiency may be obtained for the Fisher-Tropsch synthesis (Shah et al., 1982). In the present age of biotechnology, bubble columns as bioreactors have the greatest potential and have widely been used (Blenke, 1974; Schügerl et al., 1977; Chisti, 1989). An example is the biological effluent treatment plant operated by Hoechst Co., Germany, where the reactor has 5-m diameter and 22-m effective height, and the treatment ability is about two tons BOD per day (Leistner et al., 1979). As another example, bubble columns have successfully been used for the production of a single-cell protein (Sittig et al., 1979; Westlake, 1986). In gas–liquid–solid three-phase systems, some applications in which gas is used as a dispersed phase of bubble are shown in Table 1.2. An important application area of bubble columns is their use as bioreactors in which microorganism are used to

Table 1.2   Some Applications of Gas–Liquid–Solid Three-Phase Systems in Which Gas is Used as a Dispersed Phase of the Bubble

Gas–Liquid–Solid Three Phase	System	Main Products
Absorption of	SO_2 in an aqueous slurry of magnesium oxide and calcium carbonate	
Absorption of	CO_2 in lime suspension	
Desulfurization (Catalytic) of	Petroleum fractions	
Disproportionation of	Oleñn by using a catalyst consisting of molybdenum oxide and aluminum oxide	Consisting of saturated hydrocarbons and a branched chain olefins
Fischer-Tropsch synthesis		
Hydride formation and decomposition	In slurry	
Hydro-desulphurization		
Hydrogenation of	Unsaturated fatty acids	
Hydrogenation of	Benzene	Cyclohexane
Hydrogenation of	Adipic acid dinitrile	Hexamethylene Diamine
Hydrogenation of	Nitroaromatics	Amines
Hydrogenation of	Glucose	Sorbitol
Hydrogenation of	Anthraquinone	Anthrahydroquinone (in the production of hydrogen peroxide)
Hydrogenation of	Ammonium Nitrates	Hydroxyl amines
Hydrogenation of	*a*-Nitrocaprolactum	
Hydrogenation of	Edible coal	
Liquefaction (thermal) of	Coal	
Methanation	Synthesis gas	Methanol
Methanation of	CO	
Oxidation (wet) of	Waste sludge	
Oxydesulfurization of	Coal	
Oxidation of	Cyclohexane	Mixture of cyclohexanol and cyclohexanone
Oxidation of	Cyclohexene	Adipic acid
Oxidations of	*n*-Parrafins	Sec alcohols
Oxidation of	Glucose	Gluconic acid
Polymerization of	Olefins	
Upgrading of	Coal oils and heavy oil fractions by hydrogenations	
Production of	from CO and steam	Hydrocarbons
Production of		Single–cell protein
Production of		Primary and secondary metabolites
Production of		Animal cells
Treatment of	Waste water	

Table 1.3   Biochemical Application of Bubble Columns (Kantarci et al., 2005)

Bioproduct	Biocatalyst	Reference
Thienamcyn	*Streptomyces cattleya*	Arcuri et al. (1986)
Glucoamylase	*Aureobasidium pullulans*	Federici et al. (1990)
Acetic acid	*Acetobacter aceti*	Sun and Furusaki (1990)
Monoclonal	*Hybridoma cells*	Rodrigues et al. (1999)
Plant secondary metabolites	*Hyoscyamus muticus*	Bordonaro and Curtis (2000)
Taxol	*Taxus cuspidate*	Son et al. (2000)
Organic acids (acetic, butyric)	*Eubacterium limosum*	Chang et al. (2001)
Low oxygen tolerance	*Arabidopsis thaliana*	Shiao et al. (2002)
Ethanol fermentation	*Saccharomyces cerevisiae*	Ogbonna et al. (2001)

produce industrially valuable products such as enzyme, proteins, and antibiotics. Several biochemical studies using bubble columns as bioreactors are presented in Table 1.3.

Types of Bubbly Flow Devices

Bubbly flow devices are classified based on flow of phases (e.g. horizontal flow, vertical up or down flow, cocurrent, countercurrent, crosscurrent). Andreussi et al. (1999) performed experiments to analyze gas-phase distribution in horizontal bubbly flow. They measured the local void fraction and the bubble diameter and velocity using conductivity probes. With the experimental results, they developed a new correlation for the maximum bubble size in dispersed bubbly flow. Extensive studies have also been done on cocurrent up-flow bubble columns. Kantak et al. (1995) investigated the effects of gas and liquid properties on gas phase dispersion in a cocurrent upflow bubble column. The extent of liquid backmixing in gas–liquid concurrent upflow packed-bubble column reactors quantified in terms of an axial dispersion coefficient is described by Belfares et al. (2001). The backmixing of the liquid phase can be reduced in compartmentalized bubble columns (Dreher and Krishna, 2001). Partition sieve plates with open areas of 18.6% and 30.7% are used in the column. It has been pointed out in the literature that modified bubble columns are sometimes advantageous over single-staged bubble columns. Unfortunately, insufficient data are available for novel reactors for any specific design recommendations. The following novel reactors should be studied in more detail: loop reactor, jet reactor, staged bubble column, packed bubble column, countercurrent bubble column, and inverse bubble column.

Different Types of Modified Bubbly Flow Devices

There are many variations on the simple type of bubble column reactor. They are all adapted to particular practical needs. Nagel et al. (1973), Blenke (1974), and Gerstenberg (1979) all have given details on the various types of bubble column arrangements. Schügerl et al., 1977 have described the special problems associated with bioprocesses. Schügerl et al. (1977) divided various bioreactors according to energy input and provided a comparative

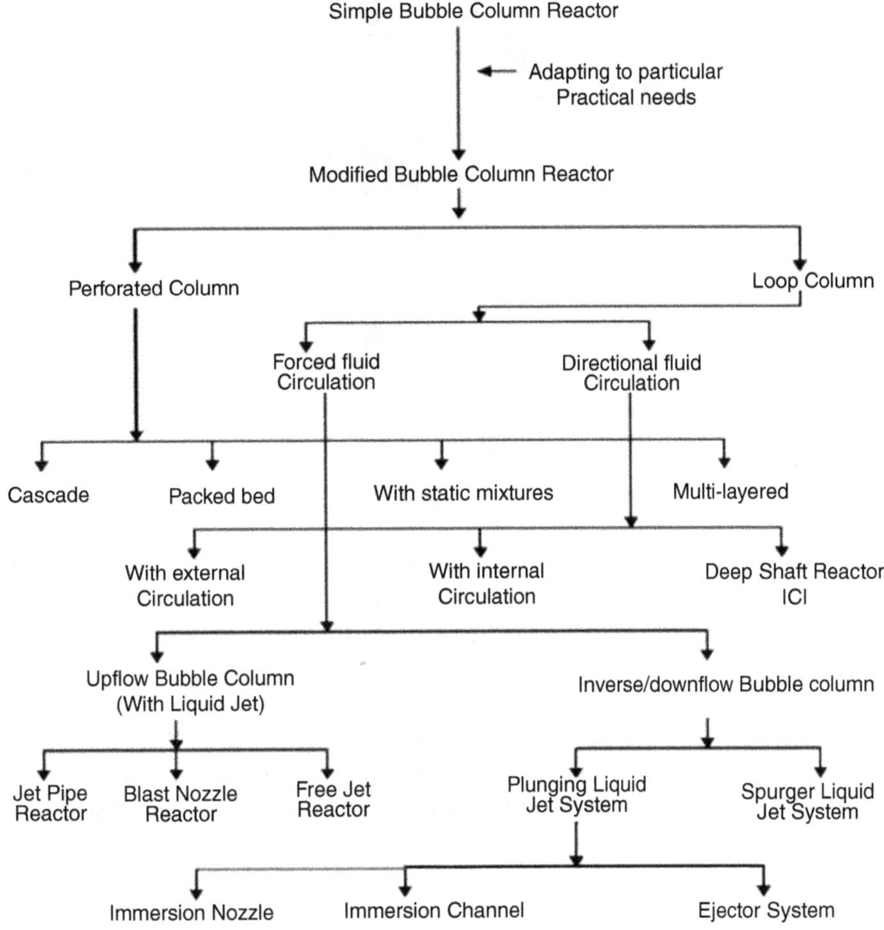

FIGURE 1.4 Schematic chart of different types of bubble column reactors.

evaluation in conjunction with the various practical requirements. Figure 1.4 depicts a few of the modifications frequently used in the field of chemical process technology. Incorporation of additional perforated plates, including the single-hole variety (Zlokarnik, 1971), transforms the simple model into a multistage cascade version. This new distribution of gas over the perforated plates intensifies mass transfer, reduces the fraction of larger bubbles, and prevents backmixing in both phases. The same effects can be obtained in packed bubble columns in which a dumped packing or static mixtures are used (Hsu et al., 1977; Maclean et al., 1977; Hofmann, 1983). Multilayer appliances, as obtained by incorporating cooling devices, prevent bulk circulation, and a uniform gas flow can be achieved throughout the reactor, provided a suitable gas distributor is fitted. Many types of bubble columns offer directional fluid circulation. The simplest case uses the major effect resulting from bubble entrainment and the difference in density between the dispersed gas and the liquid phases, the circulation being stabilized by means of an inserted loop. This eliminates

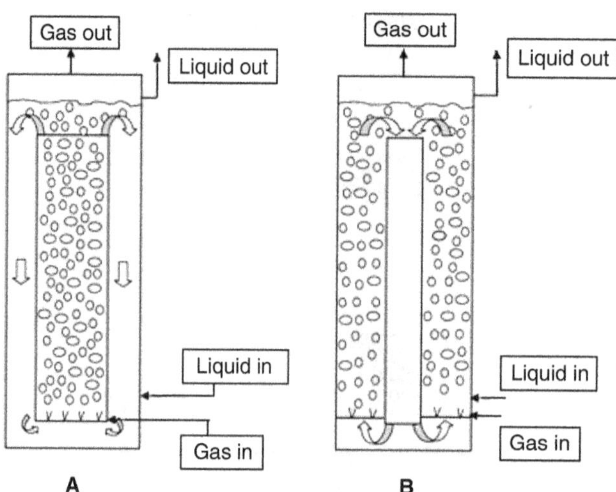

FIGURE 1.5 A, bubble column with external circulation. **B,** Bubble column with internal circulation.

complete radial transfer over the whole cross-sectional area. The large loop reactors come with either an internal or external liquid removal line, depending on the gas removal arrangement. A shaft reactor developed by ICI based on a sophisticated circulation process represents a highly significant development (Hines, 1978).

The advantage of this arrangement is that a large amount of gas reaches the lower parts of the shaft, and the surface area does not have to be very large. Bubble columns with external circulation systems are similar to loop reactors. As shown in Figure 1.5, *A*, the shaft at the center eliminates radial transfer over whole cross-sectional area and creates a homogeneous flow zone.

The dispersed phase is in center of the column, and liquid circulation occurs through the annulus region. Large amount of gas can be processed in this type of reactor. In the bubble column with internal circulation system as shown in Figure 1.5, *B*, whereas the dispersed phase exists in annulus region, liquid circulation occurs through the center of the column. These types of reactors provide a homogenous flow zone and a high rate of circulation. An inserted loop stabilizes the bulk circulation. In a multishaft bubble column, as shown in Figure 1.6, *A*, the presence of the shaft prevents the coalescence of bubbles. The shaft provides directed vertical paths for the bubbles, preventing lateral movement and bulk circulation. The shaft could also act as cooling device to control the temperature of mixture inside the column. Bubble columns with static mixers consist of motionless mixers (see Figure 1.6, *B*). They are found to be effective because they increase the rates of mass transfer and help in bubbly flow to remain homogeneous. Multistage bubble columns, as shown in Figure 1.6, *C*, allow greater variation of the retention time of the liquid phase.

The simple construction of the columns enables a cascade mode of operation to be used, with a well-defined flow path and no remixing between stages. A cascade bubble column reactor, as shown in Figure 1.6, *D*, is a vertical column with equidistantly spaced, horizontally mounted, and uniformly perforated plates fitted inside it. The backmixing

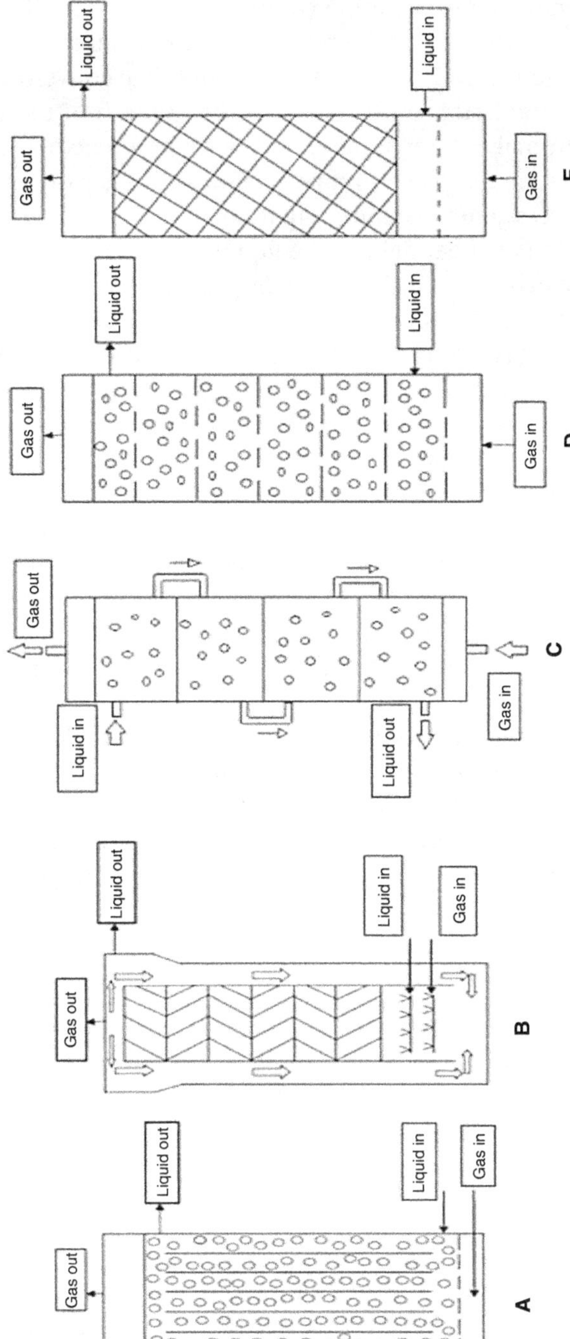

FIGURE 1.6 **A,** Multishaft bubble column. **B,** Bubble column with static mixture. **C,** Multistage bubble column. **D,** Cascade bubble column. **E,** Packed bubble columns.

of gas and liquid phases in the simple bubble column and the nonuniform distribution of gas bubbles over the cross-section can be reduced by the installation of trays. Packed bubble columns, as shown in Figure 1.6, *E,* are an offshoot of the conventional trickle bed reactors. The high holdup and better cross-sectional distribution of the liquid phase in the packed bubble column make them superior to trickle bed reactors. Bubble column reactors with forced circulation use the liquid jet as a means of gas distribution, and the energy involved in this process generates the circulation. In the case of the jet loop reactor, the gas is finely atomized in the shear field of the liquid jet, and the reactor contents are efficiently circulated by means of a conduit tube. If the gas is forced in through nozzles and a momentum exchange unit is incorporated, then it is known as a blast nozzle reactor. In the free jet configuration, the gas is efficiently distributed via two-component nozzles, and liquid circulation is strongest in the lower section of the reactor. The selection and design of the gas distributor is an important aspect of all bubble column reactors. Hebrard et al. (1999) studied axial liquid mixing in gas–liquid systems (bubble columns) and two types of gas–liquid–solid systems (the turbulent and inverse turbulent beds) for different types of gas sparger (membrane and perforated plate). In the turbulent and inverse turbulent beds, solids, larger and lighter than water and of large diameter (3–4 mm), are fluidized only by an upward gas flow. In two-phase systems, the type of gas sparger has a strong effect on the gas flow regime and consequently on the axial liquid mixing. In the gas–liquid–solid reactors, the effect of the gas sparger on axial liquid mixing can be pronounced. Heterogeneous flow behavior is observed independent of the gas sparger when the solid content of the column is sufficiently high. A spinning sparger produces smaller size bubbles with higher gas holdup and interfacial area inside a bioreactor (Fraser and Hill, 1993). Shah et al. (1983) used a ring-type sparger as the gas distributor at the top of the column to obtain a higher residence time of gas bubbles in a cocurrent inverse bubble column. Ejector type mixing devices are also good alternative for achieving efficient dispersion of one fluid into other. A wide variety of liquids, gases, and vapors may be used as either primary or secondary stream. Based on primary and secondary streams, the ejector type contacting devices may broadly be classified as:

1. *Gas–gas system:* Gases are used as both the primary and secondary fluid.
2. *Gas–liquid system:* Gas is used as the primary fluid and liquid as the secondary fluid.
3. *Liquid–liquid system:* Liquids are used as both the primary and secondary fluid.
4. *Liquid–gas system:* Liquid is used as the primary fluid and gas as the secondary fluid.

Many workers have attempted the analysis of gas–gas ejectors using the momentum, continuity, and energy equations. The effects of design variables such as projection ratio, area ratio, throat length, and diffuser angle have also been investigated. Mitra et al. (1963) reported the design and performance of single-phase air–air ejector system in wide ranges of operating conditions by means of interchangeable nozzles. Significant contribution in this area has also been made by Davies et al. (1967a), Kastner and Spooner (1950), Van der Lingen (1960), and Smith et al. (1996). Mitra and Roy (1963) investigated studies on the performance of a slurry reactor incorporating a gas-jet ejector for liquid recirculation and agitation. They carried out Fisher-Tropsch synthesis reaction with iron catalyst in a high boiling oil medium and found improved reactor performance with respect to slurry agitation, suspension of catalyst, temperature control, and conversion of carbon dioxide to liquid products. Davies et al. (1967a,b) reported studies on the performance of an ejector with air as primary fluid and various liquids and slurries as secondary fluid. Pal (1980) studied gas–liquid mixing in air-jet and water-jet ejectors and found better dispersion using water as the primary fluid. Cunningham (1957) and Cunningham et al. (1970) successfully analyzed the liquid–liquid ejector performance by momentum and continuity equations. Acharjee et al. (1978) have carried out liquid–liquid mass transfer with binary and ternary systems in vertical ejectors and reported very high values of mass transfer coefficient. Mukherjee et al. (1988) studied on mixing characteristics of two immiscible liquids in a liquidjet ejector system in which kinetic energy of a liquid jet aspirates and interdisperses the immiscible phases. They have also reported energy dissipation, pressure drop, and holdup of liquid–liquid two-phase upflow in a vertical column fitted to the ejector system. However, studies on liquid–gas systems with liquid as the primary motive fluid are found to be increasing recently. Some of the earliest reported investigations using water as the primary fluid and air as the secondary fluid are those by Kroll (1947), Silvester (1961), and Reddy and Kar (1968). Liquid jet is considered to be useful for gas–liquid contactors because it generates fine bubbles by impinging the liquid jet on the liquid surface. Another advantage of liquid jet is that it entrains certain amount of gas into the pool of liquid. Burgess and Molloy (1973) carried out experiments of gas absorption in a plunging liquid jet reactor. They reported that the reactor is analogous to a gas sparged stirred tank reactor with plunging jet acting as both the reactor agitator and gas bubble generator. Ogawa et al. (1982) studied the liquid phase mixing in the upward gas liquid jet reactor with liquid jet ejector. The longitudinal liquid phase mixing pattern was quite different between the spouting section and the calm section. In the spouting section, the liquid phase was regarded as almost completely mixed flow. By contrast, in the calm section, the value of the dispersion coefficient was fairly smaller than that in bubble columns, and backmixing of the liquid phase was suppressed. Tojo and Miyanami (1982) reported the oxygen transfer characteristics in an inverse gas–liquid system in which gas is entrained and dispersed by a liquid jet produced through a nozzle. They showed that the gas–liquid mixture with a jet is more efficient for gas–liquid mass transfer than the corresponding gas–liquid system. Kumagai and Endoh (1982) showed the characteristic regions of the gas entrainment rate of an impinging liquid jet. Bin (1993) reviewed experimental and theoretical studies on gas entrainment by a plunging jet and discussed several aspects of these phenomena. He also noted the practical applications of the plunging jet aerator in the waste treatment, fermentation, and flotation industries. Mitra-Majumdar et al. (1995) examined the mixing behavior of the two-phase air–water turbulent flow in a jet bubble column. The time evolution of the mixing behavior of a liquid tracer in a turbulent air–water flow within a jet bubble column was predicted using a model based on the fundamental governing equations of fluid motion. The predictions of the model were compared with experimental measurements and found satisfactory.

Inverse Bubbly Flow Device

Recently, inverse bubble columns have been gaining interest for chemical processes, particularly when the interfacial mass transfer area is the rate-controlling step. The inverse bubble column, in which choice of suitable liquid and gas throughput rates permits gas residence times to be adjusted up to maximum gas content within certain limits, presents a neat solution. The advantages of the inverse bubble column are:

- Bubbles are finer and more uniform in size.
- Coalescence of bubbles is negligible.
- Homogenization of the two phases in the whole column is possible.
- A large amount of liquid can be contacted with a small amount of dispersed gas efficiently.
- There is a higher residence time of the gas bubbles.

The immersion nozzle and immersion channel reactors are also important in which forced liquid rotation gives rise to a partial gas and liquid inverse system (Gerstenberg, 1979; Schügerl et al., 1977; Leuteritz et al., 1976). Ejectors are devices that use the kinetic energy of a high-velocity liquid jet to entrain and disperse the gas phase. The inverse bubble column with an ejector system is very simple in design, and no extra energy is required for gas dispersion because the gas phase is sucked and dispersed by the high-velocity liquid jet. So, from an energy point of view, it is very attractive. Hence, the cocurrent inverse bubble column with ejector-type gas distributor is getting importance because of its distinctive advantages over more conventional devices. These are:

- Lower power consumption
- Almost complete gas utilization
- Higher overall mass transfer coefficient
- Tolerance to particulates and therefore viable for slurry chemical reaction

The schematic diagram of an inverse bubbly flow column described by Majumder et al. (2005) is shown in Figure 1.7. It consists of an ejector assembly, E, for entraining gas bubble; an extended pipeline column where entrained bubbly flows against its buoyancy, PC; a gas liquid separator, SE; and other accessories such as centrifugal pumps (P_1, P_2), manometer ports (M_1–M_{10}), control valves (V_1–V_7), solenoid valves (SV_1-SV_4), rotameters (R_L, R_G), and a collector vessel (CV). The ejector assembly and extended pipeline column are made of generally transparent Perspex for visual observation of the flow and mixing patterns. The ejector assembly consists of a suction chamber, a throat, a divergent diffuser, and a forcing nozzle. The suction chamber should be large enough to avoid shock and entry losses on the one hand, and on the other, it should not be unduly large such that gas circulation occurs in the chamber. Throats of constant area and of variable area are generally used in the design of ejectors. Earlier workers (Kroll, 1947; Smith, 1951; Lapple, 1956) reported that a constant area throat produces a higher vacuum than the variable area throat. This is because the velocity distribution of the secondary fluid in

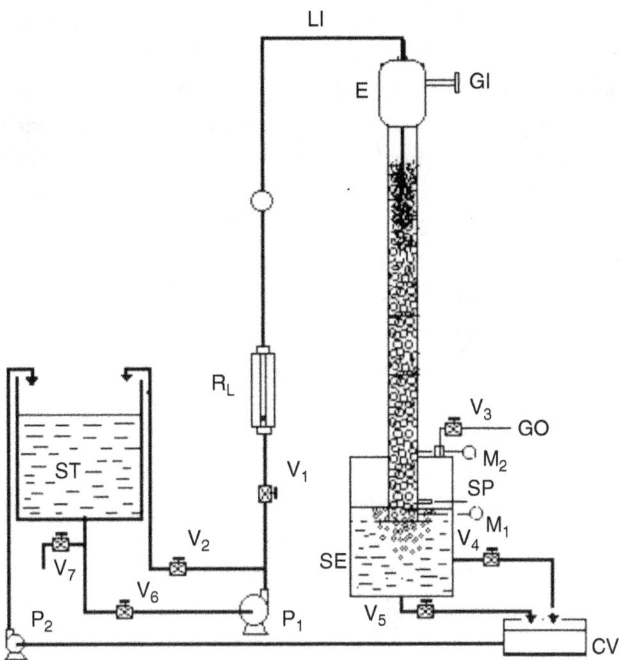

FIGURE 1.7 Schematic diagram of inverse bubbly flow in vertical column. *Al,* Air inlet; *CV,* collector vessel; *E,* ejector assembly; *P1, P2,* pump; *PC,* pipeline contactor; *RL,* rotameter for liquid; *RG,* rotameter for gas; *SE,* separator; *ST,* storage tank; *V1* to *V7,* valves.

a constant area throat is more uniform than the variable area throat. The length of the throat also plays an important role.

Dutta (1976) showed that for a liquid–gas ejector system, in which the ejector was fitted with an extended parallel contactor, a throat length of 4 to 20 times of its diameter yielded substantially the same performance. The entrance to the throat is made well rounded to minimize the entry loss for the secondary fluid at the throat. After the throat, a divergent diffuser was provided to decrease the velocity of the fluid and increase its static pressure. Abrupt divergence produces undesirable effect. Earlier workers (Kastner and Spooner, 1950; Smith, 1951) have reported that to optimize the performance, the divergent angle should be around 7 degrees, and the length of this section should be between eight and 12 times the diameter of the throat (Davies et al., 1967a, 1967b). Two types of nozzles, convergent-divergent and straight-hole nozzles, are generally used in the ejector. Straight-hole nozzles in the medium velocity range are efficient compared with the convergent-divergent types on the basis of nozzle losses (Engdahl and Holton, 1943). At low-pressure range, there is little effect of nozzle type on the performance of an ejector (Mitra et al., 1963). A gas–liquid separator is an essential part of the apparatus, in which inverse flowing of the gas–liquid mixture gets separated from each other. It should be sufficiently large to minimize the effects caused by liquid leaving the system and air–liquid separation. The important factors that characterize the efficient operation of bubbly flow in bubble column reactor are the flow regime, gas holdup, pressure drop, bubble characteristics, heat transfer coefficient, mass transfer coefficient, mixing intensity, and gas distribution. In case of inverse bubbly flow, gas is first entrained in the liquid medium, which then allows moving against its buoyancy by downward liquid momentum in the column. Gas holdup is a dimensionless parameter defined as the volume occupied by the gas phase in the form of bubbles to the total volume of bubble column. It plays an important role in the design and analysis of bubble columns. The spatial variation of gas holdup gives rise to pressure variation and thus liquid recirculation. Liquid recirculation plays an important role in mixing and heat and mass transfer; predictions of radial gas holdup profiles would lead to better understanding of these phenomena and thus more reliable bubble column scale-up. The gas holdup increases with increasing gas velocity and operating pressure, and it decreases with increasing liquid viscosity and solid concentration; adding a surface active reagent into the mixture also increases the holdup. In bubble columns, the effect of the column size on gas holdup is negligible when the column diameter is larger than 10 to 15 cm and the height is above 1 to 3 m (in other words, with height to diameter ratios [i.e., aspect ratio] >5) (Shah et al., 1982).

Bubble column reactors are particularly suited for slow reactions taking place in the liquid phase. The main resistance to the mass transfer is located in the liquid phase. The gas–liquid contact achieved is reflected by the parameter β, which is the ratio of the liquid phase volume to the volume of the film diffusion layer (Krishna et al., 1994). High-value βs are obtained with bubble column reactors and are suitable for reactions that demand high bulk liquid volume. The range for parameter is as $\beta = (1 - \varepsilon_g)/\delta_1 a = 10^3$ to 10^4 where ε_g is the gas holdup and δ_l is the thickness of liquid film.

Pressure drop is another important parameter that plays significant role in efficient operation of a bubble column. Total pressure drop is the sum of the static pressure drop and pressure drop from losses that include loss caused by friction, loss caused by bubble formation, and loss caused by drag. Pressure drop depends on various factors such as column dimension, liquid and gas flow rate, gas and liquid properties, gas holdup, and mean bubble diameter.

According to Kundu et al. (1995), prediction of pressure drop is difficult in inverse bubbly flow because of the phenomena of momentum transfer between phases, the wall friction and shear at the phase interface cannot be specified quantitatively. Bubble characteristics analysis of a bubble column mainly includes estimation of mean bubble diameter, bubble rise velocity, and bubble population. This parameter is closely related to gas holdup. The average bubble size in a bubble column is affected by gas velocity, liquid properties, gas distribution, operating pressure, and column diameter. The rise velocity of a single gas bubble depends on its size. Thus, the size and rise velocity of a bubble depend on each other and are affected by the same parameters. Gas holdup for smaller bubbles is greater than for bigger bubbles at the same superficial gas velocity because bubble rise velocity is less for smaller bubble and thus residence time is greater, leading to larger gas holdup. Bubbles are assumed to be spherical in shape in most of the cases, but it actually depends on the flow regime prevalent inside the bubble column. The basic parameters affecting the heat transfer are mainly the superficial gas velocity, particle size and concentration, liquid viscosity, particle density, axial or radial location of the heat transfer probe, and column dimensions. The heat transfer coefficients increase with increasing superficial gas velocity irrespective of the solid or liquid phase properties. The heat transfer coefficient has been found to decrease with increasing liquid viscosity in three-phase fluidized systems regardless of particle size. Increased solid concentrations increases the heat transfer coefficient values, which has been attributed to a corresponding increase of the slurry viscosity, resulting in greater bubble sizes and higher large bubble rise velocities and thus higher heat transfer rates. Heat transfer coefficients at column center is higher than near the wall because of concentration of large bubbles at the center and more turbulence at the center as compared with near the walls (Li and Prakash, 1997; Prakash et al., 2001). The heat transfer coefficient increases with increasing temperature. This could be explained by the reduced liquid viscosity and enhanced turbulence maintained at higher temperatures (Saxena et al., 1990). Chen et al. (2003) observed that the heat transfer coefficients increased with increasing pressure.

Another important factor, volumetric mass transfer coefficient, is a key parameter in the characterization and design of both industrial stirred and nonstirred gas–liquid reactors. In a bubble column as a gas–liquid reactor, the overall mass transfer rate per unit volume of the dispersions governed by the liquid–side mass transfer coefficient, $k_l a$, assuming that the gas side resistance is negligible. To determine the volumetric mass transfer coefficient, a precise knowledge of the gas holdup, mixing characteristics, and bubble size distribution is needed because volumetric mass transfer depends on both k_l (mass transfer coefficient) and a (interfacial area), which are directly related to them. The volumetric mass transfer coefficient ($k_l a$) values were found to increase with gas velocity and gas density (Ozturk et al., 1987; Behkish et al., 2002). Addition of solids and high solid concentrations caused reduced values of $k_l a$ caused by increased large bubble sizes (Vandu and Krishna, 2004). Muller and Davidson (1995) reported that with viscous media, the effect of surface active agents on the mass transfer, $k_l a$ values increase in the presence of surfactants because of formation of smaller bubbles. Behkish et al. (2002) showed that the volumetric mass transfer coefficient, $k_l a$, decreases with increasing liquid viscosity. According to Wilkinson et al., 1994 the values of $k_l a$ generally increase with pressure.

Analysis of mixing characteristics is essential for the modeling, design, and optimization of the bubble column based on mass transfer and other applications. Mixing is of great importance because homogeneity of the desired degree can be obtained by it and is used in industries to increase heat and mass transfer for a system that is undergoing a chemical change. Mixing time is a very important design parameter when the bubble columns are operated in a semibatch manner. The knowledge of mixing time gives some information regarding the liquid phase flow pattern (Pandit and Joshi, 1983). Dispersion coefficient and dimensionless Peclet number are the two quantities generally estimated for analyzing the mixing phenomena in bubble columns. Towell and Ackerman (1972) found that the axial dispersion coefficient varies with the column diameter and the superficial gas velocity. Hikita and Kikukawa (1974) stated that the viscosity of the liquid also has a strong effect on the intensity of liquid-phase dispersion. Considerable work has been reported by different authors on efficient dispersion of gas by liquid jet in gas–liquid two-phase cocurrent contactor with venturis, nozzles, and ejectors as gas–liquid mixing devices (Ohkawa et al., 1985; Bando et al., 1988; Yamagiwa et al., 1990; Kundu et al. 1997; Evans et al., 2001; Mandal et al., 2004; Majumder, 2008; Majumder et al., 2005, 2007). Dispersion not only brings about a large increase in the interfacial area available for material or heat transfer but also places the fluids in a state of motion that serves to increase the specific rates of both the above transfer process. In the design of two-phase contacting equipment, operating conditions are chosen in favor of transfer driving forces so that batch, cocurrent, or countercurrent operation is selected, and the times of exposure of the two phases are determined.

Having evaluated the minimum number of transfer units required for the proposed duty, from consideration of maximum driving force, attention must next be directed to the dispersion process whose effectiveness, in continuous flow systems, determines the height of a transfer unit. The latter should be minimal for economic design, is inversely dependent on the magnitude of the mass transfer coefficient and interfacial area, and is directly dependent on the flow velocity. Thus, if the flow velocity and output are to be high, the mass transfer coefficient and interfacial area must be as large as possible. For the proper modeling of the performance of a multiphase bubble column reactor, the residence time distributions (RTDs) of the fluid phases are of vital importance. The RTD curves allow a quantitative evaluation of the nature and degree of mixing in each fluid phase. Even when there are no extraneous mass or heat transfer effects present in the reactor, its performance depends both on the nature of intrinsic reaction kinetics and the nature of the RTD curves. The mixing or interdispersion of immiscible phases is thus an important and common operation designed to produce a high interfacial area for mass transfer, chemical reaction, or both. The specific gas liquid interfacial area (a) is related to the gas holdup (ε_g) and the Sauter mean bubble diameter (d_s) by $a_s = 6\,\varepsilon_g/d_s$. The interfacial area may be increased by increasing the holdup or by reducing the bubble diameter. In this regard, knowledge of the distributions of gas holdup and bubble size is very important.

Practical Applications of Inverse Bubbly Flow

Inverse bubbly flows have interesting potential applications, particularly in waste treatment and fermentation.

Practical examples in which the inverse bubbly flow is being used include:

- DSM Geleen in the Netherlands built its Meijet Sewage treatment plant with a capacity of 2250 m³, a depth of 4 m, and a retention time of 8.3 hours based on the plunging liquid jet aerators of inverse bubbly flow. The water sludge is recirculated between the pool and the orifice pipe with a proper pump to adjust speeds to keep the active sludge in suspension. The oxygen capacity of this unit equals 54 kg/h at the higher velocity and 21 kg/h at the lower velocity, with an oxygenation efficiency of 2 kg O2 kW^{-1} h^{-1} (based on the gross power input to the system) (Bin, 1993).

- The jet performance for the inverse bubbly flow on a small scale is reported by different authors (Bin, 1993).
- Böhnke (1970) briefly described the application of a jet-induced aerator combined with a circulation ditch to purify communal sewage. The oxygenation efficiency was reported to be dependent on the jet inclination, reaching a maximum value of 3.3 kg OZ kW^{-1} h^{-1} at about 70 degrees.
- Aerobic and anaerobic degradation by mass transfer operation in the pig slurry liquor contribute to the increase in COD removal in an inverse bubble column (Sneath, 1978). This confirms that inverse bubbly flow system has potential to enhance oxygen transfer in biological systems. More than 40% of COD reduction can be achieved; this reduction concerns only the dissolved solids by the aeration of the waste liquor in a continuous biological reactor equipped with a plunging liquid jet aerator.
- Kenyeres (1991) designed an aerator called HTPJ (high-turbulence plunging jet) which is equipped with nozzles of diameters between 50 and 100 mm ID made of abrasive resistant polyurethane on metabolite production in the activated sludge system or in fermentation processes. The bioreactor system used had a considerable potential for the biological treatment of small-scale agricultural or domestic wastewater discharged in rural or sparsely populated areas.
- Successful application of inverse bubble column with a plunging jet aerator as a small-scale activated sludge treatment unit was described by Ohkawa et al. (1986). The oxygen transfer characteristics in the plunging liquid jet aerator have been found to be superior to those of conventional aerators applied in aerobic wastewater treatment.
- The plunging liquid jet aerator is capable of treating a wastewater of high organic loading compared with that in a conventional aeration system. The plunging jet aerator is coupled with a cross-flow filtration process (Yamagiwa et al., 1991). The unit enabled removal of about 97% of total organic carbon (TOC), and the TOC removal efficiency of the cake layer deposited on the ceramic microfilter was comparable to that of the suspended microorganisms in the aeration tank.
- A plunging jet aerator can be applied in a fermentor (Jagusch and Püschel, 1968; Jagusch and Schönherr, 1972; Lafferty et al., 1978; Zaidi et al., 1991). Zaidi et al. (1991) applied a plunging liquid jet inverse bubble column reactor to xanthan production.
- Evans (1990) studied the confined plunging liquid jet system to check the range of its applicability in the flotation industry. Because in such a system, very fine bubbles can be obtained by the shearing forces generated by the jet plunging into the receiving pool, the system can be applied to improve the recovery of fine material in the flotation process.

An inverse bubbly flow column using a jet ejector for improved gas–liquid mixing would be highly suitable for many industrial processes such as absorption, desorption and scrubbing, gas–liquid reactions, aerobic fermentations, and waste treatment. However, the studies regarding hydrodynamics, mixing characteristics, bubble size distribution, and specific interfacial area are scanty. Therefore, a precise knowledge of the hydrodynamics and transport processes in inverse bubbly flow would be of considerable interest.

In this book, some hydrodynamic characteristics, including flow regimes, gas entrainment, pressure drop, holdup characteristics mixing, bubble size distribution, and interfacial area, and transport processes in the inverse bubbly flow regimes are reported. The book may be helpful for further understanding the multiphase phenomena for the scientific community for advance research and development and industry the installation of inverse bubble column to specific chemical and biochemical application.

References

Acharjee, D.K., Mitra, A.K., Roy, A.N., 1978. Co-current flow liquid–liquid binary mass transfer in ejectors. Can. J. Chem. Eng. 56, 37–42.

Andreussi, P., Paglianti, A., Silva, F.S., 1999. Dispersed bubbly flow in horizontal pipes. Chem. Eng. Sci. 54, 1101–1107.

Arcuri, E.J., Slaff, G., Greasham, R., 1986. Continuous production of thienamycin in immobilized cell systems. Biotechnol. Bioeng. 28, 842–849.

Bando, Y., Uraishi, M., Nishimura, M., Hattori, M., Asada, T., 1988. Cocurrent downflow bubble column with simultaneous gas–liquid injection nozzle. J Chem. Eng. Japan 21, 607–612.

Behkish, A., Men, Z., Inga, R.J., Morsi, B.I., 2002. Mass transfer characteristics in a large-scale slurry bubble column reactor with organic liquid mixtures. Chem. Eng. Sci. 57, 3307–3324.

Belfares, L., Cassanello, M., Bernard, P.A.G., Larachi, F., 2001. Liquid back-mixing in packed-bubble column reactors: a state-of-the-art correlation. Catal. Today 64 (3–4), 321–332.

Bin, A.K., 1993. Gas entrainment by plunging liquid jet. Chem. Eng. Sci. 48, 3585–3630.

Blenke, H., 1974. Loop reactors. Adv. Biochem. Eng. 13, 121–214.

Böhnke, B., 1970. Package treatment plant for expanding communities, in advances in water pollution research. Proceedings of the Fifth International Conference, San Francisco, Paper 1-9/2-9/7.

Bordonaro, J.L., Curtis, W.R., 2000. Inhibitory role of root hairs on transport within root culture bioreactors. Biotech. Bioeng. 70, 176–186.

Burgess, J.M., Molloy, N.A., 1973. Gas absorption in plunging jet reactor. Chem. Eng. Sci. 28, 183–190.

Chang, I.S., Kim, B.H., Lovitt, R.W., Bang, J.S., 2001. Effect of partial pressure on cell-recycled continuous co fermentation by Eubacterium limosium kist 612. Process Biochem. 37, 411–421.

Chen, W., Hasegawa, T., Tsutsumi, A., Otawara, K., Shigaki, Y., 2003. Generalized dynamic modeling of local heat transfer in bubble columns. Chem. Eng. J. 96, 37–44.

Chisti, M.Y., 1989. Airlift Bioreactors. Elsevier Applied Science, London.

Coulson, J.M., Richardson, J.F., Peacock, D.G., 2003. Coulson & Richardson's Chemical Engineering: Chemical and Biochemical Reactors and Process Control, third ed., vol. 3, Butterworth and Heinemann, Oxford, United Kingdom.

Cunningham, R.G., 1957. Jet Pump theory and performance with fluids of high viscosity. Trans. Am. Soc. Mech. Eng. 79, 1807–1820.

Cunningham, R.G., Hasen, A.G., Na, T.Y., 1970. Jet pump cavitations. J. Basic Eng., ASME 92, 483–494.

Davies, G.S., Mitra, A.K., Roy, A.N., 1967a. Momentum transfer studies in ejectors. Correlations for single-phase and two-phase systems. Ind. Eng. Chem. Process Des. Dev. 6, 293–299.

Davies, G.S., Mitra, A.K., Roy, A.N., 1967b. Momentum transfer studies in ejectors. Correlation for three-phase (air-liquid-solid) system. Ind. Eng. Chem. Process Des. Dev. 6, 299–302.

Deckwer, W.D., 1992. Bubble column reactors. Wiley, New York.

Dreher, A.J., Krishna, R., 2001. Liquid-phase backmixing in bubble columns, structured by introduction of partition plates. Catal. Today 69, 165–170.

Dutta, A.K., 1976. Effect of mixing throat length on the performance of a liquid jet ejector, M Tech Thesis. Chemical Engineering Department, IIT, Kharagpur, India.

Engdahl, R.H., Holton, W.C., 1943. Over fire air jet. Am. Soc. Mech. Eng. Trans. 65, 741–754.

Evans, G.M., 1990. A study of a plunging jet bubble column, PhD Thesis. University of Newcastle, Australia.

Evans, G.M., Bin, A.K., Machniewski, P.M., 2001. Performance of confined plunging liquid jet bubble column as a gas–liquid reactor. Chem. Eng. Sci. 56, 1151–1157.

Federici, F., Petruccioli, M., Miller, M.W., 1990. Enhancement and stabilization of the production of glucoamylase by immobilized cells of Aureobasidium pullulans in a fluidized bed reactor. Appl. Microbiol. Biotechnol. 33, 407–409.

Fraser, R.D., Hill, G.A., 1993. Hydrodynamic characteristics of a spinning sparger, external loop airlift bioreactor. Can. J. Chem. Eng. 71, 419–425.

Gerstenberg, H., 1979. Blasensaulen-reaktoren (Bubble column reactors). Chemie Ingenieur Technik 51 (3), 208–216.

Hebrard, G., Bastoul, D., Roustan, M., Comte, M.P., Beck, C., 1999. Characterization of axial liquid dispersion in gas–liquid and gas–liquid–solid reactors. Chem. Eng. J. 72 (2), 109–116.

Hikita, H., Kikukawa, H., 1974. Liquid-phase mixing in bubble columns: effect of liquid properties. Chem. Eng. J. 8, 191–197.

Hines, D.A., 1978. Proceedings of First European Congress for Biotechnology CH-Interlaken, Dechema Monographs, Verlag Chemie, Weinheim, p. 55.

Hofmann, H., 1983. Reaction engineering problems in slurry reactors in Mass transfer with chemical reaction in multiphase systems, vol II. Three Phase Systems. NATO ASI Ser. E. Appl. Sci. 73, p. 171.

Hsu, K.H., Erickson, L.E., Fan, L.T., 1977. Pressure drop, gas hold-up, and oxygen transfer in tower systems. Biotechnol. Bioeng. XIX, 247–265.

Jagusch, L., Püschel, S., 1968. Tauchstrahlbelüftung-ein neuartiges. In: VEB Kombinat "Schwartze Pumpe" erprobtes Belüftungssystem zur biologischen Reinigung hochbelasteter AbwBsser, WWT 18, pp. 160–164.

Jagusch, L., Schönherr, W., 1972. Das Tauchstrahlbegasungsverfahren-ein neues ökonomisches Verfahren mit vielen Einsatzmöiglichkeiten. Chem. Technol. (DDR) 24, 68–72.

Kantak, M.V., Hesketh, R.P., Kelkar, B.G., 1995. Effect of gas and liquid properties on gas-phase dispersion bubble columns. Chem. Eng. J. 59, 91–100.

Kantarci, N., Borak, F., Ulgen, K.O., 2005. Review: bubble column reactors. Process Biochem. 40, 2268–2283.

Kastner, L.J., Spooner, J.R., 1950. An investigation of the performance and design of the air ejector employing low-pressure air as the driving fluid. Proc. Inst. Mech. Eng. 162, 149–159.

Kenyeres, S., 1991. Plunging jet aeration technology and equipment. Brochure.

King, C.J., 1974. Separation Processes. Tata McGraw Hill Publishing Co., New Delhi.

Krishna, R., De Stewart, J.W.A., Hennephof, D.D., Ellenberger, J., Hoefsloot, H.C.J., 1994. Influence of increased gas density on hydrodynamics of bubble column reactors. AIChE J 40, 112–119.

Kroll, E.A., 1947. The design of jet pumps. Chem. Eng. Progress 1, 21–24.

Kumagai, M., Endoh, K., 1982. Effects of kinematic viscosity and surface tension on gas entrainment rate of an impinging liquid jet. J. Chem. Eng. Japan 15, 427–433.

Kundu, G., Mukherjee, D., Mitra, A.K., 1995. Experimental studies on a co-current gas–liquid downflow bubble column. Int. J. Multiphase Flow 21, 893–906.

Kundu, G., Mukherjee, D., Mitra, A.K., 1997. Ejector performance in a co-current gas–liquid downflow bubble column. Can. J. Chem. Eng. 75, 956–963.

Lafferty, R., Moser, A., Steinder, W., Saria, A., Weber, J., 1978. Gas-Fliissigkeitsstrahl-Schlaufenreaktor. VDI Bericht 315, 257–267.

Lahey, Jr., R.T., 1988. In: Afgan, N.H. (Ed.), Turbulence and Phase Distribution Phenomena in Two-Phase Flow: Transient Phenomena in Multiphase Flow. Hemisphere Publishing Corporation, New York, pp. 139–177.

Lapple, C.E., 1956. Fluid and Particle Mechanics. University of Delaware, Newark.

Leistner, G., Müller, G., Bauer, A., 1979. The bio-tower reactor—a tower-type biological waste water purification plant. Chemie Ingenieur Technik 51, 288–294.

Leuteritz, G.M., Reimann, P., Vergeres, P., 1976. Loop reactors: better gas/liquid contact. Hydrocarb. Process. 55, 99–100.

Li, H., Prakash, A., 1997. Heat transfer and hydrodynamics in a three-phase slurry bubble column. Ind. Eng. Chem. Res. 36, 4688–4694.

Maclean, G.T., Erickson, L.E., Hsu, K.H., Fan, L.T., 1977. Oxygen transfer and axial dispersion in an aeration tower containing static mixers. Biotechnol. Bioeng. 19 (4), 493–505.

Majumder, S.K., 2008. Analysis of dispersion coefficient of bubble motion and velocity characteristic factor in down and upflow bubble column reactor. Chem. Eng. Sci. 63, 3160–3170.

Majumder, S.K., Kundu, G., Mukherjee, D., 2005. Mixing mechanism in a modified co-current downflow bubble column. Chem. Eng. J. 112, 45–55.

Majumder, S.K., Kundu, G., Mukherjee, D., 2007. Energy efficiency of two phase miximg in a modified bubble column. Can. J. Chem. Eng. 85, 280–289.

Mandal, A., Kundu, G., Mukherjee, D., 2004. Gas-holdup distribution and energy dissipation in an ejector-induced downflow bubble column: the case of non-Newtonian liquid. Chem. Eng. Sci. 59, 2705–2713.

Michiyoshi, I., Serizawa, A., 1986. Turbulence in two-phase bubbly flow. Nucl. Eng. Des. 95, 253–267.

Mitra, A.K., Roy, A.N., 1963. Performance of slurry reactor for Fischer-Tropsch and related syntheses. Indian Chem. Eng. 5, 127–132.

Mitra, A.K., Guha, D.K., Roy, A.N., 1963. Studies on the performance of ejector part-i, air-air-system. Indian Chem. Eng. Trans. 5, 59–64.

Mitra-Majumdar, D., Farouk, B., Shah, Y.T., Wisecarver, K., 1995. A model for two-phase turbulent mixing in a jet bubble column. Can. J. Chem. Eng. 73, 772–778.

Mukherjee, D., Biswas, M.N., Mitra, A.K., 1988. Hydrodynamics of liquid–liquid dispersion in ejectors and vertical two phase flow. Can. J. Chem. Eng. 66, 896–907.

Muller, F.L., Davidson, F., 1995. On the effects of surfactants on mass transfer to viscous liquids in bubble columns. Chem. Eng. Res. Des. 73, 291–296.

Nagel, O., Kurten, H., Hegner, B., 1973. Criteria for the selection and design of gas liquid reactors. Int. J. Chem. Eng. 21, 161–172.

Ogawa, S., Kobayashi, M., Tone, S., Otake, T., 1982. Liquid phase mixing in the gas–liquid jet reactor with liquid jet ejector. J. Chem. Eng. Jpn 15 (6), 469–473.

Ogbonna, J.C., Mashima, H., Tanaka, H., 2001. Scale up of fuel production from sugar beet juice using loofa sponge immobilized bioreactor. Biores. Technol. 76, 1–8.

Ohkawa, A., Kusabiraki, D., Shiokawa, Y., Sakai, N., Fujii, M., 1986. Flow and oxygen transfer in a plunging water system using inclined short nozzles and performance characteristics of its system in aerobic treatment of wastewater. Biotechnol. Bioeng. 28, 1845–1856.

Ohkawa, A., Shiokawa, Y., Saki, N., Imai, H., 1985. Flow characteristics of downflow bubble columns with gas entrainment by a liquid jet. J. Chem. Eng. Jpn 18, 466–469.

Ozturk, S.S., Schumpe, A., Deckwer, W.D., 1987. Organic liquids in a bubble column: holdups and mass transfer coefficients. AIChE J. 33, 1473–1480.

Pal, S.S., Mitra, A.K., Roy, A.N., 1980. Pressure drop and holdup in vertical two-phase cocurrent flow with improved gas–liquid mixing. Ind. Eng. Chem. Process Des. Dev. 19, 67–72.

Pandit, A.B., Joshi, J.B., 1983. Mixing in mechanically agitated gas-liquid contactors, bubble columns and modified bubble columns. Chem. Eng. Sci. 38, 1189–1215.

Prakash, A., Margaritis, A., Li, H., 2001. Hydrodynamics and local heat transfer measurements in a bubble column with suspension of yeast. Biochem. Eng. J. 9, 155–163.

Reddy, Y.R., Kar, S., 1968. Theory and performance of water jet pump. J. Hydraul. Div. 94 (Hy5), 1261–1281.

Rodrigues, M.T.A., Vilaca, P.R., Garbuio, A., Takagai, M., 1999. Glucose uptake rate as a tool to estimate hybridoma growth in a packed bed bioreactor. Bioprocess Eng. 21, 543–556.

Saxena, S.C., Rao, N.S., Saxena, A.C., 1990. Heat-transfer and gas-holdup studies in a bubble column: air–water–glass bead system. Chem. Eng. Commun. 96, 31–55.

Schügerl, K., Lücke, J., Oels, U., 1977. Bubble column bioreactions. Adv. Biochem. Eng. 7, 1–84.

Serizawa, A., Kataoka, I., 1988. In: Afgan, N.H. (Ed.), Phase distribution in two-phase flow: transient phenomena in multiphase flow. Hemisphere Publishing Corporation, New York, pp. 179–224.

Serizawa, A., Kataoka, I., 1992. Dispersed flow. Proceedings of the third International Workshop on Two-Phase Flow Fundamentals, London, UK.

Shah, Y.T., Kulkarni, A.A., Wieland, J.H., 1983. Gas holdup in two- and three-phase downflow bubble columns. Chem. Eng. J. 26, 95–104.

Shah, Y.T., Kelkar, B.G., Godbole, S.P., Deckwer, W.-D., 1982. Design parameters estimations for bubble column reactors. AlChE J. 28 (3), 353–379.

Shiao, T.I., Ellis, M.H., Dolferus, R., Dennis, E.S., Doran, P.M., 2002. Overexpression of alcohol dehydrogenase or pyruvate decarboxylase improves growth of hairy roots at reduced oxygen concentrations. Biotechnol. Bioeng. 77, 455–461.

Silvester, R., 1961. Characteristics and applications of the water jet pumps, La Houlle Elanche Grenoble, France, 16, p. 45.

Sittig, W., Faust, U., Prave, P., Scholderer, J., 1978. Technologische und Wirtschaftliche Aspekte der Einzeller-Proteingewinnung. Chemische Industrie 30, p. 713.

Smith, J.S., Burns, L.F., Valsaraj, K.T., Thibodeaux, L.J., 1996. Bubble column reactor for wastewater treatment. 2. The effect of sparger design on sublation column hydrodynamics in the homogeneous flow regime. Ind. Eng. Chem. Res. 35, 1700–1710.

Smith, R.A., 1951. Some aspects of fluid flow. Institution of Physics. Edward Arnold and Co, London, p. 229.

Sneath, R.W., 1978. The performance of a plunging jet aerator and aerobic treatment of pig slurry. Water Pollut. Control 77, 408–420.

Son, S.H., Choi, S.M., Lee, Y.H., Choi, K.B., Yun, S.R., Kim, J.K., 2000. Large-scale growth and taxane production in cell cultures of Taxus cuspidate using a novel bioreactor. Plant Cell Reports 19 (6), 628–633.

Sun, Y., Furusaki, S., 1990. Effects of product inhibition on continuous acetic acid production by immobilized *Acetobacter aceti*: theoretical calculations. J. Ferment. Bioeng. 70 (1), 196–198.

Tatterson, G.B., 1991. Fluid mixing and gas dispersion in agitated tanks. McGraw-Hill Inc, New York.

Tojo, K., Miyanami, K., 1982. Oxygen transfer in jet mixer. Chem. Eng. J. 24, 89–97.

Towell, G.D. and Ackerman, G.H., 1972. Axial mixing of liquid and gas in large bubble reactors. Proceedings of the Fifth European/Second International Symposium B31.

Van der Lingen, T.W., 1960. A jet pump design theory. J. Basic Eng. 82, 947–960.

Vandu, C.O., Krishna, R., 2004. Volumetric mass transfer coefficients in slurry bubble columns operating in churn-turbulent flow regime. Chem. Eng. Process. 43, 987–995.

Wallis, G.B., 1969. One-dimensional two-phase flow. McGraw Hill, New York.

Westlake, R., 1986. Large-scale continuous production of single cell protein. Chemie Ingenieur Technik 58, 934–937.

Wilkinson, P.M., Spek, A.P., Van Dierendonck, L.L., 1992. Design parameters estimation for scale-up of high-pressure bubble columns. AIChE J. 38, 544–554.

Wilkinson, P.M., Haringa, H., Van Dierendonck, L.L., 1994. Mass transfer and bubble size in a bubble column under pressure. Chem. Eng. Sci. 49, 1417–1427.

Yamagiwa, K., Kusabiraki, D., Ohkawa, A., 1990. Gas holdup and gas entrainment rate in downflow bubble column with gas entrainment by a liquid jet operating at high liquid throughput. J. Chem. Eng. Jpn 23, 343–348.

Yamagiwa, K., Ohmae, Y., Dahlan, H.M., Ohkawa, A., 1991. Activated sludge treatment of small-scale waste water by a plunging liquid jet bioreactor with cross-flow filtration. Bioresour. Technol. 37, 215–222.

Zaidi, A., Ghosh, P., Schumpe, A., Deckwer, W.-D., 1991. Xanthan production in a plunging jet reactor. Appl. Microbiol. Biotechnol. 35, 330–333.

Zlokarnik, M., Eignung von einlochböden als gasverteiler in blasensäulen. Chemie Ingenieur Technik 43(6), 329–335.

2

Flow Regime and Its Transition

Flow Regime

Multiphase flow behaviors are affected by interfacial tension forces; wetting characteristics of the liquid on the channel wall; the contact angle; and the exchange of mass, momentum, and energy between the phases. Dictates of the flow behavior and shape of the interfaces between phases in a multiphase mixture are commonly referred to as the *flow regime* or the *flow pattern*. The flow regime in a multiphase reactor depends on several factors, including dynamic variables such as phase flow rate; geometric variables such as diameter, length, cross-sectional area of the of the device, hole size of the phase distributor, and the catalyst particle size; and the physical properties of the phases. The boundary of the two regimes is called the *flow regime transition*. The flow regime plays a significant role in the operation and performance of the bubbly flow reactor. Different hydrodynamic characteristics result in different mixing as well as heat and mass transfer rates in different flow regimes. Understanding the flow regimes is very important to reactor design and scale up. The study of flow regimes can help to resolve the complex hydrodynamics of bubbly flow devices and optimize the operating conditions of the systems. Different flow regimes affect reactor performance in many respects, such as pressure fluctuation, mass transfer, heat transfer, momentum loss, mixing, and reactor volume productivity (Nedeltchev et al., 2003). In gas–liquid chemical reactors maximum coefficients of reaction can be attained by keeping a dispersed-bubbly flow regime to maximize the total interfacial area. The efficiency of a bubbly flow device changes as the flow structures change. A heterogeneous regime is required in most industrial reactors, but a homogeneous flow regime is desired in some bioreactors (Ribeiro, 2008). Hence, further information and insight regarding the study and identification of flow structures in a bubble column under different superficial gas velocities are valuable and important. Therefore, the demarcation of flow regimes becomes an important task in the design and scale up of bubbly flow reactors and has led to considerable research efforts, which have resulted in various experimental methods and empirical, semi-empirical, and mechanistic models to identify the same. The flow regime directly impacts the interfacial area of contact and mixing between phases. The flow regime highly depends on both gas and liquid flow rates and on the diameter of the column. Bubbly flow device applications can be classified based on their flow regimes. Most biochemical applications such as cultivation of bacteria or mold fungi and production of single-cell protein, animal cell culture, and treatment of sewage are performed in bubbly flow. In addition, other examples are hydroconversion of heavy oils and petroleum feedstock and coal

Hydrodynamics and Transport Processes of Inverse Bubbly Flow. http://dx.doi.org/10.1016/B978-0-12-803287-9.00002-3

hydrogenation. For highly exothermic processes such as liquid phase methanol synthesis, Fisher-Tropsch synthesis, and catalytic hydrogenation of methyl α-acetamido cinnamate in [bmim][BF$_4$]/CO$_2$ media, the churn-turbulent flow regime is preferable (Shaikh and Al-Dahhan, 2007).

Flow Regimes in a Conventional Bubbly Flow Reactor

Three broad types of flow patterns are commonly observed in bubbly flow reactors, which are homogeneous (ideally separated bubbly flow) pattern, the heterogeneous (churn-turbulent: interacting bubbly flow, churn-turbulent bubbly flow, clustered bubbly flow) pattern, and slug flow pattern (Deckwer, 1992). There also exists a so-called "foaming pattern" that is not commonly encountered in bubble columns (Kantarci et al., 2005). In these different flow patterns, the interaction of the dispersed gas phase with the continuous liquid phase varies considerably. The various flow patterns of bubbly flow in bubble columns are shown in Figure 2.1, *I*.

Homogeneous or Dispersed Bubbly Flow Regime

This flow regime is known as a laminar, uniform, dispersed bubbly flow regime. The flow regime is obtained at low superficial gas velocities, approximately less than 5 cm/s in semi-batch columns (Hills, 1974). This flow regime is characterized by nearly spherical bubbles of relatively uniform small sizes and rise velocities traveling vertically with minor transverse and axial oscillations.

A uniform bubble distribution and relatively gentle mixing results over the entire cross-sectional area of the column. There is practically no bubble coalescence or

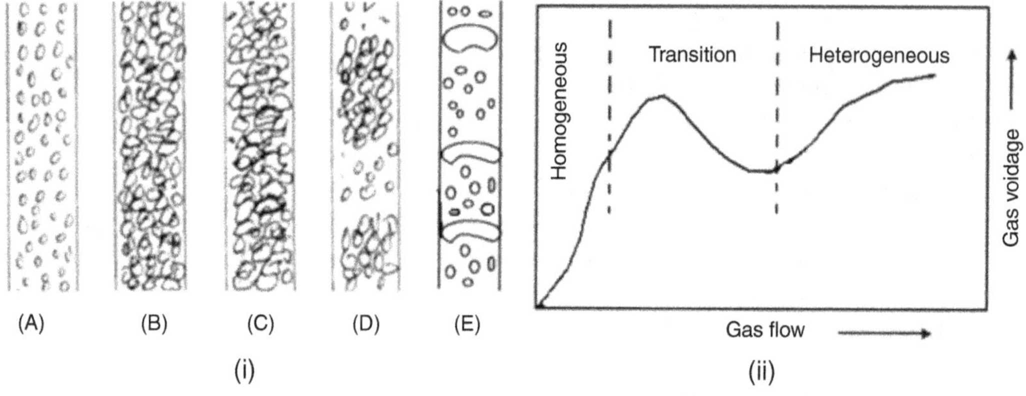

FIGURE 2.1 i, Flow regime of bubble column depicting homogenous, transition, and heterogeneous regimes: separated bubbly flow (A), interacting bubbly flow (B), churn-turbulent bubbly flow (C), clustered bubbly flow (D), and slug bubbly flow (E). ii, Transition of different flow regimes.

break-up. Hence a narrow bubble size distribution results. The size of the bubble depends mainly on the nature of the gas distribution and the physical properties of the liquid. In ideally separated bubbly flow, the bubbles do not interact with each other directly or indirectly.

Heterogeneous or Churn-Turbulent Bubbly Flow Regime

This regime is characterized by the disturbed form of the homogeneous gas–liquid system caused by enhanced turbulent motion of gas bubbles and liquid recirculation. As a result, unsteady flow patterns and large bubbles with short residence times are formed by coalescence. Because of intense coalescence and break-up, small as well large bubbles appear in this regime, leading to a wide bubble size distribution. The bubble number density becomes so large that the bubbles begin to interact with each other directly or indirectly because of collisions or the effects of wakes caused by other bubbles. With a further increase in bubble number density, the bubbles tend to coalesce to form so-called nonspherical cap bubbles, and the flow changes to interacting churn-turbulent bubbly flow. The flow contains cap bubbles formed in this way as well as smaller bubbles and is highly agitated because of the interactions between bubble motions and turbulent flow. The large bubbles churn through the liquid; thus, it is called *churn-turbulent flow*. Once in a while, the large bubbles form a clustering of bubbles, as shown in Figure 2.1, and they behave like a single gastropod. After a certain travel, they sometimes coalesce to form a gas slug, and sometimes they separate into individual bubbles. This flow regime is thus a transition from bubbly flow to slug or churn flow. Churn-turbulent flow is frequently observed in industrial-size, large-diameter columns (Hyndman et al., 1997).

Slug Bubbly Flow Regime

A slug flow regime is mainly observed in small-diameter columns at high gas flow rates (Hyndman et al., 1997). The primary characteristic of slug flow is its inherent intermittence. It is characterized by long "Taylor" bubbles, also called gas slugs, rising and nearly filling a pipe cross-section. The basic part of slug flow is a small compartment, involving the region of a long bubble plus the region of the following liquid slug. During stable slug flow, liquid is shed from the back of the slug at the same rate that liquid is picked up at the front. As a result, the slug length stays more or less constant as it travels along the tube. The liquid forms a falling annular film and accelerates as it moves downward. Bubble slugs can be observed in the column diameter up to 15 cm (Hills, 1976; Miller, 1980). The slug formed may be equal to or greater than the diameter of the column. Bubbles occupy the entire cross-section of the column and their shape resemble Taylor's bubbles. Thorat and Joshi (2004) reported that the transition gas velocity depends on column dimensions (diameter, dispersion height), sparger design, and physical properties of the system. There is no definite quantitative range for superficial velocities to characterize the flow regimes. Different studies performed with different systems and operating conditions provide different results in determination of regime boundaries and regime transitions (see Figure 2.1(ii)). From a practical point of view, slug flow occurs over a wide range of intermediate flow rates of both phases. For this reason, it presents a major interest for many industrial processes such as production of oil and gas in wells, emergency cooling of nuclear reactors, and boiling and condensation processes in power generation facilities as well as in chemical plants and refineries. The slug flow at very high liquid velocities results in very high convective heat and mass transfer coefficients for very efficient transport operations.

Flow Regimes in Inverse Bubbly Flow Reactors

The nature of the flow regime developed in the inverse flow bubble column is at least non-uniform bubbly flow in which a rather narrow heterogeneity in bubble size (d_b = 3–5 mm) can be observed. The flow regime in inverse flow bubble columns with a sintered disc, porous plate, or ring-type gas sparger varies along the flow direction in the column. Intense turbulence occurs in the vicinity of the gas distributor, where frequent coalescence and breakup of bubbles can be seen. Away from the distributor, bubble diameters become more uniform. However, bubbles with diameters larger than some critical value rise through the column to the gas distributor, where they may be split into smaller bubbles and again flow inversely into the bulk of the column. Ohkawa et al. (1986) reported that four types of flow regimes can be observed during the operation of the column using an air–water system:

- Regime A: bubble stagnant flow
- Regime B: non-uniform bubbling flow
- Regime C: uniform bubbling flow
- Regime D: churn-turbulent flow

Bando et al. (1988) observed two different flow patterns when gas and liquid were injected simultaneously into the column. A spouting section developed near the nozzle exit where the gas and liquid were violently mixed, and a calm section of uniform bubbly inverse flow appeared at some distance below the nozzle, which was similar to that observed in a concurrent upflow bubble column with a simultaneous gas–liquid injection nozzle. Relatively few studies on flow patterns in concurrent inverse bubbly flow are found in the literature. However, all of the flow patterns of concurrent upward flow may also appear in the inverse flow conditions. The gas–liquid dispersion presents regions of different flow regimes with very distinct mixing characteristics and bubble sizes along the column length. The number of those zones in the inverse bubble column differed for different researchers possibly because of the observation method used and to the configuration of the column. Two distinct flow regimes (homogeneous and churn-turbulent) are observed in an ejector-induced inverse flow bubble column with a gas–liquid system

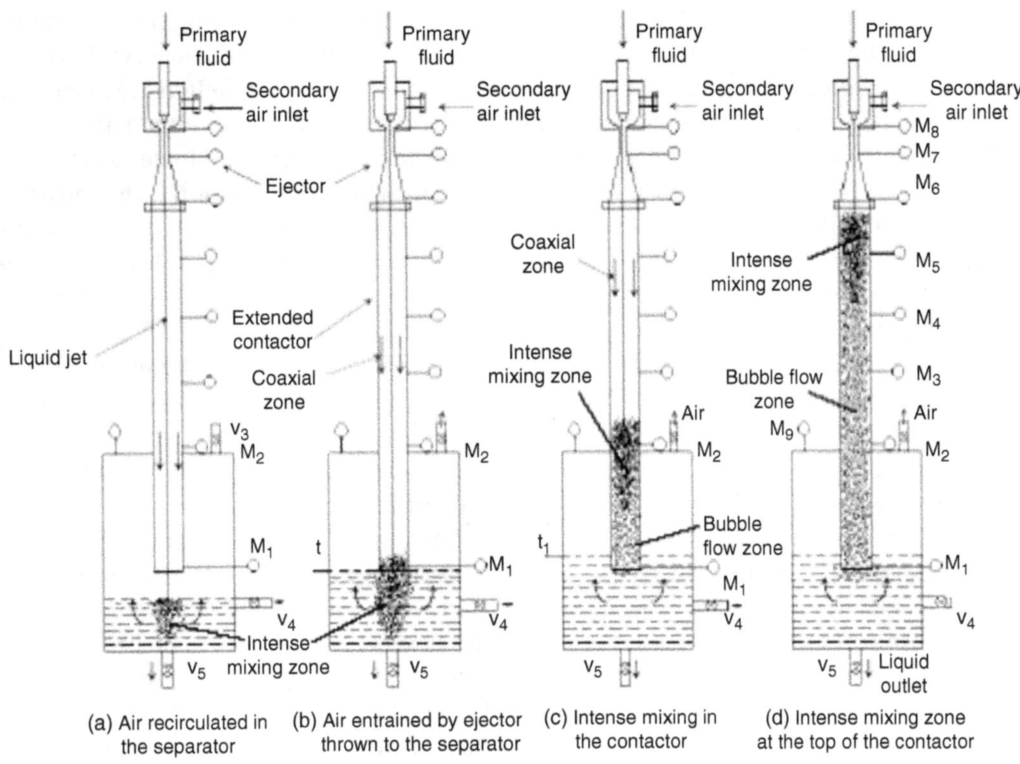

FIGURE 2.2 The zone of churn-turbulent and bubbly zone for inverse flow of bubbles.

(Majumder, 2005). Both flow regimes distingly occur for the same operating conditions in two zones of the inverse bubbly flow reactor as shown in Figure 2.2. The two zones are the intense mixing zone and the homogeneous bubble zone. In the homogeneous bubble zone, the bubbles are uniform in size. They move in an orderly fashion with little collision among bubbles, and the liquid is mildly stirred by the bubbles. As the liquid flow rate increases, there is a natural tendency of the gas bubbles to coalesce, forming large bubbles. The rate of coalescence increases rapidly after a certain limit of liquid flow rate depending on the physical properties of the liquid. Under this condition, large bubbles tend to move upward because of higher buoyant forces, and consequently, heterogeneous churn-turbulent results.

In the mixing zone, because of high momentum exchange, the bubbles are highly interacting with each other and continuously change their shape and size. The breakup of bubbles results in smaller bubbles and drag downward of the smaller bubbles by liquid momentum against their buoyancy. At high momentum (jet velocity >15 m/s), it was observed that elongated bubbles are instantly formed, which may be referred to as a *slug flow condition*. In this condition, the mixing zone increases by diminishing the bubbly zone. In a plunging liquid jet inverse bubbly flow reactor, the foaming flow pattern is observed in the presence of a foaming agent in the liquid. Evinc (1982) observed two distinctive flow regions using a column of internal diameter 0.1 m and height 1.0 m with nitrogen–water, air–water, oxygen–water, and carbon dioxide–water systems. The first one is the upper turbulent region, where small bubbles (d_b < 1.5 mm) and lower gas holdup (30%–40%) were observed in the 0.350-m top section. The second region is the lower region, where a uniform larger bubbly two-phase flow (d_b = 3 mm for carbon dioxide–water and d_b = 5 mm for oxygen–water) and 50% gas holdup were observed in the lower 0.650-m section. Wilson and Jones (1983) reported that in the highly turbulent region, bubbles are loosely packed because of a high-velocity liquid inlet stream, and a high packing bubble density is observed in less turbulent regions and the region of clearly defined bubbles with gentle fluctuations. Lu et al. (1994) studied systematically the flow regimes in a column of internal diameter 76 mm and height 760 mm using oxygen–aqueous sodium alginate, oxygen–aqueous glycerol, oxygen–aqueous propanol and oxygen–water systems and reported four distinct regions of flow patterns. They also reported that the dispersion of the of the fluid based on flow regimes in the column depends on the liquid phase properties, gas and liquid flow rates, column geometry, and liquid inlet velocity.

Gas distribution without an ejector system in case of inverse flow may also result in bubble, annular, intermittent, separated, and dispersed flow patterns with only minor modifications (Crawford et al., 1986) as shown in Figure 2.3. Crawford et al. (1986) reported that bubbly flow means small discrete bubbles of various diameters (all significantly less than the tube diameter) moving in the direction of the liquid but generally not at the liquid velocity.

In vertically inverse flow, the bubbles tend to be ellipsoidal or spherical at the high flow rates but hemispherical at the lower flow rates. The smaller, faster moving bubbles tend

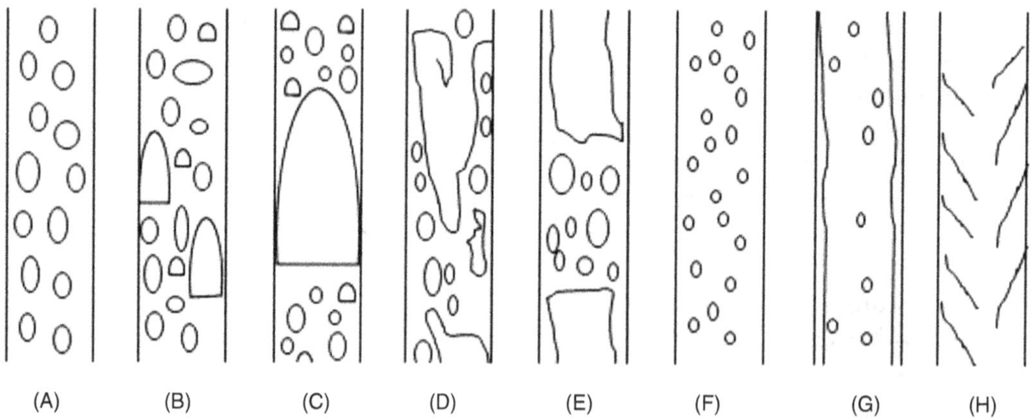

FIGURE 2.3 Flow pattern configurations in inverse flow. **(A)** Bubble. **(B)** Transition. **(C)** Plug (slug). **(D)** Churn-intermittent. **(E)** Semi-plug. **(F)** Dispersed. **(G)** Annular. **(H)** Falling film.

to move toward the center of the pipe and the larger slower moving ones tend to move near the wall. They observed the intermittent flow pattern in which there are alternating vapor and liquid packets. They found plug flow (sometimes called slug flow) in which large Taylor plugs of gas are formed. They reported that these gas bubbles with hemispherical caps nearly filled the entire tube and moved upward with respect to the liquid, but the drag and viscous forces were always sufficient to give them a net inverse velocity. This plug flow can be seen near the bubble–intermittent transition. As one moved away from the transition, the gas plugs did not fill the entire tube and did not have a hemispherical head. This region was designated the "semi-plug." The intermittent region also includes "churn" flow in which there is a chaotic mixture of vapors and liquid packets. This behavior was observed only in the vertical lines and was more prominent at the higher pressure. They pointed out that the distinction between annular and separated flow is not readily applicable in vertically inverse flow. Under these conditions, a "falling film" flow, which is characterized by liquid layers that trickledown the walls, will be encountered. The liquid layers appear to flow gently over one another as sheets. The gas flows in the center core. The annular flow can be distinguished from the "falling film" by a relatively small amount of droplet entrained in the gas core in the centerline. Generally, numerous tiny bubbles smaller than 1 mm appeared to be dispersed in the liquid layer on the wall during annular flow, they reported.

Flow Regime Map and Its Transition

The graphical representation of boundaries or transitions between different flow patterns is called the *flow pattern map*. For a quantitative analysis of the experimental conditions that lead to the transition of flow patterns, a set of parameters or parameter groups are specified to show a direct interrelations on such transitions. The flow regime transition from bubbly to churn-turbulent flow or from churn-turbulent to slug flow depends simultaneously on parameters such as superficial gas velocity, column diameter, liquid and gas phase properties, and distributor design. Figure 2.4 shows one of the few approximate flow regime maps that distinguishes among bubbly, churn-turbulent, and slug flow (Deckwer, 1992), which is valid for both bubble and slurry bubble columns with a batch (stationary) liquid phase operated with a low viscous liquid at ambient condition. The exact boundaries of the transition region shown in Figure 2.4 depend on the system properties.

There is no flow regime map available that covers a wide range of industrial conditions. To characterize the flow regimes, it is not possible to give definite quantitative ranges for superficial velocities. Different studies performed with different systems and operating conditions provide different results in determination of regime boundaries and regime transitions (Bukur et al., 1987; Deckwer, 1992; Hyndman et al., 1997). From these studies, it is found that the bubbly flow regime prevails for superficial velocities lower than 5 cm/s. The churn-turbulent flow regime prevails for gas superficial velocities between 2 and 5 cm/s. The transition from bubbly to churn-turbulent flow in a bubble column with

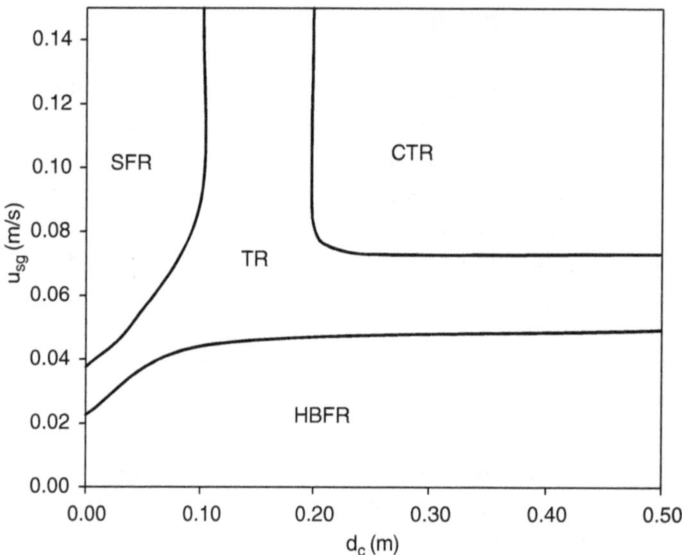

FIGURE 2.4 Flow regime map for upflow bubble columns (Deckwer, 1992). *CTR*, Churn-turbulent regime; *HBTR*, homogeneous bubbly flow regime; *SFR*, slug-flow regime; *TR*, transition ranges.

increasing superficial gas velocity is actually a gradual process (Hyndman et al., 1997). Several flow regime maps have been presented in the literature to identify the boundaries of possible flow patterns (Deckwer, 1992; Zhang et al., 1997). Quantitative criteria for the transitions from ideally separated to interacting bubbly flow and from interacting bubbly flow to churn-turbulent bubbly flow are roughly 0.01 and 0.06 in void fraction, respectively (Wallis, 1969). The bubbles ranging from 2 to 5 mm in diameter tend to collect toward the wall. If the bubble size is carefully controlled under fixed flow conditions, the transition boundaries shift toward a lower or higher gas or liquid flow rate (Serizawa and Kataoka, 1988). In case of an inverse bubbly flow reactor, it is seen that the churn-turbulent flow regime prevails at a velocity of liquid below 0.15 m/s, above which the operation is not possible (Kulkarni and Shah, 1984; Bando et al., 1988; Kundu et al., 1995; Mandal et al., 2004; Majumder, 2005). The homogeneous bubbly flow in inverse bubbly flow reactor offers the maximum gas liquid contacting area with a stable operation as shown in Figure 2.5. Kulkarni and Shah (1984) reported a flow regime map in terms of superficial gas and liquid velocities for possible operation in a concurrent inverse flow system. With a gas sparger, they found that operation is possible only in regime A. Regime C is practically undesirable, and operation is not possible in regime B. However, from Figure 2.5, it may be found that in regime B, operation is also possible with efficient gas–liquid distributor such as an ejector-nozzle system. Majumder (2005) showed that the gas-to-liquid flow ratio in his study is much higher than in Bando et al.'s study (1988) and comparable with the one in Kundu et al.'s study (1995). Relatively few studies on flow patterns of concurrent inverse bubbly flow are reported. However, all of the flow patterns of concurrent upward flow may also

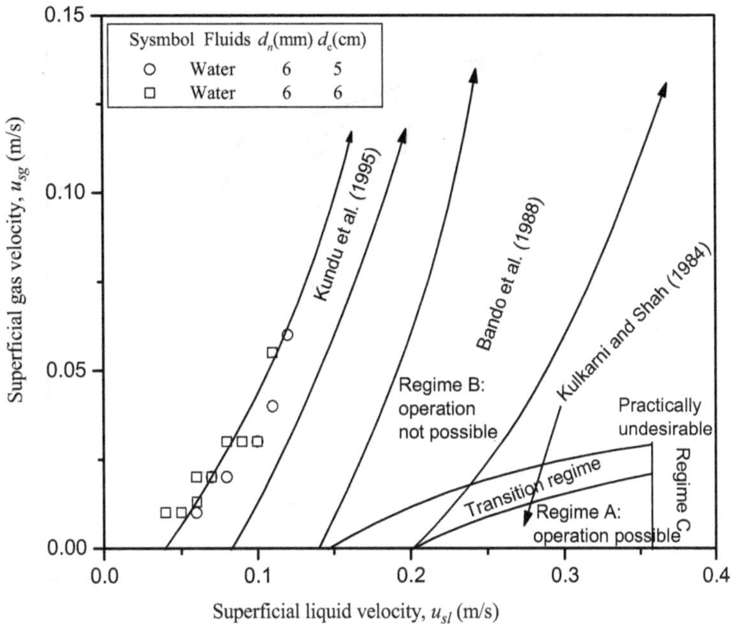

FIGURE 2.5 Flow regime map for ejector-induced inverse bubble columns (Majumder, 2005).

appear in the inverse flow situations. It is seen from the experimental results that in the ejector-induced inverse bubbly flow column, the minimum jet velocity criteria to get the visible clear bubbly and mixing zone depends on the surface tension and the jet diameter, which can be expressed as

$$u_j > \left[\frac{10.20\sigma_l}{\rho_g d_j}\right]^{1/2} \tag{2.1}$$

Crawford et al. (1986) examined the two-phase (gas–liquid) flow patterns at steady state using a refrigerant and its vapor in vertically inverse lines. They reported that the dispersed and annular flow transitions appeared essentially unchanged. The flow pattern map for the inverse flow is shown in Figure 2.6. The maps are similar in many respects to those obtained in upward or horizontal flow. Crawford et al. (1986) observed that because of increased gas density, the mass flux at which annular flow begins at 4 bars is about double that at which it begins at 2 bars.

They also observed different flow pattern in inverse flow at different downward inclinations. They noted that flow patterns at the sharp angles of 60 and 45 degrees was similar to that seen in inverse flow. However, at the less steeply declined angles of 30- and 15-degree angles, the intermittent transition line moved appreciably from the vertical, and there is observed a separated region between intermittent and annual flow. They reported that the significant expansion of the separated region in lines declined at 15 and 30 degrees. Later on Crawford et al. (1986) examined the transient flow pattern behavior of inverse

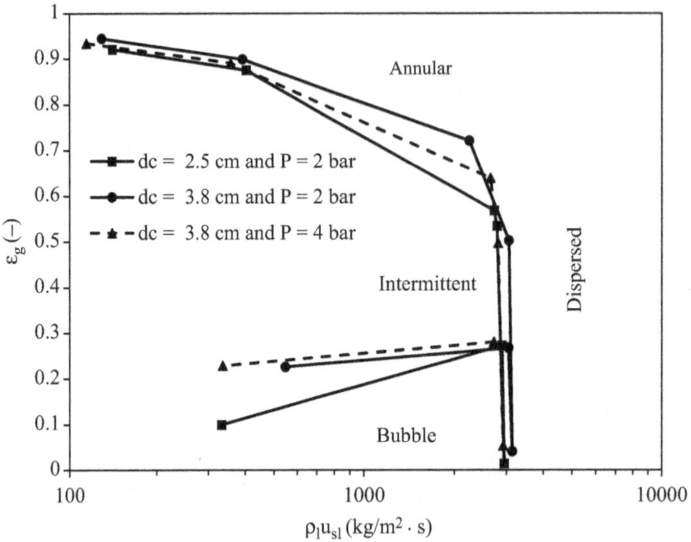

FIGURE 2.6 Steady-state flow pattern maps at 2 and 4 bar for inverse flow of refrigerant 113 in a vertical pipe. *(Adapted from Crawford et al., 1986).*

two-phase flow. They determined the relationship between void fraction and quality as a function of mass velocity for refrigerant 113 and its vapor in steady-state inverse flow. The "dynamic slip" model was used to determine the appropriate relationship between void fraction and quality during the transient, which enabled comparison of the transient flow pattern with steady-state flow maps. They concluded that the transient data approximately followed the steady-state map but that in most cases, the existing flow pattern persists slightly past the steady-state boundary.

Methods for Identification of Flow Regime Transition

There are mainly three methods that are used to identify the flow regime transitions in vertical bubble columns. They are (1) experimental methods, (2) mathematical model–based prediction, and (3) computational fluid dynamics (CFD) methods. There are various experimental methods used for identification of regime transition, which can be broadly classified in the groups of (1) visual observation, (2) evolution of global hydrodynamic parameter, (3) temporal signatures of quantity related to hydrodynamics, and (4) advanced measurement techniques.

Visual Observation

Traditionally, flow regimes have been defined according to visual observations performed by viewing the flow through transparent channels. The majority of all the reported data in the literature have been obtained in this manner. Although visual observation provides some information on the flow patterns, it is often difficult to identify the flow regime transitions without quantitative measurements, even in transparent columns, because of the relatively opaque nature of multiphase flow. The slow movement of bubbles can be observed in the homogeneous regime. However, in the heterogeneous regime, because of intense interaction of bubbles, leading to gross circulation, it is difficult to pinpoint the exact transition velocity (u_{trans}) by visual observation. It is admitted, however, that ambiguities regarding the exact nature of flow patterns may exist in the interpretation of such visual observations, particularly at high flow velocities. In this connection, still pictures by high-speed photography can be used as a useful aid. Even though such pictures may give a clear view of flow at a certain moment of time, the interpretation regarding the flow regime may be arbitrary or somewhat subjective, depending on the observer. The difficulty of obtaining a clear view of the central parts of the flow's cross-section by high-speed photography is due to the light diffraction at all gas–liquid or vapor–liquid interfaces, which, in some cases, would make the clear observations limited to only a layer of the mixture near the channel walls. However, despite the limitations, the direct visual observation approach has been used for its simplicity and inexpensiveness. It certainly is the best tool for simple experiments.

Evolution of Global Hydrodynamic Parameters

Global hydrodynamic parameters are manifestations of the prevailing flow patterns. They vary with the flow regimes. This fact has generally been used to identify the flow regime transition point. Typically, global hydrodynamics have been quantified based on overall gas holdup. The overall gas holdup increases with an increase in superficial gas velocity linearly ($\varepsilon_g \propto u_{sg}^{0.8-1}$) at low gas velocity, but because of an intense nonlinear interaction of bubbles at high gas velocities, the relationship between the overall gas holdup and superficial gas velocity deviates from linearity and obeys $\varepsilon_g \propto u_{sg}^{0.4-0.6}$. Hence, the change in the slope of the gas holdup curve can be identified as a regime transition point. Sometimes gas holdup shows an S-shaped curve, depending on operating and design conditions (Rados, 2003). In such cases, the superficial gas velocity at which maximum gas holdup is attained is identified as the transition velocity. However, when the change in slope is gradual or the gas holdup curve does not show a maximum in gas holdup, it is difficult to identify the transition point. In such cases, the Zuber and Findlay (1965) drift flux method can be used extensively. They have argued that gas holdup in two-phase flow depends on two phenomena: the gas rises locally relative to liquid because of phase density differences, and the gas holdup and velocity distribution across the column diameter cause gas to concentrate in a faster or slower region of flow, thereby affecting the average gas holdup. In the systems with down comers or in inverse bubbly flow, in which vertical plunging liquid jets act as a gas distributor to the system, a homogeneous bubbly flow is observed (Ohkawa et al., 1986; Evans, 1990; Yamagiwa et al., 1990). The following drift-flux model of Zuber and Findlay (1965) is applicable to describe the flow regime criteria whether it is upflow or inverse bubbly flow:

$$u_{sg}/\varepsilon_g = C_o(u_{sg} + u_{sl}) - u_d \tag{2.2}$$

C_0 is a distribution parameter and is a measure of the interaction of the holdup and velocity distribution. Where the gas is more concentrated in the faster region of flow, C_0 is greater than 1, and often it is taken as 1.2 for churn-turbulent flow. The parameter u_d is the weighted average drift velocity, accounting for the local slip. It represents the difference between the gas phase velocity and the average mixture velocity. Generally, it is assumed to be similar to the rise velocity of a bubble in an infinite medium. The weighted average drift velocity ranges from 0.18 to 0.24 m/s. For vertical upward flow, Lahey and Moody (1977) assumed that the drift velocity may be set as the bubble rise velocity and taken as independent of flow pattern. As flow becomes heterogeneous, intense liquid circulation sets in. According to Yamagiwa et al. (1990), the value of C_o is 1.17 in the inverse flow bubble column at higher liquid throughputs; however, Evans (1990) found that the value of this constant depends on the hydrodynamic regime in the inverse bubbly flow. For inverse bubbly flow, it rapidly decreases from about 2 at a low Reynolds number for liquid flow to a steady value of approximately unity for liquid flow Reynolds numbers greater than 10,000. In the churn-turbulent regime, this value is close to unity. Crawford et al. (1986) reported that the value of C_o depends on both the flow pattern and pressure. They found that for inverse bubbly flow, the values are 0.8 and 0.85 at pressures 2 and 4 bar, respectively. It can be noted that all the values of C_o are 1.0 or lower, which is in contrast to the values commonly used in upward flow where C_o is greater than 1.0. According to Zuber and Findlay (1965), values greater than 1 generally indicate that the gas is being transported more rapidly than the average flow velocity. In inverse bubbly flow, because gas is transported more slowly than the average flow, a value of C_o less than 1.0 is reasonable (Majumder et al., 2006). It is to be noted that for annular flow, C_o equals 1, which does not imply the same flow rate of gas and liquid velocity because the drift velocity, u_d is not zero.

Temporal Signatures of Quantity Related to Hydrodynamics

Many researchers are working in the development of detection criteria for multiphase flow regimes. One of the first research works on this subject is the one by Hubbard and Dukler (1966) in which different flow regimes were characterized through spectral signatures of temporal pressure signals. Many others are Weisman et al. (1979), Vince and Lahey (1982), Matsui (1984), Tutu (1984), Mishima and Ishii (1984), Sekoguchi et al. (1987), Sheikhi et al. (2013), and Shaban and Tavoularis (2014). Seleghim and Hervieu (1998) proposed an objective indicator for the bubbly to slug transition in vertical flow based on the quantification of the loss of stationarity through the standard deviation of Ville's instantaneous frequency. They also proposed the use of the time-frequency covariance calculated from the signal's Gabor transform as a new flow regime transition indicator. Several attempts have been made to capture the instantaneous flow behavior through an energetic parameter.

The following temporal signatures are used for flow regime transitions that use a variety of signals based on two principally different approaches:

Direct observation:

- Conductivity probe (Briens et al., 1997; Zhang et al., 1997; Julia et al., 2008)
- Local holdup fluctuations using resistive or optical probes (Bakshi et al., 1995)
- Local bubble frequency measured using an optical transmittance probe (Kikuchi et al., 1997)
- Sound fluctuations using an acoustic probe (Holler et al., 2003; Al-Masry and Ali, 2007)
- Temperature fluctuations using a heat transfer probe (Thimmapuram et al., 1992)

Indirect determination, including:

- X-ray or gamma-ray attenuation (Bennett et al., 1999; Heindel et al., 2008; Salgado et al., 2010; Saayman et al., 2013)
- Static pressure fluctuations (Drahos et al., 1992; Letzel et al., 1997; Vial et al., 2001a,b; Park et al., 2001, 2003)
- Thermal neutron scattering "noise" analysis (Stekelenburg and van der Hagen, 1993; Sunde et al., 2005; Zboray et al., 2014)
- Drag-disk signal analysis (Rouhani and Sohal, 1982)

Advanced Measurement Techniques

With advances in measurement techniques, various imaging and velocimetric techniques can be used in flow regime transition studies (Shaikh and Al-Dahhan, 2007). The reader can follow the details of the method from the reference given in the following points.

- Computer-automated radioactive particle tracking (CARPT) (Devanathan, 1990; Yang et al., 1993; Cassanello et al., 2001; Roy et al., 2002; Nedeltchev et al., 2003; Upadhyay et al., 2013)
- Electrical resistance tomography (ERT) (Dong et al., 2003; Tan et al., 2007; Sharifi and Young, 2013; Scott and McCann, 2005; Banasiak et al., 2014)
- Electrical capacitance tomography (ECT) (Bennett et al., 1999; Zhang et al., 2014)
- Laser Doppler anemometry (LDA) (Arastoopour and Shao, 1997; Olmos et al., 2003; Mudde et al., 1998; Vial et al., 2001)
- Particle image velocimetry (PIV) (Chen and Fan, 1992; Chen et al., 1994; Li and Hishida, 2009; Sathe et al., 2011)
- γ-Ray computed tomography (CT) (Shaikh and Al-Dahhan, 2005; Bieberle et al., 2013; de Mesquita et al., 2014)

Some typical literature on various experimental techniques that have been used to determine flow regime transition is summarized in Table 2.1.

Table 2.1   Different Experimental Studies Performed for Flow Regime Identification in Bubble Columns by Various Investigators

System and Mode of Flow	Column Dimension (m)	Pressure and Temperature	Measurement Technique	Investigations	Author
Air–water upflow	$d_c = 0.026$, $H = 5.2$	0.1 MPa, 298 K	Static pressure fluctuations	The standard deviation, pressure intensity, frequency function, and spectral density of static pressure pulsation showed different behavior-indifferent flow patterns.	Nishikawa et al. (1969)
Air–water upflow	$d_c = 0.04, 0.08, 0.16$	0.1 MPa, 298 K	Photographic method	Based on photographic method, three flow regimes were identified and described.	Ohki and Inoue (1970)
Air–water upflow		0.1 MPa, 298 K	X-ray	Three dominant patterns were identified: bubbly, slug, and annular regime.	Jones and Zuber (1975)
Air–water upflow	$d_c = 0.0254$	0.1 MPa, 298 K	Dual beam x-ray	Flow regime indicator based on pdf was proposed.	Vince and Lahey (1982)
Air–water, electrolyte solution upflow	$d_c = 0.1$, $H = 1.5$	0.1 MPa, 298–353 K	Drift flux and visual method	Transition velocity was found to decrease with an increase in temperature. Also, transition velocity in electrolytes was found to be higher than in water.	Grover et al. (1986)
N_2, CO_2, Ar, He, SF_6– deionized water and so on upflow	$d_c = 0.16, 0.19$	0.1–2 MPa, 298 K	Bed expansion and dynamic gas disengagement measurement	Transition velocity is significantly influenced by gas density and physical properties. A model for gas holdup incorporating influence of gas density on regime transition was proposed.	Krishna et al. (1991)
Air–water upflow	$d_c = 0.1$	0.1 MPa, 298 K	PIV, laser sheeting technique	Three flow regimes were identified: dispersed bubble, vertical-spiral, and turbulent flow. A conceptual flow four-region structure was proposed for the vertical-spiral flow regime. Similarities between two- and three-dimensional flow were pointed out.	Chen et al. (1994)
Air–water upflow	$d_c = 0.1$, $H = 2.2$	0.1 MPa, 298 K	Optical fiber probe	Multiresolution analysis of gas holdup fluctuations was performed. The intermittence of the local bubble concentration was related to the flow regime transition.	Bakshi et al. (1995)
Air–water upflow	$H = 1.6$, $W = 0.483$, $T = 0.0127$	0.1 MPa, 298 K	Flow visualization and PIV	The flow regimes were divided into dispersed bubble and coalesced bubbly flow regimes. The coalesced bubbly flow was divided into four-region and three-region flow.	Lin et al. (1999)

Table 2.1 Different Experimental Studies Performed for Flow Regime Identification in Bubble Columns by Various Investigators (*cont.*)

System and Mode of Flow	Column Dimension (m)	Pressure and Temperature	Measurement Technique	Investigations	Author
Nitrogen–water upflow	$H = 1.5$, $W = 0.56$, $T = 0.01$	0.1 MPa, 298 K	Optical transmittance probe with narrow He-Ne laser beam	Intervals between two successive bubble signals were subjected to time series analyses. The dynamics were characterized in terms of correlation dimension, Mann-Whitney statistic, and Hurst exponent.	Kikuchi et al. (1997)
Air–distilled and tap water and so on upflow	$d_c = 0.14, 0.15, 0.29$	0.1 MPa, 298 K	Bed expansion method	Transition velocity is a strong function of the type and geometry of the distributor. Also, effects of the diameter and static liquid height were studied.	Zahradnik et al. (1997)
Air–water, ethanol upflow	$d_c = 0.15$	0.1 MPa, 298 K	Gas hold-up curve, DGD	Addition of alcohol delays flow regime transition.	Krishna et al. (1999)
Air–water upflow	$d_c = 0.16$	0.1 MPa, 298 K	CARPT	Transition velocity was calculated based on Kolmogorov entropy. It was shown that the quality of mixedness in the lower and upper parts of the column is the same at transition velocity.	Nedeltchev et al. (2003)
Air–water upflow	$d_c = 0.376$, $H = 2.1$	0.1 MPa, 298 K	Pressure transducers	Wavelet analysis was applied to study bubbly and churn-turbulent flow. Based on wavelet packet table and spectrogram, it was observed that objects in bubbly flow have finer scale than in churn-turbulent flow.	Park et al. (2003)
Air–water upflow	$H = 1.2$, $W = 0.4$, $T = 0.04$	0.1 MPa, 298 K	Laser Doppler anemometry (LDA)	Various signal processing techniques were applied to LDA measurements and showed that characterization of flow regime transitions can be performed using LDA signals. Based on flow visualization, they identified two transition regimes.	Olmos et al. (2003)
Air–water upflow	$d_c = 0.15$, $H = 0.66$	0.1 MPa, 298 K	Acoustic probe	Acoustic measurements used to estimate bubble frequency, bubble size, and its distribution. The regime based study of these parameters was performed.	Al-Masry et al. (2005)
Air–Therminol LT upflow	$d_c = 0.1615$	0.1–1 MPa, 298 K	Computed tomography (CT)	A new flow regime identifier based on steepness of gas holdup radial profile was proposed and used to study the effect of pressure on flow regime transition.	Shaikh and Al-Dahhan (2005)

Table 2.1 Different Experimental Studies Performed for Flow Regime Identification in Bubble Columns by Various Investigators (*cont.*)

System and Mode of Flow	Column Dimension (m)	Pressure and Temperature	Measurement Technique	Investigations	Author
Air–water inverse flow	$d_c = 0.05, 0.06$	0.1 MPa, 298 K	Photographic method	Based on the photographic method, two flow regimes were identified and described.	Majumder (2005)
Air–water system, upflow	$d_c = 0.15$, $H = 1.5$	1 bar	Passive acoustic measurements	Using both spectral and chaos-based techniques to characterize the column flow regimes and to predict the transitions points	Ajbar et al. (2009)
Air–water upflow	$d_c = 0.0390$; $H/d_c = \sim 3.5$		Gamma-ray CT	The high-resolution gamma CT was applied to study the effect of different sparger designs for large bubble columns. The extracted holdup distributions revealed that homogeneous gas distribution was only possible using the radial sparger.	Bieberle et al. (2013)

CARPT, Computer-automated radioactive particle tracking; d_o, diameter of the column; *DGD,* dynamic gas disengagement, ***; *PIV,* particle image velocimetry; *H,* column height; *pdf,* Probability distribution function, ***.

Factors Influencing Formation and Stability of Flow Regime Transitions

The knowledge of the transition between the homogeneous bubbly flow and the churn-turbulent flow regimes is important for the design and operation of industrial reactors. The flow regime transitions depend on several factors, which are explained in this section.

Effect of Operating Pressure and Temperature

The flow regimes and the regime transition have been studied extensively under various pressure and temperature over the past 3 decades. The effect of the operating pressure on the transition has been examined by many researchers in bubble columns (Krishna et al., 1991; Wilkinson et al., 1992; Reilly et al., 1994; Lin et al., 1999). Most of the investigators have reported that an increase in pressure delays flow regime transition. The transition from the homogeneous to the heterogeneous regimes is generally found to be delayed to higher gas velocities because of the increase of overall circulation velocity with pressure. An increase in pressure increases breakup rate, reduces coalescence rate, and delays the appearance of large bubble and thereby flow regime transition changes. Most investigators have found a general trend (Figure 2.7) of an increase in transition velocity with an increase in pressure (Krishna et al., 1991; Wilkinson et al., 1992; Reilly et al., 1994; Lin et al., 1999; Shaikh and Al-Dahhan, 2005).

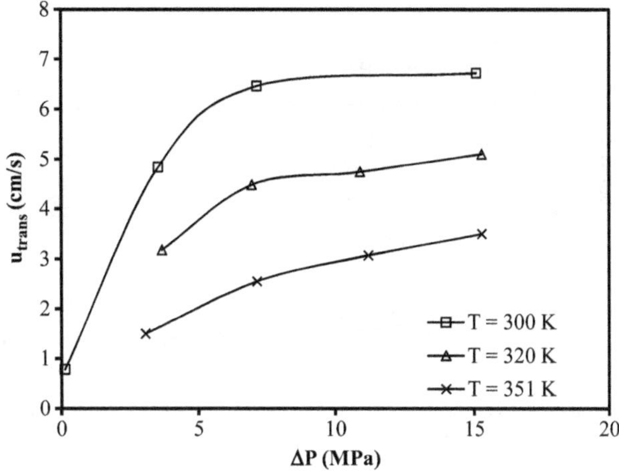

FIGURE 2.7 Effect of temperature and pressure on transition velocity. *(Reproduced from Lin et al., 1999).*

Because most industrial operations are performed at high operating pressure, it is important to investigate its effect on flow regime transition at high pressure. Increasing temperature delays the regime transition (Fan, 1989), and the rise velocity of bubbles in liquids and liquid–solid suspensions decreases with an increase in temperature. An increase in temperature increases overall gas holdup because of the formation of small bubbles. The effect of temperature can be accounted by the change in physical properties of fluid. Very few studies have investigated the effect of temperature on flow regime transition (Grover et al., 1986; Bukur et al., 1987; Lin et al., 1999, Fan, 1989). Because of generation of small bubbles, flow regimes are sustained at higher velocities with an increase in temperature. An increase in pressure and temperature has a favorable effect on flow regime stability and results in increasing transition velocity. Transition velocity does not vary as pressure exceeds a critical value, particularly at higher temperature. At 350 K, the critical value of pressure is 7 MPa as stated by Lin et al. (1999).

Effect of Physical Properties of Fluid

Effect of Liquid Viscosity

The liquid phase viscosity has a significant effect on the hydrodynamic characteristics of bubbly flow. An increase in viscosity results in a stable bubble interface and thereby results in an increase in coalescence rate and a decrease in breakup rate. This gives rise to an early appearance of large bubble and hence advances the flow regime transition. Moderate viscosities (3–22 mPa.s) destabilize the homogeneous regime, thereby reducing transition velocity. Low viscosity (1–3 mPa.s) stabilizes the homogeneous regime, hence increasing the transition velocity as viscosity increases in this narrow range (Ruzicka et al., 2003) as shown in Figure 2.8. An addition of solids increases the "pseudo-viscosity" of the liquid phase and stabilizes the interface. Hence, the coalescence rate is increased, and the breakup rate is

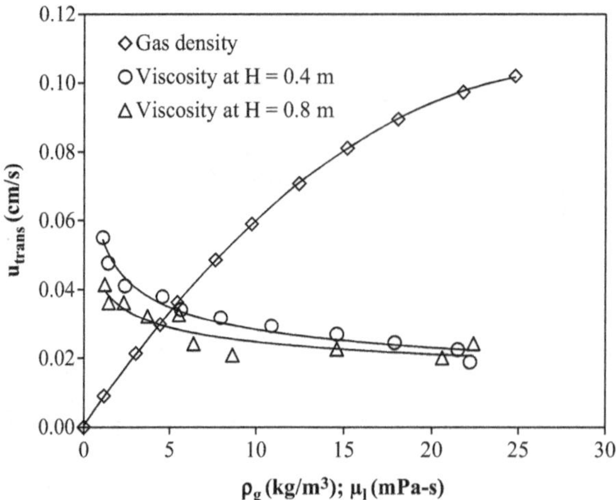

FIGURE 2.8 Variation of transition velocity of homogeneous to heterogeneous flow regimes with gas density and with viscosity. *(Reproduced from Krishna et al., 1991 and Ruzicka et al., 2001).*

reduced, resulting in an early appearance of large bubbles (Vandu et al., 2004; Ruthiya et al., 2005). Thimmapuram et al. (1992) reported that an increase in liquid viscosity lowered the bubble break-up. They observed that gas holdup increases with viscosity ranging from 1 to 3 cp and then decreases sharply until about 11 cp and then decreases slowly up to about 39 cp. The changes were relatively sharp at higher gas velocities. This was explained on the basis of the bubble coalescence phenomenon on liquid-phase viscosity. In the small-viscosity range up to about 3 cp, bubble coalescence was insignificant, and moderate drag forces contributed to more uniform distribution of bubbles and to higher gas holdups. At higher viscosity, bubble coalescence decreased the gas holdup. The change in gas holdup caused by this bubble characteristic changes the flow regimes from homogeneous to the heterogeneous. Zahradnik et al. (1997) reported the viscosity of saccharose solution on the transition regime is found to be significant. Existence of drag forces promotes the bubble coalescence in the distributor region. They used the saccharose concentration starting from 30 wt %, and this study focuses on the viscosity range from 0 to 41 wt % of glucose at different aspect ratios. Thet et al. (2006) concluded from their experimental results that with increasing liquid viscosity, a narrowing of the homogeneous regime showed at high viscosities. They observed that beyond viscosity 4 mPa.s, the bubbling was in the hetero-geneous regime for all superficial gas velocities.

They also reported that earlier transition was observed with increasing viscosity up to 3.5 mPa.s. They explained that at a high viscosity, shear stresses between the liquid and bubble phase are higher. Bubbles remain in the liquid for a longer time, and this in turn enhances the possibility of bubble coalescence and fluctuation of fluid flow patterns. Thus, higher viscosity promotes the transition from homogeneous toward heterogeneous flow by increasing the instability of the flow regime. Zahradnik et al. (1997) also found that the homogeneous bubbling column was suppressed at high liquid viscosities.

Effect of Gas Density

The gas phase plays a very important role in the design and operation of a bubble column reactor and in determining the chemical conversion achieved. However, well-defined results on the effect of gas density on the flow regimes are not plentiful. Studies by Crowley et al. (1992) with freon gas of density 32.5 kg/m^3 in a 17.8-cm pipe and by Wu et al. (1987) with natural gas of density 65.0 kg/m^3 in a 20.3-cm pipe were discussed by Hurlburt and Hanratty (2002) and by Andritsos et al. (1992), who emphasize the importance of slug stability in determining transition at high gas density. The transition from the homogeneous to the heterogeneous regimes is generally found to be delayed to higher gas velocities, and the gas holdup increases when the gas density increases because of the increase of overall circulation velocity (Letzel et al., 1997). Transition velocity increases with an increase in gas density as stated by Krishna et al. (1991) and as shown in Figure 2.8.

Effect of Surface-Active Agent

In the presence of a surface-active agent, the flow regime transition is delayed because of the suppression of the coalescing tendencies of small bubbles. The presence of alcohols in water systems reduces surface tension and hence induces no coalescing tendencies in the system (Urseanu, 2000). Figure 2.9 shows the change in transition velocity of a homogeneous to a heterogeneous regime in presence of alcohol. Reduction in the surface tension of the liquid led to a decrease in bubble stability and thus to smaller bubbles. This condition can be created by either adding surface active components to the liquid phase or using a liquid with a smaller value of surface tension. Krishna et al. (1999) and Zahradnik et al. (1997) explained that delaying the transition point by addition of alcohols is likely to be caused by the enhancement of homogeneous regime stability. Thus, addition of a surface-active agent to the viscous phase system is predicted to maintain the stability of homogeneous regime.

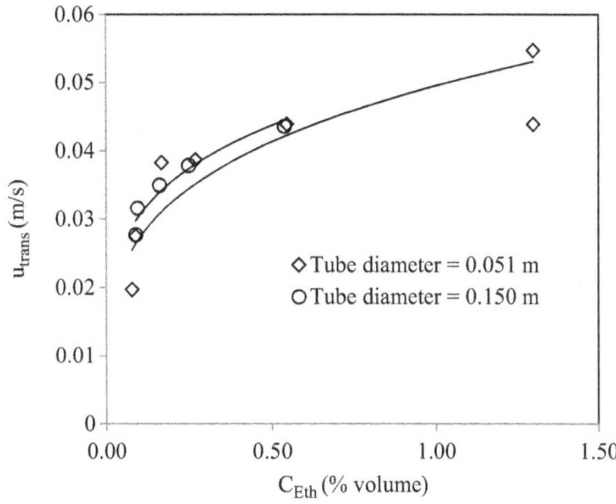

FIGURE 2.9 Effect of percent ethanol in water on the transition velocity. *(Reproduced from Urseanu, 2000).*

Effect of the Geometrical Variables

Effect of Gas Distributor

Controlling the uniformity of gas release by a gas distributor across the bubble column reactor and the initial bubble size is a challenge for efficient design of bubble column reactor. In general, the effect of the sparger is dominant in bubbly flow and diminishes as the system enters into churn-turbulent flow. The bubble size in homogeneous flow is the direct result of the nature of the distributor. The distributor design may affect the flow pattern of the bubble in the bubble column reactor.

Sarrafi et al. (1999) developed a method using the gas holdup curve to calculate the transition velocity of a flow regime. The transition velocity decreases sharply as the hole diameter of the distributor increases, up to 0.0015 m (Figure 2.10). Beyond this point, the transition velocity is independent of the hole diameter. Also as shown in Figure 2.10, the transition velocity increased with the perforation pitch (distance between the two distributor holes) up to 0.02 m. Larger orifice spacing (distance between to two orifice holes of the distributor) shows an insignificant effect on transition velocity (Shaikh and Al-Dahhan, 2007). Thorat and Joshi (2004) found that transition gas holdup increases with a decrease in the percentage of free open area and a decrease in hole diameter. Zahradnik et al. (1997) reported that an orifice diameter of 1.6 mm cannot generate the homogeneous regime and 1 mm or smaller is recommended for the homogeneous regime. The plates with small and closely spaced orifices produce uniform layers of equal-sized spherical bubbles at a low flow rate. At a high flow rate, the flow pattern changes because of the instability of homogeneous regime and form transition. For the plates with large orifices, only a heterogeneous regime can be obtained (Ruzicka et al., 2001). The small hole nozzle can produce smaller bubbles with a short distance of plunging point of which shows the uniform bubble and homogeneous flow regime; at high velocity of the liquid, because of

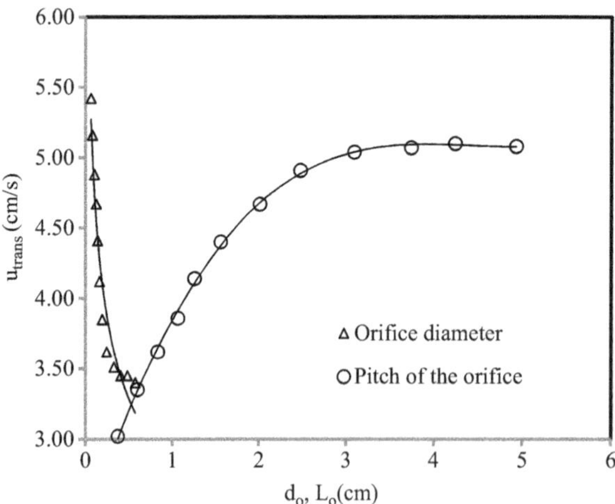

FIGURE 2.10 Effect of sparger on the transition velocity. *(Reproduced from Sarrafi et al., 1999).*

its high kinetic energy produces the heterogeneous flow of bubbles in an inverse flow bubble column (Majumder, 2005).

Effect of Column Size

The dimension of the bubble apparatus on the flow regime is also important for its scale up and design. These dimensions include the diameter of the column (d_c), unaerated liquid height or so-called column height (H), and aspect ratio (H/d_c). The diameter, d_c, should exceed 0.1 to 0.2 m (Jamialahmadi and Muller-Steinhagen, 1993); the height, H, should be larger than 1 to 3 m (Wilkinson et al., 1992); and the aspect ratio is recommended to be above 5 (Wilkinson et al., 1992; Zahradnik et al., 1997) for homogeneous operation of a bubble column reactor. Ruzicka et al. (2001) reported that the column size destabilizes the homogeneous regime. They suggest that the column aspect ratio alone cannot replace the simultaneous effect of the column height and width. The transition velocity increases with an increase of column diameter in the range of 0.04 to 0.16 m (Ohki and Inoue, 1970). Urseanu (2000) found a general trend of an increase in transition holdup with an increase in column diameter. Based on their own data and literature data in air-water systems, Sarrafi et al. (1999) found that the transition velocity increases with an increase in the column diameter. However, it becomes independent of the column diameter beyond 0.15 m (Shaikh and Al-Dahhan, 2007). At different liquid static heights, an increase in the column diameter reduced the transition velocity (Zahradnik et al., 1997; Ruzicka et al., 2001). For any particular diameter, the transition velocity showed a decreasing trend with an increase in the aspect ratio (Ruzicka et al., 2001). The authors (Ruzicka et al., 2001) concluded that aspect ratio alone is not the only physically relevant geometrical parameter and hence should not replace H and d_c in scale-up considerations. Thorat and Joshi (2004) found that an increase in the aspect ratio decreases the transition gas holdup. Thet et al. (2006) showed that increasing the height to diameter aspect ratio caused earlier transition for the range of viscosities they observed. Jamialahmadi and Muller-Steinhagen (1993) and Wilkinson et al. (1992) observed that a higher column height to diameter ratio decreases the gas hold-up because of bubble coalescence and changes the transition of the flow regime from homogeneous to heterogeneous. Ruzicka et al. (2001) and Sarrafi et al. (1999) also found that the gas velocity for the transition decreased with column size. A summary of the generalized effects of various operating and design parameters on flow regime transition is shown in Table 2.2.

Theories on Prediction of Flow Regime Transition

Each flow pattern has in some respect a particular property that either may be desirable or should be avoided in processes of different kinds. Also, the rate of exchange of mass, energy, and momentum between vapor and liquid depends on the extent of interfacial area per unit volume and hence is dependent on the flow regime. In some chemical processes, increased surface contact between gas and liquid is used to extract some chemical species

Table 2.2   Generalized Effects of Various Parameters on Flow Regime Transition

Variables	Effect on Flow Regime Transition	Reference
Pressure	In general, an increase in pressure results in an increase in transition velocity.	Krishna et al. (1991); Wilkinson et al. (1992); Reilly et al. (1994); Lin et al. (1999); Shaikh and Al-Dahhn (2005)
Temperature	An increase in temperature increases the transition velocity and delays the flow regime transition.	Bukur et al. (1987); Lin et al. (1999)
Viscosity	An increase in viscosity, in general, advances the flow regime transition.	Wilkinson (1991); Ruzicka et al. (2001)
Surface tension	A reduction in surface tension increases the transition velocity.	Grover et al. (1986); Urseanu (2000)
Sparger design	The transition velocity decreases with an increase in hole size up to a certain hole size. The transition velocity increases with perforation pitch and then remains the same after a certain critical value.	Sarrafi et al. (1999); Jamialahmadi et al. (2000)

or to promote a chemical reaction. In such processes, one needs to ensure a finely distributed bubble regime to generate the maximum contact area between the two phases. Reliable flow regime prediction capabilities would be desirable in this regard. To design bubble column reactors, one needs to know "a priori" the prevailing flow regime at the design and operating conditions. Analysis of flow regime transition has been exercised by the development of various models and approaches that include pure empirical correlations and phenomenological models, linear stability theory, artificial neural network (ANN), statistical theory, and CFD.

Empirical Method

This is a common approach in modeling two-phase phenomena. The main reason in following this approach is that minimum knowledge of the flow characteristics is required. Thus, empirical models are easy to implement, and they often provide good accuracy in the range of the database available for the development of the correlation. As a consequence, one of principal disadvantages of this approach is that they are limited by the range of their underlying database. Another important disadvantage of the empirical approach is that no single correlation is currently able to provide an acceptable accuracy for general use. Weisman and Kang (1981) developed the flow regime transition criteria based on the transition data of various investigators for vertical two-phase flow. Taitel et al. (1980) have given theoretical criteria for flow pattern transitions. They argued that as the gas flow is increased in a bubbly flow, the bubble density increases. Finally, a point can be reached where the dispersed bubbles become so closely packed that coalescence to larger bubbles increases substantially, showing transition to slug flow. The void fraction in bubbly flow rarely exceeds 0.35 in this case; according to Griffith and Wallis (1969), coalescence is rarely observed in bubbly flow for void fractions less than 0.20. Therefore, Taitel et al. (1980) took the criterion for transition from bubble to slug flow to be a void fraction of 0.25. By using the relation for the rise velocity of large bubbles, they gave the following condition for the flow to change from bubble to slug flow as

$$u_{sl} < 3.0u_{sg} - 1.15\left[\frac{g(\rho_l - \rho_g)\sigma}{\rho_l^2}\right]^{0.25} \tag{2.3}$$

If the transitional void fraction is taken to be 0.30 instead of 0.25 as taken by Taitel et al. (1980), the above transitional criterion would become

$$u_{sl} < 2.333u_{sg} - 1.071\left[\frac{g(\rho_l - \rho_g)\sigma}{\rho_l^2}\right]^{0.25} \tag{2.4}$$

If the liquid flow rate is high enough to cause turbulent flow, turbulent breakup of bubbles would take place. If the resulting bubbles are large enough and the void fraction is above 0.25, the slug flow would result because of coalescence of the large bubbles. On the other hand, if the turbulent breakup process is vigorous enough to produce small bubbles (smaller than a critical size; Brodkey, 1967), then even for void fractions above 0.25, bubbly flow would exist. When slip velocity is neglected in calculating the void fraction, it results for transition from churn-slug flow to finely dispersed bubbly flow in the following condition:

$$u_{sl} + u_{sg} > 4.0\left[\frac{d_c^{0.429}\sigma^{0.089}/\rho_l}{v_l^{0.071}}\left\{\frac{g(\rho_l - \rho_g)^{0.446}}{\rho_l}\right\}\right] \tag{2.5}$$

The bubbly flow contains deformable bubbles that move upward with a zigzag motion with Taylor-type bubbles appearing occasionally, but in finely dispersed bubbly flow, Taylor bubbles are absent. They also gave a criterion for a physical situation in which bubbly flow with Taylor bubbles cannot exist. This occurs for tubes of diameter smaller than 5 cm, and the equivalent criterion is as

$$\left[\frac{\rho_l^2 g d_c^2}{(\rho_l - \rho_g)\sigma}\right]^{0.25} \leq 4.37 \tag{2.6}$$

Jones and Zuber (1975) concluded that transition from bubble to slug flow pattern occurs near fractional gas hold-up $(\varepsilon_g) = 0.23$ for all liquid velocities. The slug flow to churn flow transition would occur when the void fraction in the liquid-slug section reaches the mean void fraction at the slug-bubble section (i.e., $\varepsilon_g \geq \varepsilon_{gm}$) (Mishima and Ishii, 1984). The limiting mean value of void fraction, ε_{gm}, is calculated on the basis of the streaming conditions around a Taylor bubble, which is expressed as

$$\varepsilon_{gm} = 1 - 0.813\times\left[\frac{0.2\{1 - \rho_g/\rho_l\}^{0.5}(u_{sg} + u_{sl}) + 0.35\{gd_c(\rho_l - \rho_g)/\rho_l\}^{0.5}}{(u_{sg} + u_{sl}) + 0.35\{gd_c(\rho_l - \rho_g)/\rho_l\}^{0.5}\{(gd_c(\rho_l - \rho_g))/(\rho_l u_{sl}^2)\}^{1/18}}\right]^{0.75} \tag{2.7}$$

The local void fraction ε_g comparing with ε_{gm} can be determined by solving the separate phase momentum equations.

Phenomenological Method

Two-phase flow pattern models following a phenomenological approach are theoretical methods as the interfacial structure is taken into account (Beattie and Whalley, 1982; Hashizume, 1985; Olujic, 1985; Hart et al., 1989; Quibe'n and Thome, 2007). Thus, they are not blind to the different flow regimes, resulting in general applicability models. Despite this obvious advantage, an important drawback of this approach is that some empiricism is still required to close the model. Another important aspect is that no general flow pattern–based model is yet available. Phenomenological models are generally suitable for individual flow patterns or flow structures. In this regard, the dynamic slip approach of McFadden et al. (1981) can be noteworthy to represent transient two-phase flow patterns. The model represents the transient two-phase flow patterns in terms of four differential equations. These are the continuity, momentum, and energy equations for the mixture and one equation to describe the dynamic behavior of the phase velocity difference ("dynamic slip" equation). The dynamic slip equation can be derived by first writing the momentum equations for the gas or vapor and liquid phases in convenient form, subtracting the resultant equations, and solving for the time derivative of the relative velocity, which can be represented as

$$
\frac{\partial u_s}{\partial t}\left(1 + \frac{C_m \bar{\rho}}{\varepsilon_g (1-\varepsilon_g)\rho_l \rho_g}\right) = -\left(\frac{1}{\rho_l} - \frac{1}{\rho_g}\right)\frac{\partial P}{\partial z} - \frac{\bar{\rho}a_{lg}F_{lg}u_s}{\varepsilon_g(1-\varepsilon_g)\rho_l\rho_g}
$$

$$
+ u_g\frac{\partial u_g}{\partial z} - u_l\frac{\partial u_l}{\partial z} + \frac{a_{wg}F_{wg}u_g}{\varepsilon_g\rho_g} - \frac{a_{wl}F_{wl}u_l}{(1-\varepsilon_g)\rho_l}
$$

(2.8)

Because there is a thin liquid layer in contact with the wall, friction between the gas and the wall can be neglected. Hence, a_{wg} is equal to zero. The liquid area in contact with the wall per unit volume is $4/d_c$, where d_c is the column diameter. The terms $u_g\,\partial u_g/\partial z$, $u_l\,\partial u_l/\partial z$ are both negligible with respect to other terms even for very rapid flow transients in the absence of heating. Equation (2.8) then becomes

$$
\frac{\partial u_s}{\partial t}\left(1 + \frac{C_m \bar{\rho}}{\varepsilon_g (1-\varepsilon_g)\rho_l\rho_g}\right) = -\left(\frac{1}{\rho_l} - \frac{1}{\rho_g}\right)\frac{\partial P}{\partial z} - \frac{\bar{\rho}a_{lg}F_{lg}u_s}{\varepsilon_g(1-\varepsilon_g)\rho_l\rho_g} - \frac{a_{wl}F_{wl}u_l}{(1-\varepsilon_g)\rho_l}
$$

(2.9)

The added mass coefficient can be calculated by (McFadden et al., 1981)

$$
\begin{aligned}
c_m &= (0.5 + \varepsilon_g)\varepsilon_g\,\bar{\rho} & &if \quad \varepsilon_g \le 0.5 \\
&= (1.5/\varepsilon_g - 1)(1-\varepsilon_g)\varepsilon_g\,\bar{\rho} & &if \quad \varepsilon_g \ge 0.5
\end{aligned}
$$

(2.10)

The $a_{lg}F_{lg}$ term can be evaluated on the basis of mechanistic models for the various flow patterns. The momentum due to friction can be expressed as

$$
F_{lg} = \frac{1}{8}C_{d,lg}\rho_l u_{lg}
$$

(2.11)

$$
F_{lw} = \frac{1}{8}f_D\rho_l u_l
$$

(2.12)

where f_D is the single-phase Darcy friction factor. For the steady-state condition, one can then write

$$a_{\mathrm{lg}}C_{d,\mathrm{lg}} = \frac{8[\varepsilon_g(1-\varepsilon_g)\rho_g\rho_l]^3}{[(1-x)\varepsilon_g\rho_g - x(1-\varepsilon_g)\rho_l]^2\,\bar{\rho}\rho_l^2}\left[\frac{-f}{2D\rho_l^2}\left(\frac{\phi_{tp}\Delta\rho}{\rho_g}+\frac{(1-x)^2}{(1-\varepsilon_g)^3}\right)+\frac{\bar{\rho}g\Delta\rho}{G^2\rho_g\rho_l}\right] \tag{2.13}$$

where ϕ_{tp} is the two-phase friction multiplier, $\rho = \rho_l\rho_g$. The single liquid and relative gas–liquid velocities for the above simplified equation are expressed respectively as

$$u_l = \frac{G(1-x)}{\rho_l(1-\varepsilon_g)} \tag{2.14}$$

$$u_s = \left(\frac{1-x}{\rho_l(1-\varepsilon_g)}-\frac{x}{\rho_g\varepsilon_g}\right)G \tag{2.15}$$

The pressure gradient dP/dz can be expressed as follows in terms of its friction and elevation components only by neglecting the acceleration component in the absence of heating.

$$\frac{dP}{dz}=\frac{\phi_{\mathrm{lg}}fG^2}{2d_c\rho_l}-\bar{\rho}g \tag{2.16}$$

In this case, it should be noted that in expressing the interfacial friction solely in terms of an interracial drag coefficient, the turbulent shearing stress can be ignored because it depends on the difference in phase velocities. The model is not valid for fully dispersed flow where u_s approaches zero. Crawford et al. (1986) pointed out that the turbulent shear stresses vary with the mass flux to density ratio, and hence a correction to be made for onset of dispersed flow if the mass velocity approaches 10×10^6 kg/m²h. They reported that the value of $a_{\mathrm{lg}}C_{d,\mathrm{lg}}$ increases with the gas void fraction at bubbly flow pattern at low gas voids but decreases with the gas void fraction at intermediate and high gas voids for the conditions of intermittent and annular flow patterns. After curves of $a_{\mathrm{lg}}C_{d,\mathrm{lg}}$ are obtained from the analysis of the experimental data, these curves may be used to calculate gas void fraction for a given x and G by use of equation (2.13) together with an iterative solution procedure.

Stability Theory

In the process industry, the presence of multiple phases poses major difficulties in both design and operation. Typically, bubbly flow systems are susceptible to flow instabilities, blockages, and pressure and temperature fluctuations (Daniels, 1995). Hence, the performance of bubbly flow equipment can be strongly enhanced when the dynamics of the flow are well understood on a fundamental level. The literature on the stability of two-phase flow shows that this type of flow is susceptible to instabilities of various kinds. The instability is caused by viscosity stratification and can result from density stratification, velocity profile curvature, or shear effects in one of the constitutive phases. By definition, instability implies the increase of kinetic energy of an initially small disturbance with time in the flow. In two-phase flow, mechanisms of different physical origin account for the production of this energy. Identification of the dominant mechanism of energy production perceives the difference between different types of instabilities or disturbances of flow. The details of fundamental of two-phase flow stability can be followed from Drazin and Reid (1981), Boomkamp (1998), and Joshi et al. (2001). A two-fluid model to study the stability of bubbly flow can be used to analyze the flow pattern of two-phase flow. Pauchon and Banerjee (1988) showed that the kinematic wave velocity based on a constant interfacial friction is weakly stable. The turbulence provides the stabilizing mechanism through axial dispersion of the void fraction. Biesheuvel and Gorisson (1990) used a one-dimensional conservation equation to describe the propagation of gas hold-up disturbances. They investigated the stability of uniform bubbly flow under planar disturbances. They reported that the transition to slug flow found to be influenced by no uniformities over the cross-section of the tube induced by the presence of the wall. The void fraction waves were found to be unstable for gas hold-up above 0.35 for homogeneous bubbly flow. The one-dimensional model of adiabatic bubbly flow in a vertical column can be derived from full three-dimensional equations by performing instantaneous space averaging by considering various interfacial momentums. The relative size of these terms can be determined based on the scale analysis. Under steady flow conditions, the critical void fraction is to be 0.42. Leon-Becerril and Line (2001) also developed one-dimensional two-fluid model for bubbly flow. They reported that instability occurs when the kinematic wave velocity is less than the slower dynamic wave velocity. They concluded that the transition gas holdup is larger for spherical bubbles because of pressure-induced forces that stabilize the flow. The transition holdups obtained by accounting bubble deformation were close to the experimental ones. Shnip et al. (1992) developed a criterion to predict flow regime transition in a two-dimensional bubble column based on the theory of linear stability. The authors considered both semi-batch as well continuous modes. The following stability criteria were proposed:

For semi-batch operation:

$$\frac{gW/u_{b\infty}^2}{(K_v + 2\varepsilon_{g0}K_T\bar{u}_s)(1-\varepsilon_{g0})^{m-1}[1-(m+1)\varepsilon_{g0}]} < \frac{\pi \sinh(\pi h)}{\cosh(\pi h)-1} \tag{2.17}$$

For continuous operation:

$$\frac{gW/u_{b\infty}^2}{[K_v + 2\varepsilon_{g0}K_T(W_0+\bar{u}_s)](w_0+u_s+\varepsilon_{g0}\bar{u}_s)} < \pi \coth(\pi h) \tag{2.18}$$

$$K_v = 0.98a_o^{1.5}/u_{b\infty}\rho_l \tag{2.19}$$

$$K_T = 0.3a_o^{1.2}/\rho_l \tag{2.20}$$

where $u_{b\infty}$ is the terminal rise velocity, m is a constant, W is the column width, and a_o is the percent open area. The maximum predicted value of the transition gas holdup was 0.42. The effects of the sparger, column height, column diameter, dispersion coefficient, and liquid phase properties on flow regime stability were studied using the proposed criteria. The authors found that coalescence in a system results in a higher local hold-up, generates local liquid recirculation, and results in higher bubble rise velocities; hence, the value of the transition hold-up reduces. The proposed transition criteria were found to be independent of liquid phase viscosity. Joshi et al. (2001) provided a generalized stability criterion using a one-dimensional model for multiphase systems based on linear stability theory. The following unified stability criterion was proposed:

$$f_1 = 1 - \frac{[A(G/F) - B/2]^2}{A(Z - C) + (B^2/4)} \tag{2.21}$$

where if f_1 is greater than 1, the system is stable; if f_1 is less than 1, the system is unstable. The parameters in the above criterion were defined for a batch bubble column as follows:

$$A = \frac{\rho_g}{\rho_l} + \frac{1 + C_m}{1 - \varepsilon_g} - 1 \tag{2.22}$$

$$B = 2\left(\frac{\rho_g}{\rho_l} + C_m\right)\frac{u_{sg}}{\varepsilon_g} \tag{2.23}$$

$$C = \left(\frac{\rho_g}{\rho_l} + C_m\right)\left(\frac{u_{sg}}{\varepsilon_g}\right)^2 \tag{2.24}$$

$$Z = \frac{\beta_0}{\rho_l}\left(\frac{1}{\varepsilon_g} + \frac{1}{1 - \varepsilon_g}\right) \tag{2.23}$$

$$G = \frac{\beta_0}{\rho_l}\frac{u_{sg}}{\varepsilon_g} + \frac{\beta_0' u_{sg}}{\rho_l} + \frac{\rho_g - \rho_l}{\rho_l}g \tag{2.24}$$

$$\beta_0' = \frac{\rho_g - \rho_l}{u_s}g \tag{2.25}$$

$$\beta_0 = \frac{(\rho_g - \rho_l)\varepsilon_g}{u_s}g \tag{2.26}$$

Bhole and Joshi (2005) developed a one-dimensional two-fluid model to study instability in uniform bubbly flow. They formulated the drag force to take into account the dependence of drag coefficient on the slip velocity considering the bubble deformation. They proposed the following equation for f transition gas hold-up at stable condition:

$$\frac{u_{b\infty}}{\sqrt{gd_b}} = \left[\sqrt{\frac{\lambda(1 - \varepsilon_{g,trans})}{C_{m0}(1 + 2\varepsilon_{g,trans}) + (1 - \varepsilon_{g,trans})^2}}\right]\left[\frac{1}{(1 - \varepsilon_{g,trans})^{m-1}}\right] \tag{2.27}$$

where $u_{b\infty}$ is the terminal bubble rise velocity, λ is proportionality constant, and C_{mo} is the virtual mass coefficient of isolated bubble. They stated that the bubble rise velocity was found to be the most important parameter affecting the transition. An increase in bubble rise velocity reduces the gas hold-up at which transition occurs.

Computational Fluid Dynamics

Generally, a population balance model (PBM) along with CFD simulations is used to predict the flow regime transition. A model with $k\text{-}\varepsilon$ turbulence in a Eulerian framework for theoretical prediction of the flow regime transition can be used. The variation of bubble size distribution with superficial gas velocity is calculated by CFD simulations to provide information regarding flow regimes. For accurate prediction of the gas hold-up radial profile, lateral forces, including the transverse lift force, wall lift force, and turbulent dispersion force, can be considered in addition to drag force (Wang et al., 2006). The bubble coalescence caused by turbulent eddies, different rise velocities, wake entrainment, and breakup caused by eddy collision and large bubble instability can be considered to model bubble characteristics. According to Wang et al. (2006), a sharp variation in the fraction of small bubbles results near the flow regime transition point as shown in Figure 2.11. In the analysis of thermohydraulic performance of many systems that contain a two-phase flow medium, such as boiling liquids or condensing vapors, a variety of computer programs are used that solve the time-dependent continuity and conservation equations for flow in different parts of the system. A particular situation of this kind, which is of special interest to nuclear engineers, is predicting two-phase flow transients in nuclear power reactors under

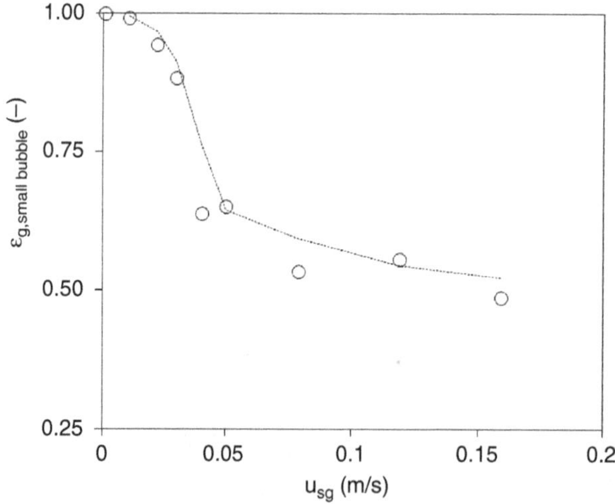

FIGURE 2.11 Prediction of variation of small bubble fraction with superficial gas velocity as per population balance model. *(Reproduced from Wang et al., 2006).*

various operational transients. The most recent generation of computer codes for the calculation of such transients is based on the use of separate flow equations for vapor and liquid. In this approach, the so-called "two-fluid" model, separate time-dependent equations for mass, momentum, and energy balance are used for gas and liquid in each element of space along the flow path. This approach is preferred because of its capability of predicting the situations in which gas and liquid are not in thermal equilibrium, such as in the case of saturated gas in contact with subcooled liquid during condensation or the case of superheated steam carrying droplets of water at saturation temperature. The use of separate continuity and conservation equations in the computer codes makes it necessary to include some equations for predicting the rates of exchange of mass, momentum, and energy between the phases. The appropriate forms of such equations, which are commonly called "constitutive relations," depend on the flow regime in actual situations. This calls for some reliable predictive capabilities for flow regime indication.

Artificial Neural Network Method

The ANNs are used for engineering purposes, such as pattern recognition, forecasting, classification, optimization function, approximation, vector quantization, and data compression. It is an alternative tool to solve many complex and nonlinear problems. Since the early 1980s, artificial ANNs have been used extensively in engineering for various applications as adaptive control, model-based control, process monitoring, fault detection, dynamic modeling, and parameter estimation. The ANN provides a nonlinear mapping between input and output variables and is useful in providing cross-correlation among these variables. The mapping is performed by the use of processing elements and connection weights. In multiphase reactor research, there have been efforts to apply neural networks for improved prediction of design and scale-up variables. Quantrille and Liu (2012) used an ANN to improve the prediction of various hydrodynamic parameters in several reactors. There are multitudes of different types of ANNs depending on how data is processed through the network such as feedforward backpropagation (FFBP) in which no loops are formed by the network, cascade feedforward backpropagation (C-FFBP) in which one or more loops are formed, generalized regression neural networks (GRNN), nonlinear autoregressive exogenous model (NARX), radial basis network (RBN), probabilistic neural network (PNN), and support vector machine (SVM). Neural networks can be hardware (neurons are represented by physical components) or software based (computer models) and can use a variety of topologies and learning algorithms. The details of ANNs applied in predicting different hydrodynamics is published in a special issue edited by Larachi and Grandjean (2003). A multilayer neural network is effective in finding complex nonlinear relationships. It has been reported that multilayer ANN models with only one hidden layer are universal approximators (Syeda et al., 2002). A multilayer feedforward neural network can be chosen as a regression model. The weighting coefficients

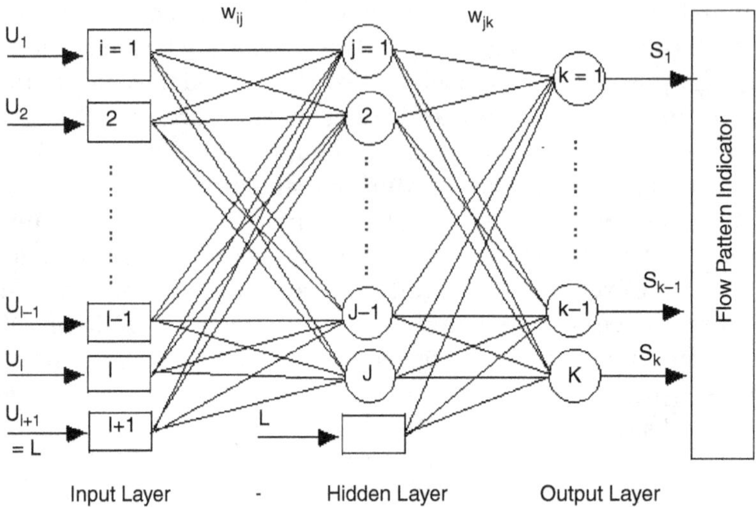

FIGURE 2.12 Architecture of the three-layered feedforward neural network.

of the neural network are calculated using the software NNFit (Cloutier et al., 1996). NNFit is a nonlinear regression software that discloses relationships between a set of normalized input variables, U_i, and a set of normalized output variable, S_k. Figure 2.12 shows a typical transformation $S = f(U)$ using a neural network with a single hidden layer. The transformation of actual variables (x, y) to normalized variables (U, S) can be expressed as

$$U_i = \frac{\log(x_i/x_{\min})}{\log(x_{\max}/x_{\min})} \tag{2.28}$$

$$S_k = \frac{\log(y_k/y_{\min})}{\log(y_{\max}/y_{\min})} \tag{2.29}$$

where x_i and y_k are raw input and output variables, respectively. The various layers are interconnected to each other by a sigmoid function through the fitted parameters w_{ij}, w_{jk} as per the following equations:

$$S_k = \left\{ 1 + \exp\left(-\sum_{J=1}^{J+1} w_{jk} H_j \right) \right\}^{-1} \tag{2.30}$$

$$H_j = \left\{ 1 + \exp\left(-\sum_{i=1}^{I+1} w_{ij} U_i \right) \right\}^{-1} \tag{2.31}$$

where the indices I, J, and K indicate the input, hidden, and output nodes of the ANN structure, respectively. H_{J+1} and U_{I+1}, as shown in Figure 2.12, are the bias constants that are set equal one. The parameters w_{ij} and w_{jk} are weighting parameters that are fitted by the NNFit regression model.

Procedure to Predict Two-Phase Flow Regime Using an Artificial Neural Network
The procedure to predict two-phase flow regimes using ANN modeling is summarized as follows:

- **Data collection:** Experimental data to be collected for identified flow regimes
- **Data preprocessing:** To get more effective, the solution of missing data has to be performed.
- **Building the network:** The investigator of the ANN identifies the number of hidden layers, neurons, and transfer functions, in each layer and training function.
- **Training the network:** The training of network involves the adjustment of associated weights to have the predicted actual outputs close the measured target output of the network. Generally, 80-percentage data are used for training for different types of training algorithms and transfer function.
- **Testing the network:** The modeled network is to be tested to predict the performance of the developed ANN model. Generally, unseen data (100-percentage data are used for training) are exposed to model.

The predicted results are then compared based on the statistical parameters, which are outlined as follows:

- **Root mean squared error (RMSE):** This shows the short-term performance information. The RMSE is defined as

$$RSME = \sqrt{\frac{1}{n}\sum_{i=1}^{n}(y_{P,i} - y_i)^2}$$

(2.32)

where $y_{p,i}$ denotes the predicted flow type value and y_i denotes the measured flow type value. The lower the value of the RMSE, the higher the accuracy of the estimation.

- **Mean bias error (MBE):** This represents the average deviation of predicted values from the corresponding measured data and can provide information on the long-term performance of the models. The error is defined as

$$MBE = \frac{1}{n}\sum_{i=1}^{n}(y_{p,i} - y_i)$$

(2.33)

The lower value of MBE is preferred because it gives the better long-term model prediction. The positive value of MBE shows overestimation in predicting the flow pattern value and vice versa.

- **Correlation coefficient (R^2):** This is a parameter to represent how well the model data is fitted with the experimental data. Its range is 0 to 1. The value of R^2 shows the overall usefulness measure of goodness of fit, which can be expressed as

$$R^2 = SSR/SST \tag{2.34}$$

where SST is the total variation, which is calculated by

$$SST = \sum (y_i - \bar{y}) \tag{2.35}$$

The parameter SSR is called the regression sum of square or the explained sum of squares, which is calculated by

$$SSR = \sum (y_c - \bar{y}) \tag{2.36}$$

where y_i is the measured target values, $\bar{y}$ is mean of measured target value, and y_c is the predicted target value.

MATLAB Algorithm to Use an Artificial Neural Network
The algorithm to predict the experimental data using MATLAB software by the ANN tool is summarized as below:

- Prepare the experimental data file in CSV format and read these data into MATLAB.
- Prepare the Mfile to target train and test the data.
- In the MATLAB command window, type nntool and import the inputs and targets data file into the nn data manager.
- Set the input data and target data.
- Hit the create button to create the neural network.
- Train the network. On the train tab of the network, network dialog, select inputs and targets; then press the train network button to start the network training.
- Obtain the results of the trained data.
- Hit the export button and test the unseen data.

A significant advance in the flow regime identification was achieved by the use of ANN (Cai et al., 1994). Using Kohonen self-organizing neural networks (SONNs), it was possible to identify the flow regimes more objectively. In the past decade, some improvements in the flow regime identification methodology have been made. Different neural network strategies to improve the flow regime identification results have also been developed (Hernandez et al., 2006; Bishop, 2007).

Nomenclature

a_{lg} Gas–liquid interfacial area per unit volume (1/m)
a_o The % open area (m²)
a_{wg} Surface area of gas per unit volume in contact with wall (1/m)
a_{wl} Surface area of liquid per unit volume in contact with wall (1/m)
C_0 Distribution parameter (-)
$C_{d,lg}$ Drag coefficient of gas–liquid flow (-)
C_{Eth} Ethanol concentration (% w)
C_m Virtual mass coefficient (-)
C_{mo} Virtual mass coefficient of isolated bubble
d_b Bubble diameter (m)
d_c Diameter of the column (m)
D_G Gas dispersion coefficient (m²/s)
d_j Liquid jet diameter (m)
D_L Liquid dispersion coefficient (m²/s)
d_o Orifice diameter
f_1 Functionality of stability criteria (-)
f_D Single phase Darcy friction factor (-)
F_{lg} Interfacial force caused by friction between gas and liquid phases (kg/m²s)
F_{wg} Force caused by friction between the wall and gas (kg/m²s)
F_{wl} Momentum caused by friction between the wall and liquid (kg/m²s)
G Mass flux (kg/m²s)
H Height (m)
L_o Pitch of orifice (m)
P Pressure (N/m²)
S_k Normalized output variable
t Time (s)
T Thickness (m)
U_i Normalized input variables
u_j Liquid jet velocity (m/s)
u_s Slip velocity (u_{sl}-u_{sg}) (m/s)
u_{sg} Superficial gas velocity (m/s)
u_{sl} Superficial liquid velocity (m/s)
u_{trans} Transition velocity (m/s)
$u_{b\mu}$ Terminal rise velocity of bubble (m/s)
u_d Weighted average drift velocity (m/s)
W Width (m)
x Vapor or gas quality (-), variable
y Variable
z Axial distance (m)

Symbol

$\bar{\rho}$ Average density of mixture (kg/m³)
β_0 Drag interaction parameter (kg/m³.s)
ε_g Gas hold-up (-)
ε_{g0} Initial gas hold-up (-)
ρ_l Density of gas (kg/m³)

ρ_l Density of liquid (kg/m^3)
ν_l Kinematic viscosity of liquid (m^2/s)
σ_l Surface tension (N/m)
ϕ_{tp} Two-phase friction multiplier (-)
μ Viscosity (mPa.s)
λ Proportionality constant (-)

References

Ajbar, A., Al-Masry, W., Ali, E., 2009. Prediction of flow regimes transitions in bubble columns using passive acoustic measurements. Chem. Eng. Process 48 (1), 101–110.

Al-Masry, W.A., Ali, E.M., 2007. Identification of hydrodynamic characteristics in bubble columns through analysis of acoustic sound measurements—influence of the liquid phase properties. Chem. Eng. Process. 46 (2), 127–138.

Al-Masry, W.A., Ali, E.M., Aqeel, Y.M., 2005. Determination of bubble characteristics in bubble columns using statistical analysis of acoustic sound measurements. Chem. Eng. Res. Des. 83 (A10), 1196–1207.

Andritsos, N., Bontozoglou, V., Hanratty, T.J., 1992. Transition to slug flow in horizontal pipes. Chem. Eng. Comm. 118, 361–385.

Arastoopour, H., Shao, S., 1997. In: Chaouki, J., Larachi, F., Dudukovic, M.P. (Eds.), Laser Doppler Anemometry: Applications in Multiphase Flow Systems. Non-Invasive Monitor Multiphase Flows. Elsevier, New York. pp. 455–494.

Bakshi, B.R., Zhong, H., Jiang, P., Fan, L.S., 1995. Analysis of flow in gas–liquid bubble columns using multiresolution methods. Chem. Eng. Res. Des. 73 (A6), 608–614.

Banasiak, R., Wajman, R., Jaworski, T., Fiderek, P., Fidos, H., Nowakowski, J., Sankowski, D., 2014. Study on two-phase flow regime visualization and identification using 3D electrical capacitance tomography and fuzzy-logic classification. Int. J. Multiphase Flow 58, 1–14.

Bando, Y., Uraishi, M., Nishimura, M., Hattori, M., Asada, T., 1988. Cocurrent downflow bubble column with simultaneous gas-liquid injection nozzle. J. Chem. Eng. Jpn 21, 607–612.

Beattie, D.R.H., Whalley, P.B., 1982. A simple two-phase frictional pressure drop calculation method. Int. J. Multiphase Flow 8 (1), 83–87.

Bennett, M.A., West, R.M., Luke, S.P., Jia, X., Williams, R.A., 1999. Measurement and analysis of flows in gas-liquid column reactor. Chem. Eng. Sci. 54, 5003–5012.

Bhole, M.R., Joshi, J.B., 2005. Stability analysis of bubble columns: predictions for regime transition. Chem. Eng. Sci. 60 (16), 4493–4507.

Bieberle, A., Härting, H., Rabha, S., Schubert, M., Hampel, U., 2013. Gamma-ray computed tomography for imaging of multiphase flows. Chem. Ing. Technik 85 (7), 1002–1011.

Biesheuvel, A., Gorissen, W.C.M., 1990. Void fraction disturbances in a uniform bubbly fluid. Int. J. Multiphase Flow 16 (2), 211–231.

Bishop, M., 2007. Pattern Recognition and Machine Learning, second ed. Springer, United Kingdom.

Boomkamp, P.A.M., 1998. Stability of parallel two-phase flow, PhD Thesis. Universiteit Twente, Netherlands.

Briens, L.A., Briens, C.L., Margaritis, A., Hay, J., 1997. Minimum liquid fluidization velocity in gas–liquid–solid fluidized bed of low-density particles. Chem. Eng. Sci. 52 (21–22), 4231–4238.

Brodkey R.S., 1967. The Phenomena of Fluid Motions, Addison-Wesley, Massachusetts.

Bukur, D.B., Petrovic, D., Daly, J.G., 1987. Flow regime transitions in a bubble column with a paraffin wax as the liquid medium. Ind. Eng. Chem. Res. 26 (6), 92–1087.

Cai, S., Toral, H., Qiu, J., Archer, J.S., 1994. Neural network based objective flow regime identification in air-water two-phase flow. Can. J. Chem. Eng. 72 (3), 440–445.

Cassanello, M., Larachi, F., Kemoun, A., Al-Dahhan, M.H., Dudukovic, M.P., 2001. Inferring liquid chaotic dynamics in bubble columns. Chem. Eng. Sci. 56, 6125–6134.

Chen, R.C., Fan, L.-S., 1992. Particle image velocimetry for characterizing the flow structure in three-dimensional gas–liquid–solid fluidized beds. Chem. Eng. Sci. 47 (13–14), 3615–3622.

Chen, R.C., Reese, J., Fan, L.-S., 1994. Flow structure in a three-dimensional bubble column and three-phase fluidized bed. AIChE J. 40 (7), 1093–1104.

Cloutier, P., Tibirna, C., Grandjean, B.P.A., Thibault, J., 1996. NNFit, non-linear regression program based on multilayered neural network models. Available from: http://www.gch.ulaval.ca/~nnfit/.

Crawford, T.J., Weinberger, C.B., Weisman, J., 1986. Two-phase flow patterns and void fractions in downward flow. Part II: void fractions and transient flow patterns. Int. J. Multiphase Flow 12 (2), 219–236.

Crowley, C.J., Wallis, G.B., Barry, J.J., 1992. Validation of a one-dimensional wave model for the stratified-to-slug flow regime transition, with consequences for wave growth and slug frequency. Int. J. Multiphase Flow 18 (2), 249–271.

Daniels, L., 1995. Dealing with two-phase flows. Chem. Eng. J. 102, 70–78.

de Mesquita, C.H., de Sousa Carvalho, D.V., Kirita, R., Vasquez, P.A.S., Hamada, M.M., 2014. Gas–liquid distribution in a bubble column using industrial gamma-ray computed tomography. Radiat. Phys. Chem. 95, 396–400.

Deckwer, W.D., 1992. Bubble Column Reactors. Wiley, New York.

Devanathan, N., Moslemian, D., Dudukovic, M.P., 1990. Flow mapping in bubble columns using CARPT. Chem. Eng. Sci. 45 (8), 2285–2291.

Dong, F., Jiang, Z.X., Qiao, X.T., Xu, L.A., 2003. Application of electrical resistance tomography to two-phase pipe flowparameters measurement. Flow Meas. Instrum. 14 (4–5), 183–192.

Drahos, J., Bradka, F., Puncochar, M., 1992. Fractal behavior of pressure fluctuations in a bubble column. Chem. Eng. Sci. 47 (15–16), 4069–4075.

Drazin, P.G., Reid, W.H., 1981. Hydrodynamic Stability. Cambridge University Press, Cambridge.

Evans, G.M., 1990. A study of a plunging jet column, PhD Dissertation. University of Newcastle, Australia.

Evinc, F., 1982. Absorption of gases in a cocurrent downflow column, MSc Thesis. University of Birmingham.

Fan, L.-S., 1989. Gas–Liquid–Solid Fluidization Engineering, Butterworth, Boston, MA.

Griffith, P., Wallis, G.B., 1961. Two-phase slug flow. Trans. ASME J. Heat Transfer 83, 307–320.

Grover, G.S., Rode, C.V., Chaudhari, R.V., 1986. Effect of temperature on flow regimes and gas hold-up in a bubble column. Can. J. Chem. Eng. 64 (3), 501–504.

Hart, J., Hamersma, P.J., Fortuin, J.M.H., 1989. Correlations predicting frictional pressure drop and liquid holdup during horizontal gas–liquid pipe flow with a small liquid holdup. Int. J. Multiphase Flow 15 (6), 947–964.

Hashizume, K., 1985. Flow pattern and void fraction of refrigerant two-phase flow in a horizontal pipe. Bull. JSME 26, 1597–1602.

Heindel, T.J., Gray, J.N., Jensen, T.C., 2008. An X-ray system for visualizing fluid flows. Flow Meas. Instrum. 19 (2), 67–78.

Hernandez, L., Julia, J.E., Chiva, S., Paranjape, S., Ishii, M., 2006. Fast classification of two-phase flow regimes based on conductivity signals and artificial neural networks. Meas. Sci. Technol. 17, 1511–1521.

Hills, J.H., 1974. Radial non-uniformity of velocity and voidage in bubble column. Ind. Eng. Chem. Process Des. Dev. 20, 540–545.

Hills, J.H., 1976. The operation of a bubble column at high throughputs. I. Gas holdup measurements. Chem. Eng. J. 12, 89–99.

Holler, V., Ruzicka, M., Drahos, J., Kiwi-Minsker, L., Renken, A., 2003. Acoustic and visual study of bubble formation processes in bubble columns staged with fibrous catalytic layers. Catal. Today 79–80, 151–157.

Hubbard, M.G., Dukler, A.E., 1966. The characterization of flow regimes for horizontal two-phase flow. In: Saad, M.A., Moller, J.A. (Eds.), Proceedings of Heat Transfer and Fluid Mechanics. Stanford University Press, Stanford.

Hurlburt, E.T., Hanratty, T.J., 2002. Prediction of the transition from stratified to slug and plug flow for long pipes. Int. J. Multiphase Flow 28 (5), 707–729.

Hyndman, C.L., Larachi, F., Guy, C., 1997. Understanding gas-phase hydrodynamics in bubble columns: a convective model based on kinetic theory. Chem. Eng. Sci. 52, 63–77.

Jamialahmadi, M., Muller-Steinhagen, H., 1993. Gas hold-up in bubble column reactors. In: Cheremisinoff, N.P. (Ed.), Encyclopedia of Fluid Mechanics. Gulf Publishing Company, Houston, pp. 387–407.

Jamialahmadi, M., Muller-Steinhagen, H., Sarrafi, A., Smith, J.M., 2000. Studies of gas holdup in bubble column reactors. Chem. Eng. Technol. 23, 919–921.

Jones, Jr., O.C., Zuber, N., 1975. The interrelation between void fraction fluctuations and flow patterns in two-phase flow. Int. J. Multiphase Flow 2, 273–306.

Joshi, J.B., Deshpande, N.S., Dinkar, M., Phanikumar, D.V., 2001. Hydrodynamic stability of multiphase reactors. Adv. Chem. Eng. 26, 1–130.

Julia, J.E., Liu, Y., Paranjape, S., Ishii, M., 2008. Local flow regimes analysis in vertical upward two-phase flow. Nucl. Eng. Des. 238, 156–169.

Kantarci, N., Borak, F., Ulgen, K.O., 2005. Review: bubble column reactors. Process Biochem. 40, 2268–2283.

Kikuchi, R., Yano, T., Tsutsumi, A., Yoshida, K., Punchochar, M., Drahos, J., 1997. Diagnosis of chaotic dynamics of bubble motion in a bubble column. Chem. Eng. Sci. 52 (21/22), 3741–3745.

Krishna, R., Ellenberger, J., Maretto, C., 1999. Flow regime transition in bubble columns. Int. Commun. Heat Mass Transfer 26 (4), 467–475.

Krishna, R., Wilkinson, P.M., Van Dierendonck, L.L., 1991. A model for gas holdup in bubble columns incorporating the influence of gas density on flow regime transitions. Chem. Eng. Sci. 46 (10), 2491–2496.

Kulkarni, A., Shah, Y.T., 1984. Gas phase dispersion in a downflow bubble column. Chem. Eng. Commun. 28, 311–326.

Kundu, G., Mukherjee, D., Mitra, A.K., 1995. Experimental studies on a co-current gas-liquid downflow bubble column. Int. J. Multiphase Flow 21, 893–906.

Lahey, R.T., Moody, F.J. 1977. The Thermal Hydraulics of a Boiling Water Nuclear Reactor. American Nuclear Society, LaGrange Park, Illinois, pp. 208.

Larachi, F., Grandjean, B.P.A., 2003. Application of neural networks to multiphase reactors. Chem. Eng. Process. Process Intens. 42 (8–9), 585–721.

Leon-Becerril, E., Line, A., 2001. Stability analysis of a bubble column. Chem. Eng. Sci. 56, 6135–6141.

Letzel, H.M., Schouten, J.C., Krishna, R., Van Den Bleek, C.M., 1997. Characterization of regimes and regime transitions in bubble columns by chaos analysis of pressure signals. Chem. Eng. Sci. 52 (24), 4447–4459.

Li, F., Hishida, K., 2009. Particle image velocimetry techniques and its applications in multiphase systems. Adv. Chem. Eng. 37, 87–147.

Lin, T.-J., Tsuchiya, K., Fan, L., 1999. On the measurements of regime transition in high-pressure bubble columns. Can. J. Chem. Eng. 77 (2), 370–374.

Lu, X.X., Boyes, A.P., Winterbottom, J.M., 1994. Operating and hydrodynamic characteristics of a cocurrent downflow bubble column reactor. Chem. Eng. Sci. 49, 5719–5733.

Majumder, 2005. Dispersion and Hydrodynamics in a Modified Bubble Column, PhD Thesis. IIT Kharagpur, India.

Majumder, S.K., Kundu, G., Mukherjee, D., 2006. Efficient dispersion in a modified two-phase non-Newtonian downflow bubble column. Chem. Eng. Sci. 61 (20), 6753–6764.

Mandal, A., Kundu, G., Mukherjee, D., 2004. Gas-holdup distribution and energy dissipation in an ejector-induced downflow bubble column: the case of non-Newtonian liquid. Chem. Eng. Sci. 59, 2705–2713.

Matsui, G., 1984. Identification of flow regimes in vertical gas-liquid two-phase flow using differential pressure fluctuations. Int. J. Multiphase Flow 10 (6), 711–720.

McFadden, J.H., Paulsen, M.P., Gose, G.C., 1981. RETRAN dynamic slip model. Nucl. Tech. 54, 287–297.

Miller, D.N., 1980. Gas holdup and pressure drop in bubble column reactors. Ind. Eng. Chem. Process. Des. Dev. 19, 371–377.

Mishima, K., Ishii, M., 1984. Flow regime transition criteria for upward two-phase flow in vertical tubes. Int. J. Heat Mass Trans. 27 (5), 723–737.

Mudde, R.F., Groen, J.S., Van Den Akker, H.E.A., 1998. Application of LDA to bubbly flows. Nucl. Eng. Des. 184 (2–3), 329–338.

Nedeltchev, S., Kumar, S.B., Dudukovic, M.P., 2003. Flow regime identification in a bubble column based on both Kolmogorov entropy and quality of mixedness derived from CARPT data. Can. J. Chem. Eng. 81, 367–374.

Nishikawa, K., Sekoguchi, K., Fukano, T., 1969. On the pulsation phenomena in gas-liquid two-phase flow. B. JSME 12 (54), 1410–1416.

Ohkawa, A., Kusabiraki, D., Kawai, Y., Sakai, N., 1986. Some flow characteristics of a vertical liquid jet system having downcomers. Chem. Eng. Sci. 41, 2347–2361.

Ohki, Yoshihiro, Inoue, H., 1970. Longitudinal mixing of the liquid phase in bubble columns. Chem. Eng. Sci. 25 (1), 1–16.

Olmos, E., Gentric, C., Poncin, S., Midoux, N., 2003. Description of flow regime transitions in bubble columns via laser Doppler anemometry signals processing. Chem. Eng. Sci. 58 (9), 1731–1742.

Olujic, Z., 1985. Predicting two-phase flow friction loss in horizontal pipes. Chem. Eng. J. 24, 45–50.

Park, S.H., Kang, Y., Kim, S.D., 2001. Wavelet transform analysis of pressure fluctuation signals in a pressurized bubble column. Chem. Eng. Sci. 56 (21–22), 6259–6265.

Park, S.H., Kang, Y., Kim, S.D., 2003. Characterization of pressure signals in a bubble column by wavelet packet transform. Korean J. Chem. Eng. 20 (1), 128–132.

Pauchon, C., Banerjee, S., 1988. Interphase momentum interaction effects in the averaged multifield model. Part II. Kinematic waves and interfacial drag in bubbly flows. Int. J. Multiphase Flow 14 (3), 253–264.

Quantrille, T.E., Liu, Y.A., 2012. Artificial Intelligence in Chemical Engineering. Elsevier.

Quibe'n, J.M., Thome, J.R., 2007. Flow pattern based two-phase frictional pressure drop model for horizontal tubes. Part II: new phenomenological model. Int. J. Heat Fluid Flow 28, 1060–1072.

Rados, N., 2003. Slurry Bubble Column Hydrodynamics, DSc. Thesis. Washington University, St. Louis, MO.

Reilly, I.G., Scott, D.S., De Bruijn, T.J.W., MacIntyre, D., 1994. The role of gas phase momentum in determining gas holdup and hydrodynamic flow regimes in bubble column operations. Can. J. Chem. Eng. 72 (1), 3–12.

Ribeiro, Jr., C.P., 2008. On the estimation of the regime transition point in bubble columns. Chem. Eng. J. 140 (1–3), 473–482.

Rouhani, S.Z., Sohal, M.S., 1982. Two-phase flow patterns: a review of research results. Prog. Nucl. Energy 11 (3), 21–259.

Roy, S., Larachi, F., Al-Dahhan, M.H., Dudukovic´, M.P., 2002. Optimal design of radioactive particle tracking experiments for flow mapping in opaque multiphase reactors. Appl. Radiat. Isot. 56 (3), 485–503.

Ruthiya, K.C., Chilekar, V.P., Warnier, M.J.F., van der Schaaf, J., Kuster, B.F.M., Schouten, J.C., van Ommen, J.R., 2005. Detecting regime transitions in slurry bubble columns using pressure time series. AIChE J. 51 (7), 1951–1965.

Ruzicka, M.C., Drahos, J., Fialova, M., Thomas, M.H., 2001. Effect of bubble column dimensions on flow regime transition. Chem. Eng. Sci. 56, 6117–6124.

Ruzicka, M.C., Drahos, J., Mena, P.C., Teixeira, J.A., 2003. Effect of viscosity on homogeneous-heterogeneous flow regime transition in bubble columns. Chem. Eng. J. 96 (1–3), 15–22.

Saayman, J., Nicol, W., Van Ommen, J.R., Mudde, R.F., 2013. Fast X-ray tomography for the quantification of the bubbling-, turbulent- and fast fluidization-flow regimes and void structures. Chem. Eng. J. 234, 437–447.

Salgado, C.M., Pereira, C.M.N.A., Schirru, R., Brandão, L.E.B., 2010. Flow regime identification and volume fraction prediction in multiphase flows by means of gamma-ray attenuation and artificial neural networks. Prog. Nucl. Energ. 52 (6), 555–562.

Sarrafi, A., Jamialahmadi, M., Muller-Steinhagen, H., Smith, J.M., 1999. Gas holdup in homogeneous and heterogeneous gas-liquid bubble column reactors. Can. J. Chem. Eng. 77 (1), 11–21.

Sathe, M.J., Mathpati, C.S., Deshpande, S.S., Khan, Z., Ekambara, K., Joshi, J.B., 2011. Investigation of flow structures and transport phenomena in bubble columns using particle image velocimetry and miniature pressure sensors. Chem. Eng. Sci. 66 (14), 3087–3107.

Scott, D.M., McCann, H., 2005. Process Imaging for Automatic Control. CRC Press Taylor and Francis Group, Boca Raton, FL.

Sekoguchi, K., Inoue, K., Imasaka, T., 1987. Void signal analysis and gas-liquid two-phase flow regime determination by a statistical pattern recognition method. JSME Int. J. 30 (266), 1266–1273.

Seleghim, P. Jr., Hervieu, E., 1994. Caractérisation des changements de configuration des éc oulements diphasiques gaz-liquide par analyse de la fréquence instantanée. C. R. Acad. Sci. Paris, t.319, série II, pp. 611–616.

Serizawa, A., Kataoka, I., 1988. In: Afgan, N.H. (Ed.), Phase Distribution in Two-Phase Flow: Transient Phenomena in Multiphase Flow. Hemisphere Publishing Corporation, New York, pp. 179–224.

Shaban, H., Tavoularis, S., 2014. Measurement of gas and liquid flow rates in two-phase pipe flows by the application of machine learning techniques to differential pressure signals. Int. J. Multiphase Flow 67, 106–117.

Shaikh, A., Al-Dahhan, M., 2005. Characterization of the hydrodynamic flow regime in bubble columns via computed tomography. Flow Meas. Instrum. 16 (2–3), 91–98.

Shaikh A., Al-Dahhan, M.H.A., 2007. Review on flow regime transition in bubble columns. Int. J. Chem. Reactor Eng. 57, 1–68.

Sharifi, M., Young, B., 2013. Electrical resistance tomography (ERT) applications to chemical engineering. Chem. Eng. Res. Des. 91 (9), 1625–1645.

Sheikhi, A., Sotudeh-Gharebagh, R., Mostoufi, N., Zarghami, R., 2013. Frequency-based characterization of liquid–solid fluidized bed hydrodynamics using the analysis of vibration signature and pressure fluctuations. Powd. Technol. 235, 787–796.

Shnip, A.I., Kolhatkar, R.V., Swamy, D., Joshi, J.B., 1992. Criteria for the transition from the homogeneous to the heterogeneous regime in two-dimensional bubble column reactors. Int. J. Multiphase Flow 18 (5), 26–705.

Stekelenburg, A.J.C., van der Hagen, T.H.J.J., 1993. Two-phase flow monitoring by analysis of in-core neutron detector noise signals—a literature survey. Ann. Nucl. Energy 20 (9), 611–621.

Sunde, C., Avdic, S., Pázsit, I., 2005. Classification of two-phase flow regimes via image analysis and a neuro-wavelet approach. Prog. Nucl. Energy 46 (3–4), 348–358.

Syeda, S.R., Afacan, A., Chuang, K.T., 2002. Prediction of gas hold-up in a bubble column filled with pure and binary liquids. Can. J. Chem. Eng. 80 (1), 44–50.

Taitel, Y., Bornea, D., Dukler, A.E., 1980. Modeling flow pattern transitions for steady upward gas-liquid flow in vertical tubes. AIChE J. 26 (3), 345–354.

Tan, C., Dong, F., Wu, M., 2007. Identification of gas/liquid two-phase flow regime through ERT-based measurement and feature extraction. Flow Meas. Instrum. 18 (5–6), 255–261.

Thet, M.K., Wang, C., Tan, R.B.H., 2006. Experimental studies of hydrodynamics and regime transition in bubble columns. Can. J. Chem. Eng. 84 (1), 63–72.

Thimmapuram, P.R., Rao, N.S., Saxena, S.C., 1992. Characterization of hydrodynamic regimes in a bubble column. Chem. Eng. Sci. 47 (13–14), 3335–3362.

Thorat, B.N., Joshi, J.B., 2004. Regime transition in bubble columns: experimental and predictions. Exp. Therm. Fluid. Sci. 28, 423–430.

Tutu, N.K., 1984. Pressure drop fluctuations and bubble-slug transition in a vertical two-phase air-water flow. Int. J. Multiphase Flow 10 (2), 211–216.

Upadhyay, R.K., Pant, H.J., Roy, S., 2013. Liquid flow patterns in rectangular air-water bubble column investigated with radioactive particle tracking. Chem. Eng. Sci. 96, 152–164.

Urseanu, M., 2000. Scaling up bubble column reactors, PhD Thesis. University of Amsterdam, Amsterdam, Netherlands.

Vandu, C.O., Koop, K., Krishna, R., 2004. Large bubble sizes and rise velocities in a bubble column slurry reactor. Chem. Eng. Technol. 27 (11), 1195–1199.

Vial, C., Laine, R., Poncin, S., Midoux, N., Wild, G., 2001a. Influence of gas distribution and regime transitions on liquid velocity and turbulence in a 3-D bubble column. Chem. Eng. Sci. 56 (3), 1085–1093.

Vial, C., Poncin, S., Wild, G., Midoux, N., 2001b. A Simple method for regime identification and flow characterization in bubble columns and airlift reactors. Chem. Eng. Proc. 40, 135–151.

Vince, M.A., Lahey, Jr., R.T., 1982. On the development of an objective flow regime indicator. Int. J. Multiphase Flow 8 (2), 93–124.

Wallis, G.B., 1969. One-Dimensional Two-Phase Flow. McGraw-Hill, New York.

Wang, T., Wang, J., Jin, Y., 2006. A CFD-PBM coupled model for gas–liquid flows. AIChE J. 52 (1), 125–140.

Weisman, J., Kang, S.Y., 1981. Flow pattern transitions in vertical and upwardly inclined tubes. Int. J. Multiphase Flow 7, 271–291.

Weisman, J., Duncan, D., Gibson, J., Crawford, T., 1979. Effects of fluid properties and pipe diameter on two-phase flow patterns in horizontal lines. Int. J. Multiphase Flow 5, 437–462.

Wilson, S.D.R., Jones, A.F., 1983. The entry of falling film into a pool and the air-entrainment problem. J. Fluid Mech. 128, 219–230.

Wilkinson, P.M., 1991. Physical aspects and scale-up of high pressure bubble columns, PhD Thesis. University of Groningen, The Netherlands.

Wilkinson, P.M., Spek, A.P., Van Dierendonck, L.L., 1992. Design parameters estimation for scaleup of high-pressure bubble columns. AIChE J. 38 (4), 54–544.

Wu, H.L., Pots, B.F.M., Hollenburg, J.F., Mehoff, R., 1987. Flow pattern transitions in two-phase gas/condensate flow at high pressure in an 8-inch horizontal pipe. Proceedings of BHRA Conference, The Hague, Netherlands, pp. 13–21.

Yamagiwa, K., Kusabiraki, D., Ohkawa, A., 1990. Gas holdup and gas entrainment rate in downflow bubble column with gas entrainment by a liquid jet operating at high liquid throughput. J. Chem. Eng. Japan 23, 343–348.

Yang, Y.B., Devanathan, N., Dudukovic, M.P., 1993. Liquid backmixing in bubble columns via computer automated radioactive particle tracking (CARPT). Exp. Fluids 16, 1–9.

Zahradnik, J., Fialova, M., Ruzicka, M., Drahos, J., Kastanek, F., Thomas, N.H., 1997. Duality of the gas–liquid flow regimes in bubble column reactors. Chem. Eng. Sci. 52 (21/22), 3811–3826.

Zboray, R., Adams, R., Cortesi, M., Prasser, H., 2014. Development of a fast neutron imaging system for investigating two-phase flows in nuclear thermal–hydraulic phenomena: a status report. Nucl. Eng. Des. 273, 10–23.

Zhang, J.-P., Grace, J.R., Epstein, N., Lim, K.S., 1997. Flow regime identification in gas–liquid flow and three-phase fluidized beds. Chem. Eng. Sci. 52 (21/22), 3979–3992.

Zhang, W., Wang, C., Yang, W., Wang, C., 2014. Application of electrical capacitance tomography in particulate process measurement—a review. Adv. Powd. Technol. 25 (1), 174–188.

Zuber, N., Findlay, J.A., 1965. Average volumetric concentration in two-phase flow systems. J. Heat Trans. 87 (4), 68–453.

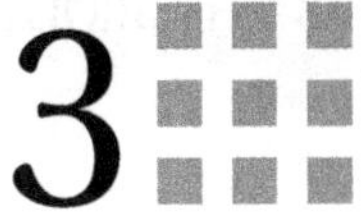

Entrainment of Gas Bubbles

Entrainment of Gas Bubbles

The gas entrainment as a dispersed phase of bubble is the prominent role to get inverse flow of the bubble. The arresting of one substance by another substance is called *entrainment,* which is commonly used in various branches of engineering (Perry and Green, 1984). Examples include, (1) the arresting of gas in a flowing liquid as a dispersed phase of bubble; (2) the holding of liquid droplets in a flowing gas as a smoke; and (3) the emulsion of droplets of one liquid into the other liquid which mutually immiscible. Gas entrainment in liquid is the process in creation of tiny gas bubbles or gas pockets in a liquid. The bubbles are introduced into the liquid by addition of a surface-active substance or by other action without a gas-entraining agent. Air entrainment by plunging water jet is an example of entrainment of gas without an agent. Bubbles are entrained locally at the intersection of the impinging jet with the surrounding waters. The gas bubble entrainment may be localized or continuous along with an interface (water jets, spillway chutes) usually parallel to the flow direction (Chanson, 1997). Gas entrainment by plunging liquid jets has various applications. It occurs during pouring and filling containers with liquids. For liquids such as molten glass, molten metals, plastics, cosmetics, paints, or food products, it is usually highly undesirable because of the presence of gas bubbles entrained in the liquid phase during these operations. On the other hand, gas entrainment may be required, for example, to achieve gas absorption coupled with good mixing in gas–liquid reactors. The purpose of gas entrainment in liquid is to increase the interfacial area for better physical or chemical absorption of gaseous component to the liquid phase. In particular, because of favorable energy requirements, jet aerators have potential application in many chemical, fermentation, and waste treatment processes. Self-purification (re-aeration) of rivers, streams, waterfalls and weirs, and jets that entrain air and provide good contact and dispersion of air bubbles into the body of water plays a significant role as a natural process. Plunging liquid jets also have an interesting application in flotation processes. The gas entrainment through the sparger or other different types of gas distributor is common in vertical upflow systems such as in bubble column reactors, fluidized bed reactors, and floatation columns. Unlike vertical upflow systems, the gas entrainment and flow of bubbles inversely in a vertical column is rather a difficult task. This is because of the gas bubbles have a natural tendency to rise because of its buoyancy force. Therefore, to get the inverse flow of gas bubbles, the carrier liquid velocity must be higher than the bubble rise velocity. Ejector-induced inverse two-phase bubble column by liquid jet conforms to a new type of concurrent inverse bubble column. This type of bubble column can be successfully

Hydrodynamics and Transport Processes of Inverse Bubbly Flow. http://dx.doi.org/10.1016/B978-0-12-803287-9.00003-5

used in chemical, fermentation, and waste treatment processes and gas-aided liquid–liquid extraction processes because of its self-aeration characteristics. Entrainment of gas in this type of bubble column occurs because of the characteristics of plunging liquid jet. The rate of entrainment is mostly controlled by the liquid jet velocity. However, entrainment depends on the two-phase gas-liquid mixing height inside the column. The entrained gas disperses into liquid as fine bubbles in the intense mixing zone (where liquid jet plunges), which are then carried inversely by high momentum of the flow of fluid. A minimum liquid flow rate is required to move the bubbles inversely by prevailing over their upward buoyant force. Significant work has been reported on gas entrainment by plunging liquid jet through nozzles, venturis, and ejector systems with and without downcomers (McCarthy et al., 1969; Van de Sande and Smith, 1973; Ohkawa et al., 1987; Yamagiwa et al., 1990; Kundu et al., 1997; Mandal et al., 2004; Majumder et al., 2007; Sivaiah et al., 2012) in a bubble column.

Mechanism of Gas Entrainment

The first significant study quantifying the entrainment of gaseous phase was carried out by Mertes (1938). Mertes described the entrainment processes as being a film of gas that was formed in the region immediately adjacent to the liquid jet. The gas film then traveled at the same velocity as the liquid jet, and after being brought below the surface of the receiving fluid, the gas film was sheared from the jet. The gas film was then distributed as bubbles throughout the liquid. De Frate and Rush (1959) measured entrainment rates for turbulent water jets with Reynolds numbers greater than 20000. They described three different mechanisms of increasing entrainment. At low velocities, the liquid jet remained as a continuous liquid stream, and entrainment occurred as bubbles detached themselves from the base of the gas film extending below the surface of the receiving liquid. When the velocity was increased, the jet becomes sinuous and eventually disintegrated into droplets at longer jet length. For higher velocities, the amplitude of the sinuous motion of the jet increased, creating a greater impact area between the jet and the receiving liquid surface. McCarthy et al. (1969) found that for rough surface jets, the basic entrainment mechanism was due to the interaction between the irregularities on the free surface of the jet and the surface of the receiving liquid. As a result of this interaction, cavities formed, which filled with gas and then dispersed as bubbles. The gas–liquid contact mechanisms proposed for inverse bubbly flow bubble column are basically the same as for an upflow bubbly flow device, although different methods of distributing the gas phase (aiming mainly to improve both the range of inverse flow bubble column stable operation and contact efficiency) have been reported. The gas can be introduced as a dispersed phase of bubble inversely into the liquid phase in an inverse flow bubble column in different methods, which are summarized as follows:

- **Sparging type:** The gas phase is dispersed into the column through a sparging device or a sintered disc, perforated plate, tube, or ring (Shah, 1979).
- **Plunging liquid jet:** The gas phase is entrained from the atmosphere by a plunging liquid jet from a nozzle (Ohkawa et al., 1987; Yamagiwa et al., 1990; Briens et al., 1992; Bando et al., 1988; Majumder et al., 2007; Sivaiah et al., 2012). As far as the entrainment of a secondary fluid by a jet of a primary liquid is concerned, some scientific issues are worth mentioning, the majority of which focus on the situation of air carry under as for the secondary fluid. Mainly it is the stability of a free jet that experiences an aerodynamic interaction with the surrounding gas. This aerodynamic interaction can even be strong enough to cause the breakup or the atomization of the jet. The following features are important to note in this case:
  - The arising of the transition from a laminar to a turbulent regime
  - The distribution of the energy across and along the jet
  - The instabilities that prefigure a breakup of the jet: varicose, sinuous waves, and Kelvin–Helmholtz instabilities for a jet that experiences a strong aerodynamic coupling with a surrounding flow of gas
  - The length of the potential core
  - The typical diameter of the droplets subsequent to atomization
- **Gas-inducing impeller:** The gas-inducing impeller plays a key role in the gas entrainment process. It induces the gas from the headspace and creates a gas–liquid dispersion. It also generates a liquid flow for the dispersion of the induced gas (Mundale and Joshi, 1995; Patwardhan and Joshi, 1999). The gas-inducing impellers are classified on the basis of the flow pattern in the impeller zone. The gas enters the impeller zone at one location and leaves at another location.
- **Bubble entrapment by the impacting drops:** Individual droplets impinging on the liquid surface entrain gas if their velocity is high enough depending on their size or if successive droplets impinge along the same trajectory. Impacting of droplets on liquid surfaces by the entrapment of an gas layer ruptures leads to the entrainment of either a multitude of tiny bubbles (Pumphrey and Elmore, 1990) or only a few individual bubbles (Thoroddsen et al., 2012). These bubble entrapment processes take place at very small time scales after impact, as opposed to that caused by the collapse of the impact crater, which happens at much longer time scales. After the gas film ruptures, the entrained gas contracts toward the bottom to create one or several primary bubbles and leaves behind those that are much smaller in size (Tran et al., 2013). This gas entrainment process is often referred to as *Mesler entrainment* (Pumphrey and Elmore, 1990). If the impact velocity is small, the volume of the small bubbles left behind may be comparable to that of the primary ones. They explained that the impact of droplets on liquid surfaces in such a way that the two approaching surfaces (of the droplet and the liquid pool) are both deformable, and the air film penetrates into the liquid pool with a penetration velocity. The gas film moves into the liquid bulk with the penetration velocity. They described the impact dynamics, including the

penetration velocity of the gas film entrapped between the drop and the pool surface. They concluded that the penetration velocity insensitive to the viscosity (from 5 to 20 cSt) can be approximated as half of the impact velocity if the densities of the drop liquid and the liquid in the pool are comparable. This approximation is used to relate the volume of the entrapped gas bubble.

- **Ejector-induced liquid jet gas pump:** This consists of a throat and diffuser section (Majumder et al., 2007). Among others, it is used to optimize the ability of a primary liquid jet to drag a surrounding fluid and to mix with it conveniently within a throat, so as to obtain a homogeneous two-phase dispersion at the outlet as shown in Figure 3.1.
- To increase the entrainment (suction) performance of the liquid-jet gas pumps, a swirling liquid jet or a lattice of liquid jets (Havelka et al., 1997) depending on a convenient internal geometry of the nozzle (made of one, two, or four orifices) can be developed. A diffuser, which is also the subject of an energetic optimization, is connected upstream of the mixing throat and allows the remaining kinetic energy of the two-phase dispersion at the exit of the throat to be transformed into potential energy. Under the effect of the pressure recovery thus obtained along the diffuser, the two-phase dispersion is able to flow within a duct located upstream. A pressure recovery ΔP along the diffuser has an advantage compared with other types because it can be estimated as the weight of a water column whose height is approximately equal to the penetration depth of the bubble swarm into the receiving pool (see Figure 3.1), which can be represented as $\Delta P = \rho_l g \delta$, where ρ_l, g, and d are the liquid density, the gravity, and the penetration depth of the bubble swarm, respectively.

Gas entrainment by a liquid jet is a very complex process that is largely controlled by the liquid jet velocity. The gas is entrained by plunging of a liquid jet on to the surface of the pool of the liquid. A coherent liquid jet plunges vertically through a liquid surface in a pool, which entrains air if the jet velocity exceeds a certain critical value. A schematic

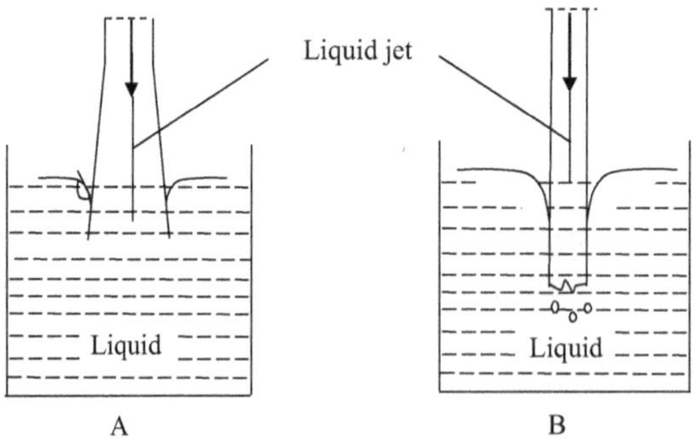

FIGURE 3.1 Liquid jet plunging through a liquid surface. **A,** Contact angle less than 180 degrees. **B,** Formation of an air film.

presentation of the mechanism of the gas entrainment is shown in Figure 3.1. For low-velocity laminar (viscous) jets, a collar-like meniscus forms around the jet at the plunging point. At slightly higher flow rates, a depression in the pool surface forms as a result of the impact pressure of the associated boundary layer of air and from the flow field induced in the pool itself. The jet where it meets the pool, enlarges to a diameter 1.5 to 2.5 times the diameter of the approaching jet. With a further increase in the flow rate, the depth of depression becomes greater, and the contact angle of the depression, measured through the liquid, increases toward 180 degrees (see Figure 3.1, *A*). The jet transfers its momentum to the pool, creating transverse and normal velocity gradients near the surface. These give rise to a viscous shear and normal stresses, which tend to pull the liquid near the surface deeper into the pool, resulting in an "inverted" meniscus around the junction. When the contact angle reaches 180 degrees, an air film is carried into the pool by the fast-moving liquid (see Figure 3.1, *B*). The lower part of this film oscillates and eventually breaks up into bubbles. This is referred to as the "stable mode" of entrainment. With lower viscosity liquids, the length of the stable air film around the plunging point during entrainment is considerably reduced. During the actual process of entrainment through the surface, the pool surface curves downward, which forms an inverted meniscus. Any disturbances moving with the jet may cause liquid particles to start outward (Van de Sande and Smith, 1973) as shown in Figure 3.2.

As the jet continues, its movement (downward the disrupted flow field) in the receiving liquid decreases and ultimately gets reversed. A toroidal hole filled with air is left in the liquid. Shear stresses then break up the captured air into bubbles. Thus, the inception of gas entrainment by low-viscosity jets depends on the magnitude of disturbances on the jet surface, which in turn depends on the velocity, diameter, and length of the jet and on the physical properties of the liquid and are also influenced by the environment as well as

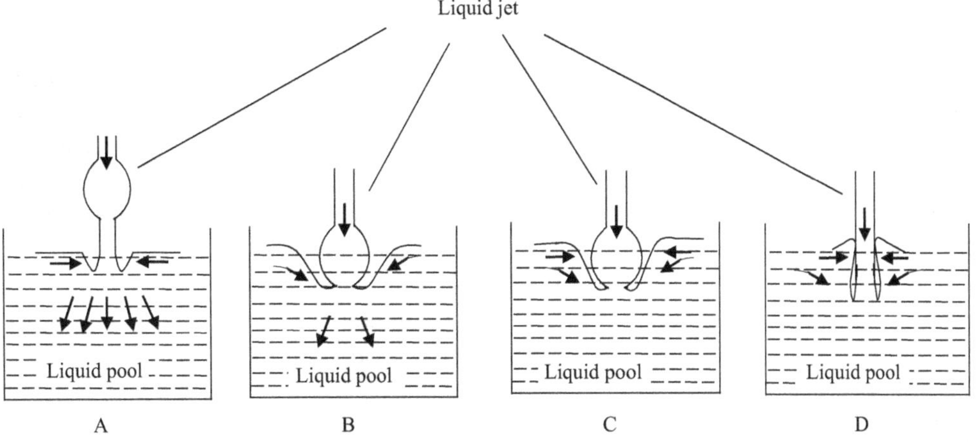

FIGURE 3.2 Mechanism of gas entrainment by liquid jet. **A** to **D,** Different subsequent phases of the mechanisms as energy transfer by the liquid jet when it moves downward.

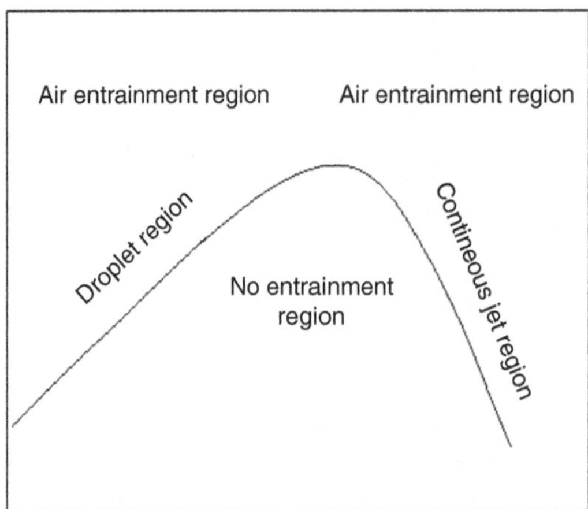

FIGURE 3.3 Threshold of air entrainment regions.

nozzle design. There are two regions of the onset of gas entrainment by a vertical plunging liquid jet: (1) then region where the jet breaks into droplets before reaching the pool surface and (2) the region where the jet is continuous (coherent). Figure 3.3 illustrates the threshold of these two entrainment regions.

In the droplet region of entrainment, as the jet velocity is increased, the distance to breakup increases until it becomes equal to the distance between the nozzle outlet and the pool surface, in which case entrainment ceases. In the continuous region, the transition is not sharp because entrainment is intermittent. It is caused mainly by irregular disturbances on the jet surface. According to Van de Sande and Smith (1973), as a liquid jet comes out of a nozzle, a thin layer of gas film is formed around it. They reported that for liquid jets with rough surfaces, some gas is entrapped into the outer boundary. When the jet impacts the pool surface, both the entrapped gas and the boundary layer gas are carried under the free liquid surface. The jet itself introduces a flow in the receiving liquid whose streamlines, at the plunging point, are directed parallel to the jet. A hole appears around the jet by the flow of liquid itself. This creates an under-pressure zone that helps the entrainment of the gas. Shear stresses then break up the captured gas into fine bubbles. The vertical jets produce an indentation in the pool surface, called an *induction trumpet,* which causes the entrainment at the plunging point (Evans et al., 1996). Entrainment from such a trumpet is very regular, and the phenomenon is analogous to a jet ejector pump, with the plunging point being the site of a free surface vortex. The entrainment for turbulent jets results from disturbances on the free surface caused by jet instability while entrainment by laminar jets occurs from the formation of a thin shell of gas at the point of entrance, the development of oscillations in the shell, and the subsequent breaking up of bubbles. Lara (1979) showed two regions of gas entrainment by a vertical plunging jet, that is, the droplet region and the continuous jet region. In the first case, entrainment occurs when the jet breaks into droplets before reaching the pool surface; in the second case, entrainment takes place by a continuous liquid jet. Lara illustrated the threshold of these two entrainment regions as a function of jet length and jet velocity.

Thomas et al. (1984) proposed another mechanism of gas entrainment for higher jet velocities. According to them, most of the entrained gas enters the main flow via the layer of foam formed on the surface of the receiving fluid. Gas enters the interstices in the foam, possibly as a result of wave action and splashing, and is then entrained into the main body of the flow along with the recirculating foam. Several other mechanisms of gas entrainment have been reported by different authors (Henderson et al., 1970; Burgess and Molloy, 1973; McKeogh and Ervine, 1981). Several other authors have clarified the effects of nozzle design on entrainment and other related phenomena, especially in the restricted geometry systems (Ohkawa et al., 1986; Bando et al., 1988; Evans et al., 1996; Kusabiraki et al., 1990; Yamagiwa et al., 1990), which occur when a vertical liquid jet plunges through the surface of liquid contained in a vertical column. Other various types of gas entrainment mechanisms in a reactor pools were proposed to apply to a number of chemical and nuclear engineering applications (e.g., fast breeder reactor based on their experimental studies, which can be summarized as follows:

- **Vortex-induced gas entrainment** (Sverak and Hruby, 1981): In this mechanism, liquid flows past a surface with high velocity, which leads to inertial dominance and vortices formation behind obstructions encountered by the flowing fluid. These vortices create dimples and lead to entrapment of gas bubbles in liquid.
- **Liquid fall–induced gas entrainment** (Madarame and Chiba, 1990): In this case, free surface velocity translates into vertical velocity near the periphery of the vessel. The liquid moves upward above the mean free surface and forms a local hump, which may fall down onto the free surface. The interaction between the falling liquid and the free surface causes entrapment of gas bubbles.
- **Drain-type vortex gas entrainment** (Takahashi et al., 1988): In this case, origination of entrainment occurs through an unstable vortex by approaching the gas core tip sufficiently close to the outlet at a certain flow rate. Bubbles are formed as they break away at the tip of the gas core and are swept into the outlet.
- **Shearing-induced gas entrainment** (Ervine and Falvey, 1987): This is caused by the shearing action at the interface between liquid and gas by generation of surface waves.
- **Dissolution-induced gas entrainment** (Cacuci, 2010): The gas gets dissolved on the free surface of liquid pool. Dissolved gas is a possible source for the formation of fine gas bubbles.
- **Rotation-induced entrainment** (Cacuci, 2010): This occurs by rotating a mechanical device (e.g., impeller, pump shaft) that is partially submerged in liquid pool. The free surface between the mechanical device and the vessel is expected to deform with a depression that forms bubbles and gets entrained along the motion.

Estimation of Entrained Gas for Inverse Bubbly Flow

Gas entrainment by plunging liquid jets occurs as a localized phenomenon at the plunging point. There are basically two methods of estimation of entraining gas are developed by Bin (1993), which are (1) measuring the amount of gas after it has been entrained into the pool liquid and (2) measuring the removal of gas from a gaseous space above the pool surface around the plunging point. The rate of gas entrainment as per the first method can be estimated corresponding to the volumetric flow rate of liquid as

$$\varepsilon_g = \frac{Q_g}{Q_g + Q_l} \tag{3.1}$$

By simplifying equation (3.1), one gets

$$Q_g = \frac{Q_l \varepsilon_g}{\left(1 - \varepsilon_g\right)} \tag{3.2}$$

where Q_g is the gas entrainment rate in the column; Q_l is the volumetric liquid flow rate; and ε_g is the overall gas hold-up in the column, which may be measured by phase isolation techniques. The second method is represented by devices used by Kusabiraki et al. (1990) and Evans et al. (1996). The gaseous space above the pool in the vicinity of the plunging point is in this case separated from the ambient, and a supplementary gas is let into this space through an appropriate flow rate device (orifice, rotameter, anemometer, or from volumetric and time readings). Traps or gas removal arrangements may interfere with the fluid flow in the pool. However, because the gas entrainment phenomenon depends mainly on the flow in the direct neighborhood of the plunging point, an appropriate submergence and geometry of the trap should not greatly affect the amount of entrained gas (Bin, 1989). Van de Donk (1981) measured entrainment by vertical jets by a collar mounted at the free liquid surface. The diameter of the collar was narrow enough to separate the rising bubbles and to allow their separate capture and measurement. The measured gas flow rates remained consistent for all collar openings from 1.2 to 3.0 times the jet diameter. Such systems of measurements prevent interaction between the rising bubbles and the jet. This is particularly relevant for vertical jets at very high entrainment rates when the surface of the pool is severely disturbed by the rising bubbles. Because gas entrainment is in this case controlled mainly by the jet roughness, it can be concluded that the pool surface disturbances are not highly significant in controlling gas entrainment. Traps are not completely satisfactory in capturing gas bubbles formed with very rough jets and for very slow rough jets that have insufficient momentum to carry the entrained gas deep into the pool. Such devices are unable to remove gas effectively. The technique may also not be satisfactory for systems that tend to foaming (e.g., more concentrated aqueous glycerol solutions). Lin and Donnelly (1966) used an indirect method of gas entrainment rate estimation from the measured bubble size and bubble frequency for aqueous glycerol solutions of viscosities of the order of 0.1 Pa s. In all of these cases, the second group of measuring techniques can be recommended. McKeogh and Elsawy (1977) experimentally studied the entrainment of gas by a jet of water falling into a pool of water at various velocities and falling distances. They measured turbulence in the jet immediately before the point of impact by laser Doppler anemometry with frequency shifts. Gas–water ratio measurements were made in the jet and the pool. The bubble size, velocity, and formation rate were investigated by cine photography.

Effect of Variables on Gas Entrainment for Inverse Bubbly Flow

In an ejector-induced bubble column of inverse bubbly flow (see Figure 1.7), the gas entrainment rates varies with the liquid flow rate, gas–liquid mixing height in the column, and nozzle diameter. The gas entrainment rate depends on the motive fluid flow rate and the resistance of liquid in the column. The resistance of liquid in the column depends on the pressure developed in the separator (P_s) because the separator pressure holds the liquid in the column. As the separator pressure increases, the entrained gas bubbles face more resistance to move inversely, and consequently, lower entrainment results. The variations of gas entrainment rate (Q_g) with separator pressure for an air–water system at different liquid flow rates are presented in Figure 3.4.

At a constant liquid flow rate, the gas entrainment rate decreases with P_s and varies approximately as $Q_g \propto P_s^{-0.631}$. Therefore, to obtain the constant rate of gas entrainment, higher energy is required at a higher separator pressure. The energy requirement as a function of separator pressure can be expressed as $E_s \propto P_s^{0.548}$. The entrainment rate increases with the liquid flow rate because of an increase in average velocity of gas film around the liquid jet (Majumder et al., 2007). A decrease in nozzle diameter and hence an increase in jet velocity at a constant liquid flow rate tends to increase the entrainment by increasing the mean linear velocity of the gas film around the jet surface. The gas entrainment rate is almost independent of the column diameter. The gas entrainment rate is dependent mainly on the kinetic energy of the liquid jet, surface roughness of the liquid jet, and contacting perimeter between the jet and the receiving liquid surface.

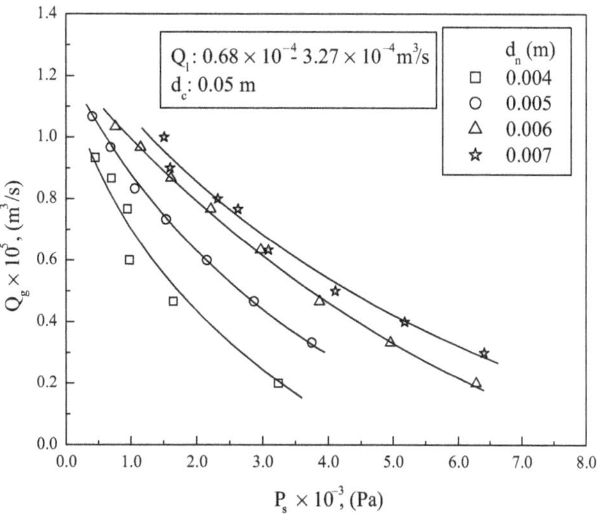

FIGURE 3.4 Variations of gas entrainment rate (Q_g) with separator pressure (P_s) at different nozzle diameters.

Another factor that is very important to note for gas entrainment by plunging liquid jets is the effect of nozzle geometry on the amount of entrained gas. The turbulence developed by the nozzle and the resultant behavior of the jet depend strongly on the nozzle geometry. To develop fully turbulent flow by the nozzle, the ratio of the length of its cylindrical section to its diameter should not be less than 50 (Van de Sande and Smith, 1975), although the entrained gas flow rate does not differ significantly for jets produced from nozzles with l/d_o ratios greater than about 15 (Bin, 1993).

Another important parameter affecting gas entrainment is the jet length. Low-velocity jets disintegrate into separate drops beyond a well-defined breakup length. For low-velocity jets near the breakup point, the total kinetic energy of the jet is related to the amount of entrained gas. For shorter jets, the jet length dependency can be found experimentally, although one can expect a similar influence of the kinetic energy of the jet. For high-velocity jets, transverse inertial forces overcome those from surface tension and change the shape of the jet, causing local enlargements and narrowing in diameter. As a result, gas is entrained within the surface enveloping the surface roughness of the jet. High-velocity jets not only carry air captured within the mean containing envelope but also entrain air in the boundary layer that develops outside that envelope into the pool. Gas entrainment rate by plunging water jets produced from long cylindrical nozzles ($l/d_o \geq 50$) has been researched by Van de Sande and Smith (1975), Kumagai and Endoh (1982), Majumder et al. (2007), and Sivaiah et al. (2012). Increasing the liquid phase viscosity affects the mechanism of entrainment because there are two controlling factors that are of relevance. Above about 6.5 mPa s, the shear between the pool liquid and the jet controls the entrainment; below this value, the roughness of the jet and the air boundary layer are the controlling parameters. As the viscosity of the liquid phase increases, laminar jets are formed at higher jet velocities than is the case for water jets. However, particularly for smaller jets, as the jet velocity increases, after a short length of apparent smoothness, the jet becomes very rough. This increasing roughness is induced by the relaxation of the steeper velocity profile in the case of more viscous jets. Addition of surface active agents decreases the static surface tension of the liquid phase. However, the actual (dynamic) surface tension of the jet at the plunging point depends on the rate of diffusion of a surfactant to the jet surface from the bulk and attaining adsorption equilibrium at this surface. This requires some finite time, and for shorter jets, the equilibrium concentration of a surfactant may not necessarily be reached before plunging to the pool liquid. Kumagai and Endoh (1982) found from their experiments with inclined jets that the gas entrainment rate decreased with increasing liquid phase kinematic viscosity in their so-called initial region of entrainment. It is increased with an increase of this parameter in the low-velocity region. For the high-velocity region of entrainment, they found no effect of this parameter on the entrainment rate. The same authors also concluded that there is a negligible effect of surface tension on the entrainment rate that is detectable only at jet velocities below 2 m/s. According to Sene (1988), at high jet speeds, the entrainment rate should be independent of the surface tension of the liquid, unlike at low speeds. For this jet, the rate of air entrainment should be reduced because the largest scales of turbulence are limited by the jet thickness. Also, at low circular jet velocities, the amplitudes of disturbances may be further reduced by surface tension effects arising from the high curvature of the jet.

Another possible reason for the surface tension effect limiting air entrainment rates with high-speed circular jets would be a reduction in the thickness of the air layer. Kusabiraki et al. (1990) reported that the gas entrainment rate can results from effect of the physical parameters as a function of the Ohnesorge number ($= \mu_l/(\rho_l \sigma d_o)^{1/2}$). It varies exponentially in such a way that the values of the power exponent on this number are from –0.58 to –0.30, depending on the jet velocity and the length to diameter ratio of the jet. There is no effect of the gas phase physical properties on the entrainment rate (Burgess and Molloy, 1973; De Frate and Rush, 1959; Lin and Donnelly, 1966). However, Sene (1988) concludes that the gas entrainment rate should be proportional to square root of the viscosity of the gas. Kumagai and Endoh (1982) observed four distinct entrainment regions for a plunging liquid jet. They found that the transition points were functions of the kinematic liquid viscosity. They reported that the entrainment rate was independent of the surface tension, which was in contrast to all previous observations. Ohkawa et al. (1986) discussed the gas entrainment characteristics between two vertical liquid jet systems with and without downcomer. They observed that the gas entrainment increases with increasing nozzle diameter and decreases with increasing height of the downcomer. Sene (1988) studied gas entrainment characteristics both theoretically and experimentally by a plunging liquid jet. He proposed that the mechanism of gas entrainment undergoes a qualitative change as the jet impact velocity (u_i) is increased. Separate models were presented that predict gas entrainment rates (Q_e) in the low- and high-velocity regimes. He proposed that at low velocities, Q_e varies with u_i^3, but at high velocities, Q_e varies with $u_i^{3/2}$. He also suggested that for high jet velocities, the entrainment rate should be independent of the surface tension.

Yamagiwa et al. (1990) have investigated the operations of an inverse flow bubble column with gas entrainment at high throughputs. For a gas–water system, they obtained a correlation for entrainment ratio (Q_e/Q_l) as a function of a dimensionless group of system parameters. They evaluated the performance of gas entrainment in terms of energy efficiency and compared their results with other gas–liquid aerators. Evans et al. (1996) developed a model of gas entrainment for a confined plunging liquid jet considering the re-entrainment of the recirculated gas. Their model was based on the effective jet diameter as a function of free jet length. Van de Sande and Smith (1975) observed entrainment mechanisms for turbulent water jets. They made a distinction between the mechanisms for low-velocity ($u_j < 5$ m/s) and high-velocity ($u_j > \{10\sigma/(\rho_g d_j)\}^{1/2}$) jets. At low velocity, they found that entrainment rate depends on the surface deformities on the jet and the physical properties of the receiving liquid. For high-velocity jets, the volumetric gas entrainment rate can be expressed by the gas contained inside the boundary layer, which was dragged along by the free surface of the jet.

Depth of Bubble Penetration Due to Gas Entrainment

Gas entrainment rate depends on the depth of the penetration of the gas bubble. The penetration depth is increased with an increase in kinetic energy supplied by the plunging liquid jet. A properly designed downcomer for inverse bubbly flow induces a velocity field of gas bubble opposite to the buoyancy force in the downcomer, which gives rise to an increased bubble penetration depth above the limiting values attainable in the ordinary jet system. Such an inverse flow bubble column with gas entrainment by a plunging jet conforms to a relatively new type of concurrent bubble column, frequently designed as a loop absorber (reactor) (Tamir, 2014), a downflow slurry reactor (Sivaiah et al., 2012). With increasing jet velocity, the bubble depth penetration increases initially, and the length of the jet above the plunging point decreases because of an effect of increased free liquid surface level in the downcomer. It will be carried out until the bubble reaches the lower end of the downcomer and is deflected by an impacting disk and then rises to the liquid surface of the pool. At this flow condition, uniform bubbling is achieved through the downcomer in a narrow central core zone. Other than this central core zone near the wall of the downcommer or column, the gas–liquid dispersion is small because gas entrainment is still poor. With a further increase in jet velocity, the jet length increases, the core zone expands, and a uniform bubble disperses through the column up to the discharging end with full occupation of the downcomer or column. During these changes, the gas phase pressure at the plunging point in the downcomer may differ from the atmospheric one because of the projection of the downcomer tube above the liquid surface in the downcomer. Bubbles entrained by a vertical plunging jet penetrate the pool liquid to some maximum depth, which is not strictly defined because the lower limit of the bubble swarm fluctuates continuously. At the maximum depth of bubble penetration, the local liquid velocity in the submerged jet at that point should be equal to the bubble free rise velocity, leading to a direct linear relationship between the maximum depth of bubble penetration and the product of the jet diameter and jet velocity (Smigelschi and Suciu, 1976). However, buoyancy forces resulting from the entrained gas complicate such an analysis. The maximum depth of bubble penetration related to the square of the jet velocity and the entrainment ratio (Bin, 1993). Van de Sande and Smith (1975) developed a correlation with nozzles of diameter ranging from 3.9 to 12 mm and for L_j less than 0.5 m as

$$H_p = 0.42u_j^{4/3}d_jQ_e^{-1/4} \tag{3.3}$$

For low jet velocity, the correlation can be represented by

$$H_p = 1.20u_j^{0.77}d_o^{0.625}L_j^{-0.094} \tag{3.4}$$

Under practical conditions, H_p will not be greater than 0.4 to 1.0 m, provided the vertical jet is not too dispersed (Bin, 1993). The effect of the jet length on H_p, is significant only for short jet lengths (Ohkawa et al., 1986, 1987). At L_j/d_o greater than 20, there is no influence of jet length on penetration depth. The depth of bubble penetration can be increased in a vertical plunging jet system by introducing a downcomer.

Minimum Entrainment Velocity

When a liquid jet strikes the pool of the liquid surface, gas entrainment occurs if the jet velocity exceeds a certain critical value. The entrainment jet velocity (u_e) is well characterized for coherent viscous laminar jets and is very complicated for turbulent jets. Lin and Donnelly (1966) developed an empirical correlation in case of laminar viscous jets within a range of jet Reynolds number of 8 to 1500, liquids of viscosities from 25 to 400 mPa s, densities from 846 to 1246 kg/m^3, and surface tensions from 0.03 to 0.063 N/m, which can be represented by

$$We_j = 10\,Re_j^{\,0.74} \tag{3.5}$$

where We_j and Re_j are related to the jet diameter and jet velocity at the plunging point. For laminar vertical jets, gravity forces affect the diameter and velocity of the jet. Theoretical treatment of free liquid jets is complicated by factors such as contraction (or expansion) of the jet at the nozzle outlet and the velocity profile relaxation after the point of efflux. An energy balance, neglecting the effects of the surrounding gas and the small contribution of the surface energy, would lead to the equation (Bin, 1993)

$$\frac{d_j}{d_r} = \left(\frac{\pi^2 g L_j d_r^4}{8 Q_l^2} + 1 \right)^{-1/4} \tag{3.6}$$

where d_r is the jet diameter at the reference point (generally, the reference point is considered the diameter of the nozzle or hole outlet). For turbulent jets and in the range of jet velocities close to the minimum entrainment velocity, the nozzle diameter is far more important than these small changes in jet diameter (Ciborowski and Bin, 1972). For simplicity, jet velocity can be calculated as

$$u_j = (u_0^2 + 2 g L_j)^{1/2} \tag{3.7}$$

For high-velocity jets, the jet is subject to violent gas friction and surface tension forces for which the jet surface becomes ill defined. The jet may spread in a dispersed, generally conical form. For high-velocity jets, Van de Sande and Smith (1973) developed a new correlation valid for $We_g \times Re_{length}$ greater than 7×10^5 and for jets produced from long cylindrical nozzles ($l/d_0 \geq 50$) as

$$\frac{d_j}{d_o} = 0.125 \left(We_g\, Re_{length} \right)^{1/6} \tag{3.8}$$

For turbulent jets produced from short cylindrical nozzles (with l/d_o, <8) Ciborowski and Bin (1972) suggested a very simple relation, $We_j = 400$. A criterion of entrainment velocity for air entrainment for circular jets is also reported by Bin (1993) as

$$Fr_j = u_j^2/(g d_o) > 10; \quad Re_j > 7000 \tag{3.9}$$

McKeogh and Ervine (1981) defined a minimum velocity to entrain gas below which jets will entrain gas only intermittently. They proposed four mechanisms characterizing the aerated region in the pool based on type of jet. They also described the minimum jet velocity to entrain gas as a function of the turbulent intensity in the jet. Bin (1988) studied vertical plunging liquid jets and developed a correlation for minimum gas entrainment velocity as a function of jet length and nozzle diameter. However, for a liquid jet with a downcomer, even after gas entrainment occurs at the plunging point, there is almost complete recirculation of the entrained gas unless certain momentum is imposed on the gas bubbles to move inverse. This is because of upward buoyant force of the gas bubbles, which tend to move the bubbles upward as stated earlier. Thus, a minimum liquid jet velocity is required to move the gas bubbles out of the column, which depends on the length of the column, the bubble size, and the physical properties of the fluids. Some theoretical attempts have been made by various authors to predict the onset conditions of entrainment. Cumming (1975), by a force balance in the plunging jet system with a few simplifying assumptions, derived an equation for the minimum entrainment velocity as

$$u_{e,min} = \left(\frac{\sigma}{C\left(\rho_l \mu_l d_j\right)^{0.5}}\right)^{\frac{2}{3}} \left(1 + C\left(\frac{\rho_g \mu_g L_j}{\rho_l \mu_l d_j}\right)^{0.5}\right)^{\frac{-2}{3}} \tag{3.10}$$

The unknown parameter C can be obtained from experiment. He reported that for an aqueous glycerol solution of viscosity 0.1 Pas, a typical value of C equals 0.0286. C is not a constant value but varies between 0.013 and 0.058, depending mainly on liquid viscosity, and actually reflects a somewhat different dependence of $u_{e,min}$, on the other relevant variables.

Energy Efficiency of Gas Entrainment

The energy efficiency of the inverse bubble column with an ejector as the gas entrainment device can be evaluated by the kinetic energy E_k of the liquid jet. The value of E_k can be calculated as follows

$$E_K = \frac{1}{2}mu_j^2 = \frac{1}{2}Q_l \rho_l u_j^2 = \frac{1}{2}a_n u_j \rho_l u_j^2 = \frac{1}{2}\left(\frac{\pi}{4}d_n^2\right)u_j \rho_l u_j^2 = \frac{\pi \rho_l d_n^2 u_j^3}{8} \tag{3.11}$$

The energy efficiency of the inverse system and other types of gas–liquid contactors with gas entraining or gas sparging are summarized in Table 3.1. The gas flow rate Q_g per unit liquid volume V_L of the apparatus has been often used as one of the measures to express the aeration intensity of the aerator. The values of Q_g/V_L are also shown in the Table 3.1. It may be seen from the table that the inverse system showed higher efficiency compared with that of other types of gas-entraining aeration systems. In estimating the

Table 3.1   Aeration Performance of Gas–Liquid Contactors

Authors	Type of Contactor	$(E_k/Q_g) \times 10^{-3}$ [kWs/m³]	$(Q_g/V_L) \times 10^2$ [s⁻¹]
Fukuda et al. (1963)	Aerated–stirred fermenter	0.08–0.14	0.10–5.00
Topiwala and Hamer (1974)	Hollow impeller	0.30–0.70	0.20–2.00
Zundelevich (1979)	Turbo aerator	0.06–0.80	1.00–13.00
Matsumura et al. (1982)	Tank-type gas entrainer	0.1–1.0	0.15–3.00
Ohkawa et al. (1987)	Jet aeration in a pool system	0.015–0.03	0.02–0.20
Majumder et al. (2005)	Inverse bubbly flow column	0.018–1.01	0.59–13.99

energy dissipation rate per unit volume of the mixing zone, $\in$, derived by Cunningham (1974) for a liquid jet gas pump, can be expressed as

$$\in \cong \frac{\rho_l d_j^2 u_j^3}{2 d_c^2 L_m}\left[1-2\left(\frac{d_j}{d_c}\right)^2-\left(\frac{d_j}{d_c}\right)^4\left\{\left(\frac{Q_g}{Q_l}\right)^2-1\right\}\right] \tag{3.12}$$

The jet–column diameter ratio is generally less than 0.223, which practically independent of the entrainment ratio.

Majumder et al. (2007) reported that gas–liquid mixing occurred in the contactor when a liquid jet of uniform initial velocity was discharged from the nozzle into a fluid. It is implicit that gas liquid bubbly mixture flows out of mixing zone while the liquid jet is detained at a certain length to the direction of flow. From the energy balances, the energy utilization efficiency for entrainment of gas in liquid phase for mixing is given by

$$\eta_m = 1 - K_m \tag{3.13}$$

$$K_m = 3.04 \, \mathrm{Re}_{\ln}^{-0.465} A_R^{0.408} H_{rn}^{-0.25} Su_n^{0.92} Mo^{0.339} \tag{3.14}$$

The liquid resistance inside the column increases with increase in concentration of the solution, which results in more energy dissipation for mixing the phases. For the same mixing intensity inside the column, more liquid jet energy required for more concentrated liquid. For this reason, the mixing dissipation coefficient K_m will be less for more concentrated liquid. There is a decrease in K_m with an increase in liquid Reynolds number, which is because of increased turbulence in the mixing zone and hence liquid velocity head. On the other hand, K_m increases with a decrease in the gas–liquid mixing height in the column. This is because for a particular liquid, the Reynolds number gas entrainment through the secondary entrance increases with decrease in the gas–liquid mixing height, which in turn increases the bubble population in the mixing zone. On an average, data reported by Majumder et al. (2007) are comparable with data reported by Ciborowski and Bin (1972), McKeogh and Ervine (1981), and Ohkawa et al. (1986) but considerably smaller than those of Van de Sande and Smith (1973). This is because Van de Sande and Smith (1973) used only plunging jet without any downcomer, but in the case of Majumder et al. (2007), a minimum height of the gas–liquid mixture is always maintained in the downcomer, which has a negative effect on gas entrainment. The energy efficiency of the inverse bubble column as a gas entrainment device can be evaluated by the rate of kinetic energy supplied by the liquid jet divided by the volumetric gas entrainment rate Q_g. The value of rate of energy is obtained by the following equation (Bin and Smith, 1982):

$$\dot{E}_s = \frac{\pi \rho_l d_n^{\,2} u_j^{\,3}}{8}$$

(3.15)

The energy used for the mixing of the two phases is then equal to the product of energy supplied to the system and the energy utilization efficiency of mixing, which is represented by

$$\dot{E}_{um} = \eta_m \dot{E}_s$$

(3.16)

Energy utilization for gas liquid mixing in the mixing zone increases with the liquid Reynolds number (Re_{ln}) at different ratios of liquid height. This is because of the energy dissipation coefficient decreases as the liquid Reynolds number increases. As the liquid Reynolds number increases, the turbulence of the gas–liquid mixture increases. As higher turbulence, the utilization of the jet energy for mixing is higher, which results increase of gas hold-up and less frictional loss (Majumder et al., 2007). This increases the efficiency of the energy utilization for gas liquid mixing. As the ratio of gas–liquid mixing height increases, the energy utilization efficiency also increases. At constant liquid Reynolds number, the energy dissipation coefficient (K_m) decreases, which results the increase of utilization efficiency of energy for mixing.

Axial Distribution of Kinetic Energy Utilization for Gas Entrainment

The axial distribution of kinetic energy utilization for gas entrainment in a downcomer of inverse bubbly flow operated by ejector system can be estimated from the energy balance between two pints of the contactor with the following assumptions:

- For a particular gas and liquid flow rate, frictional loss is considered as constant throughout the column.
- Because the column diameter is small, there is no variation of radial gas hold-up.
- The operation is isothermal and steady

Under steady operation, the energy balance equation between two points 1 and c (the gas–liquid mixing level), 2 and c, and so on can be written as

$$(E_k)_1 - (E_k)_c = P_c - P_1 + \rho_l(1 - \varepsilon_{g,1,c})(h_c - h_1)$$

(3.17)

where $(E_k)_1$ and $(E_k)_2$ are the kinetic energy at points 1 and 2, respectively; P_c is the pressure above the gas–liquid mixing level (Figure 3.5); and $\varepsilon_{g,c,1}$ is the gas hold-up between

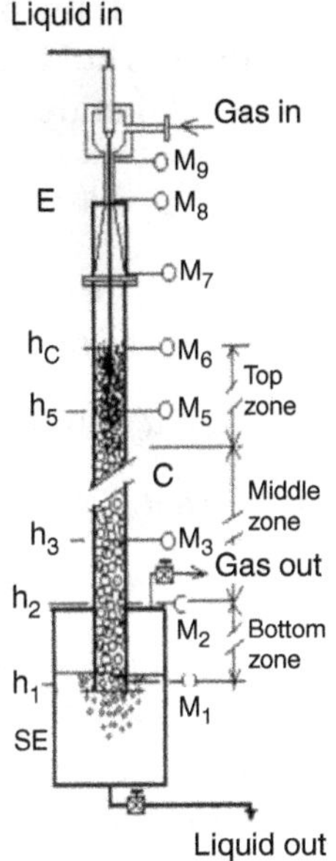

FIGURE 3.5 Layout of inverse flow of bubbles in a column.

the points 1 and c. Similarly, kinetic energy balance between points 2 and c and so on gives

$$(E_k)_2 - (E_k)_c = P_c - P_2 + \rho_l(1 - \varepsilon_{g,2,c})(h_c - h_2) \tag{3.18}$$

$$(E_k)_3 - (E_k)_c = P_c - P_3 + \rho_l(1 - \varepsilon_{g,3,c})(h_c - h_2) \tag{3.19}$$

Subtracting equations (3.18) from (3.17), (3.19) from (3.18), and so on, one gets, respectively,

$$(E_k)_1 = (E_k)_2 + (P_2 - P_1) + \rho_l[(1 - \varepsilon_{g,1,c})(h_c - h_1) - (1 - \varepsilon_{g,2,c})(h_c - h_2)] \tag{3.20}$$

$$(E_k)_2 = (E_k)_3 + (P_3 - P_2) + \rho_l[(1 - \varepsilon_{g,2,c})(h_c - h_2) - (1 - \varepsilon_{g,3,c})(h_c - h_3)] \tag{3.21}$$

Thus, in general, it can be written as

$$(E_k)_i - (E_k)_{i+1} = (P_{i+1} - P_i) + \rho_l[(1 - \varepsilon_{g,i,c})(h_c - h_i) - (1 - \varepsilon_{g,i+1,c})(h_c - h_{i+1})] \tag{3.22}$$

where, $i = 1, 2, 3, \ldots, c$

$$(E_k)_c = \eta_{ej}\rho_l u_j^2/2 \tag{3.23}$$

$\varepsilon_{g,c,i}$ is the axial cumulative gas hold-up between the points c and i, and $h_c - h_i$ is the distance between c and i. The axial cumulative gas hold-up is a function of h_i. The functionality can be represented mathematically as (Majumder et al., 2007)

$$\varepsilon_{g,c,i} = r h_{c,i}^{2} + s h_{c,i} + t \tag{3.24}$$

where r, s, and t are parameters that vary with the nozzle diameter, and the experimental results are correlated with nozzle diameter. The correlations can be expressed as

$$r = 0.12 Ln(d_n) + 0.244 \tag{3.25}$$

$$s = -0.3 Ln(d_n) - 1.103 \tag{3.26}$$

$$t = 0.0064 Ln(d_n) + 0.498 \tag{3.27}$$

Total kinetic energy supplied by the liquid jet may not be distributed in the contactor for utilization of phase mixing. The distribution of kinetic energy in the column depends on the gas hold-up distribution. The axial distribution of kinetic energy used for phase mixing in the column can be estimated from equations (3.22) to (3.24). Majumder et al. (2007) reported that the kinetic energy distribution is a polynomial function of axial length of the column. The axial length is measured from the bottom of the column. At the bottom zone of the column, the kinetic energy used in the column is relatively lower than in the other zone, and it increases with increases in height (h_i), but at the middle zone of the column, it is almost constant. The maximum amount of kinetic energy is used in the top zone of the column where the liquid jet entrains the gas. In the top zone, the kinetic energy utilization increases with the increase in axial length. The variation of energy utilization is due to the gas hold-up distribution in the column (Majumder et al., 2006). The variation of cumulative gas hold-up with respect to the location of the column is due to the variation of axial pressure drop along the column. The gas hold-ups are different in different locations of the column. The density of bubble population varies along the column, which may cause the variation of local gas hold-up. At the bottom location of the column, the bubble density is low because of higher pressure at the bottom, and less kinetic energy is distributed for gas–liquid mixing. At the bottom location, smaller bubbles coalesce, and because of buoyancy effect, coalesced bubbles move upward and accumulate in the middle location. In the middle location of the column, bubble buoyancy balances the downward kinetic force of liquid flow. Hence, at the middle zone, the density of the bubble population is higher than in the other zone of the column, which results in higher gas hold-up than in the other zone of the column. Also, a uniform bubbly zone is formed in the middle zone, which indicates the uniform energy distribution in the middle zone. In the top zone, maximum energy is used for the intense mixing of phases. The liquid jet plunges in the mixing zone, and non-uniform bubbles form because of the breakage and coalescence of bubbles. In the top zone, when the liquid jet plunges into the pool of the liquid inside the column, the maximum energy is used to overcome the maximum resistance of the liquid as well as the buoyancy force of the bubble.

Models of Entrainment Rate

The entrainment of gas through liquid jet in an inverse flow system is a very complex phenomenon; hence, it is very challenging to develop a theoretical model to predict the gas entrainment rate. The entrainment rate depends on physical properties of liquid and gas, geometric variables of the system, and dynamic variables. So, the functionality of the Q_r can be expressed as given as

$$Q_r = c_1 D_R{}^{b_1}, H_R{}^{b_2}, \mathrm{Re}_j{}^{b_3}, We_j{}^{b_4}, Fr_j{}^{b_5} \tag{3.28}$$

By the multiple linear regression analysis on experimental data, Majumder et al. (2006) developed the following generalized correlation:

$$Q_r = 1.52 \times 10^{-4} D_R{}^{-0.534} H_R{}^{0.642} \mathrm{Re}_j{}^{0.268} We_j{}^{0.295} Fr_j{}^{-0.130} \tag{3.29}$$

The correlation has been found to be satisfactory within the range of the different parameters as $0.22 < Q_r < 0.48$; $0.07 < D_R < 0.14$; $9.40 < H_R < 30.89$; $1.81 \times 10^5 < Re_j < 22.40 \times 10^5$; $5.85 \times 10^3 < We_j < 750 \times 10^3$ and $0.14 \times 10^2 < Fr_j < 18.14 \times 10^2$. Some typical ranges of gas entrainment ratio obtained by different authors are shown in Table 3.2. Ohkawa et al. (1987) and Kusabiraki et al. (1990) established three regions of the entrainment rate, which are the (1) initial or low jet velocity region, (2) transition region, and (3) high jet velocity region. They reported that a typical S-shaped entrainment rate curve is observed for the whole range of jet velocities. Van de Sande and Smith (1975) established the low jet velocity region, which extends up to about 5 m/s.

Some theoretical models are also available for predicting air entrainment rates in the low and high velocity regions (Sene, 1988). For the low-velocity region of entrainment, it is assumed in Sene's model that individual bubbles form as a result of surface tension forces. Sene (1988) derived an expression considering the turbulence intensity in the jet for this region of entrainment and showed that $Q_g \propto U_j^3$. For the high jet velocity region, he assumed that a gas layer is set into motion by shear stresses at the surface of the jet, which is broken up into a succession of bubbles by instability waves on its surface. He developed a model for the air entrainment rate for the high-velocity region based on balancing the

Table 3.2   Gas Entrainment Ratio Reported by Different Authors

Authors	$F_{rj} = (u_j^2/gd_n)$	Q_r
Ohyama et al. (1953)	400–2500	0.10–0.50
Ciborowski and Bin (1972)	64–256	0.02–0.50
Van de Sande and Smith (1975)	25–900	0.19–1.2
McKeogh and Ervine (1981)	49–400	0.14–0.45
Ohkawa et al. (1987)	25–3025	0.04–1.00
Majumder et al. (2006)	14–1814	0.22–0.48

pressure gradient in the air layer by the shear forces exerted on the air by the receiving flows, which can be represented by

$$Q_g \propto \left(\frac{\mu_g}{\rho_l g \sin\alpha}\right)^{1/2} u_j^{3/2} d_j \tag{3.30}$$

The rate of gas entrainment by plunging liquid jet of the length near the breakup length (at least 90% of this length) or beyond the breakup point is directly related to the kinetic energy (power) of the jet. For shorter jets, Sene suggested a correlation for gas entrainment rate in terms of a characteristic factor (X), which can be expressed as

$$Q_g = 0.015 X^{3/4} \tag{3.31}$$

$$X = d_o^2 V_j^3 L_j^{0.5} (\sin\alpha)^{-1.5} \tag{3.32}$$

Within a range of d_o = 2.85 to 10 mm with a large cylindrical section (l/d_o = 50), $2 < u_j < 5$ m/s, $L_j < 0.5$ m, $\alpha = 20 - 60$ degrees and $X = 10^{-4} - 10^{-2}$. For the same value of X, jets produced from nozzles with shorter cylindrical sections will entrain less gas. Equations (3.34) and (3.35) predict that $Q_g \propto (\sin\alpha)^{-1.125}$, which makes it clear that inclined jets entrain more gas than vertical ones. The power exponent at the jet inclination function $\sin\alpha$ ranges from -4.30 to -0.83 (Kumagai and Endoh, 1982; Kumagai and Imai, 1982; Ohkawa et al., 1987; Kusabiraki et al., 1990). In case of air entrainment by self-aerated flows, the entrainment ratio will be dependent on the basic system parameters such as jet velocity, jet length, nozzle diameter and design, angle of jet inclination, and liquid phase physical properties. To enunciate the effect of different parameters on the entrainment ratio, a correlation for vertical jet is proposed as the following empirical expression (Bin, 1993):

$$Q_g/Q_l = 0.04 \, Fr_j^{0.28} (L_j/d_o)^{0.4} \tag{3.33}$$

Van de Donk (1981) neglected a small effect of the jet velocity on the entrainment ratio and approximated his data for vertical jets by a simpler correlation within a range of the system parameters: u_o = 2 to 10 m/s, L_j/d_o = 2.5 to 100, d_o = 0.01 to 0.1 m as

$$Q_g/Q_l = 0.09 (L_j/d_o)^{0.65} \tag{3.34}$$

Rao and Kobus (1973) suggested that the entrainment ratio for self-aerated flows can be expressed by universal equation, which for plunging jets should take the form of $Q_g/Q_l \propto Fr_j$. As per this form, the entrainment ratio is dependent on the square of the jet velocity, which is contradictory with the results from equations (3.34) and (3.36) and with other various correlations in which the exponent of velocity is in the range 0.6 to -1.3 given by different authors (Kusabiraki et al., 1990). Ohyama et al. (1953) examined the entrainment process for turbulent water jet systems. They proposed that the roughness on the surface of the jet caused gas to be entrained at a rate given by the correlation

$$\frac{Q_g}{Q_l} = 0.75 \left(\frac{u_j^2}{gL_j} \right)^{-0.447} \left(\frac{d_n u_j \rho_l}{\mu_l} \right)^{1.53} \left(\frac{\mu_l}{(\sigma \rho_l d_n)^{1/2}} \right)^{2.18} \left(\frac{L_j}{d_n} \right)^{0.218} \tag{3.35}$$

where Q_g and Q_l are the gas and liquid volumetric flow rates, respectively; d_n is the nozzle diameter; and u_j and L_j are the velocity and free length of the jet, respectively. McKeogh and Ervine (1981) found that the turbulence level of the plunging jet determined the mechanism for gas entrainment. For gas–water systems, they found the gas entrainment ratio, which was best described by the expression

$$\frac{Q_g}{Q_l} = 1.4 \left[\left(\frac{2S}{d_n} \right)^2 + \left(\frac{S}{d_n} \right) - 0.1 \right]^{0.6} \tag{3.36}$$

where S is equal to the magnitude of the surface disturbance on the free surface of the jet. Each flow state corresponds to a critical downward liquid velocity. The uniform bubbling flow through the downcomer can be maintained if the liquid superficial velocity in the downcomer is greater than about 0.5 m/s. For longer downcomers of bubble, Ohkawa et al. (1986) developed an empirical expression for gas entrainment

$$Q_g = 0.968(u_j^3 d_o^2)^{0.8}(d_o/d_c)^{1.3}(h_c/d_c)^{-1.0} \tag{3.37}$$

which is valid in the range of $0.001 < u_j^3 d_o^2 < 0.001$ and 0.2 m⁵/s³; $7 < d_o < 13$ mm; $0.020 < d_c < 0.026$ m and $2 < h_c < 5$ m. Under the same main jet parameters (d_o, L_j, d_j, u_j), the system equipped with a downcomer allows a gas entrainment rate about 25% to 60% greater than that of system without a downcomer (Ohkawa et al., 1986). At the region of high liquid throughput in columns of diameter from 0.034 to 0.070 m, Yamagiwa et al. (1990) obtained an empirical correlation

$$Q_g = 0.040 u_j^{1.66} L_j^{0.48} d_o^{1.59} \tag{3.38}$$

with a validity for the range of values of the term $V_j^{1.66} L_j^{0.48} d_o^{1.59}$ from 0.002 to 0.03 (SI units) and for $d_o = 8$ to 13 mm and $L_j = 0.03$ to 0.15 m. The authors concluded that the inverse bubbly flow system, having a longer gas–liquid contacting section compared with that in the ordinary liquid jet system, can be operated without any concern about poorer performance in terms of gas entrainment. Evans et al. (1996) observed that the gas entrainment rate consists of two components, namely, the gas trapped within the effective of jet diameter at the plunging point and the gas contained within the film between the jet and the induction trumpet surface at the point of rupture. They developed a model that allows estimation of the volumetric rate of film wise entrainment by a liquid jet that plunges into a column of the same liquid. The model is based on thin gas film theory and assumes a laminar velocity profile in the film, where (1) the inside boundary velocity is equal to the jet velocity (u_j) and (2) the outside boundary velocity is equal to the recirculating eddy maximum velocity. They stated that a recirculating eddy in the column is induced by the plunging liquid jet, and its flow magnitude can be estimated from a correlation suggested by Liu and Barkelew (1986):

$$Q_{eddy,\max} = Q_l \left(\frac{0.37}{C_T} - 0.64 \right) \tag{3.39}$$

$$C_T = \frac{1}{\sqrt{(d_c/d_j)^2 - 1/2}} \tag{3.40}$$

where C_T is the Craya-Curtet number and, according to Barchilon and Curtet (1964) in the case of the plunging liquid jet system, where the secondary flow is zero. The value for C_T should always to be less than 0.5.

Future Scope

The following are some important issues on which future study can be focused related to the topics on present chapter:

- The detailed mechanism of entrainment of gas by other heavier gases and liquids
- The entrainment characteristics of different types of gas with a mixture of liquids
- The effect of surfactant on the entrainment characteristics
- The gas-aided extraction by entrainment of gas in inverse flow system
- The effect of non-Newtonian fluid jet on the gas entrainment characteristics
- Development of a model based on the boundary layer of gas induced by different mechanisms of gas entrainment
- Computational fluid dynamics on the entrainment characteristics to assess the flow behavior of the jet at the plunging point to produce a fine bubble by gas entrainment

Nomenclature

a_n Cross-sectional area of nozzle (m²)
C_T Craya-Curtet number
d_c Diameter of the column, (m)
d_j Diameter of the liquid jet, (m)
d_n Diameter of nozzle, (m)
d_o Diameter of nozzle outlet (m)
D_R Diameter ratio of nozzle to column (d_n/d_c) (-)
d_r Jet diameter at reference point (m)
E_s Total energy supplied (N.m/s)
E_{um} Energy utilization of mixing (N.m/s)
Fr_j Froude number $(u_j^2/d_o g)$ (-)
h_i Height of the fluid level in the column at level i (m)
H_p Penetration length (m)
H_R Ratio of gas-liquid mixing height to column diameter (h_c/d_c) (-)
H_{rn} Ratio of the mixing height to the diameter of nozzle (-)

K_m Mixing coefficient (-)
l Length of cylindrical section of nozzle (m)
L_j Jet length above gas–liquid mixing level (m)
L_{mz} Length of mixing zone (m)
Mo Morton number $(g\mu_l^4/\rho_l\sigma_l^3)$ (-)
P_1 Pressure at level 1 (N/m^2)
P_c Pressure at mixing point (N/m^2)
P_s Separator pressure (N/m^2)
Qe Gas entrainment rate (m^3/s)
Q_g Volumetric flow rate of gas (m^3/s)
Q_l Volumetric flow rate of liquid (m^3/s)
Q_r Ratio of gas to liquid flow rate (-)
Re_j Reynolds number based on jet velocity $(u_j d_j\rho_l/\mu_l$ (-)
Re_{length} Reynolds number $(\rho u_o L_j/\mu_l)$ (-)
Re_{ln} Liquid Reynolds number based on nozzle diameter, $(\rho_l u_{ln} d_n)/\mu_l$ (-)
Su_n Suratmann number based on the nozzle diameter, $(\rho_l\sigma_n d_n)/\mu_l^2$ (-)
u_e Entrainment velocity (m/s)
$u_{e,m}$ Minimum entrainment velocity (m/s)
u_i Jet impact velocity (m/s)
u_j Liquid jet velocity (m/s)
u_o Jet velocity related to nozzle outlet (m/s)
We_g Weber number based on gas $(u_o^2 d_c\rho_g/\sigma_l)$ (-)
We_j Weber number $(u_j^2 d_c\rho_l/\sigma_l)$ (-)

Greek Letters

η Efficiency (-)
ρ_g Density of the gas (kg/m^3)
ε_g Gas holdup (-)
μ_g Viscosity of the gas (N.s/m^2)
ρ_l Density of the liquid (kg/m^3)
σ_l Surface tension of the liquid (N/m)
μ_l Viscosity of the liquid (N.s/m^2)

References

Bando, Y., Uraishi, M., Nishimura, M., Hattori, M., Asada, T., 1988. Cocurrent downflow bubble column with simultaneous gas–liquid injection nozzle. J. Chem. Eng. Jpn 21, 607–612.

Barchilon, M., Curtet, R., 1964. Some details of the structure of an axisymmetric confined jet with backflow. J. Basic Engng Trans. ASME 84, 717–787.

Bin, A.K., Smith, J.M., 1982. Mass transfer in a plunging liquid jet absorber. Chem. Eng. Commun. 15, 367–383.

Bin, A.K., 1993. Gas entrainment by plunging liquid jet. Chem. Eng. Sci. 48, 3585–3630.

Bin, A.K., 1989. Comment on article "Gas holdup and gas entrainment of a plunging water jet with a constant entrainment guide." Can. J. them Engng 67, 699–701.

Briens, C.L., Huynh, L.X., Large, J.F., Catros, A., Bernard, J.R., Bergougnou, M.A., 1992. Hydrodynamics and gas–liquid mass transfer in a downward venturi-bubble column combination. Chem. Eng. Sci 47, 3549–3556.

Burgess, J.M., Molloy, N.A., 1973. Gas absorption in plunging jet reactor. Chem. Eng. Sci 28, 183–190.

Cacuci, D.G., 2010. Handbook of Nuclear Engineering. Springer Publication, USA.

Chanson, Hubert, 1997. Air Bubble Entrainment in Free-Surface Turbulent Shear Flows. Academic Press, London, UK, 401 pages.

Ciborowski, J., Bin, A., 1972. Investigation of the aeration effect of plunging liquid jets. Ini. Chem. 2, 557–577.

Cumming, I.W., 1975, The impact of falling liquids with liquid surfaces. PhD Thesis, Loughborough University of Technology, London.

Cunningham, R.G., 1974. Gas compression with the liquid pump. J. Fluids Eng. (Ser. I) Trans. ASME 3, 203–215.

De Frate, L.A., Rush, F.E., 1959. "Preprint 39D", Symposium on Selected Papers, Part II, Sixty-Fourth National Meeting of the A.I. Ch.E, New Orleans, Louisiana, pp. 16–20.

Ervine, D.A., Falvey, H.T., 1987. Behavior of turbulent water jets in the atmosphere and in plunge pools. Proc. Inst. Civil Eng. 83 (2), 295–314.

Evans, G.M., Jameson, G.J., Rielly, C.D., 1996. Free jet expansion and gas entrainment characteristics of plunging liquid jet. Exp. Ther. Fluid Sci. 12, 142–149.

Fukuda, H., Sumino, Y., Kanzaki, T., 1963. Hakko Kogaku, 46, 829.

Havelka, P., Linek, V., Sinkule, J., Zahradnik, J., Fialova, M., 1997. Effect of the ejector configuration on the gas suction rate and gas hold-up in ejector loop reactors. Chem. Eng. Sci. 53 (11), 1701–1713.

Henderson, J.B., McCarthy, M.J., Molloy, N.A., 1970. Entrainment by Plunging Jets Proceeding of Chemeca, 70, Melbourne. Section 2, 86–100.

Kumagai, M., Endoh, K., 1982. Effects of kinematic viscosity and surface tension on gas entrainment rate of an impinging liquid jet. J. Chem. Eng. Jpn 15, 427–433.

Kumagai, M., Imai, H., 1982. Gas entrainment phenomena and flow pattern of an impinging water jet. Kagaku Koguku Ronhunshu 8, 510–513.

Kundu, G., Mukherjee, D., Mitra, A.K., 1997. Ejector performance in a co-current gas–liquid downflow bubble column. Can. J. Chem. Eng. 75, 956–963.

Kusabiraki, D., Niki, H., Yamagiya, K., Ohkawa, A., 1990. Gas entrainment rate and flow pattern of vertical plunging liquid jets. Can. J. Chem. Eng. 68, 893–903.

Lara, P., 1979. Onset of air entrainment for a water jet impinging vertically on a water surface. Chem. Eng. Sci. 34, 1164–1165.

Lin, T.J., Donnelly, H.G., 1966. Gas bubble entertainment by plunging laminar liquid jets. AIChE J. 12, 563–571.

Liu, C., Barkelew, C.W., 1986. Numerical analysis of jet-stirred reactors with turbulent flows and homogeneous reactions. A. I. Ch. E. J 32, 1813–1820.

Madarame, H., Chiba, T., 1990. Gas entrainment inception at the border of a flow-swollen liquid surface. Nucl. Eng. Des. 120, 193–201.

Majumder, S.K., 2005. Dispersion and hydrodynamics in a modified bubble column, PhD Thesis. IIT Kharagpur, India.

Majumder, S.K., Kundu, G., Mukherjee, D., 2006. Efficient dispersion in a modified two-phase non-Newtonian downflow bubble column. Chem. Eng. Sci. 61 (20), 6753–6764.

Majumder, S.K., Kundu, G., Mukherjee, D., 2007. Energy efficiency of two phase mixing in a modified bubble column. Can. J. Chem. Eng. 85, 280–289.

Mandal, A., Kundu, G., Mukherjee, D., 2004. Gas-holdup distribution and energy dissipation in an ejector-induced downflow bubble column: the case of non-Newtonian liquid. Chem. Eng. Sci. 59, 2705–2713.

Matsumura, M., Sakuma, H., Yamagata, T., Kobayashi, J., 1982. Gas entrainment in a new gas entraining fermenter. J. Ferment. Technol. 60, 457–467.

McCarthy, M.J., Krichner, W.G., Molloy, N.A., Henderson, J.B., 1969. Mechanism of gas bubble entrainment by plunging jets. Trans. Inst. Min. Metal 78, C239–C241.

McKeogh, E.J., Elsawy, E.M., 1977. Self-aeration by plunging jet. Proceedings of Canadian Congress of Applied Mechanics. 6 (2), 591–592.

McKeogh, E.J., Ervine, D.A., 1981. Air entrainment rate and diffusion pattern of plunging liquid jet. Chem. Eng. Sci. 36, 1161–1172.

Mertes, A.T., 1938. United States Patent Office, Patent 2, 128, 311.

Mundale, V.D., Joshi, J.B., 1995. Optimization of impeller design for a gas inducing type mechanically agitated contactor. Can. J. Chem. Eng. 73, 161–172.

Ohkawa, A., Kusabiraki, D., Kawai, Y., Sakai, N., 1986. Some flow characteristics of a vertical liquid jet system having downcomers. Chem. Eng. Sci. 41, 2347–2361.

Ohkawa, A., Kusabiraki, D., Kawai, Y., Sakai, N., 1987. Flow characteristics of an air-entrainment type aerator having a long downcomer. Chem. Eng. Sci. 42, 2788–2790.

Ohyama, Y., Takashima, Y., Idemura, H., 1953. Induction of air by jet streams. Kagaku Hokoku 29, 344–348.

Patwardhan Ashwin, W., Joshi, Jyeshtharaj, B., 1999. Design of gas-inducing reactors. Ind. Eng. Chem. Res. 38, 49–80.

Perry, R.H., Green, D.W. (Eds.), 1984. Perry's Chemical Engineers' Handbook, sixth ed. McGraw-Hill, New York.

Pumphrey, H.C., Elmore, P.A., 1990. The entrainment of bubbles by drop impact. J. Fluid Mech. 220, 539–567.

Rao L.N.S., Kobus, H.E., 1973, Characteristics of self-aerated free surface flow. Water and Water Waste Current Research and Practice, vol. 10. Schmidt, Verlag, Berlin.

Sene, K.J., 1988. Air entrainment by plunging jets. Chem. Eng. Sci. 43, 2615–2623.

Shah, Y.T., 1979. Gas–Liquid–Solid Reactor Design. McGraw-Hill International Book Company, New York.

Sivaiah, M., Parmar, R., Majumder, S.K., 2012. Gas entrainment and holdup characteristics in a modified gas-liquid-solid down flow three-phase contactor. Powder Technol. 217, 451–461.

Smigelschi, O., Suciu, G.D., 1976. Size of the submerged biphasic region in plunging jet systems. Chem. Eng. Sci. 31, 1217–1220.

Sverak, S., Hruby, M., 1981. Gas entrainment from the liquid surface of vessels with mechanical agitators. Int. Chem. Eng. 21, 519–526.

Takahashi, M., Inouye, A., Aritomi, M., Takenaka, Y., Suzuki, K., 1988. Gas entrainment at free surface of liquid (I): gas entrainment mechanism and rate. J. Nucl. Sci. Technol. 25 (2), 131–142.

Tamir, A., 2014. Impinging-Stream Reactors: Fundamentals and Applications. Transport Processes in Engineering. Elsevier.

Thomas, N.H., Auton, T.R., Sene, K., Hunt, J.C.R., 1984. Entrapment and Transport of Bubbles by Plunging Water. In: Brutsaert, G.H., Jirka, W. (Eds.), Gas Transport at Water Surfaces. Reidel, Dordrecht, pp. 255–268.

Thoroddsen, S., Thoraval, M., Takehara, K., Etoh, T., 2012. Micro-bubble morphologies following drop impacts onto a pool surface. J. Fluid Mech. 708, 469–479.

Topiwala, H.H., Hamer, G., 1974. Mass transfer and dispersion properties in a fermenter with a gas-inducing impeller. Trans. Inst. Chem. Eng. 52, 113–120.

Tran, T., 2013. Hélène de Maleprade, Chao Sun, and Detlef Lohse, Air entrainment during impact of droplets on liquid surfaces. J. Fluid Mech. 726, R3.

Van de Donk, J.A.C., 1981. Water aeration with plunging jets, PhD Thesis. Technische Hogeschool Delft.

Van de Sande, E., Smith, J.M., 1973. Surface entrainment of air by high velocity water jets. Chem. Eng. Sci. 28, 1161–1168.

Van de Sande, E., Smith, J.M., 1975. Mass transfer from plunging water jets. Chem. Eng. J. 10, 225–233.

Yamagiwa, K., Kusabiraki, D., Ohkawa, A., 1990. Gas holdup and gas entrainment rate in downflow bubble column with gas entrainment by a liquid jet operating at high liquid throughput. J. Chem. Eng. Jpn 23, 343–348.

Zundelevich, Y., 1979. Power consumption and gas capacity of self-inducting turbo aerators. AIChE J. 25, 763–773.

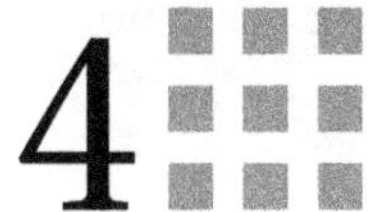

Hold-up Characteristics of Gas Bubbles

Bubble Phase Hold-up: Definition and Profile

Definition of Bubble Phase Hold-up

The volume fraction of the dispersed gas phase as a bubble is referred to as the *bubble phase hold-up*. It is of considerable practical importance because it determines the efficiency of the transport process in a gas–liquid or gas–liquid–solid reactor. It is one of the most important parameters characterizing bubble column hydrodynamics because it not only gives the volume fraction of the gas phase, but it also influences the mass transfer of gas as it determines the interfacial area for their transport. Additionally, the mixing performance of the usually used multiphase reactor is affected by the bubble phase hold-up. The bubble phase hold-up is defined as the ratio of the pipe cross-sectional area (or volume) occupied by the gas phase to the pipe cross-sectional area (or volume).

$$\varepsilon_g = \frac{\dot{A}_g}{\dot{A}} = \frac{\dot{A}_g}{\dot{A}_g + \dot{A}_{\mathrm{lg}}} \tag{4.1}$$

To characterize a two-phase flow, the slip ratio (S) is frequently used instead of bubble phase hold-up. The slip ratio is defined as the ratio of the average velocity of the gas phase flow (u_g) to the average velocity of the liquid phase flow (u_l). The bubble phase hold-up can be related to the slip ratio (S) as follows:

$$S = \frac{u_g}{u_l} = \frac{Q_g/(A\varepsilon_g)}{Q_l/(A(1-\varepsilon_g))} = \frac{Q_g(1-\varepsilon_g)}{Q_l\varepsilon_g} \tag{4.2}$$

$$S = \frac{u_g}{u_l} = \frac{Gx/(A\varepsilon_g\rho_g)}{G(1-x)/(A(1-\varepsilon_g)\rho_l)} = \frac{\rho_l x(1-\varepsilon_g)}{\rho_g(1-x)\varepsilon_g} \tag{4.3}$$

Equations (4.2) and (4.3) can be rewritten in the form:

$$\varepsilon_g = \frac{\dot{Q}_g}{S\dot{Q}_l + \dot{Q}_g} \tag{4.4}$$

$$\varepsilon_g = \frac{1}{1 + S\left(\dfrac{1-x}{x}\right)\left(\dfrac{\rho_g}{\rho_l}\right)} \tag{4.5}$$

Hydrodynamics and Transport Processes of Inverse Bubbly Flow. http://dx.doi.org/10.1016/B978-0-12-803287-9.00004-7

The volumetric quality is equivalent to the bubble phase hold-up when the slip ratio is 1. The volumetric quality is defined as:

$$\beta = \frac{\dot{Q}_g}{\dot{Q}} = \frac{\dot{Q}_g}{\dot{Q}_l + \dot{Q}_g} \tag{4.6}$$

The bubble phase hold-up is called the homogeneous bubble phase hold-up when the slip ratio is 1. When the ratio of liquid to gas density is large, the bubble phase hold-up based on the homogeneous mixture of the phases increases very rapidly when the mass quality (x = the ratio of the mass flow rate of gas phase to the total mass flow rate) increases even slightly above zero. The prediction of the void fraction using the homogeneous model is reasonably accurate for bubbly flow if the entrained phase travels at nearly the same velocity as the continuous phase. Also, when the ratio of density of liquid to gas approaches 1 (i.e., near the critical state), the bubble phase hold-up based on the homogeneous mixture of the phases approaches the mass quality (x), and the homogeneous flow model is applicable at this case. The performance of the bubbly flow device depends mostly on the bubble phase hold-up distribution. The hold-up affects the liquid recirculation, backmixing of the phases, and interface of the phases, which are the main criteria to get more yield of mass transfer in the bubbly flow devices. The bubble phase hold-up depends on several factors such as the gas distributor, the gas velocity, and the physical properties of the liquid. The various effects on the gas hold-up are discussed in a later section.

Profiles of Bubble Phase Hold-up

Radial Bubble Phase Hold-up Profile

One of the most important parameters in bubbly flow reactors is the bubble phase hold-up profile. The spatial variation of bubble phase hold-up gives rise to pressure variation, which results in liquid recirculation in the reactor. This liquid recirculation governs the rate of mixing, heat transfer, and mass transfer. The ability to predict radial bubble phase hold-up profiles in bubble reactors would help in better determining the flow regimes, liquid mixing, heat, and mass transfer. This should make bubble column scale-up more reliable. The characteristic of the heterogeneous regime of flow in bubble column that generates strong liquid recirculation results from the bubble phase hold-up distribution. The magnitude of bubble phase hold-up radial gradients and the magnitude of liquid velocity driven by the gas depend on superficial gas velocity, column diameter, the nature of the gas–liquid system, and the operating conditions (pressure and temperature of the reactor). In the long time-averaged sense, the cross-sectional bubble phase hold-up distributions in bubbly flow devices (especially in columns) are close to being axisymmetric. Therefore, it is justified in representing the cross-sectional distribution by azimuthal averaging as represented by equation (4.7).

$$\varepsilon_g(r) = \frac{1}{2\pi} \int_0^{2\pi} \varepsilon_g(r,\theta)d\theta \tag{4.7}$$

where $\varepsilon_g(r)$ represents the radial bubble phase hold-up profile. Because of the complexity of the system, no fundamental equation is available at present for prediction of the bubble phase hold-up profiles in a bubbly flow reactor. The hold-up profiles are considered in the well-developed region (region excluding the entrance and exit zones) are relatively invariant to axial position. There are a number of similar form of empirical equations that can be fitted to the observed hold-up profiles. Nassos and Bankoff (1967) proposed the following equation for the radial hold-up profile.

$$\varepsilon_g = \tilde{\varepsilon}_g \left(\frac{n+2}{n} \right) \left\{ 1 - \left(\frac{r}{R} \right)^n \right\} \tag{4.8}$$

The exponent n is a parameter that indicates the steepness of the hold-up profile, and r/R is the dimensionless radial position. When n is large, the profile is flat; when n is small, the profile is steep. If n is approximately equal to 2, the profile is parabolic. The steepness of the hold-up profile results in the intensity of liquid circulation. Ueyama and Miyauchi (1979) modified the above equation as follows to include the possibility of finite bubble phase hold-up close to the wall.

$$\varepsilon_g = \tilde{\varepsilon}_g \left(\frac{n+2}{n} \right) \left\{ 1 - c \left(\frac{r}{R} \right)^n \right\} \tag{4.9}$$

where c is an additional parameter that is indicative of the value of bubble phase hold-up near the wall. If $c = 1$, there is zero hold-up close to wall; if $c = 0$, hold-up is constant with changing r/R. The hold-up $\tilde{\varepsilon}_g$ is the radial chordal average bubble phase hold-up along the column diameter, which is related to the cross-sectional average bubble phase hold-up. The cross-sectional average bubble phase hold-up is expressed by

$$\tilde{\varepsilon}_g / \bar{\varepsilon}_g = \frac{n}{n+2-2c} \tag{4.10}$$

The radial bubble phase hold-up profile at atmospheric pressure is parabolic in nature at a gas velocity of 0.12 m/s, indicating churn turbulent flow; at higher pressure, the profile is flatter (Kemoun et al., 2001). Some values of n and c obtained by Kemoun et al. (2001) are shown in Table 4.1.

By applying equation (4.10) to data, n was found to vary from 1.4 to 11 and c from 0.5 to 1 according to different authors (Joshi et al., 1998) and based on different systems investigated. In the absence of a firm theoretical prediction of the radial bubble phase hold-up profile correlations are needed for evaluating n and c based on the knowledge of the general operating variables and physical properties of the system to estimate the bubble phase hold-up profile by equation (4.9). The model can also be applied to fit the data in case of inverse bubbly flow. Babaei et al. (2015) reported one of the best fitted radial bubble phase hold-up models (one-degree Gaussian function) in a bubble column as follows:

$$\varepsilon_g = \alpha_{gc} \exp\left(-\left(\frac{r/R}{\beta} \right)^2 \right) \tag{4.11}$$

Table 4.1   Parameters of the Fitted Radial Hold-up Profile (Kemoun et al., 2001)

P (MPa)	u_{sg} (m/s)	$\tilde{\varepsilon}_g$	n	c
0.1	0.02	0.053	3.633	0.422
	0.05	0.151	4.372	0.368
	0.12	0.145	2.228	0.605
	0.18	0.165	2.068	0.622
0.3	0.02	0.059	5.133	0.283
	0.05	0.168	4.792	0.345
	0.12	0.272	3.765	0.466
	0.18	0.27	2.955	0.487
0.7	0.02	0.062	5.424	0.307
	0.05	0.17	3.557	0.346
	0.12	0.315	3.937	0.356
	0.18	0.326	2.997	0.39

where α_{gc} and β are model parameters and r/R is the dimensionless radius. According to this equation, the maximum and minimum values of bubble phase hold-up occur at the column center ($r/R = 0$) and column wall ($r/R = 1$), respectively. The width of the Gaussian function, which is a bell curve, is determined by parameter β at the center of the column $\alpha_g = \alpha_{gc}$. At the wall, the bubble phase hold-up is

$$\varepsilon_{g,wall} = \alpha_{gc} \exp\left(-\frac{1}{\beta^2}\right) \tag{4.12}$$

The values of the α_{gc} and β parameters increase with increasing distance from the bubble distributor. For a specific system, the simultaneous increase in the α_{gc} and β values with column height leads to an increase in the wall and cross-sectional average bubble phase hold-ups depending on the different system specifications and bubble distributor geometries. Both parameters increase with an increase in the superficial gas velocity. From equation (4.12), it is obvious that at a constant α_{gc}, as the value of β decreases, the difference between central hold-up and wall hold-up increases. A decrease in the β value as for the increase in α_{gc}, indicates a higher difference between the central and wall hold-ups. This higher difference again confirms the increased bubble coalescence because larger bubbles tend to accumulate in the column center. Generally, the radial distribution of bubble phase hold-up and its radial gradient are determined by the superficial gas velocity, column dimensions, operational conditions, and physical properties of the system components. An increased radial gradient results from an increased gas velocity, increased column diameter, and decreased reactor pressure. A stronger bubble coalescing nature may result in a non-uniformity radial profile.

The Cross-Sectional Average Bubble Phase Hold-up
The cross-sectional average bubble phase hold-up can be taken as a good estimate of the overall bubble phase hold-up. The cross-sectional average bubble phase hold-up is

calculated by averaging the radial bubble phase hold-up profiles obtained from equation (4.9) as

$$\bar{\varepsilon}_g = \frac{1}{\pi R^2} \int_0^R 2\pi r \varepsilon_g(r) dr \tag{4.13}$$

Cross-sectional average bubble phase hold-ups at different column heights can also be calculated by integrating equation (4.13) with equation (4.11) over the entire cross-section as:

$$\bar{\varepsilon}_g = \alpha_{gc} \beta^2 \left(1 - \exp(-\frac{1}{\beta^2})\right) \tag{4.14}$$

Bubble Phase Hold-up Based on Liquid Velocity Profile
The mean cross-sectional hold-up, $\bar{\varepsilon}_g$, can also be determined from the liquid velocity profile by the following equation:

$$Q_l = 2\pi \int_0^R (1 - \bar{\varepsilon}_g) u_l(r) r dr \tag{4.15}$$

where Q_l is the liquid volumetric flow rate. The liquid flow and mixing behavior in bubbly flow in a column is described by means of the liquid recirculation velocity profile. A power-law liquid velocity profile is widely accepted in the literature (Garcia-Calvo and Leton, 1994; Montserrat and Garcia-Calvo, 1996; Krishna et al., 1999; Wu and Al-Dahhan, 2001). The liquid circulation velocity is characterized by the liquid moves with the bubbles in the centerline of the column (upflow or inverse flow) in the nearer wall section of the column. Wu and Al-Dahhan (2001) established a correlation for the liquid recirculation velocity profile as

$$\frac{u_l(r)}{u_{l0}} = 1 - f(n,c)\left[\frac{r}{R}\right]^{\xi(n,c)} \tag{4.16}$$

where u_{l0} is the axial liquid velocity in the center of the column. The functions are correlated as $f(n,c) = \xi(n,c) = 2.65 n^{0.44} c$. Substituting the equation (4.16) in equation (4.15) and simplifying after integration, the mean cross-sectional bubble phase hold-up can be represented theoretically as

$$\bar{\varepsilon}_g = \left[\frac{1}{2} - \frac{f(n,c)}{\xi(n,c)+2} - \frac{Q_l}{2\pi u_{l0} R^2}\right]\left[\left(\frac{1}{2} - \frac{f(n,c)}{\xi(n,c)+2} - \frac{c}{n+2} + \frac{f(n,c)c}{n+\xi(n,c)+2}\right)\left(\frac{n+2}{n+2-2c}\right)\right]^{-1} \tag{4.17}$$

The parameters n and c needed to describe the bubble phase hold-up profile can be correlated with appropriate dimensionless groups:

$$n \propto f(\text{Re}_g, Mo_l, Fr_g) \tag{4.18}$$

$$c \propto f(\text{Re}_g) \tag{4.19}$$

where $\mathrm{Re}_g = d u_{sg}(\rho_l - \rho_g)/\mu_l$, $Mo_l = g\mu_l^4/[(\rho_l - \rho_g)\sigma_l^3]$, and $Fr_g = u_{sg}^2/(gd)$. The dimensionless groups Re_g, Mo_l, and Fr_g reflect the effect of velocity and pressure, which change the density of the gas and has an effect on bubble phase hold-up profile, and the effect of gas and liquid physical properties. The cross-sectional average bubble phase hold-up can be estimated using the available correlations for overall bubble phase hold-up. It was reported (Kemoun et al., 2001; Wu et al., 2001) that the shape of hold-up profiles changes most significantly with superficial gas velocities. However, pressure affects the shapes of hold-up profiles, but it has less effect compared with superficial gas velocity within the range of pressure and superficial gas velocity that was examined. Gas distributor does affect hold-up in a certain range of gas velocities, but it has a minor effect on bubble phase hold-up profiles, particularly in the fully developed region and at high gas velocity. The shape of the bubble phase hold-up profile at different column heights seems to be unchanged at a given gas velocity when the measurement has been taken at a certain distance from the distributor (2 L/D or higher). Column diameter has been reported to have an effect on the bubble phase hold-up profile (Kumar et al., 1997). Wu et al. (2001) proposed the following correlations for n and c based on their experimental data.

$$n = 2.188 \times 10^3 \, \mathrm{Re}_g^{-0.598} \, Mo_l^{-0.004} \, Fr_g^{0.146} \tag{4.20}$$

$$c = 4.32 \times 10^{-2} \, \mathrm{Re}_g^{0.2492} \tag{4.21}$$

The agreement between the correlation and experimental data is reasonable over a wide range of operating conditions. The correlations proposed by Kemoun et al. (2001) and Majumder et al. (2006, 2007) can be used to calculate the overall bubble phase hold-up for upward bubbly flow and for inverse bubbly flow respectively. The parameter n is almost independent of Mo_l ($n \propto Mo_l^{-0.004}$). However, it is reported that the liquid physical properties affect the overall bubble phase hold-up (Luo et al., 1999) and the hold-up profile (Chen et al., 1998; Joshi et al., 1998). Therefore, Mo_l can be included in the correlation to be examined for any future necessary modification as bubble phase hold-up profiles become available for a wide enough range of liquid physical properties. Wu et al. (2001) reported that c was found to be only a function of Re_g based on the hold-up profiles developed for air–water systems. Liquid physical properties would affect the parameter c, which can be examined further experimentally. The bubble size distribution has significant effects on mass and heat transfer efficiency because it determines the interfacial area per volume of the gas phase and mixing intensity. In a given system, bubble distribution is mainly related to the local bubble phase hold-up and turbulent energy dissipation rate. Because of the non-uniform radial profiles of the bubble phase hold-up as shown in Figure 4.1, especially in the heterogeneous regime, the bubble size distribution varies not only with the superficial gas velocity but also with the radial position (Majumder, 2008).

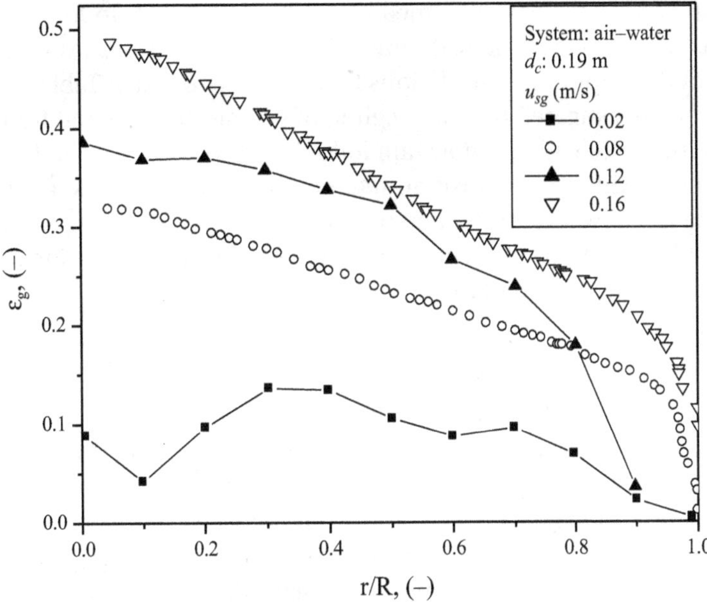

FIGURE 4.1 Radial profiles of the bubble phase hold-up in a bubble column with an air–water system. *(Lines and symbols: data from Wang and Wang, 2007; symbols: data from Sanyal et al., 1999).*

An empirical correlation based on the data in shown in Figure 4.1 developed by Majumder (2008) is:

$$\varepsilon_g = -1.32 u_{sg}^{0.28}(r/R)^3 + 1.41 u_{sg}^{0.19}(r/R)^2 + -0.95 u_{sg}^{0.24}(r/R) + 0.85 u_{sg}^{0.33} \tag{4.22}$$

The same trend of result in inverse flow bubble column can be obtained at high liquid throughput (Yamagiwa et al., 1990). In an unconstrained system, the dispersion of bubbles typically occupies a relatively small fraction of the total volume of the liquid pool. Thus, a non-uniform distribution of bubble phase hold-up is to be in the ordinary plunging liquid jet system. In the region close to the plunging point in case of plunging jet inverse bubbly flow, the gas fraction profiles have a maximum near the original free jet radius and a minimum on the jet axis. With increasing depth, the gas fraction on the jet axis increases, the maximum gas fraction becomes lower, and the gas fraction profiles become more or less Gaussian (Bin, 1993). With a similarity to the velocity profiles, the profile of gas bubble holdup can be approximated for turbulent jets (McKeogh and Ervine, 1981) by:

$$\varepsilon_g(r)/\varepsilon_{gc} = \exp(-0.69315 r^2/r_{0.5}^2) \tag{4.23}$$

$r_{0.5}$ corresponds to the radial distance for which

$$\varepsilon_g(r_{0.5}) = 0.5\varepsilon_{gc} \tag{4.24}$$

Many investigators have obtained mean bubble phase hold-up for different diameter columns larger than 0.1 m and made different forms of correlations from their experimental data. Some of the important correlations have been reported in Table 4.2. Ruzicka et al. (2001) extensively investigated the flow regime of bubble column and found that bubble phase hold-up increases with gas flow rate following a convex curve in homogeneous regime and attains maximum in transition regime; then it decreases slightly and again increases with the gas flow rate in the heterogeneous regime following a concave curve as shown in Figure 4.2. The non-uniform bubble phase hold-up distribution across the radial direction causes bulk liquid circulation.

Axial Profile of Cumulative Bubble Phase Hold-up
Estimation of axial bubble phase hold-up in two-phase gas–liquid system in a bubble column by measuring the column pressure is a well-known method (Marchese et al., 1992). The axial cumulative gas hold-up throughout the column can be obtained by pressuring pressures at different positions of the column. To incorporate this method, the assumptions were made that (1) the dynamic pressure component is negligible compared with the hydrostatic pressure, and (2) for a particular gas and liquid flow rate, frictional loss is considered as constant throughout the column. Under steady-state operation, the pressure balance between the points 1 and c (gas liquid mixing level) gives

$$P_1 - P_c = \rho_l g(h_c - h_1)(1 - \overline{\varepsilon}_{g,c,1}) \tag{4.25}$$

where P_c is the pressure above the gas–liquid mixing level and $\overline{\varepsilon}_{g,c,1}$ is the bubble phase hold-up between the points 1 and c.

Similarly, the pressure balance between points 2 and c and so on gives

$$P_2 - P_c = \rho_l g(h_c - h_2)(1 - \overline{\varepsilon}_{g,c,2}) \tag{4.26}$$

$$P_3 - P_c = \rho_l g(h_c - h_3)(1 - \overline{\varepsilon}_{g,c,3}) \tag{4.27}$$

Thus, in general, a pressure-balance equation between points at any position of the column i and c can be written as

$$P_i - P_c = \rho_l g(h_c - h_i)(1 - \overline{\varepsilon}_{g,c,i}) \tag{4.28}$$

where P_i is the pressure at any point i. At the gas–liquid mixing level, $P_i = P_c$ and $\overline{\varepsilon}_{g,c,1} = 1$.

Subtracting equation (4.28) from (4.27) and (4.29) from (4.28) and so on, one gets, respectively,

$$P_1 - P_2 = \rho_l g[(1 - \overline{\varepsilon}_{g,c,1})(h_c - h_1) - (1 - \overline{\varepsilon}_{g,c,2})(h_c - h_2)] \tag{4.29}$$

$$P_2 - P_3 = \rho_l g[(1 - \overline{\varepsilon}_{g,c,2})(h_c - h_2) - (1 - \overline{\varepsilon}_{g,c,3})(h_c - h_3)] \tag{4.30}$$

Thus, in general, it can be written as

$$P_i - P_{i+1} = \rho_l g [(1 - \overline{\varepsilon}_{g,c,i})(h_c - h_i) - (1 - \overline{\varepsilon}_{g,c,i+1})(h_c - h_{i+1})] \tag{4.31}$$

Table 4.2 Correlations for Cross-Sectional Average Bubble Phase Hold-up in an Upflow Bubble Column*

References	Gas–Liquid System	Conditions	Correlations
Hughmark (1967)	Air–water, kerosene, Na_2SO_3 aqueous solutions, glycerol, light oil, $ZnCl_2$ aqueous solutions	$u_{sg} = 0.004\text{-}0.45$, $d_c > 0.1$	$\bar{\varepsilon}_g = 1/[2 + (0.35/u_{sg})\{\rho_l\sigma/72\}^{1/3}]$
Kato and Nishiwaki (1972)	Air–water	$u_{sg} = 0\text{–}0.30$, $u_{sl} = 0\text{–}0.015$, $d_c = 0.066\text{–}0.214$, $H = 2.01\text{–}4.05$	$\bar{\varepsilon}_g = 2.51u_{sg}/[0.78 + (4.5 - 2.548d_c^{1.8})u_{sg}^{0.8}$ $(1 - \exp(717u_{sg}^{1.8}/(4.5 - 2.548d_c^{1.8}))]$
Akita and Yoshida (1973)	He/CO_2/O_2/air–H_2O/glycol/ methanol/CCl_4/Na_2SO_3/NaCl	$P = 0.1$ MPa; $T = 283\text{–}313$ K; $u_{sg} = 0.5\text{–}40$ cm/s; $d_c = 0.152, 0.301, 0.6$ m, sparger (5.0 mm)	$\dfrac{\bar{\varepsilon}_g}{(1-\bar{\varepsilon}_g)^4} = C\left(\dfrac{gd_c^2\rho_l}{\sigma_l}\right)^{1/8}\left(\dfrac{gd_c^3\rho_l^2}{\mu_l^2}\right)^{1/12}\left(\dfrac{u_{sg}}{\sqrt{gd_c}}\right)$ where $C = 0.2$ for pure liquids and non-electrolytes and $C = 0.25$ for electrolytes
Gestrich and Rahse (1975)	Air–different liquids	$u_{sg} = 0.01\text{–}0.08$, $d_c = 0.0756\text{–}0.61$, $H = 0.02\text{–}3.5$	$\bar{\varepsilon}_g = 0.89\left(\dfrac{h_i}{d_c}\right)^a\left(\dfrac{d_b}{d_c}\right)^{0.3}\left(\dfrac{u_{sg}^2}{d_b g}\right)^b K^{0.047} - 0.05$ where $K = (\rho_l\sigma^3)/(\mu_l^4 g)$; $d_b = 0.003$ m $a = 0.036(-15.7 + \log K$ $b = 0.025(2.6 + \log K)$
Hikita et al. (1980)	H_2/CO_2/CH_4/C_3H_8/$H_2 + N_2$/ air–H_2O/sucrose/methanol/ n-butanol/aniline/i-butanol/ NaCl/Na_2SO_4/$CaCl_2$/$MgCl_2$/ $AlCl_3$/KCl/K_2SO_4/K_3PO_4/ KNO_3	$P = 0.1$ MPa, $u_{sg} = 4.2\text{–}38$ cm/s, $d_c = 0.10$ m, Nozzle (1.1 cm), $H/d_c = 15$	$\bar{\varepsilon}_g = 0.672f\left(\dfrac{u_{sg}\mu_l}{\sigma_l}\right)^{0.578}\left(\dfrac{\mu_l^4 g}{\rho_l\sigma_l^3}\right)^{-0.131}$ $\left(\dfrac{\rho_g}{\rho_l}\right)^{0.062}\left(\dfrac{\mu_g}{\mu_l}\right)^{0.107}$ $f = \begin{cases} 1.0 & for \quad no-\text{electrolyte solution} \\ 10^{0.0414I} & for \quad 0 < I < 1.0 \text{kgion/m}^3 \\ 1.1 & for \quad I > 1.0 \text{kgion/m}^3 \end{cases}$
Hammer (1984)	N_2/He/Ar/CO_2/Air–H_2O/ organic liquids–glass ballotini	$P = 0.1$ MPa; $T = 293\text{–}363$ K; $u_{sg} = 0.5\text{–}13$ cm/s; $d_c = 0.106, 0.2$ m, spargers (17–40 holes of 0.5, 1, or 2 mm ID)	$\dfrac{\bar{\varepsilon}_g}{1-\bar{\varepsilon}_g} = 0.4\left(\dfrac{u_{sg}\mu_l}{\sigma_l}\right)^{0.87}\left(\dfrac{\mu_l^4 g}{\rho_l\sigma_l^3}\right)^{-0.27}\left(\dfrac{\rho_g}{\rho_l}\right)^{0.17}$

(Continued)

Table 4.2 Correlations for Cross-Sectional Average Bubble Phase Hold-up in an Upflow Bubble Column (*cont.*)

References	Gas–Liquid System	Conditions	Correlations
Idogawa et al. (1985)	Air–H_2O	$P = 0.1$–15 MPa, $H/d_c = 16.6$, $u_{sg} = 0.5$–5 cm/s, $T = 288$–293 K, $d_c = 0.05$ m Porous plate (2, 100 μm), capillary tubes (1, 3, 5 mm), perforated plate (19 holes of 1 mm)	$\dfrac{\bar{\varepsilon}_g}{1-\bar{\varepsilon}_g} = 1.44 u_{sg}^{0.58} \rho_g^{0.12} \sigma_l^{-m}$ $m = -0.16\exp(-P)$ σ_l is in mN/m
Reilly et al. (1986)	He/Ar/air–H_2O/solvent/ trichloroethylene-glass beads	$P = 0.1$ MPa, $T = 283$–323 K, $u_{sg} = 0.4$–40 cm/s, $d_c = 0.3$ m perforated plate (293 holes, 1.5 mm), single sparger, multiorifice sparger (13.4 mm)	$\bar{\varepsilon}_g = 296 u_{sg}^{0.44} \rho_l^{-0.98} \rho_g^{0.19} \sigma_l^{-0.16} + 0.009$
Idogawa et al. (1987)	H_2/He/air–H_2O/methanol/ ethanol/acetone/aqueous alcohol solution	$P = 0.1$–5 MPa, $T = 284$–293 K, $H/d_c =16.6$, $u_{sg} = 0.5$–5 cm/s, $d_c = 0.05$ m perforated plate (19 holes of 1 mm)	$\dfrac{\bar{\varepsilon}_g}{1-\bar{\varepsilon}_g} = 0.059 u_{sg}^{0.8} \rho_g^{0.17} (\sigma_l / 72)^{m'}$ $m' = -0.22\exp(-P)$ (σ_l is in mN/m)
Wilkinson et al. (1992)	N_2–H_2O/n-heptane/mono-ethylene glycol	$P = 0.1$–2.0 MPa; $H = 1.2$ m; $u_{sg} = 0$–60 cm/s; $d_c = 0.158$, 0.23 m parger ring (4 holes of 7 mm)	$\bar{\varepsilon}_g = u_{sg}/u_{sb};\quad$ for $\quad u_{sg} < u_{sg,trans}$ $\bar{\varepsilon}_g = u_{sg,trans} / u_{sb} + (u_{sg} - u_{sg,trans}) / u_{lb};\quad$ for $\quad u_{sg} > u_{sg,trans}$ where $u_{sg,trans} / u_{sb} = 0.5\exp(-193\rho_g^{-0.61}\mu_l^{0.5}\sigma_l^{0.11})$ $\dfrac{\mu_l u_{sb}}{\sigma_l} = 2.25\left(\dfrac{\sigma_l^3 \rho_l}{g\mu_l^4}\right)^{-0.273}\left(\dfrac{\rho_l}{\rho_g}\right)^{0.03}$; (sb is small bubble; lb is large buble)
Sotelo et al. (1994)	Air/CO_2–H_2O/ethanol/ saccharose/glycerin	$P = 0.1$ MPa; $H = 1.5$–2.0 m; $u_{sg} = 0$–20 cm/s; $d_c = 0.04$, 0.08 m Porous gas diffusers (30, 65, 150 μm)	$\bar{\varepsilon}_g = \left\{ 129\left(\dfrac{u_{sg}\mu_l}{\sigma_l}\right)^{0.99}\left(\dfrac{\mu_l^4 g}{\rho_l \sigma^3}\right)^{-0.123}\left(\dfrac{\rho_g}{\rho_l}\right)^{0.187} \left(\dfrac{\mu_g}{\mu_l}\right)^{0.343}\left(\dfrac{d_o}{d_c}\right)^{-0.089} \right\}$ (d_o is the inner diameter of a single nozzle, m)

Reference	System	Conditions	Correlation
Krishna and Ellenberger (1996)	Air/He/Ar/SF$_6$–H$_2$O/paraffin oil/separan/tetradecane	$P = 0.1$ MPa; $u_{sg} = 0.1$–85 cm/s; $d_c = 0.1$, 0.174, 0.19, 0.38, and 0.63 m Spider sparger, sintered glass plate (mean pore size, 150–200 μm), Polyacrylate sieve plate (2.5 mm ID), sintered bronze plate (mean pore size, 50 μm)	$\bar{\varepsilon}_g = \varepsilon_b;$ for $u_{sg} < u_{sg,trans}$ $\bar{\varepsilon}_g = \varepsilon_b + \varepsilon_{trans}(1-\varepsilon_b);$ for $u_{sg} > u_{sg,trans}$ where $\varepsilon_b = 0.268(u_{sg} - u_{trans})^{0.58} / d_c^{0.18}$ $u_{trans} = \sigma_l^{0.12}\varepsilon_{trans}(1-\varepsilon_{trans}) / (2.84\rho_g^{0.04})$ $\varepsilon_{trans} = 0.59(3.85)^{1.5}\sqrt{\rho_g^{0.96}\sigma_l^{0.12} / \rho_l}$
Kojima et al. (1997)	Air–H$_2$O/aqueous buffered solution/aqueous enzyme solution	$P = 0.1$–1.1 MPa, $T = 290$–300 K, H/d$_c$ = 20–26.7, $u_{sg} = 0.005$–0.15 cm/s, $d_c = 0.045$ m, nozzle (1.38, 2.1, 2.9, 4.03 mm)	$\bar{\varepsilon}_g = \left\{\begin{array}{l} 1.18u_{sg}^{0.679}\left(\dfrac{\sigma_l}{\sigma_{lo}}\right)^{-0.56} \\ \exp\left[1.27\times10^{-4}\left(\dfrac{\rho_l Q_g^2}{d_o^3\sigma_l}\right)\left(\dfrac{P}{P_o}\right)\right] \end{array}\right\}$ (d_o is the orifice ID in mm; P_0 is the standard atmospheric pressure)
Luo et al. (1999)	N$_2$–paratherm NF–alumina particles	$P = 0.1$–5.6 MPa, $T = 298$–351 K, $u_{sg} =$ up to 45 cm/s, $d_c = 0.102$ m Perforated plate (120 squared-pitched holes of 1.5 mm ID)	$\dfrac{\bar{\varepsilon}_g}{1-\bar{\varepsilon}_g} = 2.9\left(\dfrac{u_{sg}^4\rho_g}{g\sigma_l}\right)^{\alpha}\left(\dfrac{\rho_g}{\rho_l}\right)^{\beta} / \left[\cosh(Mo_l^{0.054}\right]^{0.41}$ where $Mo_l = \dfrac{\mu_l^4 g}{\rho_l\sigma_l^3}$; $\alpha = 0.21Mo_l^{0.0079}$; $\beta = 0.096Mo_l^{-0.011}$
Fransolet et al. (2005)	Air–tp water, aqueous solutions of xanthan	$u_{sg} = 0$–0.15, $d_c = 0.24$, $H = 2.75$	$\bar{\varepsilon}_g = 0.26u_{sg}^{0.54}\mu_l^{-0.14}$
Yifeng et al. (2008)	Air–paraffins, water, 10% K$_2$CO$_3$ and 20% K$_2$CO$_3$ solution	$u_{sg} = 0$-0.15, $d_c = 0.10$, $H = 1.5$	$\dfrac{\bar{\varepsilon}_g}{(1-\bar{\varepsilon}_g)^4} = 0.579\left(\dfrac{u_{sg}\mu}{\sigma}\right)^{0.918}\left(\dfrac{\mu^4 g}{\rho_l\sigma^3}\right)^{-0.252}$
Sal et al. (2013)	Air–water	$P = 0$–200 m Bar, $d_c = 0.33$ m in $H = 3.0$ m, hole diameter was varied in the range of 1–3 mm, while the free area was 1.0%, $u_{sg} = 0.00$–0.145 m/s	$\varepsilon_g = 0.2278\left[\dfrac{Fr^{0.7767}Ar^{0.3649}(d_o / d_c)^{0.4780}}{Eo^{0.3916}We^{0.2402}}\right],$ $We = \dfrac{d_c^4 u_{sg}^2\rho_l}{d_o^2 N^2\sigma_l}$, $Eo = \dfrac{d_o^2\rho_l g}{\sigma_l}$, $Ar = \dfrac{d_o^3\rho_l^2 g}{\mu_l^2}$, $Fr = \dfrac{u_{sg}^2}{d_o g}$ ($N =$ no. of holes)

*See the Nomenclature, Greek Letters, and Subscripts lists for abbreviations.

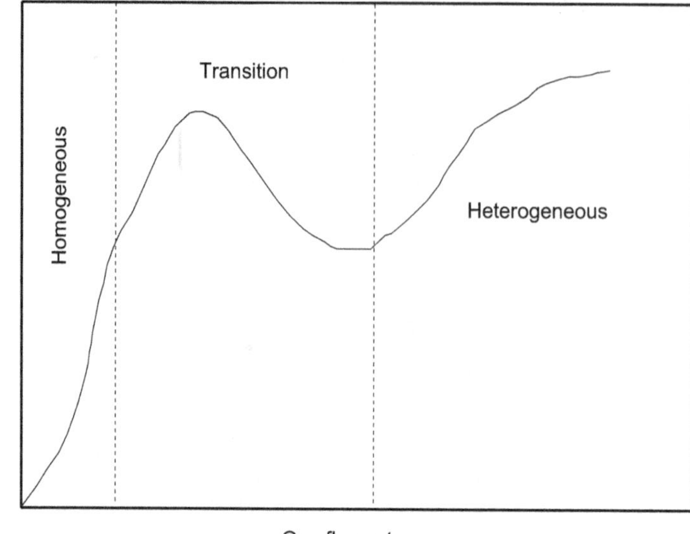

FIGURE 4.2 Flow regime-based bubble phase hold-up.

where $\varepsilon_{g,c,i}$ is the axial cumulative bubble phase hold-up between the points c and i and h_c − h_i is the distance between c and i.

From equation (4.31), $\bar{\varepsilon}_{g,i}$ can be expressed as

$$\bar{\varepsilon}_{g,c,i+1} = 1 - \left[\frac{(1-\bar{\varepsilon}_{g,c,i})(h_c-h_i)}{h_c-h_{i+1}} - \frac{P_i-P_{i+1}}{\rho_l g(h_c-h_{i+1})} \right], \quad i=1,2,3.... \tag{4.32}$$

where $\bar{\varepsilon}_{g,c,1}$ is defined as overall bubble phase hold-up and it is calculated as

$$\bar{\varepsilon}_{g,c,1} = \frac{h_c-h_l}{h_c} \tag{4.33}$$

h_l is the clear liquid height inside the column. The fractional bubble phase hold-up of the volumetric region between points 1 and 2 can be expressed as

$$\begin{aligned}
[\bar{\varepsilon}_g]_{1,2} &= \frac{\text{Volume of gas entrained between points 1 and 2}}{\text{Total volume of gas liquid mixture between points 1 and 2}} \\
&= \frac{Q_{g,c,1}-Q_{g,c,2}}{[Q_g+Q_l]_{1,2}} = \frac{\bar{\varepsilon}_{g,c,1}(\pi/4)d_c^2(h_c-h_1)-\bar{\varepsilon}_{g,c,2}(\pi/4)d_c^2(h_c-h_2)}{(\pi/4)d_c^2(h_2-h_1)} \\
&= \frac{\bar{\varepsilon}_{g,c,1}(h_c-h_1)-\bar{\varepsilon}_{g,c,2}(h_c-h_2)}{h_2-h_1}
\end{aligned} \tag{4.34}$$

where $Q_{g,c,1}$ is the rate of amount of gas occupying between the points 1 and c and $Q_{g,c,2}$ is the rate of amount of gas occupying between the points 2 and c. Similarly, the fractional bubble phase hold-up of volumetric region between points 2 and 3 is

$$[\overline{\varepsilon}_g]_{2,3} = \frac{\overline{\varepsilon}_{g,c,2}(h_c - h_2) - \overline{\varepsilon}_{g,c,3}(h_c - h_3)}{h_3 - h_2} \tag{4.35}$$

In general, the fractional bubble phase hold-up of volumetric region between points i and $i+1$ is

$$[\overline{\varepsilon}_g]_{i,(i+1)} = \frac{\overline{\varepsilon}_{g,c,i}(h_c - h_i) - \overline{\varepsilon}_{g,c,i+1}(h_c - h_{i+1})}{h_{i+1} - h_i}, i = 1, 2, ... \tag{4.36}$$

So, knowing the values of axial cumulative bubble phase hold-up, $\varepsilon_{g,c,i+1}$ from equations (4.33) and (4.36), one can calculate the value of fractional bubble phase hold-up of location between points i and $i+1$. In an ejector-induced inverse bubbly flow system, an axial distribution of cumulative bubble phase hold-up and the distribution of local bubble phase hold-ups are shown in Figures 4.3 and 4.4. The variation of cumulative bubble phase hold-up with respect to the location of the column is attributable to the variation of axial pressure drop along the column. The bubble phase hold-ups are different in different locations (top, middle, and bottom) of the column. The locations are defined as bottom as 0.1 to 0.7 m, middle as 0.7 to 1.10 m, and top as 1.10 to h_c from bottom of the column. It was observed that the local bubble phase hold-up is maximum (0.72 for $d_n = 0.004$ m, 0.69 for $d_n = 0.005$ m, 0.65 for $d_n = 0.006$ m, and 0.66 for $d_n = 0.007$ m) in the middle location ($h_i - h_{i+1} = 0.7 - 1.10$ m from the bottom of the column).

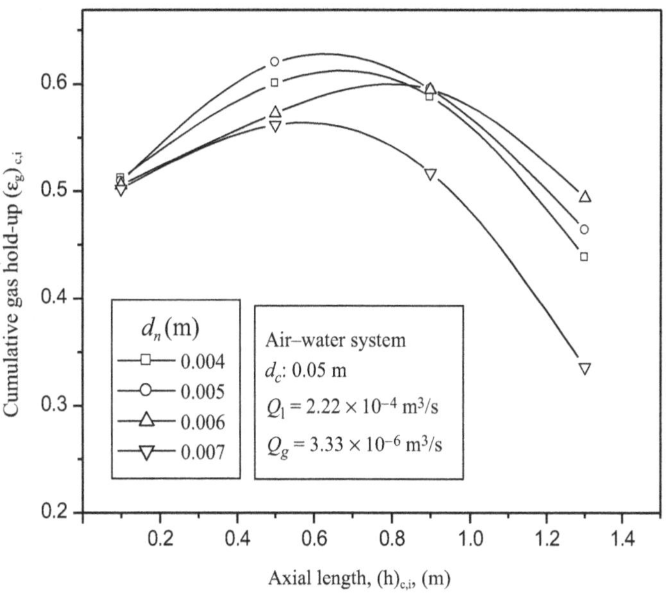

FIGURE 4.3 Axial distribution of cumulative bubble phase hold-up for jet-induced inverse bubbly flow.

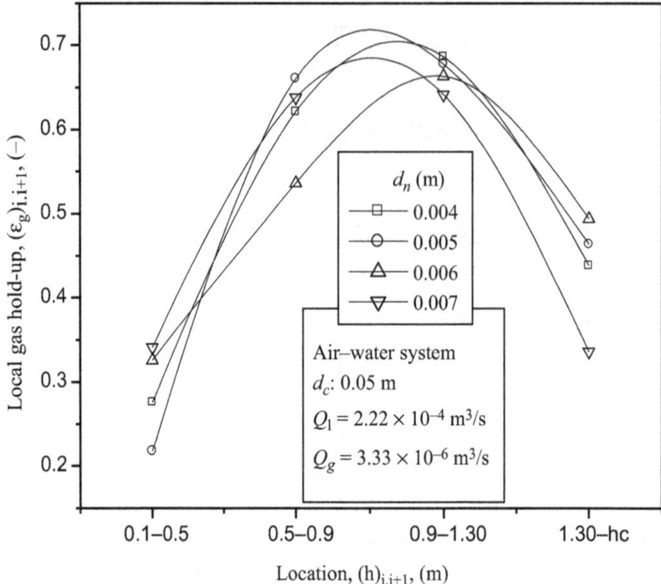

FIGURE 4.4 Distribution of local bubble phase hold-up for jet-induced inverse bubbly flow.

The density of bubble population varies along the column, which may cause the variation of local bubble phase hold-up. At the bottom location of the column, the bubble density is low because of higher pressure at the bottom. At the bottom location, smaller bubbles coalesce, and because of the buoyancy effect, coalesced bubbles move upward and accumulate in the middle location ($h_i - h_{i+1} = 0.7 - 1.10$ m). In the middle location of the column, bubble buoyancy balances the inverse kinetic force of liquid flow. Hence, at the middle location ($h_i - h_{i+1} = 0.7 - 1.10$ m), the density of bubble population is higher than the other location of the column, which results the higher bubble phase hold-up than the other location of the column.

Methods to Measure the Hold-up

Many invasive and noninvasive methods exist for the determination of bubble phase hold-up in a bubbly flow device. The numerous invasive or noninvasive techniques for two- and three-phase flow systems have been reviewed by Boyer et al. (2002). Some of the important methods to measure the bubble phase hold-up that are most suitable to measure in inverse bubbly flow conditions are described as follows.

Bubble Phase Isolation Method

In this method, first the total height of the gas–liquid mixture in the column is noted when the system is operated at steady-state condition. At a certain time, the inlet and outlet of the system are closed simultaneously with electrically operated solenoid valves, which cause an immediate termination of the fluids and the gas–liquid mixture in the column. The system is allowed to settle for some time until all the gases are separated from the liquid. Then the clear liquid height in the column is noted. The difference of these two gas–liquid mixing heights and clear liquid height gives the integral bubble phase hold-up in the column, which is expressed as

$$\varepsilon_g = \frac{h_m - h_l}{h_m} \tag{4.37}$$

Generally, in the inverse flow bubble, column air entrainment occurs because of characteristics of the plunging jet, and its rate is mostly controlled by the liquid jet. A minimum liquid flow rate is required to move the bubbles in the inverse direction by prevailing over their upward buoyancy force, which depends on the bubble size and physical properties of the liquid. The methods used so far have been based on measurements of a local or a total rise of fluid level (Ohkawa et al., 1987; Yamagiwa et al., 1990; Evans et al., 1996) on application of indirect measurements, such as γ-ray attenuation (Bin, 1974) or conductivity probes (McKeogh and Ervine, 1981; Marchese et al., 1992). The phase isolation method used by Ohkawa et al. (1987) and Yamagiwa et al. (1990) was called by them a displacement method because the gaseous phase making up the bubble phase hold-up in the system displaced water.

Differential Pressure Method

Estimating bubble phase hold-up via pressure difference measurements is a simple, low-cost, noninvasive technique to study bubble phase hold-up in a bubble reactor. Among various methods, this technique is a widely used method to estimate bubble phase hold-up. With this method, bubble phase hold-up is measured using the time-averaged static pressure drop along the column. The resulting bubble phase hold-up is an average value (both temporal and spatial) over the volume of the dispersion between the corresponding pressure taps. In applying the pressure difference method, manometers or piezoelectric pressure transducers with a data acquisition system (mounted to the column wall to minimize disturbance to the flow caused by the pressure transducers) are used to measure pressure signals. The method is also applicable to systems at high temperature and pressure. This technique does not require a transparent fluid or containment vessel, nor does it have requirements on liquid electrical properties. It can be used to measure the overall average bubble phase hold-up, as well as the average bubble phase hold-up. Thus, it can be used to probe the axial bubble phase hold-up variation in a column. As per separated flow model (Wallis, 1969; Merchuk and Stein, 1981) for vertical gas–liquid concurrent flows, bubble phase hold-up in a bubble column can be estimated as per equation follows:

$$\varepsilon_g = \left(1 + \frac{1}{\rho_l g}\frac{dP}{dz}\right) + \frac{4\tau_w}{\rho_l g d_c} + \frac{u_{sl}^2}{g(1-\varepsilon_g)^2}\frac{d\varepsilon_g}{\varepsilon_g} \tag{4.38}$$

The first term on the right hand side of equation (4.38) accounts for the hydrostatic head, the second term describes wall shear effects, and the third term represents fluid acceleration caused by void changes. The contribution of the acceleration term is typically about 1% of the total bubble phase hold-up (Merchuk and Stein, 1981). Hills (1976) has shown that in the worst case in a study with superficial liquid and gas velocities as high as 2.7 m/s and 3.5 m/s, respectively, the acceleration term amounted to less than 10% of the total bubble phase hold-up. As a result, the acceleration term is usually neglected in practice (Hills, 1976; Merchuk and Stein, 1981; Zahradnik et al., 1997; Al-Masry, 2001; Tang and Heindel, 2006). Without the acceleration term, equation (4.38) becomes

$$\varepsilon_g = \left(1 + \frac{1}{\rho_l g}\frac{dP}{dz}\right) + \frac{4\tau_w}{\rho_l g d_c} \tag{4.39}$$

To obtain the average bubble phase hold-up $\bar{\varepsilon}_g$ in a column section between two locations separated by a distance $\Delta z = z_2 - z_1\ (>0)$, average both sides of equation (4.39) from z_1 to z_2:

$$\frac{1}{\Delta z}\int_{z_1}^{z_2}\varepsilon_g\,dz = \frac{1}{\Delta z}\int_{z_1}^{z_2}\left(1 + \frac{1}{\rho_l g}\frac{dP}{dz}\right)dz + \frac{1}{\Delta z}\int_{z_1}^{z_2}\frac{4\tau_w}{\rho_l g d_c}\,dz \tag{4.40}$$

Thus,

$$\bar{\varepsilon}_g = \left(1 - \frac{1}{\rho_l g}\frac{\Delta P}{\Delta z}\right) + \frac{4\bar{\tau}_w}{\rho_l g d_c} \tag{4.41}$$

where $\Delta P = P_1 - P_2\ (>0)$ with P_1 and P_2 the pressures at location z_1 and z_2, respectively, and $\bar{\tau}_w$ representing the average wall shear stress in the same column section. The two-phase pressure drop can also be expressed based on single-phase pressure drop as

$$\Delta P = (1 - \bar{\varepsilon}_g)\Delta P_{l,s} \tag{4.42}$$

where $\Delta P_{l,s}$ is the single liquid phase pressure difference between z_1 and z_2 when $\bar{\varepsilon}_g = 0$ or $u_{sg} = 0$ and $\bar{\tau}_{w0}$ is the wall shear stress for single-phase liquid flow with the same superficial liquid velocity corresponding to ΔP. For single-phase liquid flows (i.e., $\bar{\varepsilon}_g = 0$), using equation (4.41), one can get

$$0 = \left(1 - \frac{1}{\rho_l g}\frac{\Delta P_{l,s}}{\Delta z}\right) + \frac{4\bar{\tau}_{w0}}{\rho_l g d_c} \tag{4.43}$$

Hence,

$$\Delta P_{l,s} = \rho_l g \Delta z + \frac{4\bar{\tau}_{w0}}{d_c}\Delta z \tag{4.44}$$

Substituting equation (4.44) into (4.42),

$$\bar{\varepsilon}_g = 1 - \left(\frac{1}{\rho_l g}\frac{\Delta P}{\Delta z}\right)\bigg/\left(1 + \frac{4\bar{\tau}_{w0}}{\rho_l g d_c}\right) \tag{4.45}$$

The single-phase flow wall shear stress can be estimated by

$$\bar{\tau}_{w0} = \frac{1}{8} f \rho_l u_{sl}^2 \tag{4.46}$$

$$f_{ls} = \frac{16}{\mathrm{Re}_n} \qquad \text{for laminar flow} \tag{4.47}$$

$$f_{ls} = \frac{0.079}{n^5 (\mathrm{Re}_n)^{\frac{2.63}{10.5n}}} \qquad \text{for turbulent flow} \tag{4.48}$$

where Re_n is the Reynolds number for non-Newtonian fluid flow, which is defined as (Chhabra and Richardson, 1999):

$$\mathrm{Re}_n = \frac{d_c^n V_{sl}^{2-n} \rho}{8^{n-1} K} \left(\frac{4n}{3n+1} \right)^n \tag{4.49}$$

For transition flow, the friction factor f can be deduced from the generalized pressure loss equation as (Desouky and E1-Emam, 1989):

$$f_{ls} = 0.125 \left[n^{\sqrt{n}} \left(0.0112 + \mathrm{Re}_n^{-0.3185} \right) \right] \qquad \text{for transition flow} \tag{4.50}$$

For Newtonian fluid, $n = 1$ and $K = \mu$. The factor $4\bar{\tau}_{w0}/(\rho_l g d_c) \ll 1$ is applicable for most bubble columns, which usually have diameters at least on the order of several centimeters and operate at superficial liquid velocities lower than 1 m/s (Tang and Heindel, 2006). When $4\bar{\tau}_{w0}/(\rho_l g d_c) \ll 1$, a Taylor (1953) series expansion can be used to estimate equation (4.45), thus

$$\bar{\varepsilon}_g \approx 1 - \frac{1}{\rho_l g} \frac{\Delta P}{\Delta z} + \left(\frac{1}{\rho_l g} \frac{\Delta P}{\Delta z} \right) \left(\frac{4\bar{\tau}_{w0}}{\rho_l g d_c} \right) \tag{4.51}$$

In systems with downcomers or in inverse bubble columns for which gas is fed using entrainment action of vertical plunging liquid jets, concurrent flow of both phases is obtained. In this case, the manometric method gives good agreement with the more accurate displacement method of bubble phase hold-up determination (Ohkawa et al., 1987; Yamagiwa and Ohkawa, 1989; Yamagiwa et al., 1990). The total pressure drop (ΔP_{tp}) is equal to the sum of the frictional pressure drop (ΔP_{ftp}), hydrostatic head (ΔP_h), and pressure drop caused by acceleration (ΔP_a), which is represented as

$$\Delta P_{tp} = \Delta P_{ftp} + \Delta P_h + \Delta P_a \tag{4.52}$$

where ΔP_h is the pressure drop caused by the change in the potential energy of liquid and the bubble gas. The hydrostatic pressure drop can be evaluated as

$$\Delta P_h = h_m \rho_l \left(1 - \varepsilon_g \right) g \tag{4.53}$$

When the liquid jet impinges on the pool surface, the liquid suffers an instantaneous deceleration, which enhances to raise the dynamic pressure inside the column. The dynamic pressure (ΔP_a) pushes the column contents in the downward direction. It is equal to the ratio of force exerted by the nozzle and the cross-sectional area of the column, which can be expressed as

$$\Delta P_a = \frac{F_j}{A_c} = \frac{ma_j}{A_c} = \frac{Q_l \rho_l \left(u_j - u_{sl}\right)}{A_c} \tag{4.54}$$

where u_j is the jet velocity, u_{sl} is the interstitial liquid velocity in the column, and Q_l is the volumetric flow rate of the liquid. From the continuity equation, one can write

$$u_j A_j = u_l A_c (1 - \varepsilon_g) = u_l A \tag{4.55}$$

where A_j and A_c are the areas of the nozzle and the column, respectively. From equation (4.55), one gets:

$$u_j = u_l \frac{A_c}{A_j} \tag{4.56}$$

and

$$u_l = \frac{u_{sl}}{(1 - \varepsilon_g)} \tag{4.57}$$

By substituting equations (4.56) and (4.57) into equation (4.54), one gets:

$$\Delta P_a = u_l^2 \rho_l \left[\frac{d_c^2}{d_j^2} - \frac{1}{\left(1 - \varepsilon_g\right)} \right] \tag{4.58}$$

From equations (4.52), (4.53), and (4.58), the frictional pressure drop (ΔP_{ftp}) can be represented as

$$\Delta P_{ftp} = \Delta P_{tp} - \left(h_m \rho_l (1 - \varepsilon_g) g \right) - \left(u_l^2 \rho_l \left[\frac{d_c^2}{d_j^2} - \frac{1}{\left(1 - \varepsilon_g\right)} \right] \right) \tag{4.59}$$

The two-phase frictional pressure drop based on superficial slurry velocity can also be expressed as

$$\frac{\Delta P_{ftp}}{g \rho_l h_m} = \frac{2 f_{tp} u_l^2}{g d_c} \tag{4.60}$$

By substituting equation (4.60) into equation (4.59), the bubble phase hold-up can be obtained as

$$\varepsilon_g = 1 - \left[\frac{\sqrt{K^2 + 4C} - K}{2} \right] \tag{4.61}$$

where

$$K = \frac{f_{tp} - AB + ACD}{A}$$

(4.62)

$$A = \frac{gd_c}{2u_l^2}; \ B = \frac{\Delta P_{tp}}{g\rho_l h_m}; \ C = \frac{u_l^2}{gh_m}; \ D = \frac{d_c^2}{d_j^2}$$

(4.63)

Conductometric Method

In the conductometric method, the bubble phase hold-up is measured by the classical potential theory to relate the measured conductivity to the bubble phase hold-up. Maxwell (1892) reported that the effective conductivity of dispersion ($k_{l\text{-}d}$) is related to the volume fraction (ε_d) of a dispersed nonconductive phase by

$$k_{l-d} = k_l \left(\frac{1-\varepsilon_d}{1+0.5\varepsilon_d} \right)$$

(4.64)

Gas bubble hold-up in two-phase bubbly flow is calculated based on this principle as:

$$\varepsilon_g = \left(\frac{k_l - k_{l-g}}{k_l + 0.5k_{l-g}} \right)$$

(4.65)

where k_l and $k_{l\text{-}g}$ are the electrical conductivities of liquid and liquid–gas mixture, respectively. A schematic representation of the bubble phase hold-up measurement by electrical conductivity is shown in Figure 4.5.

The electrical conductivities of liquid and liquid–gas mixture can be measured by electrical conductivity meter or online conductivity probe with data acquisition system. The cell in the conductivity meter is an adaptation of a section of the "ideal" cell consisting

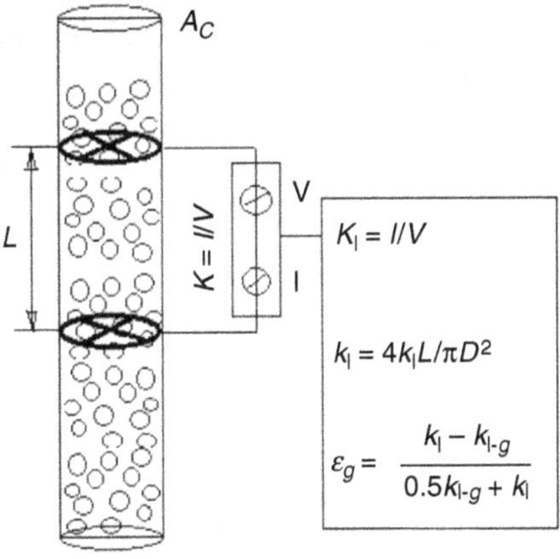

FIGURE 4.5 Schematic representation of bubble phase hold-up measurement by electrical conductivity.

of two infinite and parallel plate electrodes. The electrical conductance profaned by the electrodes is described by the equation

$$K_l = \kappa \frac{A}{L_e} \tag{4.66}$$

where K_l is the electrical conductance (inverse of resistance); κ is the electrical conductivity; A and L_e are the area (m^2); and D is the diameter of electrode and the separation (m) of the electrodes, respectively. A/L_e is referred to as the cell constant. Such a cell has been used to measure effective conductivities of dispersions. Alternating current (AC) of sufficiently high frequency ($\sim$1000 Hz) and low voltage ($\sim$1.5 V) is used to avoid polarization of the electrodes.

Electrical Resistance Tomography

Electrical resistance tomography (ERT) is an advanced and noninvasive flow visualization technique that is applicable to opaque fluids and has the capability of performing online measurements in different axial and radial locations within the reactor. ERT has been used successfully for bubble phase hold-up characterization in several two- and three-phase reactors, including bubble column reactors. An ERT system is used to obtain the conductivity distribution across different sensor planes. The ERT system consists of three components: electrodes, data acquisition system (DAS), and image reconstruction system. A number (m) of sensor planes (according to experimental design) can be located along the reactor height, and for each plane, n (as per design) equally spaced stainless steel rectangular electrodes can be noninvasively fitted to the internal wall of the reactor. The dimensions of the electrodes are a function of the different factors such as the reactor diameter, range of the fluid conductivity, fluid velocity, and required image speed. The dimensions of the electrodes can be defined as per recommendation by the tomography machine manufacturer. Figure 4.6 shows the details of the ERT system. The electrodes can be connected to the DAS using co-axial cables. The DAS uses an adjacent measurement strategy by applying a current between two adjacent electrodes and measuring the corresponding voltage between all other adjacent electrode pairs. An excitation frequency and an injection current are required in the experiments. A suitable spatial resolution (%) of the reactor diameter and a temporal resolution (frames per second) according to the design of experiments should be maintained. The DAS is to be connected to a host computer for data processing using a suitable image reconstruction algorithm (Pakzad et al., 2008) to compute the conductivity distribution across the planes based on the raw voltage measurements. By applying the algorithm, a pixel array ($m \times n$) can be created for each sensor plane, and after excluding the pixels (i) that lie outside of the reactor circumference, the remaining pixels ($m \times n - k$) can be used for the reconstruction of the circular image. If there are p number of measuring planes, a total of ($m \times n - k$) p noninvasive conductivity measurements can be obtained for each frame.

Using the conductivity distribution, which is determined by ERT, the distribution of the gas phase can be obtained at different cross-sections of the reactor. The Maxwell equation

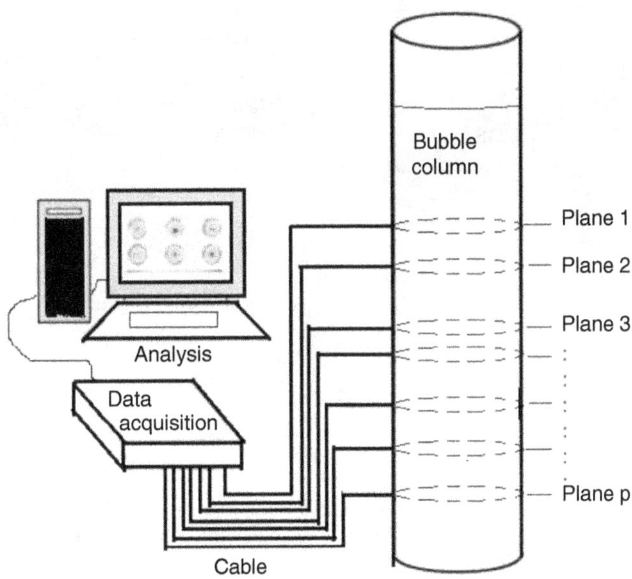

FIGURE 4.6 The schematic diagram of the electrical resistance tomography measurement system (Babaei et al., 2015).

can be applied to convert the conductivity data to the concentration of dispersed phase (bubble phase hold-up):

$$\varepsilon_g = \frac{2k_1 + k_2 - 2k_{rc} - k_{rc}k_2 / k_1}{k_{rc} - k_{rc}k_2 / k_1 + 2(k_1 - k_2)} \tag{4.67}$$

where k_1 and k_2 are the conductivities of the continuous phase and dispersed phase, respectively. k_{rc} is the reconstructed conductivity determined by ERT measurements. If the dispersed phase in nonconductive, k_2 is considered as zero. Thus, the simplified Maxwell equation can be expressed as follows:

$$\varepsilon_g = \frac{2(k_1 - k_{rc})}{k_{rc} + 2k_1} \tag{4.68}$$

Capabilities of ERT in performing point measurements enable to determine the radial distribution of bubble phase hold-up at different heights of the bubbly flow column reactor. An appropriate regionization scheme for sensor planes should be used in this regard. Based on ERT measurements, the mean value of bubble phase hold-up for each cell can be calculated by averaging the pixel values within that cell. These mean values can be fitted to several mathematical models.

Dynamic Gas Disengagement Technique

This technique was introduced by Sriram and Mann (1977). The basic principle of this technique is to stop the bubbling in a bubble column and to measure the liquid level or

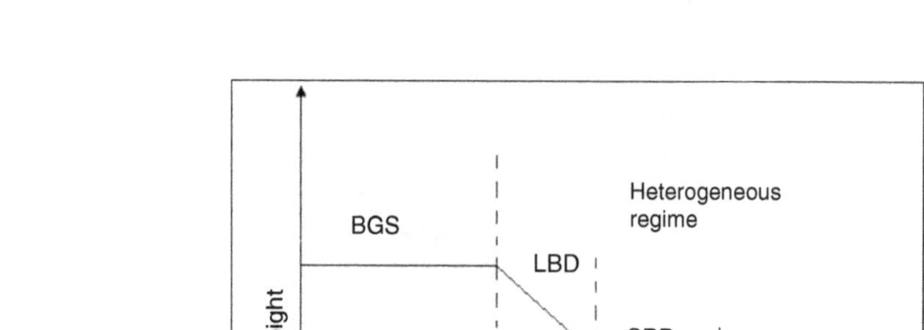

FIGURE 4.7 Schematic of regime transitions determination in bubble columns. *BGS,* Before gas shutoff; *LBD,* large bubble disengagement; *LL,* liquid level; *SBD,* small bubble disengagement. *(Adapted from Vial et al., 2001.)*

the pressure at different levels in the column as a function of time. This technique can be used either to determine the global bubble phase hold-up or to determine the structure of the bubble phase hold-up in bubble columns (as per the amount of gas bubbles separated in classes according to their size). The method requires accurate measurement of the decaying height of a gas–liquid dispersion level in a column upon cessation of gas flow. The decrease of the level can be observed using various techniques, including x-rays (Beinhauer, 1971), visual observation, and pressure profiles (Daly et al., 1992). This decaying height with respect to time is called the *dynamic gas disengagement profile.* The pressure measurement device is connected to a few centimeters below the nonaerated liquid height to measure the pressure change with changing dispersion height caused by cessation of gas. The measured disengagement profile enables the estimation of the hold-up structure and allows the evaluation of the rise velocities of bubbles in the dispersion before gas flow interruption. A typical schematic disengagement profile is shown in Figure 4.7. The basic assumptions in the analysis are (1) the dispersion is axially homogeneous at $t = 0$ when the gas feed is interrupted, (2) there are no bubble–bubble interactions (i.e., coalescence and breakup) happening during disengagement, (3) a constant rate disengagement process occurs (i.e., the bubble disengagement of each bubble class is not influenced by the other bubble classes), and (4) the cross-sectional area occupied by bubbles rising at velocity u_{bi} remains constant throughout the disengagement. The uncertainty in the direct measurement of the liquid level originates from the waviness of the liquid surface. Therefore, the measurement accuracy can be improved if the average level can be obtained reliably. One approach is to measure continuously the liquid level with a suitable sensor such as a resistance probe and obtain a time-average value. However, because a resistance measures only the local value, multiple probes have to be used to obtain the "surface-averaged" mean liquid level.

Another approach to damp the surface waviness by some mechanical means (Lee et al., 1985). Lee et al. (1999) checked experimentally the assumptions of dynamic gas disengagement (DGD) in a two-dimensional bubble column by particle image velocimetry measurements. They showed that the assumptions 2 and 3 of the classical disengagement technique are not valid, especially in the coalesced bubbly flow (heterogeneous or transition) regime. The limits of the application of the DGD technique indicated by these authors have still to be checked in case of three-dimensional columns.

Several other methods are also available to measure the hold-up characteristics, which are mentioned in Chapter 2.

Effect of Different Variables on Bubble Phase Hold-up

It was pointed out earlier that entrained gas, after getting dispersed as fine bubbles into liquid in the plunging zone, moves through the column inversely as homogeneous gas–liquid bubbly flow and finally gets separated from liquid in the separator. Kantarci et al. (2005) presented an extensive survey of the results of possibly all available experimental and theoretical studies on bubble phase hold-up characteristics by upflow and inverse flow bubble columns. In concurrent inverse flow, the gas bubbles move against their buoyant force, having a higher slip velocity compared with the upflow system. Thus, the bubbles have a larger residence time and hence higher bubble phase hold-up. Briens et al. (1992) studied the venturi bubble column with both inverse flow and upflow mode and obtained much higher bubble phase hold-up (0.15–0.40) in inverse flow compared with upflow (0.08–0.12). The gas distributor and physical properties of fluids also have some effect on bubble phase hold-up (Deckwer, 1992). It is well known that surface-active solutes, organic or inorganic, hinder coalescence of bubbles in water. This in turn alters the gas hold-up characteristics in bubble columns. The gas hold-up increases in water in the presence of different electrolytes. Syeda and Reza (2011) observed enhancement of the bubble phase hold-up for four electrolytes (namely, NaCl, $MgSO_4 \cdot 7H_2O$, Na_2SO_4, and $CaCl_2 \cdot 2H_2O$) at concentrations less than 0.1 mol/L. With the increase in concentration, the gas hold-up showed two different trends in Na_2SO_4 and $CaCl_2 \cdot 2H_2O$ solutions. The gas hold-up formed a sharp peak after the enhancement and leveled off at a value somewhat higher than that in water, but in the NaCl and $MgSO_4 \cdot 7H_2O$ solutions, gas hold-up leveled off immediately after the enhancement without forming any peak. They have found a strong relation between the gas hold-up enhancement and the change of surface tension with the addition of electrolyte. They also observed that the concentration at which maximum value of C $(d\sigma/dC)^2$ (i.e., Concentration × Surface tension gradient with respect to concentration^2) is obtained corresponds to the concentration at which maximum gas hold-up enhancement occurs. Bubble phase hold-up in a bubbling regime is strongly related to the coalescence tendency of bubbles in the respective gas–liquid system. As the suppression of bubble coalescence leads to gas hold-up enhancement, the concentration at which minimum bubble coalescence is achieved is likely to correspond to the concentration at which maximum gas hold-up occurs. About a 75% increase in gas hold-up at a concentration corresponding to minimum bubble coalescence in different electrolyte solutions can be observed (Zahradnik et al., 1995). Generally, two different approaches, namely, the dynamic surface tension theory of Andrew (1960) and the liquid phase diffusion theory of Marucci (1969), are often used to explain the coalescence behavior of the solutions.

Syeda et al. (2002) investigated the effect of addition of alcohols on bubble phase hold-up enhancement of water in a bubble column and successfully used the dynamic surface tension theory to predict the hold-up trends. It is well accepted that gas hold-up will be high when the number of bubbles is high and their diameters small. At the transition between the homogeneous and heterogeneous regime, many authors observed a decrease in bubble phase hold-up for the air–water system at ambient temperature and pressure. This is supposed to be an effect of the apparition bubbles and the disappearance of small bubbles in the transition regime. The bubbly flow is also characterized by the dispersion of the bubbles and the free space available between the bubbles. Most of the bubble column works in the heterogeneous regime. In the homogeneous regime, bubble phase hold-up increases faster than in the heterogeneous regime, mostly because bubbles remain small, and their number increases. In the case of an organic liquid, the profile may not show a maximum, probably because of viscosity or interfacial tension effects. If the bubbles are small enough (d_b <1 mm) and there is sufficient free space, bubble population increases with increase in gas flow rate, and thus bubble phase hold-up increases. But beyond a certain limit, there is little effect of gas flow rate on bubble population and coalescence of bubbles increases significantly, and hence bubble phase hold-up remains constant (Godbole et al., 1982; Schumpe and Deckwer, 1982; Zahradnik et al., 1997).

In an ejector-induced concurrent inverse flow (Majumder et al., 2006), the bubbles are comparatively smaller in size (d_b ~3–5 mm) and spend more time in the column; thus, higher bubble phase hold-up (0.34–0.61) results. At a constant liquid flow rate, bubble phase hold-up increases with an increase in the gas flow rate because of increased bubble population. The bubble phase hold-up increases with the increase in liquid jet velocity (u_j). This is because of increase of kinetic energy of the liquid jet, surface roughness of the jet, and contacting perimeter between the jets and receiving liquid surface (Kumagai and Endoh, 1982). The bubble phase hold-up increases with decreasing nozzle diameter because of the increase of bubble residence time by decrease of superficial liquid velocity in the column.

The overall gas hold-up in a bubble column is influenced by both the column diameter and dispersion height. The influence of the dispersion height at a constant column diameter has been studied by Zahradnik et al. (1997). They observed that the lower the height of dispersion, the higher the gas hold-up in the case of the homogeneous regime. In the heterogeneous regime, it has no influence. The decrease in the height of dispersion extends the homogeneous regime. They observed that the height of dispersion has no more influence for a ratio H_c/D_c higher than 5. A decrease of gas hold-up increases the column diameter caused by wall effects at a low column diameter, which decreases the bubble rise velocity. In the homogeneous regime, the same trend occurs along with an increase of the liquid recirculation while increasing column diameter. In the heterogeneous system, the presence of large-diameter spherical caps subject to wall effect, especially at atmospheric pressure, is reported (Krishna et al., 1999). The effect of column diameter for viscous fluid is higher. Most authors (Kantarci et al., 2005) report that the column diameter has no longer influence for diameters higher than 0.15 m in either well established homogeneous or heterogeneous system. This is validated by many experimental results (Zahradnik et al., 1997; Ruzicka et al., 2001; Forret et al., 2003; Chilekar et al., 2010). However, Urseanu et al. (2003) observed the decrease of gas hold-up for a range of column diameters from 0.15 to 0.23 m ($H_c/d_c >5$) while working with viscous fluids. The velocity corresponding to the transition from the homogeneous regime to the heterogeneous regime is also affected by column dimensions. Both bubble coalescence and break-up are significant in large bubble columns. Small columns, because of short residence time, impede bubble break-up and provide better understanding of the collective coalescence process. Bubble phase hold-up increases with an increase in column diameter. The increase of bubble phase hold-up with increasing column diameter at a constant liquid jet velocity is considered to be attributable to the decrease of the inverse liquid velocity in the column, which results in decreasing in inverse velocity of bubbles. The bubble phase hold-up in a bubble column with a porous plate distributor is affected by pressure in the range 0.1 to 1.6 MPa when the superficial gas velocity, evaluated at the pressure in the column, is less than 0.03 m/s (Kölbel et al., 1971). The bubble phase hold-up in a slurry bubble column with a porous plate distributor containing fine particles at pressures up to 1.1 MPa and with superficial gas velocities below 0.04 m/s does not significantly change because of pressure in that range of operating variables (Deckwer, 1992). A pressure increase results in an increase of the bubble phase hold-up. The influence of pressure seems to be higher in the heterogeneous regime than in the homogeneous regime. An increased number of bubbles while increasing pressure may result in an increase of bubble phase hold-up. Urseanu et al. (2003) reported a minor effect of pressure on bubble phase hold-up while working with viscous fluids (μL >70 mPa/s), which is attributed to the opposite effect of pressure and viscosity on bubble size.

There is a decrease of gas hold-up while increasing temperature (Pohorecki et al., 2001). Lin et al. (1998) observed an increase of the gas hold-up with temperature at a pressure of 15.2 MPa, which can be related to their bubble size distribution: it becomes narrower when temperature is increased. A temperature increase leads to a decrease in surface tension and thus leads to a decrease of the film drainage speed, a decrease of primary bubble size, and a decrease of maximum stable bubble size. A temperature increase also leads to a decrease of viscosity, which can lead to a decrease of the primary bubble diameter and a decrease of maximum stable bubble size but has two effects on bubble coalescence. On the one hand, it increases the film drainage speed and collision frequency and thus promotes coalescence. It is likely that the increase of bubble phase hold-up with the increase of temperature is linked to the concomitant decrease of viscosity and surface tension.

Viscosity and surface tension can have contradictory effects on bubble size via the effect on coalescence. The increase of bubble phase hold-up is then linked to the effect on primary bubble size, maximum bubble size, and contact time for coalescence. Olivieri et al. (2011) reported that as the viscosity of the liquid increases the bubble phase hold-up increases up to a certain limit. A further increase of viscosity leads to a destabilization of the homogeneous regime and a decreaseof the gas hold-up at transition. However, García-Abuín et al. (2012) reported the opposite result. They reported that increasing viscosity by adding a polymer leads to a decrease of bubble phase hold-up and interfacial until a certain concentration. A further increase of the viscosity results in an increase of the bubble phase hold-up. Furthermore, they observed that increasing viscosity results in a slight increase of the bubble mean diameter while the bubble size distribution becomes larger: small and large bubbles are produced. The heterogeneous regime appears. Biń et al. (2001) studied the effect of the working mode on bubble phase hold-up. They measured gas hold-up in counter, cocurrent, and semibatch mode for a small superficial gas velocity range(up to 1.6 cm/s). They showed that the bubble phase hold-up increases with superficial liquid velocity in the countercurrent mode and decreases or remains constant in the cocurrent mode. The effect is more pronounced at a high gas velocity. The difference in bubble phase hold-up between the cocurrent and countercurrent mode is around 10%. The hold-up at transition is higher for the countercurrent operation.

A Comparative Picture of Bubble Phase Hold-up

In two-phase gas–liquid flow, the bubble phase hold-up generally depends on the operation mode of the fluid flow. Higher bubble phase hold-up was observed in concurrent inverse flow compared with upflow caused by a higher residence time of gas bubbles. A comparative picture of bubble phase hold-up obtained by different authors over a range of gas and liquid flow is given in Table 4.3. Briens et al. (1992) obtained a maximum bubble phase hold-up of 0.40 in an inverse venturi–bubble column combination, while in upward flow, the bubble phase hold-up achieved was only 0.12 under similar conditions. Bubble size has the opposite effect on the bubble phase hold-up, depending on whether the system is upflow or inverse flow. For an inverse flow system, large bubbles spend more time in the column; thus, higher bubble phase hold-up is obtained. In the present work of an inverse flow system, a relatively higher bubble phase hold-up was obtained compared with other reported work even at lower gas flow rates. Also, in this modified inverse flow bubble column reactor, bubble phase hold-up obtained in air–carboxymethyl cellulose (CMC) solution (0.45–0.61) are higher than those for air–water system (0.34–0.50) because of the highly viscous behavior of CMC solutions.

Models to Analyze Bubble Phase Hold-up Characteristics

The entrained gas subsequently gets dispersed into liquid in the plunging zone as fine bubbles and then moves inversely as homogeneous gas–liquid two-phase flow. A large number of studies have been reported on the measurement and analysis of dispersed

Table 4.3 Comparative Picture of Bubble Phase Hold-up Obtained by Different Authors Over a Range of Gas and Liquid Flow

Reference	Type of Flow Column	Column Diameter (m)	Liquid Used	Liquid Flow Rate Range, u_l (m/s)	Gas Flow Rate Range, u_l (m/s)	Bubble Phase Hold-up (–)
Schumpe and Deckwer (1982)	Concurrent upflow	0.10–0.14	CMC (0–1.8 wt. %)	—	0.003–0.025	0.03–0.20
Khatib and Richardson (1984)	Concurrent upflow	0.039	Kaolin suspension	0.305–0.61	0.30–3.5	0.15–0.55
Ohkawa et al. (1987)	Concurrent inverse flow	0.02–0.026	Water	0.05–0.20	0.05–0.20	0.01–0.40
Bando et al. (1988)	Concurrent inverse flow	0.07	Water	0.10–0.20	0.01–0.10	0.01–0.32
Yamagiya et al. (1990)	Concurrent inverse flow	0.034–0.070	Water	0.4–0.912	0.1–0.50	0.15–0.40
Briens et al. (1992)	Inverse flow venturi column	0.095	Water	0.2–0.5	0.01–0.10	0.15–0.40
Huynh et al. (1991)	Upflow venturi column	0.095	Water	0.2–0.5	0.01–0.10	0.08–0.12
Das et al. (1992)	Concurrent upflow	0.019	CMC (0.5–1.0 kg/m^3)	0.296–1.00	0.17–1.60	0.12–0.45
Zahradnik et al. (1997)	Concurrent upflow	0.29	Water	0.008–0.029	0.004–0.076	0.05–0.24
Majumder et al. (2006)	Concurrent inverse flow	0.05, 0.06	Air–water	0.05–0.14	0.001–0.013	0.34–0.61
Sivaiah et al. (2012)	Concurrent inverse flow	0.05, 0.06	Air–water–soild	0.04–0.14	0.0085–0.0255	0.01–0.61

CMC, carboxymethyl cellulose.

phase hold-up of gas–liquid two-phase flow in different types of gas liquid contactors. Various models and correlations have also been proposed by several authors. Some of the important models and correlations are as follows.

Homogeneous Flow Model

In this model, it is assumed that the fluids behave like a homogeneous mixture. The velocity and density of the mixture are considered constant across the tube, and the bubble phase hold-up is predicted from the flow rate of the fluids entering the system by the following relationship:

$$\varepsilon_g = \frac{Q_g}{Q_g + Q_l}$$

(4.69)

Isbin et al. (1957) proposed that the homogeneous model gives reasonable prediction of the bubble phase hold-up if pressure losses are negligible. This model leads to larger errors when pressure losses are significant.

Variable Density Model

Bankoff (1960) reported a modification of the homogeneous flow model to consider the effect of radial non-uniformity and concentration. He suggested that although there is no slip between the gas and the liquid in the direction of flow, a gradient in the bubble concentration exists in the axial direction. By assuming a power law distribution for both the velocity and the concentration profiles in the radial direction, Bankoff proposed the following relationship between the liquid hold-up and the input flow rates:

$$\varepsilon_l = K \frac{Q_l}{Q_l + Q_g} \tag{4.70}$$

where K is the Bankoff factor. Although the Bankoff model was developed for vertical flow, Scott (1963) and Chawla (1972) suggested that this model can also be used for horizontal flow. For single-component two-phase flow, the Bankoff model correlates the experimental data satisfactorily, but for a two-component system, it does not give good results.

Momentum Exchange Model

Levy (1966) assumed that when the sum of the frictional and hydrostatic losses in one phase equals the sum of the same losses in the other phase, there would be a rapid exchange of momentum between the phases to maintain the equilibrium condition. He proposed the momentum exchange model on this assumption. According to this model, the bubble phase hold-up fraction ε_g is given by the expression

$$\frac{1}{\varepsilon_g (1+R_m)^2} + \frac{\rho_l}{2(1+R_m)\rho_g} - \frac{1}{2(1+R_m)^2 (1-\varepsilon_g)^2} = 0 \tag{4.71}$$

where R_m is the gas–liquid mass flow ratio. The Levy model gives reasonable agreement with experimental data in horizontal tubes for high void volumes.

Lockhart–Martinelli Correlations

Lockhart and Martinelli (1949) proposed graphical correlations for the analysis of the hold-up data in the horizontal cocurrent two-phase gas–Newtonian liquid systems for four different combinations—laminar–laminar, laminar–turbulent, turbulent–laminar, and turbulent–turbulent-—of the gas–liquid flow. They made a graphical relation between the liquid hold-up ε_l and the parameter X, defined by the following equation:

$$X^2 = \left(\frac{\Delta p_f}{\Delta z}\right)_l \Big/ \left(\frac{\Delta p_f}{\Delta z}\right)_g \tag{4.72}$$

Lockhart–Martinelli correlations predict excellent results for high void fractions and for liquid superficial Reynolds numbers below 8000. Butterworth and Hewitt (1977) presented the Lockhart–Martinelli relation for bubble phase hold-up based on quality (x) and fluid properties such as density (ρ) and viscosity (μ) as

$$\frac{1-\varepsilon_g}{\varepsilon_g} = K\left(\frac{1-x}{x}\right)^p \left(\frac{\rho_g}{\rho_l}\right)^q \left(\frac{\mu_l}{\mu_g}\right)^r \tag{4.73}$$

The values of constants K, p, q, and r in equation (4.73) depends on the experimental operating variables. The quality x is equal to the ratio of the mass flow rate of gas and the mass flow rate of gas–liquid mixture.

$$x = \frac{\dot{m}_g}{\dot{m}_m} = \frac{\rho_g Q_g}{\left(\rho_g Q_g + \rho_l Q_l\right)} \tag{4.74}$$

The volumetric flow rate of the gas (Q_g) can be calculate from the volumetric flow rate of the liquid (Q_l) and the bubble phase hold-up as

$$Q_g = \frac{\varepsilon_g Q_l}{\left(1-\varepsilon_g\right)} \tag{4.75}$$

By substituting equation (4.75) into equation (4.74), the bubble phase hold-up as a function of quality can be expressed as

$$\varepsilon_g = \frac{1}{\left(1+\dfrac{\rho_g}{\rho_g}\dfrac{(1-x)}{x}\right)} \tag{4.76}$$

Davis (1963) extended the applicability of the Lockhart–Martinelli approach for vertical flow through the modification of the parameter X by incorporating the Froude number Fr_{tp} to allow the effects of gravity and velocity. Thus the modified parameter X' is expressed as:

$$X' = \frac{0.19X}{Fr_{tp}^{-0.185}} \tag{4.77}$$

Drift-Flux Model

Bankoff (1960) proposed a model considering the differing distribution of void fraction and fluid velocity. Later Zuber and Findlay (1965) developed their well-known "drift flux model" on the basis of the existence of local slip. This model takes into account the effect of non-uniform flow and concentration distribution across the duct as well as the effect of local relative velocity between the two phases. To analyze the bubble phase hold-up data by drift flux model, the average gas velocity, $\bar{u}_g$, is plotted against the gas–liquid mixture velocity, u_m, as given by the following equation:

$$\bar{u}_g = C_o u_m + u_d \tag{4.78}$$

where C_o is the "profile constant," which accounts for interaction of the velocity and bubble phase hold-up distributions, and u_d is a weighted average drift velocity term. The drift velocity accounts the effect of local relative velocity between the bubbles and liquid. The values of the distributed parameter and the weighted average drift velocity can be obtained by the curve-fitting method. The value of the distribution parameter can be obtained from the slope of equation (4.78), and the intercept can be interpreted as the weighted average drift velocity. The Zuber–Findlay model provides a simple technique to analyze and correlate the experimental hold-up data and can be applied to any two-phase flow regime. Besides these models, a large number of empirical and semiempirical correlations have been proposed by different investigators. Some of them are by Akita and Yoshida (1973), Nguyen and Spedding (1977a, 1977b), Schumpe and Deckwer (1982), Radhakrishnan and Mitra (1984), Yamagiwa et al. (1990), and Krishna and Ellenberger (1996). After reviewing the literature, it was found that all the models discussed here deal with bubble phase hold-up for different systems. However, the inverse system deviates from the Lockhart–Martinelli (1949) correlation along with the modified forms suggested by the Davis (1963) and Eisenberg and Weinberger (1979) models. This is because of enhanced gas dispersion in the inverse system. The C_o values for the inverse bubbly flow varies from 1.60 to 1.81, and u_d varies from –0.23 to –0.09 with the ranges of parameters: $0.033 \times 10^{-4} < Q_g < 0.267 \times 10^{-4}\,\mathrm{m}^3/\mathrm{s}$, $1.39 \times 10^{-4} < Q_l < 2.78 \times 10^{-4}$; $0.004 < d_n < 0.007\,\mathrm{m}$ and $d_c = 0.05\,\mathrm{m}$ and $0.06\,\mathrm{m}$. Clark and Flemmer (1985) obtained average values of C_o as 1.07 and 1.17 for upflow and inverse flow, respectively. They also obtained the value of drift velocity as 0.25 m/s for upflow and 0.25 m/s for inverse flow of a gas water system. The values of C_o and u_d obtained by Yamagiwa et al. (1990) in concurrent inverse flow with an air–water system were 1.17 and 0.19 m/s, respectively. Kelkar et al. (1983) made studies on different alcohols in concurrent upflow systems and obtained the values as $C_o = 1.24$ to 2.41 and $u_d = 0.06$ to 0.083 m/s. Although current two-phase flow research is concerned with the measurement of local flow properties, most engineers still use models requiring only average quantities in a design. The Zuber–Findlay drift flux model is used to predict bubble phase hold-up in vertical bubbly flow, but some authors disagree as to whether the profile constants C_o and u_d are invariant with respect to operating parameters. Petrick and Kudirka (1956), Nassos and Bankoff (1967), Govier and Aziz (1972), and Mudde and Saito (2001) concluded that C_o varies with gas and liquid flow rates. Clark et al. (1990) reported that drift flux parameters change with low liquid input flow. The authors applied a stress balance to compute the liquid velocity and showed that with a parabolic void fraction distribution in a 10-cm-diameter gas–water column, the distribution parameter reach values of up to $C_0 = 5$ because of a liquid flow at the wall for low input conditions. Ishii (1977) developed a simple correlation for the distribution parameter in upward two-phase flow by considering a fully developed bubbly flow and assuming that the distribution parameter would depend on the density ratio, ρ_g/ρ_l, and on the Reynolds number, Re. Recently, Hibiki and Ishii (2003) modified Ishii's (1977) correlation for the distribution parameter in vertical bubbly flow based on the column diameter and the bubble dynamics. Clark and Flemmer (1986) demonstrated that C_o varies with column diameter. A list of published correlation for distribution parameter

Table 4.4 Recommended Correlations for Distribution Parameter and Drift Velocity for Inverse Flow*

Reference	Distribution Parameter (C_0)	Drift Velocity (u_d)
Hirao et al. (1986)	$C_0 = 1.2 - 0.2\sqrt{\dfrac{\rho_g}{\rho_l}}$ for $\bar{u}_m < -3.5\,\text{m/s}$ $C_0 = 0.9 + 0.1\sqrt{\dfrac{\rho_g}{\rho_l}} - 0.3\left(1 - \sqrt{\dfrac{\rho_g}{\rho_l}}\right)(2.5 + \bar{V}_m)$ for $-3.5 \leq \bar{u}_m \leq -2.5$ m/s	$u_d = \sqrt{2}\left(\dfrac{g\sigma(\rho_l - \rho_g)}{\rho_l^2}\right)$
Yamagiwa et al. (1990)	$C_0 = 1.17$	$u_d = 0.19$
Goda et al. (2003)	$C_0 = (-0.0214V_m + 0.772)$ $+(0.0214V_m + 0.228)\sqrt{\rho_g/\rho_l}$ for $-20 \leq \bar{u}_m < 0$ m/s,	$u_d = \sqrt{2}\left(\dfrac{g\sigma(\rho_l - \rho_g)}{\rho_l^2}\right)$
Majumder et al. (2006)	$Ln(C_0) = 0.054Ln(D_R) - 0.72Ln(\rho_R)$ $\quad\quad - 0.19Ln(\text{Re}_m) - 0.097Ln(Su_c)$ $\quad\quad - 0.65$ where $D_R = d_n/d_c$, $\text{Re}_m = (\rho_l u_m d_c)/\mu_l$, $\rho_R = \rho_g/\rho_l$, $Su_c = (\rho_l \sigma_l d_c)/\mu_l^2$; $0.05 \leq u_m < 0.155$ m/s	$Ln(-u_d) =$ $-0.042Ln(D_R) - 7.97Ln(P_R)$ $+0.82Ln(\text{Re}_m) - 1.39Ln(Su_c)$ -17.75

*See the Nomenclature, Greek Letters, and Subscripts lists for abbreviations.

and drift flux velocity in inverse flow is given in Table 4.4. Kawanishi et al. (1990) studied steam–water inverse two-phase flows for 19.7- and 102.3-mm-diameter round tubes and proposed a correlation based on their experimental data. They assumed that the drift velocity for inverse two-phase flow would be regardless of the flow regime. The distribution parameter was empirically determined with respect to the mixture volumetric flux. Goda et al. (2003) also developed a correlation for distribution parameter taking into account the effect of two-phase mixture volumetric flux.

Analysis by the Slip Velocity Model

The slip velocity model was first introduced by Behringer in 1936. The slip velocity of bubbles relative to surrounding liquid is defined as:

$$u_S = \pm[u_g - u_l] = \pm\left[\frac{u_{sg}}{\varepsilon_g} - \frac{u_{sl}}{1 - \varepsilon_g}\right] \tag{4.79}$$

Where a negative sign has been incorporated to consider the inverse flow system. Thus, bubble phase hold-up data can be correlated as a function of u_g and u_l. In the churn-turbulent flow regime, Lapidus and Elgin (1957) found evidence of bubble interactions that characterize the slip velocity within the column. To account for the interaction between bubbles within the column, the following Lapidus and Elgin relationship has been frequently used for slip velocity:

$$u_S = u_b f(\varepsilon_g) \tag{4.80}$$

where u_b is the bubble terminal rise velocity and $f(\varepsilon_g)$ shows the effect of the interaction of neighboring bubbles within the column. Hills (1976) obtained the following correlation for a concurrent upflow system:

$$u_s = 0.24 + 4.0\varepsilon_g^{1.72} \text{ for } u_l \leq 0.3\,\text{ms}^{-1} \tag{4.81}$$

with u_b = 0.24 m/s. Clark and Flemmer (1985) performed a regression analysis on their results using the equation proposed by Wallis (1969):

$$u_s = u_b (1 - \varepsilon_g)^n \tag{4.82}$$

and found best fit for $n = 0.702$ and $u_b = 0.25$ m/s. Various models are available to describe the slip velocity as shown in Table 4.5. Majumder et al. (2006) studied the bubble phase hold-up behavior extensively in an inverse bubble column. They observed that slip velocity is not only a function of bubble phase hold-up but also a function of the nozzle-to-diameter ratio. So, a correlation has been developed for slip velocity as $u_s = u_b \varepsilon_g^\alpha D_R^\beta$, and it has been found that the correlation holds good for the experimental data of the inverse system compared with the other models as shown in Table 4.5.

Kawase and Moo-Young (1987) Model

The model is based on the momentum change caused by liquid motion in the column. The model is developed based on the assumption that the flow behavior of the liquid can be represented by a power-law model. It is therefore applicable to non-Newtonian fluids as well as Newtonian fluids. The equation of motion for the liquid phase in a bubble column may be written as

$$-\frac{1}{r}\frac{d}{dr}(r\tau) = -(\bar{\varepsilon}_g - \varepsilon_g)\rho_l g \tag{4.83}$$

$$\tau = -(v_l + v_{t,l})\rho_l \frac{du}{dr} \tag{4.84}$$

$$\varepsilon_g = \bar{\varepsilon}_g \left(\frac{2+n}{n}\right)\left\{1 - \left(\frac{r}{R}\right)^n\right\} \tag{4.85}$$

Table 4.5   Slip Velocity Model in an Inverse Bubble Column* (Majumder et al., 2006)

Model Equations	Results by Investigators	Goodness of Fit
$u_s = u_b(1-\varepsilon_g)^{n-1}$	$u_b = 0.03;\ n = -1.28$	$R^2 = 0.43,\ \chi^2 = 0.0013$
$u_s = u_b[\varepsilon_g(1-\varepsilon_g)]^n$	$u_b = 6.42 \times 10^{-9},\ n = -12.28$	$R^2 = 0.387,\ \chi^2 = 0.0013$
$u_s = u_b(1-\varepsilon_g)^k(1+l\varepsilon_g^m)$	$u_b = 1.0;\ k = 1.0;\ l = -13.57;\ m = -7.28$	$R^2 = 0.570,\ \chi^2 = 0.0067$
$u_s = u_b(1-\varepsilon_g)/(\alpha - \beta\varepsilon_g^\gamma)$	$u_b = 1.0,\ \alpha = 1.0;\ \beta = -24.83;\ \gamma = 1/2$	$R^2 = 0.605,\ \chi^2 = 0.0061$
$u_s = a + b\varepsilon_g^c$	$a = 0.09;\ b = 2.38;\ c = 5.25$	$R^2 = 0.431,\ \chi^2 = 0.00124$
$u_s = u_b(1-\varepsilon_g^m)^n$	$u_b = 0.03,\ m = 1,\ n = -2.28;$	$R^2 = 0.43,\ \chi^2 = 0.00123$
$u_s = \lambda\varepsilon_g^\alpha D_R^\beta$ and $u_b = \lambda D_R^\beta,\ D_R = d_n/d_c$	$u_b = 0.73\text{–}1.80,\ \alpha = 3.05,\ \beta = 0.58,\ \lambda = 4.51$	$R^2 = 0.890,\ SE = 0.060$

*See the Nomenclature, Greek Letters, and Subscripts lists for abbreviations.

The flow behavior of the liquid can be represented by a power-law model as

$$\tau = K\dot{\gamma}^n \tag{4.86}$$

When the liquid is turbulent, the molecular kinematic viscosity, v_l, is negligible compared with the turbulent kinematic viscosity, v_t. Many reports on the gas hold-up profile in the literature indicate a value for N is 2 (Ueyama and Miyauchi, 1979). Therefore, a general solution of equation (4.83) is written as

$$u = \frac{\bar{\varepsilon}_g g}{v_{l,t}}\left(\frac{1}{8}\frac{r^4}{R^2} - \frac{1}{4}r^2\right) + A_1 \ln(r) + A_2 \tag{4.87}$$

The constants of integration A_1 and A_2 are determined using appropriate boundary conditions. Despite the fluctuations in the local liquid velocity, relatively stable velocity profiles can be found in bubble columns. In case of upward flow of bubble in a bubble column, the liquid rises with the bubbles in the center of the column and flows downward in the outer annular region. The liquid velocity is maximum at the center of the column and decreases in a radial direction. At high Reynolds numbers, a transition point at which the time average velocity is zero occurs at around $r/R = 0.7$, and large velocity, the magnitude of which is 50% to 100% of the velocity at the column axis, is found very close to the column wall (Kawase and Moo-Young, 1987). On the basis of these observations, they have formulated the following boundary conditions:

$$u = u_{l0} \quad at \quad r = 0 \tag{4.88}$$

$$u = 0 \quad at \quad r = 0.7R \tag{4.89}$$

$$u = (-0.5 \sim -1.0)u_{l0} \quad at \quad r = R \tag{4.90}$$

They also pointed out that profile in non-Newtonian fluids at high Reynolds numbers is similar to that in Newtonian fluids described earlier. The constants of equation (4.87) are determined by using only boundary condition equation (4.88), and the solution can be written as

$$u = \frac{\bar{\varepsilon}_g g}{v_{l,t}}\left(\frac{1}{8}\frac{r^4}{R^2} - \frac{1}{4}r^2\right) + u_{l0} \tag{4.91}$$

It is noted by them that condition equation (4.89) is better than condition equation (4.90) in order to represent a liquid velocity profile in the core region of the column, which may characterize the mixing in a whole bubble column. Substitution of equation (4.89) into equation (4.91) yields an expression for the average gas hold-up as

$$\bar{\varepsilon}_g = 43\frac{v_{l,t}u_{l0}}{d_c^2 g} \tag{4.92}$$

To calculate gas hold-up by using the above equation, expressions for the turbulent kinematic viscosity, v_t, and the liquid velocity at the column axis, u_o, must be known. The turbulent kinematic viscosity may vary with the distance from the wall. A characteristic kinematic viscosity can be defined in a bubble column on the basis of the Prandtl mixing length theory:

$$\bar{v}_{l,t} = \frac{\chi^2}{2} u_{l0} R \tag{4.93}$$

$$\chi = 0.4n \tag{4.94}$$

It is advised that the model can be applied in case of inverse bubbly flow system only after the measurements of liquid velocity profile in an inverse bubbly flow column and considering the proper boundary condition to validate the model.

Model of Axial Bubble Phase Hold-up Profiles

This model describes the changes in bubble phase hold-up with the column length where bubble is flowing inversely against its buoyancy by using a force balance method. This model is referred as the Zhou and Egiebor (1993) model. The model is applicable only for the bubble size range where the bubble Reynolds number range as $1 \ll Re_b < 500$, so the bubbles can be approximately regarded as spherical and moving straightly in an unpurified water. At this moderate Reynolds number range, viscous flow cannot satisfactorily describe the bubble motion. For simplicity, the potential flow condition is assumed, in which case the viscous forces are negligible in comparison with the inertia forces. The virtual mass of a bubble is considered, and the density of the gas is ignored compared with that of liquid. Thus, the virtual mass of a bubble, M_{vb}, can be expressed as

$$M_{vb} \approx \alpha M_l = \alpha(4\pi/3)R_b^3\rho_l \tag{4.95}$$

where α is the virtual mass coefficient, M_l is the liquid mass displaced by the bubble, R_b is the bubble radius, and ρ_l is the density of liquid. Therefore, the buoyancy, F_b, is reduced to

$$F_b = \beta(1-\alpha)gM_l) \tag{4.96}$$

where β is a coefficient related to the reduction of buoyancy by the virtual mass. The bubble phase hold-up can be defined as the volumetric ratio of a spherical gas bubble with the radius R_b, to a cubic liquid envelope surrounding the bubble with the lateral length $2l$, that is,

$$\varepsilon_g = (4\pi/3)R_b^3/((2l)^3 = 0.523R_b^3/l^3 \tag{4.97}$$

When a bubble is moving in the column, the bubble changes at a rate of dR_b/dH. This implies that the virtual mass, drag force, buoyancy, bubble phase hold-up, and superficial gas velocity also change with the height. In case of inverse flow of bubble, the change of bubble size only for those who are flow inverse against their buoyancy. In both cases (up and inverse), the bubble changes its size because of bubble–bubble interaction. In the bubbly flow regime at a given height of the column, there is approximately a linear relationship

$(\varepsilon_g = ku_{sg})$ between the superficial gas velocity and bubble phase hold-up. Based on these assumptions, the following force balance equation on a bubble motion can be written:

$$d(M_{vb}u_b)/dt = M_{vb}du_b/dt + u_b dM_{vb}/dt = \beta(1-\alpha)gM_l \tag{4.98}$$

$$dM_{vb} = 4\pi\alpha R_b^2 \rho_l dR_b \tag{4.99}$$

If the bubble velocity change along the column is unchanged, equation (4.98) becomes

$$dt = 3u_b\alpha dR_v / \left[\beta(1-\alpha)gR_b\right] \tag{4.100}$$

So, the distance traveled by the bubble in the column, *H*, can be given by

$$H = \int u_b dt = \int u_b \left\{3u_b\alpha / \left[\beta(1-\alpha)gR_b\right]\right\}dR_b$$
$$= \left\{3u_b^2\alpha / \left[\beta(1-\alpha)g\right]\right\}\ln\left[\beta(1-\alpha)gR_b\right] + C_o \tag{4.101}$$

where C_0 is the constant of integration. The constant can be evaluated from the boundary condition, $R_b = R_{b0}$ when $H = H_0$. The distance traveled by the bubble can then be expressed as

$$H = H_0 + \left\{3u_b^2\alpha / \left[\beta(1-\alpha)g\right]\right\}\ln\left[R_b / R_{b0}\right] \tag{4.102}$$

If the number of bubbles per unit volume, N_b, along the column height, is same, the following relationship applies

$$R_b = R_{b0}(\varepsilon_g / \varepsilon_{g0})^{1/3} \tag{4.103}$$

Substituting equation (4.103) into equation (4.102) gives the relationship between the bubble phase hold-up and column height as

$$\varepsilon_g = \varepsilon_{g0} \exp\left\{\left[(H - H_0)\beta(1-\alpha)g\right]/(u_b^2\alpha\right\} \tag{4.104}$$

This equation shows that there exists an exponential relationship between bubble phase hold-up and column height. At a given column height, the bubble phase hold-up is mainly related to bubble motion and initial bubble phase hold-up. The virtual mass coefficient α ranges from 0.5 to 11/16. The relationship between β and other operating parameters can be determined experimentally. The parameter β is a function of the column diameter and gas flow rates. To establish the detailed relationship between β and other operating parameters, a systematic experimental measurement is needed.

General Empirical Correlation Model of Bubble Phase Hold-up

A correlation has been developed by dimensional analysis to predict the bubble phase hold-up in terms of the physical, dynamic, and geometric variables of the system in the inverse bubble column. The bubble phase hold-up fraction can be expressed as a function of following parameters:

$$\varepsilon_g = f_2(Q_l, Q_g, P_s, \rho_l, \mu_l, \sigma_l, \rho_g, \mu_g, d_c, d_n, h_c, g) \tag{4.105}$$

In the inverse bubbly flow system, because the experiments have been carried out (Figure 1.4 in Chapter 1) with gas aspiration, it was found that Q_g was a direct function of Q_l, and the gas–liquid mixing height in the column, h_c, is dependent on the pressure developed in the gas–liquid separator (i.e., P_s). Furthermore, because gas was used as a secondary fluid, the effects of ρ_g and μ_g compared with ρ_l and μ_l were neglected. Hence, equation (4.105) may be rewritten as:

$$\varepsilon_g = f_2(Q_l, d_c, d_n, h_c, \rho_l, \mu_l, \sigma_l, g) \tag{4.106}$$

The effect of all of these parameters independently on bubble phase hold-up is very complicated; thus, dimensional analysis of the parameters has been incorporated to analyze the hold-up data. Combining the relevant dimensionless groups obtained by the dimensional analysis in a similar manner discussed earlier, the following functional relationship was obtained

$$\varepsilon_g = c_2 D_R^{b_6} H_R^{b_7} \mathrm{Re}_j^{b_8} We_j^{b_9} Fr_j^{b_{10}} \tag{4.107}$$

The multiple linear regression analysis of the experimental data yielded the following correlation for bubble phase hold-up a comprising gas–water system:

$$\varepsilon_g = 1.49 \times 10^{-3}\, D_R^{-1.185} H_R^{1.087} \mathrm{Re}_j^{-0.529} We_j^{1.453} Fr_j^{-0.964} \tag{4.108}$$

The correlation has been found to be satisfactory within the range of the different parameters: $0.34 < \varepsilon_g < 0.61$; $0.07 < D_R < 0.14$; $9.40 < H_R < 30.89$; $1.81 \times 10^5 < Re_j < 22.40 \times 10^5$; $5.85 \times 10^3 < We_j < 750 \times 10^3$; $0.14 \times 10^2 < Fr_j < 18.14 \times 10^2$.

Nomenclature

A Cross-sectional area (m²)
c Coefficient (-)
d Diameter (m)
D Dispersion coefficient (m²·s)
D_R Diameter ratio of nozzle to column (d_n/d_c) (-)
E Energy (N-m)
E_z Axial dispersion coefficient (m²/s)
f Function
Fr Froude number (-)
G Mass flux (kg/m²·s)
h Height
H Height of column (m)
H_R Ratio of gas–liquid mixing height to column diameter (h_c/d_c) (-)
I Ionic strength (kg ion/m³)
Mo Morton number (-)
n A parameter defined in equation (4.8)
n Flow behavior index
P Pressure (N/m²)
Q Volumetric flow rate (m³/s)
r Radial distance (m)
R Radius (m)
Re Reynolds number (-)

Re_g Gas Reynolds number (-)
Re_j Liquid Reynolds number based on nozzle diameter, $(\rho_l u_j d_n)/\mu_l$ (-)
S Slip ratio (-)
T Temperature (K)
u Velocity (m/s)
We Weber number (-)
We_g Weber number based on gas $(u_o^2 d_c \rho_g / \sigma_l)$ (-)
We_j Weber number $(u_j^2 d_c \rho_l / \sigma_l)$ (-)
x Mass quality (-)

Greek Letters

ε Bubble phase hold-up (-)
$\tilde{\varepsilon}$ Radial chordal average bubble phase hold-up (-)
$\bar{\varepsilon}$ Cross-sectional average bubble phase hold-up (-)
β Model parameter
$\varepsilon(r)$ Radial bubble phase hold-up (-)
α Constant (-)
ρ Density (kg/m³)
ξ Function
τ Shear stress (N/m²)
σ Surface tension (N/m)
μ Viscosity (N·s/m²)
β Volumetric quality (-)

Subscripts

0 Zero bubble phase hold-up condition
b Bubble
c Column, center, level
d Dissipation
d Drag
g Gas
gc Column center
j Jet
l Liquid
$l0$ Center line
lb Large bubble
lg Gas–liquid
ls Liquid–solid
m Mixing
m Motion
n Non-Newtonian
s Supplied, slip
sb Small bubble
sg Superficial gas
sl Superficial liquid
$trans$ Transition
w Wall

References

Akita, K., Yoshida, F., 1973. Gas holdup and volumetric mass transfer coefficient in bubble columns. Ind. Eng. Process Des. Dev. 12, 76–80.

Al-Masry, W.A., 2001. Gas holdup in circulating bubble columns with pseudo plastic liquids. Chem. Eng. Technol. 24, 71–76.

Andrew, S.P.S., 1960. Frothing in two-component liquid mixtures. International Symposium on Distillation, IChemE 73–78.

Babaei, R., Bonakdarpour, B., Ein-Mozaffari, F., 2015. The use of electrical resistance tomography for the characterization of gas holdup inside a bubble column bioreactor containing activated sludge. Chem. Eng. J. 268, 260–269.

Bando, Y., Uraishi, M., Nishimura, M., Hattori, M., Asada, T., 1988. Cocurrent inverse flow Bubble column with simultaneous gas–liquid injection nozzle. J. of Chem. Eng. Jpn 21, 607–612.

Bankoff, S.G., 1960. A variable density single fluid model for two phase flow with particular reference to steam–water flow. ASME J. Heat Trans. 82, 265–272.

Behringer, H., 1936. The flow of liquid–gas mixtures in vertical tubes. Zeit. Ges. Kalte-Ind. 43, 55–58.

Beinhauer, R., 1971. Dynamische Messungen des relativen Gasgehalts in BlasensJaulen mittels Absorption von Rontgenstrahlen. Dissertation TU Berlin.

Bin, A.K., 1993. Gas entrainment by plunging liquid jet. Chem. Eng. Sci. 48, 3585–3630.

Bin, A., 1974. Gas holdup in the system with air entraining liquid plunging jets. Ini. Chem. 4, 363–381.

Biń, A.K., Duczmal, B., Machniewski, P., 2001. Hydrodynamics and ozone mass transfer in a tall bubble column. Chem. Eng. Sci. 56 (21–22), 6233–6240.

Boyer, C., Duquenne, A., Wild, G., 2002. Measuring techniques in gas–liquid and gas–liquid–solid reactors. Chem. Eng. Sci. 57, 3185–3215.

Briens, C.L., Huynh, L.X., Large, J.F., Catros, A., Bernard, J.R., Bergougnou, M.A., 1992. Hydrodynamics and gas–liquid mass transfer in a downward venturi-bubble column combination. Chem. Eng. Sci. 47, 3549–3556.

Butterworth, D., Hewitt, G.F., 1977. Two-phase flow and heat transfer. Oxford University Press, Oxford.

Chawla, J.M., 1972. Frictional pressure drop on flow of liquid/gas mixtures in horizontal pipes. Chemie Ingenieur Technik 44, 58–63.

Chen, J.W., Gupta, P., Sujatha, D., Al-Dahhan, M.H., Dudukovic, M.P., Toseland, B.A., 1998. Gas holdup distribution in large diameter bubble columns. Flow Meas. Instrum. J. 9 (2), 91–101.

Chhabra, R.P., Richardson, J.F., 1999. Non-Newtonian Flow in the Process Industries: Fundamentals and Engineering Applications. Butterworth-Heinemann, Oxford.

Chilekar, V.P., Van der Schaaf, J., Kuster, B.F.M., Tinge, J.T., Schouten, J.C., 2010. Influence of elevated pressure and particle lyophobicity on hydrodynamics and gas–liquid mass transfer in slurry bubble columns. AIChE J. 56, 584–596.

Clark, N.N., Flemmer, R.L., 1985. Predicting the holdup in two-phase bubble upflow and inverse flow using the Zuber and Findlay Drift Flux Model. AIChE J. 31, 500–503.

Clark, N.N., Flemmer, R.L., 1986. The effect of varying gas voidage distributions on average holdup in vertical bubbly flow. Int. J. Multiphase Flow 12 (2), 299–302.

Clark, N.N., van Egmond, J.W., Nebiolo, E.P., 1990. The drift–flux model applied to bubble columns and low velocity flows. Int. J. Multiphase Flow 16, 261–279.

Daly, J.G., Patel, S.A., Bukur, D.B., 1992. Measurement of gas holdups and sauter mean bubble diameters in bubble column reactors by dynamics gas disengagement method. Chem. Eng. Sci. 47 (13–14), 3647-3654.

Das, S.K., Biswas, M.N., Mitra, A.K., 1992. Holdup for two phase flow gas-non-Newtonian liquid mixtures in horizontal and vertical pipes. Can. J. Chem. Eng. 70, 431–437.

Davis, W.J., 1963. The effect of Froude number in estimating vertical two phase gas–liquid friction losses. Bri. Chem. Eng. 8, 462–468.

Deckwer, W.D., 1992. Bubble Column Reactors. Wiley, New York.

Desouky, S.E.M., El-Emam, N.A., 1989. Program designs for pseudoplastic fluids. Oil Gas J. 87 (51), 48–51.

Eisenberg, F.G., Weinberger, C.B., 1979. Annular two phase flow of gases and non-Newtonian liquids. AIChE J. 25, 240–246.

Evans, G.M., Jameson, G.J., Rielly, C.D., 1996. Free jet expansion and gas entrainment characteristics of plunging liquid jet. Exp. Ther. Fluid Sci. 12, 142–149.

Forret, A., Schweitzer, J.M., Gauthier, T., Krishna, R., Schweich, D., 2003. Influence of scale on the hydrodynamics of bubble column reactors: an experimental study in columns of 0.1, 0. 4 and 1 m diameters. Chem. Eng. Sci. 58, 719–724.

Fransolet, E., Crine, M., Marchot, P., Toye, D., 2005. Analysis of gas holdup in bubble columns with non-Newtonian fluid using electrical resistance tomography and dynamic gas disengagement technique. Chem. Eng. Sci. 60, 6118–6123.

García-Abuín, A., Gómez-Díaz, D., Losada, M., Navaza, J.M., 18 June 2012. Bubble column gas–liquid interfacial area in a polymer + surfactant + water system. Chem. Eng. Sci. 75, 334–341.

Garcia-Calvo, E., Leton, P., 1994. Prediction of fluid dynamics and liquid mixing in bubble columns. Chem. Eng. Sci. 49 (21), 3643–3649.

Gestrich, W., Rahse, W., 1975. Der relative Gasgehalt von Blasenschichten (The relative gas content in bubble layers). Chem. -Ing. -Tech. 17 (1), 8–13.

Goda, H., Hibiki, T., Kim, S., Ishii, M., Uhle, J., 2003. Drift-flux model for downward two-phase flow. Int. J. Heat Mass Trans. 46, 4835–4844.

Godbole, S.P., Honath, M.F., Shah, Y.T., 1982. Holdup structure in highly viscous Newtonian and non-Newtonian liquids in bubble columns. Chem. Eng. Commun. 16, 119–134.

Govier, G.W., Aziz, K., 1972. The flow of complex mixture in pipes. Van Nostrand-Reinhold Co., New York.

Hammer, H., 1984. Frontiers in chemical reaction engineering. Halsted Press, New Delhi, India.

Hibiki, T., Ishii, M., 2003. One dimensional drift–flux model for two-phase flow in a large diameter pipe. Int. J. Heat Mass Transfer 46, 1773–1790.

Hikita, H., Asal, S., Tanigawa, K., Segawa, K., Kitao, M., 1980. Gas holdup in bubble column. Chem. Eng. J. 20, 59–67.

Hills, J.H., 1976. The operation of a bubble column at high throughputs I. Gas measurements. Chem. Eng. J. 12, 89–99.

Hirao, Y., Kawanishi, K., Tsuge, A., Kohriyama, T., 1986. Experimental study on drift flux correlation formulas for two-phase flow in large diameter tubes. Tokyo, Japan. Proceedings of Second International Topical Meeting on Nuclear Power Plant Thermal Hydraulics and Operations, pp. 1-88-1-94.

Hughmark, G.A., 1967. Holdup and mass transfer in bubble columns. Ind. Eng. Chem. Process Des. Dev. 6 (2), 218–220.

Huynh, X., Briens, C.L., Catros, A., Bernard, J.R., Bergougnou, M.A., 1991. Hydrodynamics and mass transfer in an upward venturi/bubble column combination. Can. J. Chem. Eng. 69, 711–722.

Idogawa, K., Ikeda, K., Fukuda, T., Morooka, S., 1985. Effect of gas and liquid properties on the behavior of bubbles in a bubble column under high pressure. Kag. Kog. Ronb. 11, 432–437.

Idogawa, K., Ikeda, K., Fukuda, T., Morooka, S., 1987. Effect of gas and liquid properties on the behavior of bubbles in a column under high pressure. Int. Chem. Eng. 27, 93–99.

Isbin, H.S., Sher, Neil C., Eddy, K.C., 1957. Void fractions in two-phase steam–water flow. AIChE J. 3, 136–142.

Ishii, M., 1977. "One-dimensional drift-flux model and constitutive equations for relative motions between phases in various two-phase flow regimes." ANL Report ANL-77–47.

Joshi, J.B., Veera, U.P., Parasad, C.V., Phanikkumar, D.V., Deshphande, N.S., Thakre, S.S., Thorad, B.N., 1998. Gas holdup structure in bubble column reactors. PINSA 64A (4), 441–567.

Kantarci, N., Borak, F., Ulgen, K.O., 2005. Review: bubble column reactors. Process Biochem. 40, 2268–2283.

Kato, Y., Nishiwaki, A., 1972. Longitudinal dispersion coefficient of a liquid in a bubble column. Int. Chem. Eng. 12, 182.

Kawanishi, K., Hirao, Y., Tsuge, A., 1990. An experimental study on drift flux parameters for two-phase flow in vertical round tubes. Nucl. Eng. Des. 120, 447–458.

Kawase, Y., Moo-Young, M., 1987. Theoretical prediction of gas hold-up in bubble columns with Newtonian and non-Newtonian fluids. Ind. Eng. Chem. Res. 26 (5), 933.

Kelkar, B.G., Godbole, S.P., Honath, M.F., Shah, Y.T., Carr, N.L., Deckwer, W.D., 1983. Effect of addition of alcohols on gas holdup and backmixing in bubble columns. AIChE J. 29, 361–369.

Kemoun, A., Rados, N., Li, F., Al-Dahhan, M.H., Dudukovic', M.P., Mills, P.L., Leib, T.M., Lerou, J.J., February 2001. Gas holdup in a trayed cold-flow bubble column. Chem. Eng. Sci. 56 (3), 1197–1205.

Khatib, Z., Richardson, J.F., 1984. Vertical co-current flow of air and shear thinning suspensions of kaolin. Chem. Eng. Res. Des. 62, 139–154.

Kojima, H., Jun, S., Hideyuki, S., 1997. Effect of pressure on volumetric mass transfer coefficient and gas holdup in bubble column. Chem. Eng. Sci. 52, 4111–4116.

Kölbel, H., Klötzer, D., Hammer, H., 1971. Zur Reaktionstechnik von Blasensäulen-Reaktoren mit suspendiertem Katalysator bei erhöhtem Druck. Chemie Ingenieur Technik 43, 103–111.

Krishna, R., Ellenberger, J., 1996. Gas holdup in bubble column reactors operating in the churn-turbulent flow regime. AIChE J. 42, 2627–2634.

Krishna, R., Urseanu, M.I., van Baten, J.M., Ellenberger, J., 1999. Influence of scale on the hydrodynamics of bubble columns operating in the churn-turbulent regime: experiments vs. eulerian simulations. Chem. Eng. Sci. 54, 4903–4911.

Kumagai, M., Endoh, K., 1982. Effects of kinematic viscosity and surface tension on gas entrainment rate of an impinging liquid jet. J. Chem. Eng. Jpn 15, 427–433.

Kumar, S.B., Moslemian, D., Duducovic, M.P., 1997. Gas-holdup measurements in bubble columns using computed tomography. AIChE J. 43, 1414–1425.

Lapidus, L., Elgin, J.C., 1957. Mechanics of vertical-moving Fluidized systems. AIChE J. 3, 63–68.

Lee, Y.H., Kim, Y.J., Kelkar, B.G., Weinberger, C.B., 1985. A Simple digital sensor for dynamic gas holdup measurements in bubble columns. Ind. Eng. Chem. Fund. 24, 105–107.

Lee, D.J., Luo, X., Fan, L.S., 1999. Gas disengagement technique in a slurry bubble column operated in the coalesced bubble regime. Chem. Eng. Sci. 54 (13–14), 2227–2236.

Levy, S., 1966. Prediction of two-phase annular flow with liquid entrainment. Intern. J. Heat Mass Transfer 9, 171–188.

Lin, T.J., Tsuchiya, K., Fan, L.S., 1998. Bubbly flow characteristics in bubble columns at elevated pressure and temperature. Am. Inst. Chem. Eng. J. 44, 545–560.

Lockhart, R.W., Martinelli, R.C., 1949. Proposed correlation of data for isothermal two-phase, two-component flow in pipes. Chem. Eng. Prog. 45, 39–48.

Luo, X., Lee, D.J., Lau, R., Yang, G., Fan, L.S., 1999. Maximum stable bubble size and gas holdup in high-pressure slurry bubble columns. AIChE J. 42, 665–680.

Majumder, S.K., 2008. Efficiency of non-reactive isothermal bubble column based on mass transfer. Asia-Pacific J. Chem. Eng. 3 (4), 440–451.

Majumder, S.K., Kundu, G., Mukherjee, D., 2006. Efficient dispersion in a modified two-phase non-newtonian inverse flow bubble column. Chem. Eng. Sci. 61 (20), 6753–6764.

Majumder, S.K., Kundu, G., Mukherjee, D., 2007. Energy efficiency of two phase mixing in a modified bubble column. Can. J. Chem. Eng. 85, 280–289.

Marchese, M.M., Uribe-Salas, A., Finch, J.A., 1992. Measurement of gas holdup in a three-phase concurrent inverse flow column. Chem. Eng. Sci. 47, 3475–4382.

Marucci, G., 1969. Theory of coalescence. Chem. Eng. Sci. 24, 975–985.

Maxwell, J.C., 1892, A Treatise of Electricity and Magnetism, third ed., Vol. 1 Oxford University Press, London, Part II, Chapter IX, 435449.

McKeogh, E.J., Ervine, D.A., 1981. Air entrainment rate and diffusion pattern of plunging liquid jet. Chem. Eng. Sci. 36, 1161–1172.

Merchuk, J.C., Stein, Y., 1981. Local hold-up and liquid velocity in air-lift reactors. AIChE J. 27, 377–388.

Montserrat, T., Garcia-Calvo, Eloy, 1996. Prediction of hydrodynamic behavior in bubble columns. J. Chem. Tech. Biotechnol. 66, 199–205.

Mudde, R.F., Saito, T., 2001. Hydrodynamical similarities between bubble column and bubbly pipe flow. J. Fluid Mech. 437, 203–228.

Nassos, G., Bankoff, S.G., 1967. Slip velocity ratios in an air-water systems under steady state and transient conditions. Chem. Eng. Sci. 22, 681–688.

Nguyen, V.T., Spedding, P.L., 1977a. Holdup in Two-Phase, Gas-Liquid Flow. I. Theoretical Aspects. Chem. Eng. Sci. 32, 1003–1014.

Nguyen, V.T., Spedding, P.L., 1977b. Holdup in two-phase, gas–liquid flow. II. Experimental results. Chem. Eng. Sci. 32, 1015–1021.

Ohkawa, A., Kusabiraki, D., Kawai, Y., Sakai, N., 1987. Flow characteristics of an air-entrainment type aerator having a long downcomer. Chem. Eng. Sci. 42, 2788–2790.

Olivieri, G., Elena Russo, M., Simeone, M., Marzocchella, A., Salatino, P., 2011. Effects of viscosity and relaxation time on the hydrodynamics of gas–liquid systems. Chem. Eng. Sci. 66, 3392–3399.

Pakzad, L., Ein-Mozaffari, F., Chan, P., 2008. Using electrical resistance tomography and computational fluid dynamics modeling to study the formation of cavern in the mixing of pseudoplastic fluids possessing yield stress. Chem. Eng. Sci. 63 (9), 2508–2522.

Petrick, M., Kudirka, A.A., "On the relationship between the phase distributions and relative velocities in two-phase flows," Proceedings of Third International Heat Transfer Conference, AIChE, 4, pp. 184–191 (1956).

Pohorecki, R., Moniuk, W., Zdrójkowski, A., Bielski, P., February 2001. Hydrodynamics of a pilot plant bubble column under elevated temperature and pressure. Chem. Eng. Sci. 56 (3), 1167–1174.

Radhakrishnan, V.R., Mitra, A.K., 1984. Pressure drop, holdup and interfacial area in vertical two-phase flow of multi-jet ejector induced dispersion. Can. J. Chem. Eng. 62, 170–178.

Reilly, J.G., Scott, D.S., Bruijn, T., Jain, A., Diskorz, J., 1986. Correlation for gas holdup in turbulent coalescing bubble columns. Can. J. Chem. Eng. 64, 705–717.

Ruzicka, M.C., Zahradnik, J., Drahoš, J., Thomas, N.H., 2001. Homogeneous–heterogeneous regime transition in bubble columns. Chem. Eng. Sci. 55, 4609–4626.

Şal, S., Güla, Ö.F., Özdemirb, M., 2013. The effect of sparger geometry on gas holdup and regime transition points in a bubble column equipped with perforated plate spargers. Chem. Eng. Process. 70, 259–266.

Sanyal, J., Vasquez, S., Roy, S., 1999. Numerical simulation of gas–liquid dynamics in cylindrical bubble column reactors. Chem. Eng. Sci. 54, 5071–5083.

Schumpe, A., Deckwer, W.D., 1982. Gas holdups, specific interfacial areas, and mass transfer coefficients of aerated carboxymethyl cellulose solutions in a bubble column. Ind. Eng. Chem. Process Des. Dev. 21, 706–711.

Scott, D.S., 1963. Properties of cocurrent gas-liquid flow. Adv. Chem. Eng. 4, 199–277.

Sivaiah, M., Parmar, R., Majumder, S.K., 2012. Gas entrainment and holdup characteristics in a modified gas-liquid-solid down flow three-phase contactor. Powder Technol. 217, 451–461.

Sotelo, J.L., Benitez, F.J., Beltran-Heredia, J., Rodriguez, C., 1994. Gas holdup and mass transfer coefficients in bubble columns. 1. porous glass-plate diffusers. Int. Chem. Eng. 34, 82–91.

Sriram, K., Mann, R., 1977. Dynamic gas disengagement: a new technique for assessing the behavior of bubble columns. Chem. Eng. Sci. 32, 571–580.

Syeda, S.R., Reza Md., J., 2011. Effect of surface tension gradient on gas hold-up enhancement in aqueous solutions of electrolytes. Chem. Eng. Res. Des. 89 (12), 2552–2559.

Syeda, S.R., Afacan, A., Chuang, K.T., 2002. Prediction of gas hold-up in a bubble column filled with pure and binary liquids. Can. J. Chem. Eng. 80 (1), 44–50.

Tang, C., Heindel, T.J., 2006. Estimating gas holdup via pressure difference measurements in a concurrent bubble column. Int. J. Multiphase Flow 32 (7), 850–863.

Taylor, G.I., (1953) Dispersion of soluble matter in solvent flowing slowly through a tube. Proceedings of the Royal Society of London A 219, pp. 186–203.

Ueyama, K., Miyauchi, T., 1979. Properties and recirculating turbulent two phase flow in gas bubble columns. A.I. Ch.E. J. 25, 258–266.

Urseanu, M.I., Guit, R.P.M., Stankiewicz, A., van Kranenburg, G., Lommen, J.H.G.M., 2003. Influence of operating pressure on the gas hold-up in bubble columns for high viscous media. Chem. Eng. Sci. 58, 697–704.

Vial, C., Poncin, S., Wild, G., Midoux, N., 2001. A Simple method for regime identification and flow characterization in bubble columns and airlift reactors. Chem. Eng. Proc. 40, 135–151.

Wallis, G.B., 1969. One Dimensional Two Phase Flow. McGraw-Hill, New York.

Wang, T., Wang, J., 2007. Numerical simulations of gas–liquid mass transfer in bubble columns with a CFD–PBM coupled model. Chem. Eng. Sci. 62 (24), 7107–7118.

Wilkinson, P.M., Haringa, H., van Dierendonck, L.L., 1994. Mass transfer and bubble size in a bubble column under pressure. Chem. Eng. Sci. 49 (9), 1417–1427.

Wilkinson, P.M., Spek, A.P., Van Dierendonck, L.L., 1992. Design parameters estimation for scaleup of high-pressure bubble columns. AIChE J. 38 (4), 54–544.

Wu, Y., Al-Dahhan, M.H., 2001. Prediction of axial liquid velocity profile in bubble columns. Chem. Eng. Sci. 56 (3), 1127–1130.

Wu, Y., Cheng Ong, B., Al-Dahhan, M.H., 2001. Predictions of radial gas holdup profiles in bubble column reactors. Chem. Eng. Sci. 56 (3), 1207–1210.

Yamagiwa, K., Ohkawa, A., 1989. A technique for measuring gas holdup in a inverse flow bubble column with gas entrainment by a liquid jet. J. Ferment. Bioeng. 68, 160–162.

Yamagiwa, K., Kusabiraki, D., Ohkawa, A., 1990. Gas holdup and gas entrainment rate in downflow bubble column with gas entrainment by a liquid jet operating at high liquid throughput. J. Chem. Eng. Jpn 23, 343–348.

Yifeng, S., Yifei, W., Qinghua, Z., Jian-hui, L., Guangsuo, Y., Xin, G., Zunhong, Yu., 2008. Influence of liquid properties on flow regime and back mixing in a special bubble column. Chem. Eng. Process. 47 (12), 2296–2302.

Zahradnik, J., Fialova, M., Kastanek, F., Green, K.D., Thomas, N.H., 1995. The effect of electrolytes on bubble coalescence and gas hold-up in bubble column reactors. Chem. Eng. Res. Des. 73A, 341–346.

Zahradnik, J., Fialova, M., Linek, V., Sinkule, J., Reznickova, J., Kastanek, F., 1997. Dispersion efficiency of ejector type gas distributor in different operating modes. Chem. Eng. Sci. 52, 4499–4510.

Zhou, Z.A., Egiebor, N.O., 1993. Prediction of axial gas holdup profiles in flotation columns. Miner. Eng. 6 (3), 307–312.

Zuber, N., Findlay, J.A., 1965. Average volume concentration in two-phase flow systems. J. Heat Transfer, Ser. C. 87, 453–468.

5

Pressure Drop in Bubbly Flow

Pressure Drop in Bubbly Flow

Pressure drop in bubbly flow is one of the main design variables that governs the flow energy required to transport multiphase fluids in a certain flow system. Bubbly flow in a pipe is rendered a more complex phenomenon when the movement of bubbles in bubbly flow constitutes the several change in flow behavior to influence the transport processes in the pipe either the direction of flow in vertical or horizontal. The pressure losses suffered in bubbly flow through pipes are influenced by a number of parameters, and no generalized method is available to calculate them accurately. The usual practice is to multiply the single-phase pressure losses by a factor known as the two-phase multiplier, which is empirically correlated. Pressure drop or head loss occurs in all piping systems because of elevation changes, turbulence caused by abrupt changes in direction, and friction within the pipe and fittings.

Why Knowledge on Pressure Drop Is Required

Pressure drop is an important design parameter in gas–liquid two-phase bubbly flow devices. The knowledge of pressure drop gives the pattern of energy dissipation, helps in modeling the system, and forms the basis of assessment of performance of the equipment. The influence of pressure seems to be higher on the heterogeneous regime than in the homogeneous regime. This tendency could be linked to a decrease of the bubble size. The bubble diameter decreases when the pressure increases. A pressure increase results in an increase in gas hold-up. In the heterogeneous regime, this positive effect of pressure on gas hold-up is commonly observed in the literature (Pjontek et al., 2014). However, in the homogeneous regime or for low superficial gas velocity (typically below 0.05 m/s), an appreciable increase is reported in the literature (Shaikh and Al-Dahhan, 2005), some observe an increase less than in the heterogeneous regime (Kang et al., 2000; Hashemi et al., 2009), and some observe no effect at all (Kemoun et al., 2001; Chilekar et al., 2010; Kumar et al., 2012). The overall mass transfer coefficient increases while increasing pressure in bubbly flow device (Wilkinson et al., 1994; Letzel et al., 1999; Lau et al., 2004; Jin et al., 2014). The mass transfer increases with the gas density by increasing pressure (Jordan and Schumpe, 2001). Wilkinson et al. (1994) explained that in the homogeneous regime, pressure would have a negligible effect because breakage and coalescence are not the governing mechanisms of the bubble column, which is instead the formation of the primary bubble. However,

Hydrodynamics and Transport Processes of Inverse Bubbly Flow. http://dx.doi.org/10.1016/B978-0-12-803287-9.00005-9

Maalej et al. (2003) reported a decrease of the overall volumetric mass transfer coefficient with pressure at constant mass flow rate. In fact, fixing a mass flow rate and increasing pressure result in a decrease of the superficial gas velocity because of the increase of the gas density. This results in a decrease of the overall volumetric mass transfer coefficient. A pressure increase leads to an increase of the interfacial area (Han and Al-Dahhan, 2007). Han and Al-Dahhan (2007) observed the decrease of individual mass transfer coefficient of liquid (k_l) while increasing the pressure. The mass transfer coefficient (k_l) decreases by 20% between 0.1 and 0.4 MPa but remains constant between 0.4 and 1 MPa. The authors explained this trend with the Higbie's theory and reported that the increase in pressure reduces the size of bubbles. The authors argued that the small bubbles have a slower slip velocity, so the contact time would decrease. This would mean that the k_l decreases when the bubble size decreases. This is contradictory with Kulkarni (2007) and his interpretation of the theory of Calderbank, which affirms exactly the opposite. In fact, if the slip velocity decreases when the diameter decreases, it is difficult to assess the evolution of the ratio d_b/u_s when both terms decrease. Yang et al. (2001) reported no influence of pressure on the mass transfer coefficient k_l at different temperatures (from 293 to 473 K) in the case of a quite viscous liquid paraffin ($\mu_l = 0.26\text{--}3$ Pa s). The authors explained that the gas solubility increases with pressure. This increasing solubility results in more gas dissolved in the liquid, which leads to a decrease in viscosity. They therefore concluded that the k_l decreases when the bubble size decreases, but in parallel, it is favored by the decrease in viscosity reduction that would increase the molecular diffusivity. Ultimately, the two effects cancel each other out and result in a constant k_l in this pressure range. A decrease of the liquid axial dispersion coefficient with an increase of pressure in a wide range of superficial gas and liquid velocity is reported by Yang and Fan (2003). The effect is higher at high superficial gas velocity, that the effect of pressure decreases when increasing pressure. The authors explained it by using fluid circulation model proposed by Joshi and Sharma (1979) when the turbulence associated with liquid circulation results the dispersion in case of heterogeneous regime. However, the effect of pressure would be linked to several contradictory effects on fluid circulation velocity reported by Therning and Rasmuson (2001) and Chilekar et al. (2010). In fact, liquid recirculation increases when increasing pressure. Houzelot et al. (1983) observed no effect of pressure on the liquid axial dispersion coefficient for pressures between 0.1 and 0.3 MPa in the homogeneous regime. Holcombe et al. (1983) also reported no effect of pressure in continuous mode at higher superficial gas velocities and for a limited range of pressure (0.3–0.71 MPa). This disagrees with Wilkinson et al. (1993), who observed an increase of the liquid axial dispersion coefficient with pressure between 0.1 and 1.5 MPa for the same nitrogen–water system but with a higher column diameter, although they observed an increase of the liquid axial dispersion coefficient with pressure despite the decrease of bubble size. The influence of pressure on liquid axial dispersion coefficient also depends on liquid phase properties. For the more viscous fluids, mixing may be dramatically affected by pressure because liquid recirculation turbulence is lower for these systems, and mixing is more dependent on bubble-induced turbulence.

Models to Analyze Pressure Drop

In the bubble column, estimation of bubbly flow pressure drops is of paramount importance to the design and optimization of the process in the bubble column. The optimal use of the bubbly flow pressure drop to obtain the maximum heat and mass transfer performance is one of the primary design goals. If the column is inaccurately designed with a bubbly flow pressure drop only half of the real value, then the system efficiency will suffer accordingly from the larger than expected fall in saturation temperature and pressure through the reactor. The bubbly flow models are divided into two categories, homogenous flow models and separated flow models. The homogenous models are simpler, assuming that the liquid and bubble phases move at the same velocity. In the separate flow models, it is assumed that different equations are valid for each phase, indicating the different velocities of the two phases. These equations are one form of empirical correlations. The separate flow models are considered to give a more reliable result in the pressure drop calculations. However, the empirical correlations are usually correlated inside laboratories at a small scales compared with the size of actual unit for specific application and do not necessarily give an accurate result when used in simulations.

Three different approaches can be used to model pressure drop in bubbly flow: semi-analytic models, numerical models, and empirical models. The empirical models are simplified relationships based on a series of experiments. Semianalytic methods or mechanistic methods approximate a physical phenomenon by taking into consideration the most important processes and neglecting other less important effects that add to uncertainty. The third one, the numerical method, is a derivation from numerical analysis, such as distribution of the phases, dynamic flow regime transition, and turbulence effects. This method is a mixture of the first two types. Some of the input values used in the models are from empirical correlations, and the model is fitted to measured data by using the semi-analytic method. One of many obstacles in bubbly flow modeling in the bubble column is limited knowledge of the subject. Therefore, all of the bubbly flow models depend on empirical correlations to some extent.

Homogeneous Flow Model

This model, which considers the fluid mixture as a homogeneous mixture of gas bubble and liquid that obeys the conventional design equations for single-phase fluids, is characterized by suitably average properties of the liquid and gas phase. Pressure gradients in vertical two-phase flow can be evaluated by using either an energy balance or momentum balance approach. The total pressure drop is composed of three factors:

- **Hydrostatic:** Pressure changes caused by difference in height. The hydrostatic head term is based on the true average density of the mixture in the pipe, which is evaluated with a precise knowledge of the hold-up of liquid and gas in the pipe section.

- **Momentum:** The different phases do not have the same velocity, which causes a change in momentum and thus the pressure.
- **Friction:** This is the friction between the two phases and between the phases and the column wall.

The total pressure drop for the bubble–liquid flow across a section of a vertical column of height Δz can be expressed as

$$\Delta P_t = \Delta P_s + \Delta P_{mom} + \Delta P_f \tag{5.1}$$

Hydrostatic Pressure

The column hydrostatic pressure, ΔP_s depends on the bubble–liquid mixture density, which in turn is a function of the composition of the mixture. There are two methods for calculating the hydrostatic component. In the first method, the two-phase bubble–liquid mixture is assumed to be homogeneous, and its composition is assumed to be the same as that at the inlet to the system. According to this method,

$$\Delta P_s = g\Delta z \left(\frac{\dot{m}_l + \dot{m}_g}{\dot{Q}_l + \dot{Q}_g} \right) \tag{5.2}$$

In the second method, the actual density of the bubble–liquid system is used for the calculation of the hydrostatic pressure, and this gives the relation

$$\Delta P_s = \rho_{tp} g H \sin\theta \tag{5.3}$$

where H is the vertical height of the bubble–liquid mixture and θ is the angle with respect to horizontal. $Sin\theta$ is equal to 1 for a vertical bubble column. The homogenous mixture density in the equation above is usually calculated as follows:

$$\frac{1}{\rho_{tp}} = \frac{x}{\rho_g} + \frac{1-x}{\rho_l} \tag{5.4}$$

The homogeneous mixture density ρ_{tp} can also be calculated as follows:

$$\rho_{tp} = \rho_l(1-\varepsilon_g) + \rho_g\varepsilon_g \tag{5.5}$$

The homogeneous bubble phase fraction (ε_b) can be calculated from the quality (x) as

$$\varepsilon_g = \left(1 + \left(\frac{u_g(1-x)}{u_l x} \times \frac{(1-x)}{x} \frac{\rho_g}{\rho_l}\right)\right)^{-1} \tag{5.6}$$

where x is the mass quality of gas, which is defined as the mass flow rate of gas divided by the total mass flow rate, and u_g/u_l is the velocity ratio or slip ratio (S), which is equal to

1.0 for a homogeneous flow. Dukler et al. (1964) proposed the two-phase mixture density quality weighted two-phase mixture density at no slip conditions as

$$\rho_{tp} = \rho_l \lambda + \rho_g (1 - \lambda)$$
(5.7)

$$\lambda = \left[\frac{\dot{Q}_l}{\dot{Q}_l + \dot{Q}_g} \right] = \frac{1}{1 + \left(\dfrac{x}{1-x} \right) \dfrac{\rho_l}{\rho_g}}$$
(5.8)

Later all researchers followed the same equation to calculate the mixture density for all of their research.

Momentum Pressure
The momentum pressure gradient is effectively the difference in momentum flux between the inlet and the outlet of the conduit. Andeen and Griffith (1968) reported that its accurate estimation might be best achieved by assuming a homogeneous two-phase mixture:

$$\Delta P_{mom} = \dot{M}_t^2 \int_{z_1}^{z_2} \left[\frac{x}{\rho_g} + \frac{1-x}{\rho_l} \right] dz$$
(5.9)

where $\dot{M}_t$ is the total mass flux and z_1 and z_2 are the axial positions. However, the momentum pressure losses, ΔP_{mom}, can often be neglected in absence of interphase mass transfer or in absence of gas expansion caused by large pressure differences. Hughmark and Pressburg (1961) reported that the momentum component is insignificant compared with the total pressure drop in a tube of uniform cross-section.

Frictional Pressure
For single-phase flow, Fanning's friction factor, f_{sp}, is related to the frictional pressure drop by the following equation:

$$\Delta P_{f,sp} = \frac{2 f_{sp} \rho_{sp} u_{sp}^2 \Delta z}{d_c}$$
(5.10)

In the same fashion, a two-phase friction factor based on the homogeneous mixture velocity in the column can be used and the frictional pressure drop can be expressed as

$$\Delta P_{f,tp} = \frac{2 f_{tp,hm} \rho_{tp,hm} u_{tp,hm}^2 \Delta z}{d_c}$$
(5.11)

Friction Factor in Homogeneous Two-Phase Flow
The friction factor in homogeneous mixture flow can be expressed as

$$f_{tp,hm} = \frac{16}{\mathrm{Re}_{tp,hm}} \quad \text{for laminar flow}$$
(5.12)

$$f_{tp,hm} = \frac{0.079}{\mathrm{Re}_{tp,hm}^{0.25}} \quad \text{for turbulent flow} \tag{5.13}$$

For transition flow, the friction factor can be deduced from the generalized pressure loss equation as (Desouky and El-Emam, 1989):

$$f_{tp,hm} = 0.125\left[\left(0.0112 + \mathrm{Re}_{tp,hm}^{-0.3185}\right)\right] \quad \text{for transition flow} \tag{5.14}$$

where $Re_{tp,hm}$ is the Reynolds number based on homogeneous mixture flow, which is defined as

$$\mathrm{Re}_{tp,hm} = \frac{\rho_{tp} u_{tp} d_c}{\mu_{tp}} \tag{5.15}$$

The viscosity of the homogeneous mixture for calculating the Reynolds number is generally taken as quality average viscosity. Cicchitti et al. (1960) developed a homogeneous two-phase correlation for upward vertical tube for adiabatic and nonadiabatic flows by defining two-phase viscosity as

$$\mu_{tp} = (1-x)\mu_l + x\mu_g \tag{5.16}$$

This equation is valid for mass velocities greater than 2000 kg/m²s in the case of frictional pressure drop calculations and for mass velocities less than 2000 kg/m²s and $\rho l/\rho g$ less than 10 for gravitational pressure drop calculations. This equation should be used for high mass velocities and at high reduced pressure. Awad and Muzychka (2008) used an analogy between thermal conductivity in porous media and viscosity in two-phase flow to develop four new two phase viscosity definitions within a range of mass quality 0 to 1. Different definitions of two-phase viscosity are shown in Table 5.1. Later Beggs and Brill (1973) developed a correlation for the no-slip condition of homogeneous flow mixture to calculate the friction factor at different inclination. The no-slip Reynolds number will approach the Reynolds number for gas or liquid as the input liquid content approaches zero or one, respectively. The correlation can be expressed as

$$f_{tp,hm} = \left[4\log\left(\frac{\mathrm{Re}_{tp,ns}}{4.5223\log(\mathrm{Re}_{tp,ns}) - 3.8215}\right)\right]^{-2} \tag{5.17}$$

Dukler et al. (1964) suggested a correlation to calculate the two-phase friction factor by using an analogy between single phase and two-phase flows for the no-slip condition in homogeneous two-phase flow as

$$f_{tp,hm} = a + \frac{b}{\mathrm{Re}_{tp}^c} \tag{5.18}$$

The coefficients a, b, and c depends on the experimental conditions. They obtained the values for a, b, and c, respectively, for upward bubbly flow as 0.0014, 0.125, and 0.32. The

Table 5.1   Different Definitions of Two-Phase Viscosity

Author	Viscosity
Arrhenius (1887)	$\mu_{tp} = \mu_l^{1-\varepsilon_g}\mu_g^{\varepsilon_g}$
Bingham (1906)	$\dfrac{1}{\mu_{tp}} = \dfrac{\varepsilon_g}{\mu_g} + \dfrac{1-\varepsilon_g}{\mu_l}$
Einstein (1909)	$\mu_{tp} = \mu_l(1+k\varepsilon_l)$
Hatschek (1928)	$\mu_{tp} = \mu_l\left(1-\sqrt[3]{\varepsilon_l}\right)^{-1}$
Taylor (1932)	$\mu_{tp} = \mu_l\left[1+2.5\left(\dfrac{\mu_g+0.4\mu_l}{\mu_g+\mu_l}\right)\varepsilon_l\right]$
Richardson (1933)	$\mu_{tp} = \mu_l\exp(k\varepsilon_l)$
McAdams et al. (1942)	$\dfrac{1}{\mu_{tp}} = \dfrac{x}{\mu_g} + \dfrac{1-x}{\mu_l}$
Davidson et al. (1943)	$\mu_{tp} = \mu_l\left[1+x\left(\dfrac{\rho_l}{\rho_g}-1\right)\right]$
Vermeulen et al. (1955)	$\mu_{tp} = \dfrac{\mu_l}{\varepsilon_g}\left[1+\left(\dfrac{1.5\mu_g(1-\varepsilon_g)}{\mu_l+\mu_g}\right)\right]$
Akers et al. (1959)	$\mu_{tp} = \dfrac{\mu_l}{1-x+x\left(\dfrac{v_g}{v_l}\right)^{0.5}}$; v_g and v_l are specific volume of gas and liquid respectively.
Hoogendoorn (1959)	$\mu_{tp} = \mu_g^{1-\varepsilon_l}\mu_l^{\varepsilon_l}$; ε_l is the fractional liquid holdup
Cicchitti et al. (1960)	$\mu_{tp} = x\mu_g+(1-x)\mu_l$
Bankoff (1960)	$\mu_{tp} = \varepsilon_l\mu_l+(1-\varepsilon_l)\mu_g$
Owens (1961)	$\mu_{tp} = \mu_l$
Dukler et al. (1964)	Definition 1: $\mu_{tp} = \mu_l\lambda+\mu_g(1-\lambda)$ where $\lambda = \left[\dfrac{\dot{Q_l}}{\dot{Q_l}+\dot{Q_g}}\right] = \left[1+\left(\dfrac{x}{1-x}\right)\dfrac{\rho_l}{\rho_g}\right]^{-1}$, Definition 2: $\mu_{tp} = \dfrac{xv_g\mu_g+(1-x)v_l\mu_l}{xv_g+(1-x)v_l}$
Cengel (1967)	$\mu_{tp} = \mu_l(1+2.5\varepsilon_l-11.01\varepsilon_l^2+52.62\varepsilon_l^3)$
Soot (1971)	$\mu_{tp} = \mu_l\left[1+\varepsilon_l\left(\dfrac{\rho_g}{\rho_l}-1\right)\right]^{1-1/n}$
Oliemans (1976)	$\mu_{tp} = \mu_l(1-\varepsilon_g)+\mu_g\varepsilon_g$
Oglesby (1979)	$\mu_{tp} = \mu_l\exp\left[\left(\dfrac{\mu_g+0.4\mu_l}{\mu_g+\mu_l}\right)\left(\varepsilon_l+\varepsilon_l^{1.667}+\varepsilon_l^{3.66}\right)\right]$
Beattie and Whalley (1982)	$\mu_{tp} = \mu_l(1-\beta)(1+2.5\beta)+\mu_g\beta$ where $\beta = \dfrac{\rho_l x}{\rho_l x+\rho_g(1-x)}$, β is called volumetric flow quality.
Olujic (1985)	$\mu_{tp} = \left[1-x\left(1-\dfrac{\mu_l}{\mu_g}\right)\right]^{-1}$

Table 5.1   Different Definitions of Two-Phase Viscosity (*cont.*)

Author	Viscosity
Lin et al. (1991)	$\mu_{tp} = \dfrac{\mu_g \mu_l}{\mu_g + x^{1.4}(\mu_l - \mu_g)}$
Fourar and Bories (1995)	$\mu_{tp} = \rho_m \left(\sqrt{x \dfrac{\mu_g}{\rho_g}} + \sqrt{(1-x)\dfrac{\mu_l}{\rho_l}} \right)^2$
Garcia et al. (2003)	$\mu_{tp} = \mu_l \left(\dfrac{\rho_{tp}}{\rho_l} \right) = \dfrac{\mu_l \rho_g}{\rho_l x + \rho_g (1-x)}$
Awad and Muzychka (2008)	Definition 1: $\mu_{tp} = \mu_l \left(\dfrac{2\mu_l + \mu_g - 2(\mu_l - \mu_g)x}{2\mu_l + \mu_g + 2(\mu_l - \mu_g)x} \right)$
	Definition 2: $\mu_{tp} = \mu_g \left(\dfrac{2\mu_g + \mu_l - 2(\mu_g - \mu_l)(1-x)}{2\mu_g + \mu_l + 2(\mu_g - \mu_l)(1-x)} \right)$
	Definition 3: $\mu_{tp} = \dfrac{1}{4}\left[\omega + \sqrt{\omega^2 + 8\mu_l \mu_g} \right]$ where $\omega = (3x-1)\mu_g + (3(1-x)-1)\mu_l$
	Definition 4: $\mu_{tp} = \dfrac{1}{2}\left[\mu_l \left(\dfrac{2\mu_l + \mu_g - 2(\mu_l - \mu_g)x}{2\mu_l + \mu_g + 2(\mu_l - \mu_g)x} \right) + \mu_g \left(\dfrac{2\mu_g + \mu_l - 2(\mu_g - \mu_l)(1-x)}{2\mu_g + \mu_l + 2(\mu_g - \mu_l)(1-x)} \right) \right]$
Muzychka et al. (2011)	Definition 1: $\mu_{tp} = \left[\mu_l \left(\dfrac{2\mu_l + \mu_g - 2(\mu_l - \mu_g)x}{2\mu_l + \mu_g + (\mu_l - \mu_g)x} \right) \times \mu_g \left(\dfrac{2\mu_g + \mu_l - 2(\mu_g - \mu_l)(1-x)}{2\mu_g + \mu_l + (\mu_g - \mu_l)(1-x)} \right) \right]^{0.5}$
	Definition 2: $\mu_{tp} = \left[2\mu_l \left(\dfrac{2\mu_l + \mu_g - 2(\mu_l - \mu_g)x}{2\mu_l + \mu_g + (\mu_l - \mu_g)x} \right) \times \mu_g \left(\dfrac{2\mu_g + \mu_l - 2(\mu_g - \mu_l)(1-x)}{2\mu_g + \mu_l + (\mu_g - \mu_l)(1-x)} \right) \right] \Bigg/ \left[\mu_l \left(\dfrac{2\mu_l + \mu_g - 2(\mu_l - \mu_g)x}{2\mu_l + \mu_g + (\mu_l - \mu_g)x} \right) + \mu_g \left(\dfrac{2\mu_g + \mu_l - 2(\mu_g - \mu_l)(1-x)}{2\mu_g + \mu_l + (\mu_g - \mu_l)(1-x)} \right) \right]$

correlation can be applied for inverse homogeneous bubbly flow based on the velocity of bubbly flow relative to bubble rise velocity.

Separated Flow Models

In the separated flow model, the two phases are artificially considered to be separated into two streams, each flowing in its own column. The areas of the two columns are proportional to the bubble phase hold-up. The bubble phase hold-up can be used those are given in previous chapter. The restriction on equal phase velocities, that is, slip ratio $(S) = 1$, is relaxed when the densities between the phases are quite different. The buoyancy effects tend to induce a drift velocity of the lighter phase in the heavier phase because of the density difference. The momentum exchange between the phases and the column separately with different velocities (e.g., bubble and liquid velocities). In this model, the total pressure drops are also considered to be the sum of the three contributions:

the hydrostatic pressure drop (ΔP_s), the momentum pressure drop (ΔP_{mom}), and the frictional pressure drop (ΔP_f) as

$$\Delta P_t = \Delta P_f + \Delta P_s + \Delta P_{mom} \tag{5.19}$$

The hydrostatic pressure drop is given by

$$\Delta P_s = \rho_{tp} g H \sin\theta \tag{5.20}$$

The momentum pressure drop reflects the change in kinetic energy of the flow and is for the case of separated flow model given by

$$\Delta P_{mom} = \dot{m}_t^2 \left[\left\{ \frac{(1-x)^2}{\rho_l(1-\varepsilon_g)} + \frac{x^2}{\rho_g \varepsilon_g} \right\}_{out} - \left\{ \frac{(1-x)^2}{\rho_l(1-\varepsilon_g)} + \frac{x^2}{\rho_g \varepsilon_g} \right\}_{in} \right] \tag{5.21}$$

where $\dot{m}_t$ is the total mass velocity of bubble phase and liquid and x is thye bubble phase quality. The momentum pressure drop can be calculated by the input of the inlet and outlet bubble phase qualities. Because the density of the bubble phase is less than the liquid phase, the momentum pressure drop results in a lower pressure at the exit than at the inlet.

Based on the separated flow model, the basic equations for the frictional pressure drop in the bubbly flow are not dependent on the particular flow configuration adopted. It is assumed that the velocities of each phase are constant. The separated flow models can be categorized as asymptotic, empirical, mechanistic, and numerical to predict the frictional pressure drop. Some common models that can be used to predict the frictional pressure drop are given as follows.

Asymptotic Models

The asymptotic model relates the two-phase frictional pressure gradient to the single-phase frictional pressure gradients of the liquid and gas flowing alone.

MARTINELLI ET AL. (1944) MODEL

A method of using a two-phase frictional pressure drop multiplier is a very popular method of developing a separated flow model pressure drop correlation. This type of analysis was found to be appealing for most researchers because single-phase flow techniques and results are analogically related to two-phase flows by this method. This instance has a benefit of avoiding ambiguity over which physical property of the phases to use, such as which viscosity of either of the phases to use during calculation of two-phase pressure drop. There are two ways of modeling the two-phase friction multiplier. The first one is assuming all the flow to be as one of the single phases such as all flow as liquid or all flow as gas. The implication here is to use the total mass flux (the sum of the mass fluxes for each phase) instead of the mass flow for each phase. The concept of two-phase friction multiplier was first introduced by Martinelli et al. (1944) and is represented as

$$\phi_{l,M}^2 = \left(\frac{\Delta p_f}{\Delta z}\right)_{tp} \bigg/ \left(\frac{\Delta p_f}{\Delta z}\right)_l \qquad (5.22)$$

$$\phi_{g,M}^2 = \left(\frac{\Delta p_f}{\Delta z}\right)_{tp} \bigg/ \left(\frac{\Delta p_f}{\Delta z}\right)_g \qquad (5.23)$$

The frictional pressure drop, $(\Delta p_f/\Delta z)_l$ and $(\Delta p_f/\Delta z)_g$, for the flow of liquid and gas alone over a distance, Δz of the column can be evaluated by Fanning's equation. Usually using the liquid two-phase friction multiplier is preferred because the liquid density generally does not vary too much in most of the applications as compared with the gas density. The same concept can be used to predict the frictional pressure drop in upward and inverse bubbly flow where the gas phase can be considered as the bubble phase. Many investigators explained the two-phase frictional pressure drop by two-phase frictional multipliers. However, the Lockhart–Martinelli correlation is unable to predict the experimental data of two-phase frictional pressure gradient properly in the system with enhanced gas–liquid mixing. In an ejector-induced inverse bubbly flow system, the frictional pressure drop may be much higher because of higher gas-hold-up and significant backmixing of the gas bubbles. Relatively bigger bubbles are unable to move downward at a particular liquid flow rate because of higher buoyant forces and leave from the top of the column for which higher frictional pressure losses results. In this case, the modification of the parameter is suggested.

MARTINELLI AND NELSON (1948) MODEL

Martinelli and Nelson (1948) proposed a graphical pressure drop correlation for forced circulation system. It is an empirical correlation based on experimental data for the flow combination of air and various liquids. The authors assumed the static pressure drop of the liquid phase and the vapor phase to be the same. The authors provided a graph on which the two-phase pressure drop can be determined when values of exit mass flow quality, system pressure, and the single-phase pressure drop of the liquid are known. The model can be used to predict the inverse bubble column when forced circulation of the systems prevails.

LOCKHART AND MARTINELLI (1949) MODEL

Lockhart and Martinelli (1949) proposed their correlation based on the assumptions that the static pressure drop of the liquid phase and the gas phase must be equal for all the flow patterns when there is no appreciable radial static pressure difference, and the sum of the volume occupied by each phase must be equal to the total volume of the pipe. The model can be used for bubbly flow systems.

Four types of flow mechanisms were assumed during the development of their correlation as:

- Viscous–viscous (vv): when the flow of both the liquid and the gas is laminar
- Viscous–turbulent (vt): when the flow of the liquid is laminar and the gas is turbulent

- Turbulent–viscous (tv): when the flow of the liquid is turbulent and the gas is laminar
- Turbulent–turbulent (tt): when the flow of both the liquid and the gas is turbulent

The authors introduced a new parameter called X in addition to multipliers introduced by Martinelli et al. (1944). The parameter X relates the single-phase pressure drops for liquid and gas as if each fluid is flowing alone in the pipe, which is defined by

$$X = \sqrt{\left(\frac{\Delta p_f}{\Delta z}\right)_l \bigg/ \left(\frac{\Delta p_f}{\Delta z}\right)_g} \tag{5.24}$$

This parameter X is a function of the ratios of the mass fluxes, densities, and viscosities of the liquid and the gas phase in addition to the diameter of the pipe.

$$X^2 = \lambda \left(\frac{\dot{m}_l}{\dot{m}_g}\right)^a \left(\frac{\rho_g}{\rho_l}\right)^b \left(\frac{\mu_l}{\mu_g}\right)^c \tag{5.25}$$

The constant λ and the coefficients a, b, and c can be obtained by regression analysis of experimental data. The Lockhart and Martinelli (1949) correlation is generally presented graphically as plot of the two-phase friction multiplier versus the dimensionless parameter X. Chenoweth and Martin (1955) argued that the accuracy of Lockhart and Martinelli (1949) correlation deteriorates for larger diameter pipes and higher system pressures.

KATO (1958) MODEL

Kato (1958) proposed a correlation for the Lockhart–Martinelli parameter, ϕ_l in the vertical two-phase air-water flow, as

$$\phi_{lK}^2 = 36.8 \times 10^{-3} \left(\frac{\dot{m}_l}{\dot{m}_g}\right)^{1.27} \frac{\mathrm{Re}_g^{0.25}}{X^2} \tag{5.26}$$

The value of ϕ_{lK} for the inverse system is smaller than ϕ_l predicted by the Kato correlation.

BANKOFF (1960) MODEL

Bankoff (1960) proposed a pressure drop based on phenomenology of interfacial shear stress of bubbly flow. The authors reported that gradient of bubbles number density where the maximum density is located at the center of the pipe or column can be assumed to develop the two-phase model. The author stated that the relative velocity of the bubbles with respect to the surrounding liquid is negligible compared with the stream velocity. Therefore, in a similar manner to homogeneous models, they developed a correlation for a two-phase frictional pressure gradient in terms of a two-phase friction multiplier ϕ_{Bf}, which can be represented as

$$\left(\frac{dp}{dz}\right)_{f,tp} = \phi_{Bf}^{7/4} \left(\frac{dp}{dz}\right)_l \tag{5.27}$$

The Bankoff multiplier is expressed as

$$\phi_{Bf} = \frac{1}{1-x}\left[1-\gamma\left(1-\frac{\rho_g}{\rho_l}\right)\right]^{3/7}\left[1+x\left(\frac{\rho_l}{\rho_g}-1\right)\right] \tag{5.28}$$

where

$$\gamma = \left[c_1 + c_2\left(\frac{\rho_g}{\rho_l}\right)\right]\left[1+\left(\frac{1-x}{x}\right)\left(\frac{\rho_g}{\rho_l}\right)\right]^{-1} \tag{5.29}$$

The coefficients c_1 and c_2 depend on the experimental data. The same model can be used to analyze the frictional pressure drop for inverse bubbly flow. Mandhane et al. (1977) proposed a composite (slug + bubbly flow) by improving the Lockhart and Martinelli (1949) equation for slug and bubbly flow. According to the author, the parameter can be a function of ratios of velocities, densities, and viscosities of liquid to gas, which can be expressed as

$$X \propto \left(\frac{u_{sl}}{u_{sg}}\right)^{c_3}\left(\frac{\rho_l}{\rho_g}\right)^{c_4}\left(\frac{\mu_l}{\mu_g}\right)^{c_5} \tag{5.30}$$

The constants are to be obtained from the experimental data of frictional pressure drop. For horizontal flow, the typical values are 0.5, 0.375, and 0.1 for c_3, c_4, and c_5, respectively for slug flow; for bubbly flow, the values are 0.875, 0.375, and 0.125, respectively. The multiplier for liquid and gas can then be related to the parameter X as per experimental design.

DAVIS (1963) MODEL

Davis (1963) extended the applicability of the Lockhart–Martinelli approach through the modification of the parameter X by incorporating the Froude number of the gas–liquid mixture, Fr_{tp}, to allow the effects of gravity and velocity. Thus, the modified parameter X' is expressed as

$$X' = 0.19 Fr_{tp}^{0.185} X \tag{5.31}$$

In the inverse bubble column, the data deviate remarkably from the Lockhart–Martinelli correlation based on Davis modification (Mandal et al., 2004a). The experimental value of the frictional pressured drop for a particular value of X' is higher than that given by this approach. This may be attributed to the better dispersion of gas in the liquid in the inverse system. The improved dispersion reduces the actual flow area available for liquid flow in the column, thereby increasing the actual liquid velocity and thus increasing the pressure drop. The enhanced gas–liquid mixing and hence higher $(\Delta p_f/\Delta z)_{tp}$, the frictional multipliers, ϕ_l of the Lockhart–Martinelli correlation can be modified by multiplying a factor. Mandal et al. (2004a) reported that the factor is 0.15 in case of inverse bubble column for the air–non-Newtonian system.

AOKI AND INOUE (1965) MODEL

Aoki and Inoue (1965) developed a correlation for the prediction of the Lockhart–Martinelli parameter, ϕ_l, for the vertical two-phase air-water flow as

$$\phi_{lA}^2 = 1 + 250 \left(\frac{Q_r}{1 - Q_r} \right)^{0.8} \tag{5.32}$$

The values of ϕ_l are faintly higher than the values predicted by the Aoki correlation, and this deviation implies an enhanced mixing and higher frictional loses in the inverse bubbly flow system.

BAROCZY (1966) MODEL

Baroczy (1966) introduced a term called the property index (I_p), which is a function of viscosity and density of each phases represented as

$$I_p = \left(\frac{\mu_l}{\mu_g} \right)^{0.2} \Bigg/ \left(\frac{\rho_l}{\rho_g} \right) \tag{5.33}$$

The two-phase pressure drop correlation is expressed graphically as a plot of the two-phase multiplier versus the property index. The author stated that when each phase flowing alone at the total mass flow rate is turbulent, the reciprocal of the property index is equal to the ratio of the pressure drop gradient for all gas flow to that for all liquid flow, which is given by

$$\left(\frac{\Delta p_f}{\Delta z} \right)_g \Bigg/ \left(\frac{\Delta p_f}{\Delta z} \right)_l = \frac{1}{I_p} \tag{5.34}$$

CHISHOLM (1967) MODEL

Chisholm (1967) proposed a very simplified form of the Lockhart–Martinelli correlation with one arbitrary constant, the value of which depends on the laminar or turbulent nature of each phase and the gas-to-liquid density ratios. He presented a theoretical analysis of the Lockhart and Martinelli (1949) correlation by including the effect of interfacial shear forces. Considering the interfacial shear force between the phases while developing the correlation enabled Martinelli to predict the hydraulic diameters of the phases more accurately than Lockhart and Martinelli (1949). The equation recommended by Chisholm is as follows:

$$\phi_{l,Ch}^2 = 1 + \frac{C}{X} + \frac{1}{X^2} \tag{5.35}$$

$$\phi_{g,Ch}^2 = 1 + CX + X^2 \tag{5.36}$$

The values of the Chisholm constant, C, vary as per experimental operating variables. For two-phase flow in a horizontal tube, the values are given in Table 5.2.

The experimental data in an inverse bubble column show a much higher pressure drop than those predicted by Chisholm's correlation. He also analyzed the frictional pressure

Table 5.2 Values of the Constant C in Chisholm (1967) Correlation for Two-Phase Flow in Horizontal Pipe

Flow Mechanism (Liquid–Gas)	Value of C
Viscous–viscous	5
Turbulent–viscous	10
Viscous–turbulent	12
Turbulent–turbulent	20

drop in a different way. For fully turbulent flow at Re greater than 2000, he defined a parameter Y, which is obtained from the ratio of the frictional pressure gradient as

$$Y^2 = \left(\frac{\Delta p_f}{\Delta z}\right)_g \Big/ \left(\frac{\Delta p_f}{\Delta z}\right)_l \tag{5.37}$$

His two-phase multiplier is then determined as

$$\phi_{l,Ch}^2 = 1 + (Y^2 - 1)\left\{\beta_{Ch} x^{(2-n)/2}(1-x)^{(2-n)/2} + x^{2-n}\right\} \tag{5.38}$$

where n is the exponent from the Blasius friction factor expression ($n = 0.25$). Chisholm's parameter β_{Ch} is a function of inverse of mass velocity of the two-phase mixture. According to him, the parameter can be calculated as

For the range of $0 < Y < 9.5$:

$$\beta_{Ch} = 55\left(\frac{1}{\dot{m}_t}\right)^{0.5} \quad \text{for} \quad \dot{m}_t \geq 1900\,\text{kg/m}^2\text{s} \tag{5.39}$$

$$\beta_{Ch} = 2400\left(\frac{1}{\dot{m}_t}\right) \quad \text{for} \quad 500 < \dot{m}_t < 1900\,\text{kg/m}^2\text{s} \tag{5.40}$$

$$\beta_{Ch} = 4.8 \quad \text{for} \quad \dot{m}_{total} < 500\,\text{kg/m}^2\text{s} \tag{5.41}$$

For the range of $9.5 < Y < 28$:

$$\beta_{Ch} = \frac{520}{Y}\left(\frac{1}{\dot{m}_t}\right)^{0.5} \quad \text{for} \quad \dot{m}_t \leq 600\,\text{kg/m}^2\text{s} \tag{5.42}$$

$$\beta_{Ch} = \frac{21}{Y} \quad \text{for} \quad \dot{m}_t > 600\,\text{kg/m}^2\text{s} \tag{5.43}$$

For $Y > 28$, β_{Ch} is to be calculated as

$$\beta_{Ch} = \frac{15000}{Y^2}\left(\frac{1}{\dot{m}_t}\right)^{0.5} \quad \text{for} \quad \dot{m}_t \leq 600\,\text{kg/m}^2\text{s} \tag{5.44}$$

The functionality can be obtained from the experimental data for inverse bubbly flow with different coefficients according to the experimental design. Other following correlations can also be used for interpretation of frictional pressure drop in inverse bubbly flow in a column.

WALLIS (1969) MODEL

Wallis (1969) reported that the value of liquid friction factor multiplier decreases with increasing system pressure at a given value of the flow quality x. He proposed a correlation and reported good prediction results for bubbly flow using his correlation, which is expressed in terms of the liquid friction factor multiplier. The proposed correlation is

$$\phi_W^2 = \left(1+x\frac{\rho_l-\rho_g}{\rho_g}\right)\left(1+x\frac{\mu_l-\mu_g}{\mu_g}\right)^{-0.25} \tag{5.45}$$

However, the correlation predicts the frictional pressure gradient well for the annular flow pattern. The correlation can be modified and applied in bubbly flow system by incorporating experimental data.

CHAWLA (1972) MODEL

Chawla (1972) proposed the following methods based on the gas pressure gradient:

$$\left(\frac{dp}{dz}\right)_{f,tp} = \phi_{Chawla}\left(\frac{dp}{dz}\right)_g \tag{5.46}$$

The vapor phase frictional gradient is calculated as

$$\left(\frac{dp}{dz}\right)_g = \frac{2f_g \dot{m}_t^2}{d_c \rho_g} \tag{5.47}$$

The Chawla two-phase multiplier depends on the slip ratio (S), and the multiplier is expressed as

$$\phi_{Chawla} = x^{1.75}\left[1+S\left(\frac{1-x}{x}\right)\frac{\rho_g}{\rho_l}\right]^{2.375} \tag{5.48}$$

where the slip ratio can be expressed by the following correlation:

$$S = \frac{1}{9.1\left[\dfrac{1-x}{x}(\mathrm{Re}_g\,Fr_m)^{-0.167}\left(\rho_l/\rho_g\right)^{-0.9}\left(\mu_l/\mu_g\right)^{-0.5}\right]} \tag{5.49}$$

where $Fr_m = \dot{m}_t^2/(gd_c\rho_m^2)$ and $\mathrm{Re}_g = \dot{m}_g d_c/\mu_g$.

CHISHOLM (1973) MODEL

Chisholm (1973) studied the influence of pipe surface roughness and proposed an equation to extrapolate the frictional pressure drop for rough surface pipes from his correlations to smooth pipes.

$$\frac{\beta_R}{\beta_S} = \left[\frac{1}{2}\left\{1+\left(\frac{\mu_g}{\mu_l}\right)^2 + 10^{-600(e/d)}\right\}\right]^{\frac{0.25-n}{0.25}}$$

(5.50)

β_s and β_r are the β coefficients discussed by Chisholm (1973) for smooth and rough pipes, respectively, as defined in the earlier section. The degree of roughness depends on the Blasius exponent, n, is evaluated from the experimental data. Generally, its value lies between 0 and 0.25.

GRÖNNERUD (1979) MODEL

The correlation of Grönnerud (1979) was developed specially for refrigerants However, the concepts can be modified to analyze the frictional pressure drop for two-phase bubble–liquid flow. The correlation is as follows:

$$\left(\frac{dp}{dz}\right)_f = \phi_{Gd}\left(\frac{dp}{dz}\right)_l$$

(5.51)

The two-phase multiplier denoted by Grönnerud (1979) is

$$\phi_{Gd} = 1+\left(\frac{dp}{dz}\right)_{Fr}\left[\frac{\rho_l/\rho_g}{(\mu_l/\mu_g)^{0.25}}-1\right]$$

(5.52)

$$\left(\frac{dp}{dz}\right)_l = \frac{2f_l\dot{m}^2_{total}}{d_c\rho_l}$$

(5.53)

According to his observation, the frictional pressure gradient depends on the Froude number, which can be expressed as

$$\left(\frac{dp}{dz}\right)_{Fr} = f_{Fr}\left[x+4(x^{1.8}-x^{10}\sqrt{f_{Fr}})\right]$$

(5.54)

He reported that if the liquid Froude number, $Fr_l \geq 1$, then the friction factor $f_{Fr} = 1$ or if $Fr_l < 1$, it can be expressed as:

$$f_{Fr} = Fr_l^a + b\left(\ln\frac{1}{Fr_l}\right)^2$$

(5.55)

where $Fr_l = \dot{m}^2_{total}/(gd_c\rho_l^2)$. The coefficient can be obtained from experimental data. He reported that typical values for a and b, respectively, are 0.3 and 0.0055.

EISENBERG AND WEINBERGER (1979) MODEL

Eisenberg and Weinberger (1979) suggested modifications in the Lockhart–Martinelli parameters, ϕ_l and X, for two-phase gas–non-Newtonian liquid flow. According to them, because for two-phase flow, the average shear rate in the liquid phase will be higher than in the single-phase flow at the same flow rate of the liquid phase, the average viscosity of the non-Newtonian liquid will be lower in two-phase flow case. To account for this effect of lower effective viscosity, they incorporated a correction in the frictional pressure drop, which is given as

$$\left(\frac{\Delta p_f}{\Delta z}\right)_l' = \frac{(\mu_{eff})_{tp}}{(\mu_{eff})_l}\left(\frac{\Delta p_f}{\Delta z}\right)_l \tag{5.56}$$

where

$$\frac{(\mu_{eff})_{tp}}{(\mu_{eff})_l} = \left[\frac{(n+1)}{3n+1}\right]^{n-1}\left[1-\varepsilon_g-\left(\frac{2n}{3n+1}\right)\left(1-\varepsilon_g^{(3n+1)/2n}\right)\right]^{(1-n)} \tag{5.57}$$

Hence, the modified parameters in the Lockhart–Martinelli correlation become

$$\phi_{l,E-W}^2 = \left(\frac{\Delta p_f}{\Delta z}\right)_{tp}\Bigg/\left(\frac{\Delta p_f}{\Delta z}\right)_l' \tag{5.58}$$

and

$$X_{E-W}^2 = \left(\frac{\Delta p_f}{\Delta z}\right)_l'\Bigg/\left(\frac{\Delta p_f}{\Delta z}\right)_g \tag{5.59}$$

The Eisenberg and Weinberger parameters, X_{E-W} and $\phi_{l,E-W}$, do not agree well with Lockhart–Martinelli parameters in case of inverse bubbly flow. The deviation may be because the correction factor obtained by equation (5.57) has been developed for annular two-phase flow of gas–non-Newtonian liquid, and no correction has been incorporated for vertical flow.

FRIEDEL (1980) MODEL
The correlation is also based on the two-phase multiplier concept. The correlation can be expressed as

$$\left(\frac{dp}{dz}\right)_f = \phi_{Frd}\left(\frac{dp}{dz}\right)_l \tag{5.60}$$

The correlation for the Friedel two-phase multiplier can be expressed as

$$\phi_{Frd}^2 = E + \frac{3.24\,FG}{Fr_m^{0.045}We_l^{0.035}} \tag{5.61}$$

where

$$E = (1-x)^2 + x^2\,\frac{\rho_l}{\rho_g}\frac{f_g}{f_l} \tag{5.62}$$

$$F = (1-x)^{0.224} x^{0.78} \tag{6.63}$$

$$G = \left(\frac{\rho_l}{\rho_g}\right)^{0.91} \left(\frac{\mu_g}{\mu_l}\right)^{0.19} \left(1 - \frac{\mu_g}{\mu_l}\right)^{0.70} \tag{5.64}$$

$$Fr_m = \frac{\dot{m}_t^2}{g d_c \rho_m^2} \tag{5.65}$$

$$Fr_m = \frac{\dot{m}_t^2 d_t}{\sigma \rho_m} \tag{5.66}$$

$$\rho_m = \left(\frac{x}{\rho_g} + \frac{1-x}{\rho_l}\right)^{-1} \tag{5.67}$$

The correlation is applicable only the ratio of liquid density to the gas density is less than 1000.

THEISSING (1980) MODEL

Theissing (1980) proposed an improvement for the Lockhart–Martinelli (1949) correlation theory based on the interaction between the phases and suggested a unique friction multiplier technique.

$$\left(\frac{dp}{dz}\right)_f = \left[\left(\frac{dp}{dz}\right)_l^{1/(N\varphi)} (1-x)^{1/\varphi} + \left(\frac{dp}{dz}\right)_g^{1/(N\varphi)} x^{1/\varphi}\right]^{N\varphi} \tag{5.68}$$

where

$$\varphi = 3 - 2\left(\frac{2\sqrt{\rho_l/\rho_g}}{1 + \rho_l/\rho_g}\right)^{0.7/n} \tag{5.69}$$

$$N = \frac{N_1 + N_2\left[\dfrac{dp/dz)_g}{(dp/dz)_l}\right]^{0.1}}{1 + \left[\dfrac{dp/dz)_g}{(dp/dz)_l}\right]^{0.1}} \tag{5.70}$$

$$N_1 = \frac{\ln\left(\dfrac{(dp/dz)_l}{(dp/dz)_{l0}}\right)}{\ln(1-x)} \tag{5.71}$$

$$N_2 = \frac{\ln\left(\dfrac{(dp/dz)_g}{(dp/dz)_{g0}}\right)}{\ln(x)} \tag{5.72}$$

CHEN AND SPEDDING (1981) MODEL

Chen and Spedding (1981) made analytical study of the Lockhart and Martinelli (1949) correlation and developed a semiempirical correlation to predict the two-phase gas friction multiplier. They proposed a simple correlation as a function of superficial Reynolds number of gas and liquid as

$$\phi_g^2 = c_6 \, \mathrm{Re}_g^{c_7} \, \mathrm{Re}_l^{c_8} \tag{5.73}$$

For a horizontal tube, the values of c_6, c_7, and c_8 are 4050, –0.91, and 0.44, respectively, for an air–water system. Although the correlation is developed for horizontal flow, it can be modified based on experimental data for a vertical column under bubbly flow conditions.

MÜLLER-STEINHAGEN AND HECK (1986) MODEL

Müller-Steinhagen and Heck (1986) developed a correlation for the two-phase frictional pressure gradient that covered a large database of air–oil, air–water, water–steam, and several refrigerents, which can be expressed as

$$\left(\frac{dp}{dz}\right)_f = G(1-x)^{1/3} + Bx^3 \tag{5.74}$$

where the factor G is

$$G = \left(\frac{dp}{dz}\right)_l + 2\left(\left(\frac{dp}{dz}\right)_g - \left(\frac{dp}{dz}\right)_l\right)x \tag{5.75}$$

GHARAT AND JOSHI (1992) MODEL

Gharat and Joshi (1992) developed a coherent model to predict the pressure drop in homogeneous (bubbly flow) and heterogeneous (churn-turbulent) regimes. They pointed out that the liquid phase hydrodynamics and the frictional pressure drop are closely related to the hold-up structure. For instance, whereas internal liquid circulation is an integral part of the flow pattern in bubble columns, it is absent in pipe flows. Furthermore, they reported that the turbulence intensities are much higher in bubble columns compared with those in two-phase flows. They explained that the principal reason for the differences is the range of superficial liquid velocity. In upward bubble columns, the values of liquid velocity are relatively low. The liquid flow is upward in the central region and inverse near the column wall. With an increase in liquid velocity, the inverse flow is reduced. They have added that additional turbulence in presence of bubble movement (the level of turbulence is much higher than that in single-phase flow) increases the friction factor. The overall pressure drop is the combined effect of the increased friction factor and the velocity and is given by the following equation:

$$\Delta P_{f,tp} = \Delta P_{f,sp} + \Delta P_{f,AT} \tag{5.76}$$

$$\Delta P_{f,sp} = \frac{2 f_{f,sp} u_{sp}^2 \rho_l \Delta z}{d_c} \tag{5.77}$$

$$\Delta P_{f,AT} = \frac{2 f_{f,AT} u_s^2 \rho_l \Delta z}{d_c} \tag{5.78}$$

and then the overall friction factor is expressed in terms of the additional friction factor (f_{AT}) as

$$f_{tp} = \frac{f_{sp}}{\varepsilon_l^2} + f_{AT} \tag{5.79}$$

The additional friction factor depends on the transverse component of turbulence intensity (u_y') of the flow and slip velocity of the bubble, which is represented as

$$f_{AT} = 2 \left(\frac{u_y'}{u_s} \right)^2 \tag{5.80}$$

where

$$u_y' = \frac{1}{\sqrt{3}} \left[gl \left(u_{sg} - \frac{\varepsilon_g u_{sl}}{\varepsilon_l} - \varepsilon_g u_s \right) \right]^{1/3} \tag{5.81}$$

In the normal direction, the fluctuating velocity component practically equals the friction velocity (u^*) (Davies, 1972). The parameter l is the scale of turbulence in single-phase pipe flow, which is given by

$$l = \chi d_c \tag{5.82}$$

where the parameter m depends on the operating conditions given by Gharat and Joshi (1992):

$$\chi = 0.17 - 0.19 u_{sl} - 0.24 \varepsilon_g + 0.015 u_g \tag{5.83}$$

For homogeneous flow,

$$u_y' = 1.5 \varepsilon_g u_s \tag{5.84}$$

where

$$u_s = \pm \frac{u_{sg}}{\varepsilon_g} \mp \frac{u_{sl}}{1 - \varepsilon_g} \tag{5.85}$$

(The upper signs are for gas–liquid upflow and the lower for gas–liquid inverse flow.) The two-phase multiplier can then be expressed as

$$\phi_l^2 = \frac{\Delta P_{f,tp}}{\Delta P_{f,sp}} = \frac{\Delta P_{f,sp} + \Delta P_{f,AT}}{\Delta P_{f,sp}} \tag{5.86}$$

or

$$\phi_l^2 = \frac{1}{\varepsilon_l^2} + \frac{f_{AT}}{f_{sp}}\left(\frac{u_s}{u_{sl}}\right)^2 \tag{5.87}$$

or

$$\phi_l^2 = \frac{1}{\varepsilon_l^2} + \frac{2}{f_{sp}}\left(\frac{u_y'}{u_{sl}}\right)^2 \tag{5.88}$$

According to Clark and Flemmer (1985a,b), within a range of superficial liquid velocity 0 to 0.8 m/s, the two-phase multiplier in upflow and inverse flow of bubble are, respectively,

$$\phi_l^2 = \frac{1}{\varepsilon_l^2} + \frac{46.5\varepsilon_g u_s}{\varepsilon_l u_{sl}} \quad \text{for upflow} \tag{5.89}$$

$$\phi_l^2 = \frac{1}{\varepsilon_l^2} + \frac{5.45\varepsilon_g u_s}{\varepsilon_l u_{sl}} \quad \text{for inverse flow} \tag{5.90}$$

AWAD AND MUZYCHKA MODEL

Awad and Muzychka (2004) proposed a correlation that relates the two-phase frictional pressure gradient to the single-phase frictional pressure gradients of the liquid and gas flowing alone. The authors recommended Churchill's (1977) correlation to calculate the fanning friction factor for the single-phase pressure drop calculations. Their correlations for liquid and gas multipliers are

$$\phi_{l,A-M} = \left[1 + \left(\frac{1}{X^2}\right)^q\right]^{1/q} \tag{5.91}$$

and

$$\phi_{g,A-M} = \left[1 + \left(X^2\right)^q\right]^{1/q} \tag{5.92}$$

The values for the parameter X are to be evaluated using the Lockhart–Martinelli (1949) correlation. The value of the parameter q is to be determined from experimental data. The parameter q depends on the size of the pipe or column. They also proposed an expression for the two-phase friction multiplier as a function of friction factors and physical properties as

$$\phi_{l,A-M}^2 = \left(\frac{f_{tp}}{f_{lo}}\right)\left(1 + x\frac{\Delta\rho}{\rho_g}\right) \tag{5.93}$$

The frictional pressure drop can also be expressed in terms of the interfacial pressure drop. Awad and Muzychka (2010) stated that the total frictional pressure drop gradient can be expressed as the sum of the frictional pressure drop of each single phase and the interfacial pressure drop gradient, which is represented as

$$\left(\frac{dp}{dz}\right)_{f,tp} = \left(\frac{dp}{dz}\right)_l + \left(\frac{dp}{dz}\right)_i + \left(\frac{dp}{dz}\right)_g \tag{5.94}$$

They suggested that the two-phase liquid multiplier in the Lockhart–Martinelli (1949) correlation can then be expressed in terms of interfacial friction multiplier $\phi_{l,i}$ as

$$\phi_{l,S-M}^2 = 1 + \phi_{l,i}^2 + \frac{1}{X^2} \tag{5.95}$$

$$\phi_{l,i}^2 = \frac{C_{A-M}}{X^m} \tag{5.96}$$

The values of C_{A-M} and m can be obtained from experimental data, which are dependent on the mass flow rates, flow patterns, and flow mechanisms.

SUN AND MISHIMA (2009) MODEL
Sun and Mishima (2009) represented a new correlation based on the Chisholm parameter (C). They reported that the parameter (C) can be expressed as a function of the flow quality and superficial Reynolds number of the liquid and the gas phases. According to the Sun and Mishima (2009) correlation, the liquid multiplier can be expressed as

$$\phi_{l,S-M}^2 = 1 + \frac{C_{S-M}}{X^{1.19}} + \frac{1}{X^2} \tag{5.97}$$

where

$$C_{S-M} = \lambda \left(\frac{Re_g}{Re_l}\right)^a \left(\frac{1-x}{x}\right)^b \tag{5.98}$$

The constants λ, a, and b can be obtained from the experimental data.

Empirical Model
The phenomenon such as momentum transfer between the phases, the wall friction, and the shear at the phase interface cannot be quantified directly from the experiments. Therefore, the measurement of frictional pressure drop by theoretical analysis alone is not possible. During the early stage of two-phase flow study, a number of correlations have been developed by curve fitting of data based on experimental pressure drop measurements. Correlating the experimental data by using some carefully selected variables is a convenient way of developing a correlation with a minimum analytical knowledge of the problem. Because the two-phase flow problem involves several independent physical quantities, analyzing their relationship and developing dimensionless parameters is a tedious job. Some dimensionless groups play a dominant role in determining pressure drop in variety of applications. By this method, the prediction capability heavily relies on the quality of the data and vastness of the experimental data. Several workers

have developed correlations for the two-phase frictional pressure drop by dimensional or similarity analysis as a function of the directly measurable parameters. The factors that affect the two-phase frictional pressure drop are

1. **System variables:** physical properties of liquid, namely, density (ρ_l) effective viscosity (μ_l), and surface tension (σ_l), and physical properties of air, namely, density (ρ_g) and viscosity (μ_g)
2. **Operating variables:** liquid and gas mass flow rates (Q_l and Q_g, respectively) and gas–liquid mixing height in the column (h_c)
3. **Geometric variables:** the diameter of the column (d_c) and diameter of the nozzle (d_n) or distributor orifice diameter

With the variables and by the use of dimensional analysis (either by the Raleigh's method or Buckingham Pi theorem), this functional relationship between frictional pressure drop and the relevant physical and dynamic parameters can be reduced to the following equivalent relationship:

$$\frac{\Delta p_f}{\rho u^2} = \lambda \left(\frac{\rho d_c u}{\mu}, \frac{g \mu^4}{\rho \sigma^3}, \frac{d_c}{d_o}, \frac{h_m}{d_c} \dots \dots \right) \tag{5.99}$$

The functionalities of the dimensionless groups are can be expressed as

$$Gr_1 = c_9 Gr_2^{c_{10}} Gr_3^{c_{11}} Gr_4^{c_{12}} \dots \dots \dots \dots \dots \tag{5.100}$$

The coefficients of the functional relationship can be obtained by regression analysis with the experimental data. Kundu et al. (1995) developed a correlation for inverse bubbly flow in a vertical column by similarity analysis, which is given by

$$Eu = 0.11 \times 10^{-8} \, Re_l^{-2.15} \, Su^{-1.71} Mo^{-1.33} A_r^{-0.07} H_r^{0.09} \tag{5.101}$$

The two-phase friction factor can be evaluated from the experimental data of frictional pressure drop as

$$f_{tp} = \frac{d_c}{2 \rho_{tp} u_{tp}^2} \left(\frac{\Delta P_{ftp}}{\Delta z} \right)_{measured} = \frac{d_c}{2 \rho_{tp} u_{tp}^2} \left(\frac{\Delta P_{tot} - \Delta P_h - \Delta P_{mom}}{\Delta z} \right)_{measured} \tag{5.102}$$

The friction factor depends on the experimental operating variables. The correlation model can be developed for the friction factor by dimensional analysis with the variables. The correlation can be represented the functional relationship typically for the ejector induced inverse bubbly flow system as

$$f_{tp} = f(Re_l, Mo, A_r, H_r) \tag{5.103}$$

The multiple linear regression analysis for the inverse flow with experimental data (Majumder et al., 2007) yielded the following correlation:

$$f_{lg} = 1.465 \times 10^{-10} \, Re_l^{-2.036} \, Mo^{-0.709} A_r^{-0.013} H_r^{-6.709} \tag{5.104}$$

The range of variation of the different parameters of the two-phase friction factor for the inverse bubbly flow is $4424.73 < Re_l < 8849.47$; $51.02 < A_r < 156.25$ and $27.6 < H_r < 32.80$. Dukler et al. (1964) reported that in heterogeneous mixture of gas–liquid two-phase flow, the friction factor is enhanced from the single-phase friction factor by the factor depending on the slip ratio and the mass quality, which can be expressed as

$$f_{tp} = f_0\left[1 + \frac{-\ln\lambda}{S}\right]$$

(5.105)

$$S = 1.281 - 0.478(-\ln\lambda) + 0.444(-\ln\lambda)^2 - 0.094(-\ln\lambda)^3 + 0.00843(-\ln\lambda)^4$$

(5.106)

$$f_l = 0.0014 + \frac{0.125}{Re_l^{0.32}}$$

(5.107)

$$\lambda = \left[\frac{Q_l}{Q_l + Q_g}\right] = \left[1 + \left(\frac{x}{1-x}\right)\frac{\rho_l}{\rho_g}\right]^{-1}$$

(5.108)

Garcia et al. (2003) proposed a single equation concept (called composite) that can be developed to predict the mixture friction factors for a wide range of gas and liquid flow rates, viscosity values, and different flow patterns. The equation can be represented by

$$f_{tp} = F_2 + \frac{F_1 - F_2}{\left[1 + \left(\frac{Re}{t}\right)^c\right]^d}$$

(5.109)

F_1 and F_2 depend on the two-phase Reynolds number as

$$F_1 = 13.98\,Re_{tp}^{-0.9501}$$

(5.110)

$$F_2 = 0.1067\,Re_{tp}^{-0.2629}$$

(5.111)

The parameters c, d and t can be obtained by fitting equation (5.109) to the experimental data points. For slug and dispersed bubbly flow, they reported the values of c, d, and t are 293, 3.577, and 0.2029 and 304, 2.948, and 0.2236, respectively.

Mechanistic Models
CLARK AND FLEMMER (1985a) MODEL

Clark and Flemmer (1985a) developed a theory to explain pressure losses using classical mixing length theory; it is verified with data gained in two-phase bubbly flow. They have extended single-phase mixing length theory to the two-phase case to take into account additional Reynolds stresses generated by flow around the bubbles. Their analysis showed that transverse velocity components induced by bubbles moving in the flow give rise to "excess shear" stresses that are of significant magnitude at low flow velocities. They have reported the

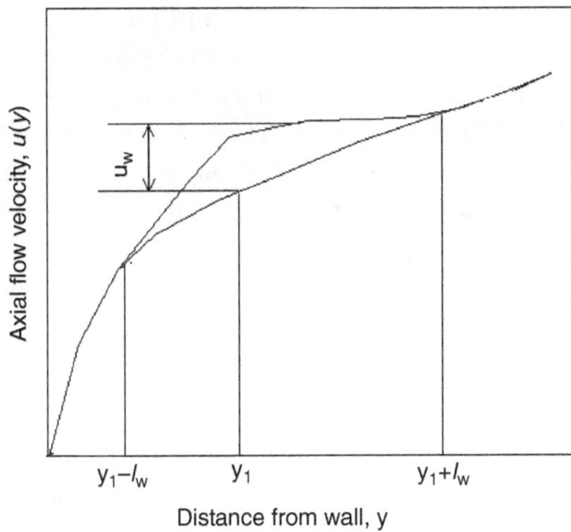

FIGURE 5.1 Effect of bubbles on axial flow velocity.

fundamental explanation of the mechanism causing the unexpectedly high shear stresses in presence of bubbly flow. A simpler working model was also developed in part 2 of their other paper (Clark and Flemmer, 1985b) based on average properties of the flow for evaluation of the pressure loss. They validated the model with experimental results for inverse bubbly flow to confirm the form of the model. According to them, a reduction of shear stress results in the presence of bubbles that is proportional to the square of the voidage because shear stress is proportional to the product of velocity fluctuation caused by bubble movement. They reported that the presence of the bubble causes a distortion of the velocity profile by an amount that may be smaller than bubble rise velocity. A graphical nonscaled representation of an axial velocity profile in presence of bubble movement is shown in Figure 5.1. They pointed out that there must be transverse velocity components feeding into the wake to replace fluid below the bubble nadir. These may be of the same order of magnitude of axial liquid velocity because of the presence of a bubble wake. The stronger the bubble wake, the smaller and more diffuse the transverse components. In the final analysis, they suggested to consider some average values for velocity components of liquid caused by the presence of a bubble wake. According to Prandtl mixing length theory, based on the average value of axial liquid velocity in the presence of bubbles (average in some small volume of the flow, or a time-average value), instead of the local value the axial liquid velocity, they derived the shear stress profile for two-phase bubbly flow as

$$\tau_{tp} = \rho_l l^2 \left(\frac{du}{dy}\right)^2 + \rho_l C_{13} l_w u_b \varepsilon_g \left(\frac{du}{dy}\right) \tag{5.112}$$

Because the excess shear arises from the transverse components induced above and below the bubble and not the axial components, this analysis is applicable in both upward and inverse bubbly flow. The larger bubbles may surmount turbulent eddies by these bubbles. In this case, equation (5.112) is to be modified to account for turbulent eddies in low-velocity flow. For small uniform bubbles, the pressure drop may be higher. The uniform size results in ideal bubbly flow, but bubbly flow of different sizes may appear to be as churn-turbulent bubbly flow and incorporates strong waking and training or channeling (the following of one bubble by another), so that transverse velocity components may be small, with a consequent reduction in excess shear. Moreover, they reported that the bigger bubbles may have some capacity to reduce the turbulence. In this case, their modified equation can be represented as

$$\tau_{tp} = (1 - c_{14}\varepsilon_g)^2 \rho_l l^2 \left(\frac{du}{dy}\right)\left[\left(\frac{du}{dy}\right) + c_{15}u_b\varepsilon_g/l^2\right] \tag{5.113}$$

The mixing length, l, was expressed in the original analysis by Prandtl as increasing linearly from the wall, where its value was zero. Clark and Flemmer (1985a) reported some criteria for the mixing length, including that it must tend to zero at the wall; it must not be zero at the pipe center because mixing certainly occurs there; and the derivative of it with respect to distance from the pipe wall, dl/dy, should be zero at the pipe center.

They suggested a simple function meeting these criteria:

$$\frac{l}{R} = c_{16}\left(\frac{y}{R} - \left(\frac{y}{R}\right)^n/n\right) \tag{5.114}$$

where c_{16}, and n are constants. Nikuradse (Schlichting, 1966) suggested another empirical relationship for mixing length given by

$$\frac{l}{R} = 0.14 - 0.08\left(1 - \frac{y}{R}\right)^2 - 0.06\left(1 - \frac{y}{R}\right)^4 \tag{5.114}$$

which shows favorable agreement with relationships given by Clark and Flemmer for a value of $n = 1.3$ in equation (5.114). The two-phase mixing length theory can be used to develop the model related to the shear stress to the velocity gradient, voidage, and degree of bubble waking. Clark and Flemmer developed the following equation for two-phase shear stress based on the two-phase mixing length theory, which can be represented as

$$\tau_{tp} = \rho_l\left(\frac{c_{16}u_{sl}}{1 - \varepsilon_g}\right)\left(\frac{c_{17}u_{sl}}{1 - \varepsilon_g} + c_{18}u_b\varepsilon_g\right) \tag{5.115}$$

The shear stress at a fixed point in the bulk flow a distance y from the wall can be related to the shear stress at the wall by

$$\tau_{tp} = \tau_{tp}^w\left(1 - \frac{y}{R}\right) \tag{5.116}$$

where y is by definition constant at the fixed point. Hence, at that fixed point in the flow, the shear stress varies directly proportional to the wall shear stress. The wall shear stress is given by

$$\tau_{tp}^{w} = B_1 \frac{u_{sl}^{2}}{(1-\varepsilon_g)^2} + B_2 \frac{u_b u_{sl} \varepsilon_g}{1-\varepsilon_g} \tag{5.117}$$

where B_1 and B_2 are dimensional with units of $N/(m^4\, s^2)$. For inverse flow of the bubble, the values of B_1 and B_2 are 2.65 and 57.8, respectively.

MAJUMDER ET AL. (2007) MODEL

This model describes the gas–liquid two-phase frictional pressure drop in vertical flow by considering dynamic interaction between the two phases. The dynamic interaction between liquid and gas is taken into account in modeling the flow by introducing the rate of energy dissipation for formation and breakage of bubbles. On the basis of mechanical energy, a unified treatment is presented to describe the two-phase pressure drop. The treatment is presented with the following assumptions:

1. There is isothermal steady flow of phases through the column.
2. The hold-up is considered as average for particular gas and liquid flow rates.
3. Acceleration effects are negligible because of no interphase mass transfer.
4. For a particular gas and liquid flow rate, frictional loss is considered as uniform throughout the column.

The mechanical energy balance equation for the gas is given by

$$\Delta P_{tp} \varepsilon_g A_c u_{sg} + g \Delta z A_c \varepsilon_g \rho_g u_{sg} + E_g = 0 \tag{5.118}$$

In the above equation, the first term is "energy due to two-phase pressure," the second term is "potential energy," and the third term is "energy dissipation due to friction." Similarly, for the liquid,

$$\Delta P_{tp} (1-\varepsilon_g) A_c u_{sl} + g \Delta z A_c (1-\varepsilon_g) \rho_l u_{sl} + E_l = 0 \tag{5.119}$$

Adding equations (5.118) and (5.119), one gets

$$\Delta P_{tp}[(1-\varepsilon_g)u_{sl} + \varepsilon_g u_{sg}] + [(1-\varepsilon_g)\rho_l u_{sl} + \varepsilon_g \rho_g u_{sg}]g\Delta z = -(E_l + E_g)/A_c \tag{5.120}$$

The energy dissipation in gas–liquid two-phase flow occurs from the frictional losses of individual phase. The energy dissipation not only occurs because of frictional losses but also for the formation and breakage of bubbles. Therefore, total energy dissipation of the two phases is equal to the sum of frictional loss caused by each gas and liquid phase and the loss caused by bubble formation. Therefore, an expression for $(E_l + E_g)$ can be obtained as

$$\Delta P_{fl} A_c (1-\varepsilon_g) u_{sl} + \Delta P_{fg} A_c \varepsilon_g u_{sg} + E_b = (E_l + E_g) \tag{5.121}$$

where ΔP_{fl} is the frictional pressure loss caused by liquid in two-phase flow and ΔP_{fg} is the frictional pressure loss caused by gas in two-phase flow. E_b is the energy loss caused by bubble formation. The frictional energy dissipation per unit area can be conveniently written in terms of mass velocities, L and G. Therefore, equation (5.121) can be written as

$$\Delta P_{fl}L/\rho_l + \Delta P_{fg}G/\rho_g + E_b/A_c = (E_l + E_g)/A_c \tag{5.122}$$

where L and G are the mass flux of liquid and gas, respectively, and these have been defined as $L = (1 - \varepsilon_g)u_{sl}\rho_l$ and $G = \varepsilon_g u_{sg}\rho_g$, respectively. Substituting equation (5.122) for $(E_l + E_g)$ in equation (5.120), one gets

$$\Delta P_{tp}[L/\rho_l + G/\rho_g] + [L + G]g\Delta Z + \Delta P_{fg}G/\rho_g + \Delta P_{fl}L/\rho_l + E_b/A_c = 0 \tag{5.123}$$

To determine the frictional losses attributable to each gas and liquid phase in two-phase flow in the column, a model can be formulated on the basis of the following assumptions: (1) the friction factor for each phase in two-phase flow is a constant multiple, α', of that if only flow of single phase takes place in the column, and (2) the area of contact of each phase with wall in two-phase flow is α'' times that of only flow of single phase in the column. In case of the gas phase, the values of α' and α'' depend on the number of bubbles coming in contact with the wall of the column. From these assumptions and from a simple overall momentum balance, it follows

For liquid phase:

$$\frac{\Delta P_{fl}(\text{Cross-sectional area})_{tp}}{\Delta P_{fl,sp}(\text{Cross-sectional area})_{l,sp}} = \frac{(\text{Wall shear stress})_{tp}}{(\text{Wall shear stress})_{l,sp}} \times \frac{(\text{Area of contact with wall})_{tp}}{(\text{Area of contact with wall})_{l,sp}} \tag{5.124}$$

implies

$$\begin{aligned}
\frac{\Delta P_{fl}A_c(1-\varepsilon_g)}{\Delta P_{fl,sp}A_c} &= \frac{(0.5f\rho_l u_{sl}^2)_{tp}}{(0.5f\rho_l u_{sl}^2)_{l,sp}} \times \frac{(\text{Area of contact with wall})_{tp}}{(\text{Area of contact with wall})_{l,sp}} \\
&= \frac{f_{tp}}{f_{l,sp}} \cdot \frac{(u_{sl}^2)_{tp}}{(u_{sl}^2)_{l,sp}} \times \frac{(\text{Area of contact with wall})_{tp}}{(\text{Area of contact with wall})_{l,sp}} \\
&= \alpha_l' \cdot \frac{1}{(1-\varepsilon_g)^2} \cdot \alpha_l'' = \frac{\alpha_l}{(1-\varepsilon_g)^2}
\end{aligned} \tag{5.125}$$

Hence,

$$\frac{\Delta P_{fl}}{\Delta P_{fl,sp}} = \frac{\alpha_l}{(1-\varepsilon_g)^3} \tag{5.126}$$

where the subscript sp refers to single phase; tp refers to liquid–gas two phase; sl refers to liquid superficial, $(u_{sl})_{tp} = (u_{sl})_{l,sp}/(1-\varepsilon_g)$, $\alpha_l = \alpha_l'\alpha_l''$; and cross-sectional area for the flow of liquid in two-phase is equal to $(1-\varepsilon_g)$ times the cross-sectional area of the column. α_l' is defined as $f_{tp}/f_{l,sp}$, and α_l'' is defined as $(\text{Area of contact})_{tp}/(\text{Area of contact})_{l,sp}$. $\Delta P_{fl,sp}$ is the frictional pressure drop caused by liquid when single liquid phase flows through the column. It is assumed that the area of contact of liquid with the wall in two-phase flow is the same as the area of contact with the wall in single-phase liquid flow. Therefore, α_l''

is equal to 1.0. Therefore, α_l can be interpreted as α_l', which is defined as $f_{tp}/f_{l,sp}$. Similarly, for the gas phase,

$$\frac{\Delta P_{fg}}{\Delta P_{fg,sp}} = \frac{\alpha_g}{\varepsilon_g^3} \tag{5.127}$$

where $\Delta P_{fg,sp}$ is the frictional pressure drop due to gas when only single gas phase flows through the column and $\alpha_g = \alpha_g' \alpha_g''$.

Pressure drop caused by form drag: The form drag represents the interaction forces between the continuous and the dispersed phase. The forces acting on a motionless bubble in a motionless liquid are pressure and gravity. Because there is usually a relative motion between the bubbles and liquid, the liquid flow around individual bubbles leads to local variations in pressure (Joshi, 2001). The force caused by gravity of gas bubbles is negligible. A bubble imbedded in a flowing fluid is influenced by a number of mechanisms that act on it through the traction at the gas–liquid interface. The bubbles moving through the fluid experience pressure forces that affect their motion. The formulation of the pressure force is given in the following section. For a single bubble at steady state, the force balance is given by

$$F_d = C_d \frac{\pi d_b^2}{4} \cdot \frac{1}{2} \rho_l u_s^2 = \frac{\pi}{6} d_b^3 (\rho_l - \rho_g) g \tag{5.128}$$

Equation (5.129) gives the drag force for a single bubble. C_d is the drag coefficient, d_b is the bubble diameter, and u_s is the slip velocity. The drag force for a single bubble depends on the size and shape of the bubble and the nature of the gas–liquid interface. Because of homogeneity of the inverse bubbly flow, it is assumed that all bubbles are of equal size and spherical in shape. For N number of bubbles, the total drag force can be estimated on the basis of the left-hand side of equation (5.129):

$$F_{td} = N F_d = \frac{\pi/4 d_c^2 h_m \varepsilon_g}{\pi/6 d_b^3} C_d \frac{\pi d_b^2}{4} \frac{1}{2} \rho_l u_s^2 \tag{5.129}$$

where

$$N = \frac{(\pi/4) d_c^2 h_m \varepsilon_g}{\pi/6 d_b^3} \tag{5.130}$$

$$u_s = \pm \frac{u_{sg}}{\varepsilon_g} \mp \frac{u_{sl}}{1 - \varepsilon_g} \tag{5.131}$$

The upper signs are for gas–liquid upflow and the lower for gas–liquid inverse flow. According to Lain et al. (1999), the drag coefficient C_d can be calculated by using the following empirical correlations:

$$C_d = \frac{16}{\mathrm{Re}_b}, \quad \mathrm{Re}_b < 1.5, \tag{5.132}$$

$$C_d = 14.9 \, \mathrm{Re}_b^{-0.78}, \quad 1.5 < \mathrm{Re}_b < 80, \tag{5.133}$$

$$C_d = \frac{48}{\mathrm{Re}_b}(1.0 - 2.21 \, \mathrm{Re}_b^{-0.5}), \quad 80 < \mathrm{Re}_b < 700 \tag{5.134}$$

where

$$\mathrm{Re}_b = \frac{\rho_l d_b u_s}{\mu_l} \tag{5.135}$$

Therefore, the pressure drop attributable to form drag can be calculated as the total form drag (F_d) divided by the total interfacial area ($N\pi d_b^2$) of N number of bubbles and can be obtained as follows:

$$\Delta P_{FD} = \frac{\text{Total form drag}}{\text{Total interfacial area}} = \frac{N F_d}{N \pi d_b^2} = \frac{N C_d (\pi/4) d_b^2 (1/2) \rho_l u_s^2}{N \pi d_b^2} = \frac{1}{8} C_d \rho_l u_s^2 \tag{5.136}$$

Additional pressure drop caused by formation of bubble: The rate of energy loss caused by bubble formation is the product of the rate of bubble formation and the energy loss in forming a single bubble. The rate of bubble formation is given by the volumetric gas flow rate divided by the single bubble volume. In the ejector-induced inverse bubble column, the bubbles are more uniform in size (Kundu et al., 1995; Majumder et al., 2007) Assuming that all the bubbles are of equal size, the number of bubbles formed per unit time can be calculated as

$$\dot{N} = \left(\frac{G}{\rho_g} A_c\right) \Big/ \left(\frac{\pi}{6} d_b^3\right) \tag{5.137}$$

Energy loss in forming a single bubble is $\pi d_b^2 \sigma$. Therefore, the rate of energy loss for the formation of N number of bubbles is $E_b = \dot{N} \pi d_b^2 \sigma$, which on substitution in equation (5.138) gives

$$E_b = \left(\frac{G}{\rho_g} A_c\right) \frac{6\sigma}{d_b} \tag{5.138}$$

Therefore, additional pressure drop caused by bubble formation can be obtained by dividing the aforementioned rate of energy loss (E_b) by the volumetric flow rate of gas, and this can be obtained as

$$\Delta P_b = \frac{6\sigma}{d_b} \tag{5.139}$$

The size of bubbles formed is determined by the forces acting on the gas–liquid interphase. In low-viscosity liquids, the bubbles are deformed by forces arising from liquid velocity fluctuations acting over distances of the order of bubble diameter, d_b. The restoring force resisting the deformation of the bubble is caused by surface tension acting on the gas–liquid interface. The ratio of these two forces is known as the Weber number, *We*, and can be written as $We = \rho_l u^2 d/\sigma_l$, where u^2 is the average value of the squares of the velocity differences acting over length scales of the order d. σ_l and ρ_l are the surface tension and density of the liquid, respectively. The Weber number can be used to predict a maximum stable bubble diameter, d_b, by assuming that a bubble will split when a critical Weber number (We_c) is reached (i.e., $We_c = \rho_l u^2 d_b/\sigma_l$). In this case, u^2 is the average value of the squares of the velocity differences acting over length scales of the order d_b. For isotropic flow fields, the critical Weber number can be related to the average energy dissipation rate per unit volume, $\bar{\epsilon}$, within the liquid (Hinze, 1955). Applying this assumption to the mixing zone at the top of the column, the corresponding bubble diameter is given by (Evans et al., 1992)

$$d_b = (We_c \sigma_l/2)^{3/5}\, \rho_l^{-1/5}\, \bar{\epsilon}^{-2/5} \tag{5.140}$$

where $\bar{\epsilon}$ is the average energy dissipation rate per unit volume. The critical Weber number can be chosen from a number of theoretical and numerical studies relating to the breakup of bubbles. Hinze (1955) gave a value of 1.18 based on the breakup of a droplet or bubble in a viscous shear flow. Sevik and Park (1973) obtained a value of 1.24 based on the assumption that a bubble will oscillate violently and break up when the characteristic frequency of the turbulence is equal to one of the resonant frequencies of the bubble. Miksis et al. (1981) predicted a value of 3.23 for a bubble in a uniform flow; Lewis and Davidson (1982) found that a cylindrical bubble became unstable in an axisymmetric inviscid shear flow at a critical Weber number of 4.7. Ryskin and Leal (1984) predicted that the critical Weber number was a function of the Reynolds number and gave values in the range of 0.95 to 2.76 for bubbles in a uniaxial extensional Newtonian flow. According to the work of Evans et al. (1992), a value of 1.2 is a reasonable assumption for the critical Weber number for inverse bubbly flow when break-up of bubble is related to the energy dissipation rate. It has been pointed out earlier that in the inverse bubbly flow, gas–liquid mixing occurs in the extended contactor. After coming out from the divergent diffuser, the liquid jet can be arrested in the contactor by adjusting the liquid column height inside the column. It is well known that when a liquid jet of uniform initial velocity is discharged from a nozzle into a fluid, a tangential separation surface is created between the jet and the surrounding medium. The instability of this surface creates eddies that move both in the main direction of flow and across it. This causes exchange of momentum with the surrounding medium. As a result, a region of finite thickness with a velocity is formed on the border of the jet and surrounding fluid. When this jet impinges on the liquid surfaces, it entrains the surrounding gas along with it. The entrained gas undergoes subsequent breakdown and disintegrates as fine bubbles within a finite length, called the *intense mixing zone.* In the intense mixing zone, the jet energy is used to form fine gas bubbles, which are carried in the inverse direction by the bulk liquid motion and leave the mixing zone. The region below the mixing zone is referred to as *homogeneous bubbly zone.* The rate of energy dissipation that forms the bubble in the mixing zone can be obtained from energy balance in the mixing zone.

The specific energy dissipation for the formation of bubble in the mixing zone can be calculated as follows (Kundu et al., 1997):

$$\rho_l e_m = K_m(\rho_l u_{\text{in}}^2/2) \tag{5.141}$$

where K_m is the mixing loss coefficient. A correlation has been developed by Mandal et al. (2004b) in the same system in terms of physical properties of liquid and gas and geometric and dynamic variables of the system as

$$K_m = 1.426 \times 10^{-2}\, \text{Re}_l^{0.604}\, A_r^{0.590}\, H_r^{-0.072}\, Mo^{0.144} \tag{5.142}$$

The specific energy, e_m, can be converted to the average energy dissipation rate per unit volume ($\bar{\in}$) by multiplying by the liquid mass flow rate, $\dot{m}_l$, and dividing by the volume of the column of length, Δz:

$$\bar{\in} = \frac{e_m \dot{m}_l}{(\pi/4)d_c^2 \Delta Z} \tag{5.143}$$

Substituting e_m from equation (5.142) and replacing $\dot{m}_l$ with $\rho_l Q_l$ and Q_l with $(\pi/4)d_n^2 u_j$, equation (5.144) can be written as

$$\bar{\in} = \frac{1}{2}\frac{K_m \rho_l u_j^3 d_n^2}{d_c^2 \Delta z} \tag{5.144}$$

The values for mixing loss coefficient, K_m were calculated from equation (5.143). The stable bubble diameter can be calculated from equation (5.141) by knowing the values of K_m from equation (5.143) and $\bar{\in}$ from equation.

Determination of model parameters: The total gas phase frictional pressure drop is the sum of pressure drop due to flow of gas and due to form drag and can be written as

$$\Delta P_{fg} = \frac{\alpha_g \Delta P_{fg,sp}}{\varepsilon_g^3} + \frac{1}{8}C_d \rho_l u_s^2 \tag{5.145}$$

The liquid phase pressure drop was calculated using Fanning's equation:

$$\Delta P_{fl,sp} = 2 f_{l,sp} \rho_l u_{sl}^2 \Delta z/d_c \tag{5.146}$$

Similarly, for the gas phase, the pressure drop was calculated as

$$\Delta P_{fg,sp} = 2 f_{g,sp} \rho_g u_{sg}^2 \Delta z/d_c \tag{5.147}$$

Substituting the model equations (5.127) for ΔP_{fl} and (5.146) for ΔP_{fg} and (5.139) for E_b into equation (3.35), one gets the final equation for ΔP_{tp} as

$$-\Delta P_{tp} = \left[\frac{\alpha_g \Delta P_{fg,sp}}{\varepsilon_g^3} + \frac{1}{8}C_d \rho_l u_s^2\right]\frac{[G/\rho_g]}{[L/\rho_l + G/\rho_g]} + \left[\frac{6\sigma}{d_b}\right]\frac{[G/\rho_g]}{[L/\rho_l + G/\rho_g]}$$
$$+ \left[\frac{\alpha_l \Delta P_{fl,sp}}{(1-\varepsilon_g)^3}\right]\frac{L/\rho_l}{[L/\rho_l + G/\rho_g]} + \frac{[L+G]g\Delta Z}{[L/\rho_l + G/\rho_g]} \tag{5.148}$$

As discussed in an earlier section, in case of gas phase, the parameter α_g defined as $\alpha_g = \alpha_g' \alpha_g''$ depends on the parameter α_g''. The parameter α_g'' depends on the number of bubbles coming in contact with the wall of the column. Because there is a thin layer of liquid through the wall of the column, no bubbles practically come in contact with the wall of the column. So, α_g'' and hence the value of α_g can be considered as zero. Therefore, in equation (5.149), the value of $\alpha_g \Delta P_{fg,sp} / \varepsilon_g^3$ can be neglected. Therefore, equation (5.149) can be written as

$$
\begin{aligned}
-\Delta P_{tp} = &\left[\frac{1}{8} C_d \rho_l u_s^2\right] \frac{[G/\rho_g]}{[L/\rho_l + G/\rho_g]} + \left[\frac{6\sigma}{d_b}\right] \frac{[G/\rho_g]}{[L/\rho_l + G/\rho_g]} \\
&+ \left[\frac{\alpha_l \Delta P_{fl,sp}}{(1-\varepsilon_g)^3}\right] \frac{L/\rho_l}{[L/\rho_l + G/\rho_g]} + \frac{[L+G]g\Delta Z}{[L/\rho_l + G/\rho_g]}
\end{aligned}
\tag{5.149}
$$

Equation (5.150) can be rewritten in dimensionless form as

$$
-\delta_{tp} = (\lambda + \psi)(1-\gamma_l)\rho_R + \frac{\alpha_l \delta_{fl,sp}\gamma_l}{(1-\varepsilon_g)^3} + [\gamma_l + (1-\gamma_l)\rho_R]
\tag{5.150}
$$

where, $\delta_{lg} = \Delta P_{lg}/\rho_l g \Delta z$, $\delta_{fl,sp} = \Delta P_{fl}/\rho_l g \Delta z$, $\delta_{fg,sp} = \Delta P_{fg,sp}/\rho_g g \Delta z$, $\lambda = 6\sigma/\rho_l g \Delta z d_b$

$\psi = (\frac{1}{8} C_d \rho_l u_s^2)/(\rho_l g \Delta z)$, $\gamma_l = (L/\rho_l)/(L/\rho_l + G/\rho_g)$, and $\rho_R = \rho_g/\rho_l$. In equation (5.151), except for α_l, all parameters are known. α_l can be obtained from the equation (5.151) as

$$
\alpha_l = [-\delta_{lg} - (\lambda + \psi)(1-\gamma_l) - \{\gamma_l + (1-\gamma_l)\rho_R\}]\frac{(1-\varepsilon_g)^3}{\delta_{fl,sp}\gamma_l}
\tag{5.151}
$$

One can calculate the two-phase friction factor, f_{lg} from the definition of α_l as follows:

$$
f_{tp} = \alpha_l f_{l,sp}
\tag{5.152}
$$

Using the experimental pressure drop and gas hold-up, the corresponding values of α_l can be calculated for different operating variables from equation (5.152). Then two-phase friction factor can be calculated by equation (5.153). The drag coefficient can be calculated from equation (5.135).

The bubble–liquid interfacial shear stress depends on the two-phase friction factor, f_{lg}, which that can be calculated as

$$
\tau_i = \frac{1}{2} f_{lg} \rho_l u_{sl}^2
\tag{5.153}
$$

A correlation is developed for the interfacial shear stress for the inverse flow, which can be represented by

$$
\frac{\tau_i}{\rho_l u_{sl}^2} = 2.63 \times 10^3 \frac{A_r^{0.23}}{Re_l^{1.61} Mo^{0.42} H_r^{1.03}}
\tag{5.154}
$$

within a range of $4424.73 < Re_l < 11723.78$; $41.6 < A_r < 156.25$ and $12.21 < H_r < 33.91$. Many other authors (Govier and Aziz, 1972; Fernandes et al., 1983; Couët et al., 1991; Vassallo, 1999; Woerlee et al., 2001) developed pressure gradient correlation for an air–water system based on two-phase mechanical energy balance in vertical pipe flow. Martindale and Smith (1981) examined the behavior of the interface friction factor or drag coefficient as a means for extending the modeling of separated two-phase flows through the separate consideration of each phase. Kabir and Hasan (1990) presented application of a recently developed method for predicting two-phase gas–oil pressure drop in vertical oil wells. Five principal flow regimes—bubbly, dispersed bubbly, slug, churn, and annular—were recognized while developing appropriate correlations for predicting pressure drop in each flow regime. Joye (2003) reported that pressure drop is significantly affected by heat transfer in mixed convection situations. He developed a semiempirical correlation for pressure drop from a theoretical base by first making use of the momentum integral solution to the heat and momentum equations for natural convection on a vertical surface with constant wall temperature. Wen and Kenning (2004) investigated two-phase pressure drop during flow boiling of water in a vertical channel. They reported that the Lockhart–Martinelli (1949) correlation for pressure drop in large tubes with a modified C value is applicable with uncertainty of 35%. Kulov et al. (1979) reported pressure drop, mean film thickness, and entrainment in inverse two-phase flow with a water–air system at room temperature. They proposed relationships for calculation of the measured parameters as functions of gas and liquid flow rates and the height of the tube. Blok et al. (1983) described the transition from gas-continuous to pulsing flow, the liquid hold-up, and the pressure drop in cocurrent gas–liquid inverse flow packed column. They reported that pressure drop is linearly dependent on the pulse frequency.

Klausner et al. (1990) studied the frictional pressure drop and vapor volume fraction in vertical upflow and inverse flow boiling. They observed that all values of two-phase multipliers evaluated from their measured frictional pressure drop data in vertical upflow fall in a range bounded by the predictions of the Chisholm correlation and the homogeneous model. For vertical inverse flow, the frictional pressure drop exhibits an anomalous behavior similar to that reported by previous investigators.

Kundu et al. (1995) reported that two-phase cocurrent vertical inverse flow systems offer some distinct advantages, such as uniform and finer bubbles, greater residence time, and negligible coalescence of the bubbles. They carried out experimental studies to evaluate the total pressure gradient in an inverse flow bubble column fitted with an ejector system. They also developed correlations to predict pressure drop as a function of different physical and dynamic variables.

Fu and Klausner (1997) developed a separated flow model that is applicable to vertical annular two-phase inverse flow in the convective heat transfer regime. They used conservation of mass, momentum, and energy to solve for the pressure drop. They specified relationships for the interfacial friction factor, liquid film eddy viscosity, turbulent Prandtl number, and entrainment rate in annular inverse flow system. Mandal et al. (2004a) exclusively investigated the two-phase frictional pressure drop with air–non-Newtonian liquid (carboxymethyl cellulose (CMC) solutions) system in a concurrent inverse flow bubble column. They analyzed their experimental two-phase frictional pressure drop data by modifying the Lockhart–Martinelli correlation and Aoki correlation. Also, they developed correlations for two-phase friction factor in terms of dimensionless groups of the operating and system variables in inverse bubbly flow regime. The typical effects of gas flow rate on gas–liquid two-phase pressure drop, $\Delta P_{tp}/\rho_l g\Delta Z$, in the inverse bubble column is shown in Figure 5.2.

The two-phase pressure drop increases with increasing gas flow rate, Q_g, at a constant liquid flow rate. An increase of Q_g causes a higher population of gas bubbles, which in turn increases the true liquid velocity. Lower pressure drops are observed at higher liquid flow rates for the same gas flow rate. At a higher liquid flow rate, comparatively bigger bubbles are formed because of coalescence, which causes a decrease in the true liquid velocity because of an increase in the liquid flow area. Moreover, because of bigger bubble size, some of the liquid get entrapped in the wake of the bubble (Collins, 1965; Pal et al., 1980); as a result, the bulk liquid velocity reduces, which causes decreasing two-phase pressure drop at an increasing liquid flow rate. The pressure drops because of bubble formation and form drag are significant in the dispersed bubbly flow in the column (Majumder et al., 2006, 2007). As the fluid flow rate increases, the frictional pressure drop because of form drag increases. The increase of form drag is influenced by the local gas phase distribution and bubble interactions because of momentum exchange of the fluids. The pressure drop caused by form drag is a function of slip velocity. As the liquid flow rate is increased, the slip velocity (u_s) increases, which contributes to the increase in pressure drop caused by

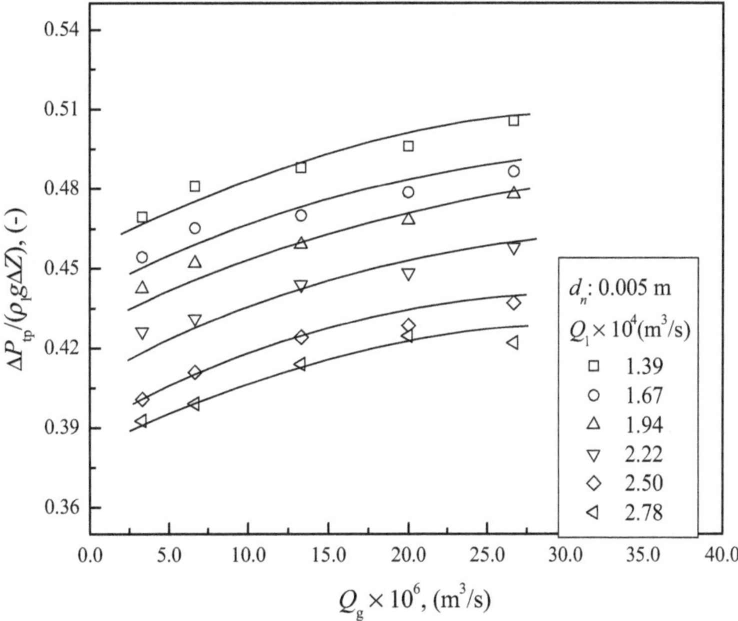

FIGURE 5.2 Effect of gas flow rate on gas–liquid two-phase pressure drop.

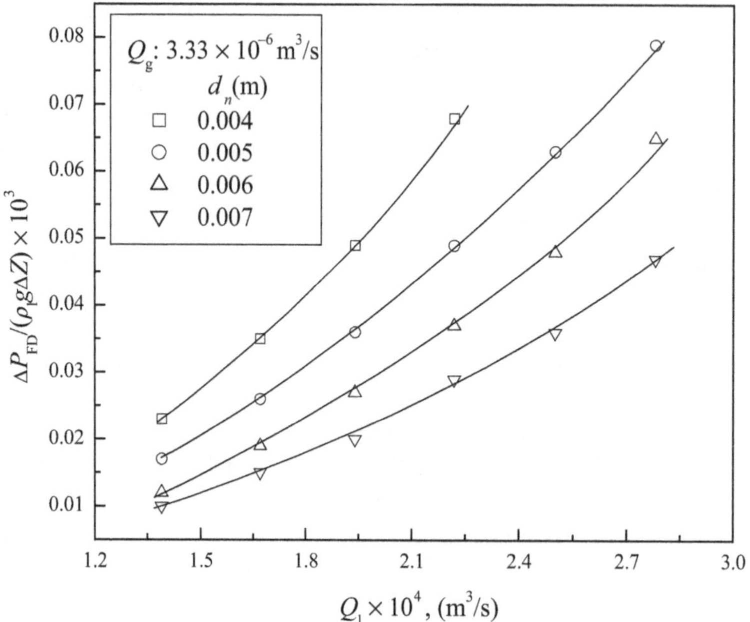

FIGURE 5.3 Variation of pressure drop with liquid flow rate.

form drag as shown in Figure 5.3. Also, the drag force takes into account the interaction force between the liquid and bubbles in a uniform flow field under a nonaccelerating condition. As the gas flow rate increases, the fractional gas hold-up increases, which enhances intimate contact between the liquid and bubbles inside the column. This increases the interfacial drag force between the phases. Also, the bubble number density increases with the increase in gas hold-up. As the bubble number density increases, the interfacial drag force increases because of the increase in interfacial area. This results in an increase in pressure drop caused by form drag with an increase in the gas flow rate (Figure 5.4). Bubble formation also contributes to the variation of pressure drop. Table 5.3 shows the values of pressure drop caused by bubble formation and phase interaction that contribute to the total frictional pressure drop for typical flow rates of the phases.

As the fluid flow rate increases, the frictional pressure drop caused by phase interaction increases. The increase of pressure drop caused by phase interaction is influenced by the local gas phase distribution and bubble interactions caused by momentum exchange. As the liquid flow rate increases, the pressure drop caused by bubble formation increases as shown in Figure 5.5. With an increase in the liquid flow rate, because of higher momentum transfer, entrainment of gas increases, and the number density of bubble increases as smaller bubbles are formed at a higher liquid flow rate. The pressure drop caused by bubble formation is a function of bubble diameter. A smaller bubble diameter contributes to a higher pressure drop. This results in an increase in pressure drop caused by bubble formation with

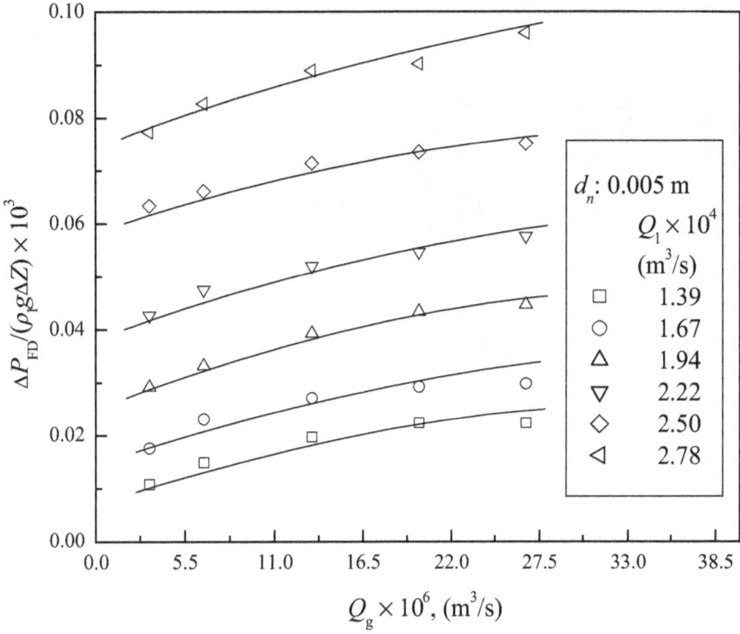

FIGURE 5.4 Variation of pressure drop caused by form drag with gas flow rate.

an increase in the liquid flow rate. But at the same liquid flow rate, the pressure drop caused by bubble formation does not change significantly with the gas flow rate. At a constant liquid flow rate, if the gas flow rate is increased, the population of the bubble increases, but the size of the stable bubble is almost the same at steady state. This causes the pressure drop caused by bubble formation to remain almost the same with the increasing gas flow rate. Also, increasing the gas hold-up may increase the frictional pressure drop. Frictional pressure drop increases with gas hold-up because of decreasing flow area of the liquid.

Numerical Models

For modeling purposes, one needs to know the classification of the two-phase flows. The two phases are is mainly classified based on the physical state of the two phases and on the flow structure.

As per the physical state, four combinations can be considered: gas–solid, gas–liquid, liquid–solid, and liquid–liquid mixtures. For the flow structure, especially the interface topology, it is more difficult to establish a set of combinations because changes in the flow may be occurring continually in a system. However, according to Ishii and Hibiki (2011), two-phase flows are generally classified into three categories according to the shape of the interface: separated flows, mixed flows, and dispersed flows. In separated two-phase flows, each phase plays a symmetrical role, and the • topology of the interface is not assumed. No hypothesis can be proposed for the shape of each phase, and the volume of

Table 5.3   Typical Values of Contribution of Different Components to the Total Pressure Drop

System	Superficial Liquid Velocity, $u_{sl} \times 10^2$, (m/s)	Superficial Gas Velocity, $u_{sg} \times 10^2$, (m/s)	Bubble–Liquid Interfacial Shear Stress, τ, (N/m^2)	Dimensionless Pressure Drop		
				Friction Caused by Liquid Phase, $\Delta P_{fl}/(\rho_l g \Delta Z)$, (-)	Contribution from Phase Interaction, $\Delta P_{FD}/(\rho_l g \Delta Z) \times 10^4$, (-)	Contribution from Bubble Formation, $\Delta P_b/(\rho_l g \Delta Z) \times 10^2$, (-)
Air–water	10.18	2.12	6.51	0.125	0.364	1.803
	11.66	2.38	13.40	0.233	0.376	1.856
	2.57	2.08	14.43	0.251	0.447	1.953
	13.44	1.91	17.90	0.360	0.460	1.965
Air–CMC-1	17.05	1.61	21.30	0.375	1.364	2.280
	17.92	1.70	25.60	0.361	1.508	2.367
	17.92	1.88	27.00	0.359	1.518	2.538
Air–CMC-2	10.18	1.91	11.31	0.193	1.509	2.509
	10.89	1.70	12.22	0.217	1.746	2.383
	11.66	2.00	15.75	0.253	1.758	2.719
	12.57	1.70	20.33	0.316	2.010	2.884
Air–CMC-3	13.34	2.20	20.50	0.215	2.588	2.500
	14.20	2.04	22.00	0.286	2.527	2.539
	15.02	2.16	23.10	0.290	3.084	2.748
	15.93	2.25	24.10	0.296	3.238	2.807
Air–CMC-4	15.93	2.04	25.20	0.260	3.997	3.117
	16.60	2.04	26.20	0.279	4.598	3.288
	17.31	1.61	27.20	0.320	4.654	3.308
	17.97	1.91	27.90	0.339	4.559	3.350

CMC, carboxymethyl cellulose.

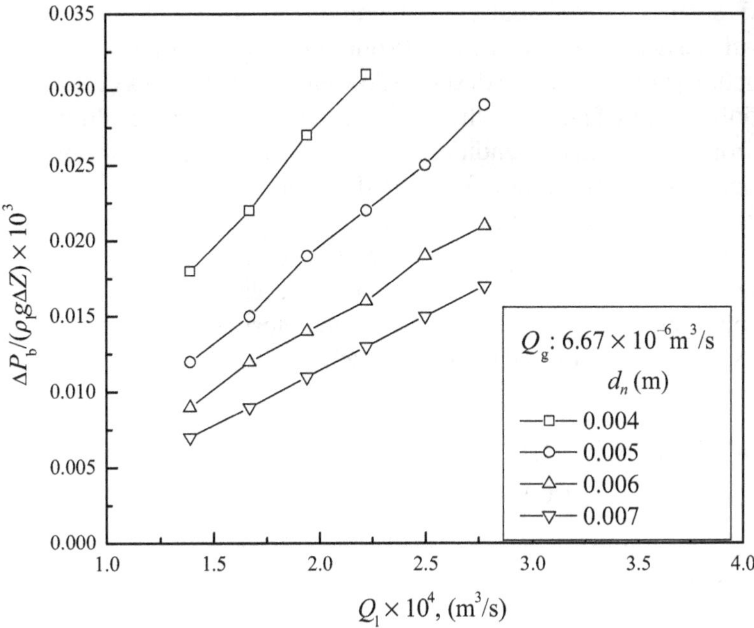

FIGURE 5.5 Variation of pressure drop caused by bubble formation with liquid flow rate.

one of the phases cannot be neglected. There are two approaches for the mathematical description of such flows:

1. **Tracking the interface:** By tracking the interface between both fluids and solving Navier-Stokes equations in each one by volume of fluid (VOF) (Hirt and Nichols, 1981), level-set methods (LSMs) (Sussman et al., 1994)
2. **Averaging phase flow:** By using an averaging procedure to derive a set of conservation equations describing the two-phase flow as a whole and assuming that each phase may be simultaneously present at any point. Using the averaging procedure, the smallest scales of the interface cannot be resolved but modeled (Ishii and Hibiki, 2011).

In the case of dispersed flows, one phase is assumed to be dispersed inside the other phase. The flow around and inside the dispersed phase does not have to be computed. The influence of the dispersed phase on the continuous phase is taken into account with the introduction of source terms in the Navier-Stokes equations. Direct solution to the Navier-Stokes equations is becoming available for relatively low Reynolds numbers in case of single-phase turbulent pipe flows. However, in case of two-phase flows, the problem becomes more complicated. The presence of a second phase with a different transport properties and the fact that the second phase is usually not distributed across the interior of the pipe makes solution very challenging. Even if attempts to modify the single-phase energy correlations to suit gas–liquid flow has been made by some authors, rigorous numerical models for two-phase flow do not contribute in most practical applications today. The two-phase flow problem can be solved using methods such as integral analysis, differential analysis, computational fluid dynamics (CFD), and artificial neural network (ANN). In one-dimensional integral analysis, some forms of certain functions are assumed to describe flow parameters such as velocity or concentration distribution in a pipe first. Then the functions are made to satisfy fluid mechanics equations and appropriate boundary conditions in integral form. Differential analysis is a method by which velocity and concentration fields are deduced from suitable differential equations. Time-averaged quantities are usually used to write the equations such as single-phase turbulence theories. CFD solutions are usually challenged by stability, convergence, and accuracy. The importance of well-placed mesh is also crucial in implementing CFD methods. ANN uses the prior acquired knowledge to respond to a new information rapidly and automatically. Large data sets are usually required to develop ANN methods.

Estimation of Frictional Pressure Drop in a Plunging Liquid Jet Inverse Bubbly Flow Column

When the liquid jet impinges on the pool surface, the liquid sustains an instantaneous deceleration, which increases the dynamic pressure inside the column (Majumder et al., 2007). The momentum pressure (ΔP_{mom}) pushes the column contents in the inverse direction of bubble buoyancy. It is equal to the ratio of force exerted by the nozzle and the cross-sectional area of the column, which can be expressed as

$$\Delta P_{mom} = \frac{F_j}{A_c} = \frac{ma_j}{A_c} = \frac{Q_l \rho_l \left(u_j - u_l\right)}{A_c} \tag{5.155}$$

where u_j is the jet velocity, u_l is the interstitial liquid velocity in the column, and Q_l is the volumetric flow rate of the liquid. From the continuity equation, one can write

$$u_j A_j = u_l A_c (1 - \varepsilon_g) = u_l A_c \tag{5.156}$$

where A_j and A_c are the areas of the nozzle and the column, respectively. From the equation (5.157), one gets

$$u_j = u_l \frac{A_c}{A_j} \tag{5.157}$$

and

$$u_l = \frac{u_{sl}}{(1 - \varepsilon_g)} \tag{5.158}$$

By substituting equations (5.158) and (5.159) into equation (5.156), one gets

$$\Delta P_{mom} = u_l^2 \rho_l \left[\frac{d_c^2}{d_j^2} - \frac{1}{\left(1 - \varepsilon_g\right)} \right] \tag{5.159}$$

The frictional pressure drop (ΔP_{ftp}) can then be represented as

$$\Delta P_{ftp} = \Delta P_{tp} - \left(h_m \rho_l \left(1 - \varepsilon_g\right) g \right) - \left(u_l^2 \rho_l \left[\frac{d_c^2}{d_j^2} - \frac{1}{\left(1 - \varepsilon_g\right)} \right] \right) \tag{5.160}$$

The two-phase frictional pressure drop based on superficial liquid velocity can also be expressed as

$$\frac{\Delta P_{ftp}}{g \rho_l h_m} = \frac{2 f_{tp} u_l^2}{g d_c} \tag{5.161}$$

By substituting equation (5.162) into equation (5.161), the two-phase friction factor (f_{tpl}) can be obtained as

$$f_{tp} = \left(\frac{g d_c}{2 u_l^2} \right) \left(\frac{\Delta P_{tp}}{g \rho_l h_m} - \left(1 - \varepsilon_g\right) - \left(\frac{u_l^2}{g h_m} \left[\frac{d_c^2}{d_j^2} - \frac{1}{\left(1 - \varepsilon_g\right)} \right] \right) \right) \tag{5.162}$$

Nomenclature

A Cross-sectional area (m²)
A_r Area ratio (-)
c Parameter/coefficient (-)
d Diameter (m)
e Energy per unit mass (N·m/kg)
E Energy (Nm)
f Friction factor (-)
F Force (N)
Fr Froude number (-)
g Gravitational acceleration (m/s²)
G Gas mass velocity (kg/m²s)
Gr Dimensionless group (-)
H Height (m)
H Height of gas–liquid mixture (m)
H_r Height ratio (-)
I Index (-)
l Scale of turbulence or Prandtl's mixing length (m)
L Liquid mass velocity (kg/m²s)
m Mass flow rate (kg/s)
M Mass flux (kg/m²s)
Mo Morton number (-)
P Pressure drop (N/m²)
q Parameter (-)
Q Volumetric flow rate (m³/s)
R Radius of pipe (m)
Re Reynolds number (-)
S Slip ratio (-)

Su Suratman number (-)
u Velocity (m/s)
We Weber number (-)
x Mass quality (-)
X Martinelli parameter (-)
z Length (m)

Greek Letters

θ Angle of inclination (degree)
$\in$ Energy per unit volume (N/m^2)
ε Hold-up (-)
λ Parameter (-)
α Parameter (-)
α' Parameter (-)
ρ Density (kg/m^3)
Δ Difference
σ Surface tension (N/m)
ϕ Two-phase friction multiplier (-)
μ Viscosity (Pa-s)
τ Shear stress (N/m^2)
β Volumetric flow quality (-)
β_{Ch} Chisholm parameter (-)
υ Specific volume (m^3/kg)
χ Parameter (-)
ω Quality viscosity (Pa-s)

Suffixes

a Acceleration
A Aoki
A-M Award and Muzychka
AT Additional turbulence
b bubble
Bf Bankoff
c Column/critical
Ch Chisholm
d Drag
E-W Eisenberg and Weinberger
eff Effective
f Frictional
FD Form drag
Fr Froude
Frd Friedel
g Gas
Gd Gronneoud
hm Homogeneous
j Jet

K Kato
l Liquid
m Mixture/mixing
M Martinelli
mom Momentum
n Nozzle
p Property
s Hydrostatic/slip/superficial
sl Superficial liquid
S-M Sun and Mishima
sp Single phase
sg Superficial gas
t Total
tp Two-phase
td Total drag
w wall
W Wallis
y Transverse/y-coordinate

References

Aoki, S., Inoue, S., 1965. Fundamental studies on pressure drop in an air–water two phase flow in vertical pipes. RRR of 2nd SSSS of Heat Transfer Society of Japan 137.

Akers, W.W., Deans, H.A., Crosser, O.K., 1958. Condensing heat transfer within horizontal tubes. Chem. Eng. Prog. 54, 89–90.

Akers, W.W., Deans, H.A., Crosser, O.K., 1959. Condensation heat transfer within horizontal tubes. Chem. Eng. Prog. S. Ser. 55 (29), 171–176.

Andeen, G.B., Griffith, P., 1968. Momentum flux in two-phase flow. J. Heat Transfer 90 (2), 211–222.

Arrhenius, S., 1887. On the internal friction of solutions in water. ZeitschriftfürPhysikalischeChemie (Leipzig) 1, 285–298.

Awad, M.M., Muzychka, Y.S. 2004, A simple asymptotic compact model for two-phase frictional pressure gradient in horizontal pipes, RRR of the International Mechanical Engineering Conference and Exposition, 675-685.

Awad, M.M., Muzychka, Y.S., 2008. Effective property models for homogeneous two-phase flows. Exp. Ther. Fluid Sci. 33, 106–113.

Awad, M.M., Muzychka, Y.S., 2010. Asymptotic generalizations of the Lockhart–Martinelli method for two phase flows. J Fluids Eng. 132, 1–12.

Bankoff, S.G., 1960. A variable density single-fluid model for two-phase flow with particular reference to steam–water flow. J. Heat Transfer 82 (4), 265–272.

Baroczy, C.J., 1966. A systematic correlation for two-phase pressure drop. Chem. Eng. Progress SSSS 62 (64), 232–249.

Beattie, D.R.H., Whalley, P.B., 1982. A simple two-phase frictional pressure drop calculation method. Int. J. Multiphase Flow 8, 83–87.

Beggs, H.D., Brill, J.P., 1973. A study of two-phase flow in inclined pipes. JPT 607 Trans. AIME 255.

Bingham, E., 1906. Viscosity and fluidity. Am. Chem. J. 35, 95–217.

Blok, J.R., Varkevisser, J., Drinkenburg, A.A.H., 1983. Transition to pulsing flow, holdup and pressure drop in packed columns with cocurrent gas-liquid downflow. Chem. Eng. Sci 38 (5), 687–699.

Cengel, J., 1967. Viscosity of liquid–liquid dispersions in laminar and turbulent flow, PhD Thesis. Oregon State University, Oregon. Abstract No. 67-11805.

Chawla, J.M., 1972. Frictional pressure drop on flow of liquid/gas mixtures in horizontal pipes. Chemie Ingenieur Technik 44, 58–63.

Chen, J.J.J., Spedding, P.L., 1981. An extension of the Lockhart–Martinelli theory of two phase pressure drop and holdup. Int. J. Multiphase Flow 7 (6), 659–675.

Chenoweth, J.M., Martin, M.W., 1955. Turbulent two-phase flow. Petrol. Refin. 34 (10), 151–155.

Chilekar, V.P., van der Schaaf, J., Kuster, B.F.M., Tinge, J.T., Schouten, J.C., 2010. Influence of elevated pressure and particle lyophobicity on hydrodynamics and gas–liquid mass transfer in slurry bubble columns. AIChE J. 56, 584–596.

Chisholm, D., 1967. A theoretical basis for the Lockhart–Martinelli correlation for two-phase flow. Int. J. Heat Mass Transfer 10 (12), 1767–1778.

Chisholm, D., 1973. Pressure gradients due to friction during the flow of evaporating two phase mixtures in smooth tubes and channels. Int. J. Heat Mass Transfer (16), 347–358.

Churchill, S.W., 1977. Friction factor for equations spans all fluid flow regimes. Chem. Eng. 84 (24), 91–92.

Cicchitti, A., Lombaradi, C., Silversti, M., Soldaini, G., Zavattarlli, R., 1960. Two-phase cooling experiments—pressure drop, heat transfer, and burnout measurements. Energia Nucleare 7 (6), 407–425.

Clark, N.N. Flemmer, R.L.C., 1985a. Two-phase pressure loss in terms of mixing length theory. 1. Derivation for the general case of dispersed flow. Ind. Eng. Own. Fundam. 24, 412–418.

Clark, N.N., Flemmer, R.L.C., 1985b. Two-phase pressure loss in terms of mixing length theory. 2. Comparison of model with upflow and downflow data. Ind. Eng. Chem. Fundam. 24 (4), 418–423.

Couët, B., Brown, P., Hunt, A., 1991. Two-phase bubbly-droplet flow through a contraction: Experiments and a unified model. Int. J. Multiphase Flow 17 (3), 291–307.

Collins, R., 1965. Structure and behaviour of wakes behind two-dimensional air bubbles in water. Chem. Eng. Sci. 20, 851–853.

Davies, J.T., 1972. Turbulence phenomena. Academic Press, New York.

Davis, W.J., 1963. The effect of Froude Number in estimating vertical two phase gas–liquid friction losses. Bri. Chem. Eng. 8, 462–468.

Davidson, W.F., Hardie, P.H., Humphreys, C.G.R., Markson, A.A., Mumford, A.R., Ravese, T., 1943. Studies of heat transmission through boiler tubing at pressures from 500-3300 Lbs. Trans. ASME 65 (6), 553–591.

Desouky, S.E.M., El-Emam, N.A., 1989. Program designs for pseudo plastic fluids. Oil Gas J. 87 (51), 48–51.

Dukler, A.E., Wicks, M., Cleaveland, R.G., 1964. Pressure drop and hold up in two-phase flow. AIChE J. 10, 38–51.

Einstein, A., 1909. A new definition of molecular dimensions. Annalen der Physik 4, 289–306.

Eisenberg, F.G., Weinberger, C.B., 1979. Annular two phase flow of gases and non-Newtonian liquids. AIChE J. 25, 240–246.

Evans, G.M., Jameson, G.J., Atkinson, B.W., 1992. Prediction of the bubble size generated by a plunging liquid jet bubble column. Chem. Eng. Sci. 47, 3265–3272.

Fernandes, R.C., Semiat, R., Dukler, A.E., 1983. Hydrodynamic model for gas–liquid slug flow in vertical tubes. AIChE Journal 29 (6), 981–989.

Fourar, M., Bories, S., 1995. Experimental study of air–water two-phase flow through a fracture (Narrow Channel). Int. J. Multiphase. 21 (4), 621–637.

Friedel, L., 1980. Pressure drop during gas/vapour–liquid flow in pipes. Int. Chem. Eng. 20, 352–367.

Fu, F., Klausner, J.F., 1997. A separated flow model for predicting two-phase pressure drop and evaporative heat transfer for vertical annular flow. Int. J. Heat Fluid Flow 18 (6), 541–549.

García, F., García, R., Padrino, J.C., Mata, C., Trallero, J.L., Joseph, D.D., 2003. Power law and composite power law friction factor correlations for laminar and turbulent gas–liquid flow in horizontal pipelines. Int. J. Multiphase Flow 29, 1605–1624.

Govier, G.W., Aziz, K., 1972. The flow of complex mixture in pipes. Van Nostrand-Reinhold Co., New York.

Gharat, S.D., Joshi, J.B., 1992. Transport phenomena in bubble column reactors II: pressure drop. Chem. Eng. J. 48, 153–166, 153.

Grönnerud, R. 1979, Investigation of liquid hold-up, flow resistance and heat transfer in circulation type evaporators, part iv: two-phase flow resistance in boiling refrigerants, Annexe 1972-1, Bull. de l'Inst du Froid.

Han, L., Al-Dahhan, M.H., 2007. Gas–liquid mass transfer in a high pressure bubble column reactor with different sparger designs. Chem. Eng. Sci. 62, 131–139.

Hashemi, S., Macchi, A., Servio, P., 2009. Gas–liquid mass transfer in a slurry bubble column operated at gas hydrate forming conditions. Chem. Eng. Sci. 64, 3709–3716.

Hatschek, E., 1928. The viscosity of liquids. Bell & Sons Ltd., London.

Hinze, J.O., 1955. Fundamentals of the hydrodynamic mechanism of splitting in dispersion processes. AIChE J. 1, 289–295.

Hirt, C.W., Nichols, 1981. B.D., Volume of fluid (VOF) method for the dynamics of free boundaries. J. Comput. Phys. 39, 201–225.

Holcombe, N.T., Smith, D.N., Knickle, H.N., O'Dowd, W., 1983. Thermal dispersion and heat transfer in nonisothermal bubble columns. Chem. Eng. Commun. 21, 135–150.

Hoogendoorn, C.J., 1959. Gas–liquid flow in horizontal pipes. Chem. Eng. Sci. 9, 205–217.

Houzelot, J.L., Thiebaut, M.F., Charpentier, J.C., Schiber, J., 1983. Contribution à l'étude hydrodynamique des colonnes à bulles. Entropie (Paris) 19, 121–126.

Hughmark, G.A., Pressburg, B.S., 1961. Holdup and pressure drop with gas–liquid flow in a vertical pipe. AIChE J. 7 (4), 677–682.

Ishii, M., Hibiki, T., 2011. Thermo-Fluid Dynamics of Two-Phase Flow, second ed., Springer, New York.

Jin, H., Yang, S., He, G., Liu, D., Tong, Z., Zhu, J., 2014. Gas–liquid mass transfer characteristics in a gas–liquid–solid bubble column under elevated pressure and temperature. Chinese J. Chem. Eng. 22, 955–961.

Jordan, U., Schumpe, A., 2001. The gas density effect on mass transfer in bubble columns with organic liquids. Chem. Eng. Sci. 56, 6267–6272.

Joshi, J.B., 2001. Computational flow modeling and design of bubble column reactors. Chem. Eng. Sci. 56, 5893–5933.

Joshi, J.B., Sharma, M.M., 1979. A circulation cells model for bubble column. Trans. Inst. Chem. Eng. 57, 244–251.

Joye, D.D., 2003. Pressure drop correlation for laminar, mixed convection, aiding flow heat transfer in a vertical tube. Int. J. Heat Fluid Flow 24 (2), 260–266.

Kabir, C.S., Hasan, A.R., 1990. Performance of a two-phase gas/liquid flow model in vertical wells. J. Petrol. Sci. Eng. 4 (3), 273–289.

Kang, Y., Cho, Y.J., Woo, K.J., Kim, K.I., Kim, S.D., Bubble properties and pressure fluctuations in pressurized bubble columns. Chem. Eng. Sci., 55 2000, 411–419.

Kato, Y., 1956, 1958. Report Faculty Eng., Yamanashi Univ., Japan 7:105; 9:29.

Kemoun, A., Cheng, B.O., Gupta, P., Al-Dahhan, M.H., Dudukovic, M.P., 2001. Gas holdup in bubble columns at elevated pressure via computed tomography. Int. J. Multiphase Flow 27, 929–946.

Klausner, J.F., Chao, B.T., Soo, S.L., 1990. An improved method for simultaneous determination of frictional pressure drop and vapor volume fraction in vertical flow boiling. Exp. Ther. Fluid Sci. 3 (4), 404–415.

Kulkarni, A.A., 2007. Mass transfer in bubble column reactors: effect of bubble size distribution. Ind. Eng. Chem. Res. 46, 2205–2211.

Kulov, N.N., Maksimov, V.V., Maljusov, V.A., Zhavoronkov, N.M., 1979. Pressure drop, mean film thickness and entrainment in downward two-phase flow. Chem. Eng. J. 18 (2), 183–188.

Kumar, S., Munshi, P., Khanna, A., 2012. High pressure experiments and simulations in cocurrent bubble columns. Proc. Eng. 42, 842–853.

Kundu, G., Mukherjee, D., Mitra, A.K., 1995. Experimental studies on a co-current gas–liquid downflow bubble column. Int. J. Multiphase Flow 21, 893–906.

Kundu, G., Mukherjee, D., Mitra, A.K., 1997. Ejector performance in a co-current gas–liquid downflow bubble column. Can. J. Chem. Eng. 75, 956–963.

Lain, S., Broder, D., Sommerfeld, M., 1999. Experimental and numerical studies of the hydrodynamics in a bubble column. Chem. Eng. Sci. 54, 4913–4920.

Lau, R., Peng, W., Velazquez-Vargas, L.G., Yang, G.Q., Fan, L.-S., 2004. Gas–liquid mass transfer in high-pressure bubble columns. Ind. Eng. Chem. Res. 43, 1302–1311.

Lewis, D.A., Davidson, J.F., 1982. Bubble splitting in shear flows. Trans. Inst. Chem. Eng. 60, 283–291.

Letzel, H.M., Schouten, J.C., Krishna, R., Van den Bleek, C.M., 1999. Gas holdup and mass transfer in bubble column reactors operated at elevated pressure. Chem. Eng. Sci. 54, 2237–2246.

Lin, S., Kwok, C.C.K., Li, R.Y., Chen, Z.H., Chen, Z.Y., 1991. Local frictional pressure drop during vaporization of R-12 through capillary tubes. Int. J. Multiphase Flow 17, 95–102.

Lockhart, R.W., Martinelli, R.C., 1949. Proposed correlation of data for isothermal two-phase, two component flow in pipes. Chem. Eng. Progress 45 (1), 39–48.

Maalej, S., Benadda, B., Otterbein, M., 2003. Interfacial area and volumetric mass transfer coefficient in a bubble reactor at elevated pressures. Chem. Eng. Sci. 58, 2365–2376.

Mandal, A., Kundu, G., Mukherjee, D., 2004a. Studies on frictional pressure drop of gas-non-Newtonian two-phase flow in a cocurrent downflow bubble column. Chem. Eng. Sci. 59 (18), 3807–3815.

Mandal, A., Kundu, G., Mukherjee, D., 2004b. Gas-holdup distribution and energy dissipation in an ejector-induced downflow bubble column: the case of non-Newtonian liquid. Chem. Eng. Sci. 59 (13), 2705–2713.

Mandhane, J.M., Gregory, G.A., Aziz, K., 1977. Critical evaluation of friction pressure-drop prediction methods for gas–liquid flow in horizontal pipes. J. Petrol. Technol. 29 (10), 1348–1358.

Majumder, S.K., Kundu, G., Mukherjee, D., 2006. Prediction of pressure drops in a modified gas–liquid downflow bubble column. Chem. Eng. Sci. 61 (12), 4060–4070.

Majumder, S.K., Kundu, G., Mukherjee, D., 2007. Pressure drop and bubble-liquid interfacial shear stress in a modified gas-non-Newtonian-liquid downflow bubble column. Chem. Eng. Sci. 62, 2482–2490.

Martinelli, R.C., Boelter, L.M.K., Taylor, T.H.M., Thomsen, E.G., Morrin, E.H., 1944. Isothermal pressure drop for two-phase two-component flow in a horizontal pipe. Trans. ASME 66, 139–151.

Martinelli, R.C., Nelson, D.B., 1948. Prediction of pressure drop during forced-circulation boiling water. Trans. ASME 70, 695–702.

Martindale, W.R., Smith, R.V., 1981. Two-phase two-component interface drag coefficients in separated phase flows. Int. J. Multiphase Flow 7 (2), 211–219.

McAdams, W.H., Woods, W.K., Heroman, L.C., 1942. Vaporization inside horizontal tubes, II. Benzene-oil mixture. Trans. ASME 64, 193–200.

Miksis, M., Vanden-Broeck, J.M., Keller, J.B., 1981. Axisymmetric bubble or drop in a uniform flow. J. Fluid Mech. 108, 89–100.

Müller-Steinhagen, H., Heck, K., 1986. A simple friction pressure drop correlation for two-phase flow in pipes. Chem. Eng. Technol. 20 (6), 291–308.

Muzychka, Y.S., Walsh, E., Walsh, P., Egan, V., 2011. Non-boiling two phase flow in microchannels. In: Mitra, S.K., Chakraborty, S. (Eds.), Microfluidics and Nanofluidics Handbook: Chemistry, Physics, and Life Science Principles. CRC Press Taylor & Francis Group, Boca Raton, FL.

Oglesby, K., 1979. An experimental study on the effects of oil viscosity mixture velocity and water fraction on horizontal oil-water flow, TEGM The University of Tulsa.

Olujic, Z., 1985. Predicting two phase flow friction loss in horizontal pipes. J. Chem. Eng. 92 (13), 45–50.

Oliemans, R., 1976. Two phase flow in fas-transmission pipelines, ASME paper 76-Pet-25, Petroleum Division ASME meeting, Mexico, September 19–24.

Owens, W.L., 1961. Two-phase pressure gradient. ASME Int. Dev. Heat Transfer, Part II, 363–368.

Pal, S.S., Mitra, A.K., Roy, A.N., 1980. Pressure drop and holdup in vertical two-phase cocurrent flow with improved gas-liquid mixing. Ind. Eng. Chem. Proc. Des. Dev. 19, 67–72.

Pjontek, D., Parisien, V., Macchi, A., 2014. Bubble characteristics measured using a monofibre optical probe in a bubble column and freeboard region under high gas holdup conditions. Chem. Eng. Sci. 111, 153–169.

Richardson, E., 1933. On the viscosity of emulsions. Kolloid Zeitschrift 65, 32–37.

Ryskin, G., Leal, L.G., 1984. Numerical solution of free boundary problems in fluid mechanics Part 3. Bubble deformation in an asymmetric straining flow. J. Fluid Mech. 148, 37–43.

Schlichting, H., 1966. Boundary Layer Theory. McGraw-Hill: New York.

Sevik, M., Park, S.H., 1973. The splitting of drops and bubbles by turbulent fluid flow. J. Fluid Eng. Trans. ASME 95, 53–60.

Shaikh, A., Al-Dahhan, M., 2005. Characterization of the hydrodynamic flow regime in bubble columns via computed tomography. Flow Meas. Instrum. 16, 91–98.

Soot, P., 1971. A study of tow-phase liquid–liquid flow in pipes, PhD Thesis. Oregon State University, Oregon. Abstract No. 71-2479.

Sun, L., Mishima, K., 2009. Evaluation analysis of prediction methods for two-phase flow pressure drop in mini-channels. Int. J. Multiphase Flow 35 (1), 47–54.

Sussman, M., Smereka, M., Osher, S., 1994. A level-set approach for computing solutions to incompressible two-phase flows. J. Comput. Phys. 114, 146–159.

Taylor, G., 1932. The viscosity of fluid containing small drops of another fluid. Proc. R. Soc. Lond. 138A (834), 41–48.

Theissing, P., 1980. A generally valid method for calculating frictional pressure drop in multiphase flow. Chem. Eng. Tech. 52, 344–355.

Therning, P., Rasmuson, A., 2001. Liquid dispersion, gas holdup and frictional pressure drop in a packed bubble column at elevated pressures. Chem. Eng. J. 81, 331–335.

Vassallo, P., 1999. Near wall structure in vertical air–water annular flows. Int. J. Multiphase Flow 25 (3), 459–476.

Vermeulen, T., Williams, G.M., Langlois, Y.G.E., 1955. Interfacial area in liquid–liquid and gas–liquid agitation. Chem. Eng. Progress 51, 85–94.

Wallis, G.B., 1969. One dimensional two-phase flow. McGraw-Hill Company, New York.

Wen, D.S., Kenning, D.B.R., 2004. Two-phase pressure drop of water during flow boiling in a vertical narrow channel. Exp. Therm. Fluid Sci. 28 (2–3), 131–138.

Wilkinson, P.M., Haringa, H., Stokman, F.P.A., Van Dierendonck, L.L., 1993. Liquid mixing in a bubble column under pressure. Chem. Eng. Sci. 48, 1785–1791.

Wilkinson, P.M., Haringa, H., Van Dierendonck, L.L., 1994. Mass transfer and bubble size in a bubble column under pressure. Chem. Eng. Sci. 49, 1417–1427.

Woerlee, G.F., Berends, J., Olujic, Z., Jan de, G., 2001. A comprehensive model for the pressure drop in vertical pipes and packed columns. Chem. Eng. J. 84 (3–15), 367–379.

Yang, W., Wang, J., Jin, Y., 2001. Mass transfer characteristics of syngas components in slurry system at industrial conditions. Chem. Eng. Technol. 24, 651–657.

Yang, G.Q., Fan, L.-S., 2003. Axial liquid mixing in high-pressure bubble columns. AIChE J. 49, 1995–2008.

6

Mixing in Inverse Bubbly Flow

Mixing

Mixing is a term that implies homogenization or reduction in the variability of concentration, temperature, and so on. It is important to know the scale at which this variability exists and the various mixing mechanisms that reduce it. One of the most important tasks facing the designer of any industrial chemical equipment is the prediction of the flow conditions in the vessel. For the need of scale-up and operation optimization, the mixing mechanism is necessary. The understanding of mixing is desired for the purpose of optimized operation and design of industrial-scale reactors. The investigation of mixing based on local bubble motion would be beneficial for optimization of operation, scale-up, and design of efficient bubble column reactors and their applications. In many cases, several competing reactions occur simultaneously with the time scale of the mixing process. The progress of the reactions, the product distribution, and the yield are governed by the mixing of the components in the mixture. By controlling the rate of mixing, it is possible to find out the effect of concentration difference on yield. The effect of phase mixing on the behavior of a chemical process can be used to control and optimize the distribution of the products that are formed in the processing unit. Generally, mixing time is one of the most important hydrodynamic parameters in different types of chemical engineering reactors for design and scale-up purposes.

The intensity of mixing in a process unit can be characterized using the mixing or residence time (t_m). To improve the yield of various commercial chemical and bioprocesses, it is important to estimate the changes to the design and operation of process unit that lead to a reduction in mixing time. Factors that may affect the mixing time are the column dimensions, gas distributor design, superficial gas velocity, and level of the fluid mixture inside the column. These factors determine the overall rate of the reactions if the mixing times are much larger than the reaction times. For a very slow chemical reaction relative to the time scale for mixing of the reactants, the extent of reaction is determined by the reaction kinetics and the stoichiometric ratio of the reactants. When the reaction is relatively fast compared with the accomplished mixing rate, physical factors (e.g., sparging design, column diameter, rate of stirring, solvent viscosity) also contribute to determining the yield and selectivity.

Rao and Joshi (1998) have reported that the knowledge of mixing is required when the rate of mixing is almost same to or slower than that of chemical reaction or mass transfer. In bubbly flow systems, the rapid mixing of gas can result in uniform temperature profiles, even in very large industrial-scale bubble column reactors for very exothermic reactions. It may also result in lateral (and axial) mixing. The deviation from the ideal plug-flow behavior by this lateral or axial mixing clearly affects the concentration profiles and significantly

Hydrodynamics and Transport Processes of Inverse Bubbly Flow. http://dx.doi.org/10.1016/B978-0-12-803287-9.00006-0

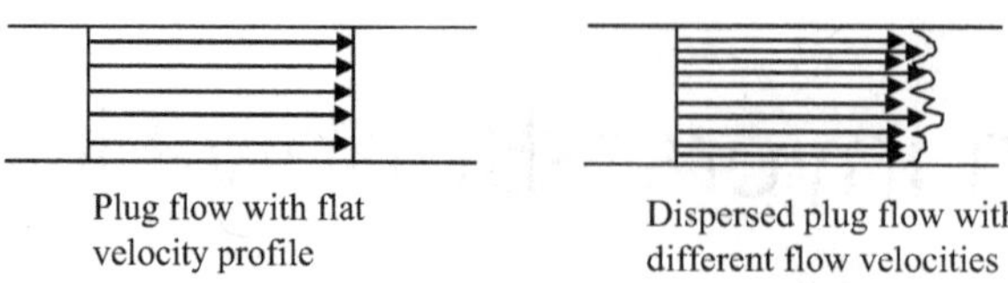

FIGURE 6.1 Ideal and non-ideal velocity profiles.

varies the conversion and possibly the product selectivities depending on the type of the chemical reaction, the corresponding reaction rate parameters, and the desired extent of reaction. In a bubble column reactor, the dispersion has the effect of reducing conversion and hence selectivity. The degree of backmixing is considered as a measure of the deviation from the ideal plug flow system. Figure 6.1 shows schematic of the ideal and non-ideal velocity profiles. Mixing of the phases is characterized by using dispersion coefficients based on Fick's law of diffusion. *Axial mixing, axial dispersion,* and *backmixing* are all terms used to describe a phenomenon that causes a distribution of residence time. The intensity of backmixing is mainly expressed in terms of an axial dispersion coefficient (ADC). Unlike diffusion, dispersion results from convective motion of fluid caused by relative movement of the gas and liquid phases, bubble coalescence and break-up, and the accumulation of fluid in wakes behind the moving gas bubbles. The dispersion and bubble phase hold-up in bubble columns are two paired phenomena because bubble movement results in mixing of the continuous liquid phase.

If a bubble moves in a liquid, a wake forms behind the bubble, which entrains the liquid element. The dimensions of the wake are heavily dependent on the bubble shape and size, which intuitively vary with operating variables and the physical properties of liquid. A random movement of the bubbles causes turbulence because of its bubble wake, accreting the liquid according to the direction of the bubble movement. Small, slow-moving or temporarily stagnant bubbles, especially for an inverse bubbly flow system in which upward bubble balancing their buoyancy with inverse liquid momentum, may not induce any serious turbulence. Interactions between fast- and slow-moving bubbles occur in bubble column reactors because of high bubble density and non-uniform bubble size distributions. This rarifies the explanation of the backmixing phenomena that occurs in bubble columns. In case of inverse bubbly flow, smaller bubbles move in an inverse direction because of liquid momentum, and at the same time, larger bubbles move in the opposite direction because of their buoyancy; hence interactions, are more intensive relative to upward bubbly flow. Radial dispersion in bubble columns has generally been grouped with the ADC. Although substantial information exists on axial dispersion of fluid in bubble columns (Ohki and Inoue, 1970, Towell and Ackerman, 1972; Deckwer et al., 1974; Field and Davidson, 1980; Shah et al., 1982; Majumder et al., 2005), radial mixing in these reactors has been ignored almost completely. The radial dispersion coefficient is always less than one-tenth the value of the axial coefficient (Deckwer, 1992). Recently, inverse bubble column reactors for improved gas-liquid mixing have been developed to adapt for many industrial processes such as absorption and scrubbing, gas–liquid reactions, waste treatment, aerobic fermentations, and so on, although the different hydrodynamic characteristics such as flow regimes and mixing characteristicsof inverse bubbly flow systems are not yet fully understood. Therefore, a precise knowledge of the mixing characteristics in inverse bubble columns is of considerable interest.

Importance of Understanding Mixing in Bubbly Flow Devices

Fluid mixing in bubble columns is deemed an a complex process to obtain a uniform mixture of ingredients distributed among each other. Some important points can be highlighted as follows to understand why mixing is so important in bubbly flow devices:

- In many industrial applications of bubble column reactors, because mixing is closely related to the rates of mass, heat, and moment transfers, the mixing of reactants and products very often is the controlling factor in the reactor's performance.
- Knowledge of good mixing among reactors is essential to avoid hot zones caused by the heat released by the highly exothermic reactions in the bubble column reactor.
- Poor homogeneity of the gas bubble can lower the efficiency of the overall heat and mass transfer processes.
- The mixing mechanism, as a qualitative feature, characterizes method of intermixing of components, which significantly influences the rate of the process and the desired homogeneity, which are important to enhance the performance of operations.
- The mixing process happening during the practical operation heavily depends on the operating conditions. Therefore, it is important to know mixing mechanism at different operating conditions, which helps to control the operation.
- In catalytic reactions in slurry bubble column reactors, large-scale vertical movement of porous particles can carry large amounts of adsorbed reaction components up and down the bed, which is referred as *backmixing*. This type of backmixing usually lowers conversion and selectivity. So knowing the extent of mixing of solids that occurs in the reactor is essential.
- To improve the selectivity and conversion, it is required to (2) control temperature, (2) reduce bulk circulation of solids, (3) reduce bubble size in the bubble column, (4) increase the gas bubble voidage, and (5) increase the overall residence time of the reactant gas bubbles in the bubble column reactor.
- The estimation of radial dispersion of gas or liquid is important in bubble columns if it is used as a column photobioreactor. In the reactor, the suspended photosynthesizing microorganisms (microalgea and cyanobacteria) cause a radial decline in irradiance from a high value near the externally illuminated transparent wall to a low value in the center of the photobioreactor (Grima et al., 2000).

Therefore, ways should be found for quantifying the extent of radial mixing and its dependence on the bubbling rate.

Methods to Quantify the Intensity of Mixing

Recent literature suggest that two different methods are available to interpret the mixing characteristics. The first method makes the use *of population balance* over the whole vessel. This has led to the postulation of various mixing models, which describe the effect of mixing of material of different ages in a vessel of known *residence time distribution (RTD)*. Most authors have used simplified mechanical models in colligation with experimental data of residence time to correlate or predict the results. In the second method, *advanced mathematical techniques* are used to solve the conservation equations of mass, momentum, and energy for specific industrial applications. These methods provide a fundamental understanding of the flow pattern of the fluid in the vessel. For the past decade, the computational fluid dynamics (CFD) approach increased in popularity as a means to model phase flow in multiphase reactors. For this approach, microscopic momentum balance (with microscopic mass balance used as a constraint) or force balance on discrete phases is solved to find many hydrodynamic parameters, including local velocity field. However, the CFD approach is time consuming and is difficult to use at industrial scale. The mixing modeling approach is less fundamental than CFD, but the former is actually more easily applicable to industrial scale. The axial dispersion model (ADM) is the most popular mixing model used. However, some authors have doubted that the ADM could adequately represent phase mixing and proposed different models. For the prediction of conversion and selectivity, the use of phenomenological mixing models is required for rapid kinetics of interfacial reactive mass transfer. Generally, three basic ideas are used to characterize and model the non-ideal reactors (Levenspiel, 1972; Wen and Fan, 1975; Fogler, 1992):

- The distribution of residence times in the system
- The intensity of mixing
- The model used to describe the system

Theoretically, the deviations from ideal plug flow by measuring fluid velocities throughout the reactor to obtain a complete distribution profile can be estimated. In this regard, residence time measurement is an important technique for the prediction of the exact flow behavior and intensity of mixing in process vessel with non-ideal flow.

Basic Concepts of Residence Time Distribution

The parameters that are important to quantify the dispersion characteristics in the non-ideal flow reactor are RTD, the mean residence time and mixing time, and the axial or radial dispersion coefficient. Flow behavior in continuous process units is usually too complex to be experimentally measured or theoretically predicted numerically by the

Navier-Stokes equation or statistical mechanical theory (Wen and Fan, 1975). The real flow pattern is often aberrant from the ideal flow patterns obtained from various models. Real flow patterns were first approached by Danckwerts (1953) on the basis of RTD analysis. The residence time of an element of fluid is defined as the time elapsed from its entry into the system until it reaches the outlet. The distribution of these times is called the RTD function of the fluid, and it represents the fraction of fluid leaving the system at each time, having units of time^{-1}:

$$E(t)=\frac{C(t)}{\int\limits_{0}^{\infty} C(t)\,dt} \tag{6.1}$$

Where $C(t)$ is the exit concentration (e.g., of a tracer) at a certain time t. It is often convenient to use a dimensionless time θ and the corresponding RTD $E(t)$ (Levenspiel, 1972):

$$\theta=\frac{t}{\tau} \quad and \quad E(\theta)=\tau E(t) \tag{6.2}$$

The cumulative curve that may also be used to characterize the RTD is called the F-curve, which corresponds to the fraction of fluid that has spent less than a time t in the system. The F-curve is defined as:

$$F(t)=\int\limits_{0}^{t} E(t)\,dt \tag{6.3}$$

Thus, the fraction of the fluid that leaves the reactor with an age greater that t is:

$$1-F(t)=1-\int\limits_{0}^{t} E(t)\,dt = \int\limits_{t}^{\infty} E(t)\,dt \tag{6.4}$$

One can also define an internal age distribution $I(t)$ that describes the reactor contents. The fraction of fluid within the reactor with an age of t is $I(t)dt$. The relation between $E(t)$ and $I(t)$ can be found from the mass balance as shown by Danckwerts:

$$I(t)=\frac{1}{\tau}\left(1-\int\limits_{0}^{t} E(t)\,dt\right); \quad E(t)=-\tau\frac{dI(t)}{dt} \tag{6.5}$$

The moments, the F-curve, do not depent on the time units. For first-order reactions under isothermal conditions, RTD information is sufficient to predict conversion in the system. In this case, the final average concentration in the exit stream of the system is (Levenspiel and Bischoff, 1963):

$$\bar{C}_e = \int\limits_{0}^{\infty} C_e(t)E(t)\,dt \tag{6.6}$$

where $C_e(t)$ is the exit concentration at a certain time t. Transformation of RTD into the Laplace domain is another useful concept in RTD modeling and data fitting to experimental RTD curves. The Laplace transform of a RTD or concentration curve is expressed as (Hopkins et al., 1969):

$$C(s) \equiv L\{C(t)\} \equiv \int_0^\infty e^{-st} C(t)\,dt \tag{6.7}$$

The system response for any input function in the Laplace domain, more commonly called the system transfer function $F(s)$, is defined as

$$F(s) \equiv \frac{C_e(s)}{C_i(s)} \tag{6.8}$$

which is used to get the transfer function for several flow models (Hopkins et al., 1969; Wen and Fan, 1975). The transfer function $F(s = k)$, for a first-order reaction with rate constant k corresponds to the remaining fraction of reactant in the exit stream. In addition, the ordinary moments $M_i = \int_0^\infty t^i E(t)\,dt$ (descibed in later section) can be directly obtained for any transfer function of a reactor flow model from the following equation (Wen and Fan, 1975):

$$M_i = (-1)^i \lim_{s \to 0} \frac{d^{(i)} F}{ds^{(i)}} \tag{6.9}$$

Models for Residence Time Distribution

Various theoretical models for fluid flow in tubular systems at isothermal conditions have been used to analyze the RTD, which are summarized in Table 6.1. These models can be divided into two classes: mechanistic (derived from fluid mechanics theory) and stochastic (based on probability theory) models.

Non-ideal models (single and multiparameter) are usually developed from ideal models that account for the deviations from real systems. One parameter model is often adequate to represent tubular systems (Levenspiel and Bischoff, 1963). The most common models corresponding RTD and relative moments are summarized in Table 6.2.

The selection of the particular model to be used depends largely on the engineering judgment of the person carrying out the analysis. The following guidelines are suggested by Fogler (1992) when developing models for non-ideal reactors:

- The model must be mathematically easily managed. The equations that are to be used to describe a chemical reactor should be able to be solved quickly.
- The model must realistically describe the characteristics of the non-ideal reactor physically, chemically, and mathematically.
- The model must not have more than two adjustable parameters because an equation with more than two adjustable parameters can be fitted to a great variety of

Table 6.1   Theoretical Models in the Literature for Fluid Flow in Pipes

Model	Mixing Conditions	Reference
Mechanistic	Laminar flow without diffusion	Bosworth (1948)
	Laminar flow with radial or axial diffusion	
	Laminar flow, power law fluid	Lin (1980)
	Laminar flow, Bingham-plastic fluid	Guariguata et al. (1979)
	Laminar flow, Prandtl-Eyring fluid	Sawinsky and Simandi (1982)
	Turbulent flow without diffusion	Bosworth (1949)
	Laminar and turbulent flow with radial or axial diffusion	Bosworth (1949)
Stochastic, single parameter	Axial dispersion model	Danckwerts (1953)
	Tanks-in-series model	Buffham and Gibilaro (1968)
	Backflow model	Deckwer (1992)
	Interstage recirculation model	Mecklenburgh and Hartland (1976)
	Bypass model	Michell and Furzer (1972)
Stochastic, multiparameter	Dispersion models	Wen and Fan (1975)
	Two-parameter mixing cell model	Deans (1963); Deans and Lapidus (1960)
	Two-parameter cross-flow model	Hoogendoorn and Lips (1965)
	Two-parameter time-delay model	Buffham et al. (1970)
	Three-parameter PDE model	van Swaaij et al. (1969)
	Three-parameter mixing cell model	van Swaaij et al. (1969)
	Four-parameter model	Raghuraman and Verma (1974)
Stochastic, combined	Combined models	Levenspiel and Bischoff (1963)

PDE, Partial differential equation.

experimental data, and the modeling process in this circumstance is nothing but an exercise in curve fitting. A one-parameter model is superior to a two-parameter model if the one-parameter model is sufficiently realistic. However, in complex systems (e.g., internal diffusion, conduction, mass transfer limitations) in which other parameters may be measured independently, more than two parameters are quite acceptable.

The dispersion model and the tanks-in-series models are the examples of one-parameter models for tubular reactors. For the dispersion model, the parameter is the dispersion coefficient, Dz, and for the the tanks-in-series model, it is the number of tanks, N. Knowing the parameter values, the conversion or solute concentrations for the reactor and the intensity of the mixing are determined. It is important to point out that small deviations from plug flow are better depicted by an axial dispersion and small deviations from perfect backmixing with a tanks-in-series model. In this regard, it is worth noting that empty bubble columns are most likely to be expected to be perfectly backmixed.

Dispersion Model

Most work relating to the study of phase dispersion coefficients in bubble column reactors have attempted to characterize the mixing in terms of an overall axial dispersion for the whole reactor (Carleton et al., 1967; Towell and Ackerman, 1972; Pilhofer et al., 1978;

Table 6.2   Most Common One-Parameter Models Describing Residence Time Distribution in Pipes with Statistical Parameters

Model	Submodel	Parameter	E-curve	Mean (t/τ)	Variance (σ^2/τ^2)
Ideal RTD models	Plug flow	τ	$E(t) = 1/\tau \; \delta(t-\tau)$	1	0
	Continuous stirred tank	τ	$E(t) = 1/\tau \exp(-t/\tau)$	1	1
Mechanistic ideal flow models	Newtonian fluid	τ	$E(t) = \tau^2/2t^3; t > 0.5\tau$	1	1
	Power law fluid	$n.\tau$	$E(t) = \left(\dfrac{1}{2}\right)\dfrac{\tau^2}{t^3} \; ; \; t \geq \tau/2$ $$E(t) = \left(\frac{2n}{3n+1}\right)\frac{\tau^2}{t^3}\left[1-\frac{n+1}{3n+1}\cdot\frac{\tau}{t}\right]^{\frac{n-1}{n+1}}$$ $$; \; t \geq \left(\frac{n+1}{3n+1}\right)\tau$$ where the power-law index n is related to the shape of the radial velocity profile in tube flow: $$u(r) = u_b\left(\frac{3n+1}{n+1}\right)\left[1-\left(\frac{r}{R}\right)^{(n+1)/n}\right]$$	—	—
Non-ideal RTD models	Dipersed plug flow ($Pe > 500$)	$Pe.\tau$	$E(t) = \dfrac{1}{\tau}\sqrt{\dfrac{Pe}{4\pi}}\exp\left\{-\dfrac{Pe(t-\tau)^2}{4\tau^2}\right\}$	1	$2/Pe$
	Dipersed plug flow (open boundary condition)	$Pe.\tau$	$E(t) = \dfrac{1}{\tau}\sqrt{\dfrac{Pe}{4\pi t}}\exp\left\{-\dfrac{Pe(t-\tau)^2}{4\tau^2}\right\}$	$1+2Pe$	$2/Pe$ $+8/Pe^2$
	Dipersed plug flow (closed boundary condition)	$Pe.\tau$	$E(\theta) = \exp\left[\dfrac{Pe}{2}\right]\displaystyle\sum_{i=1}^{\infty}(-1)^{i+1}$ $$\frac{8\alpha_i^2}{4\alpha_i^2+4Pe+Pe^2}\exp\left[\frac{-\theta(Pe^2+\alpha_i^2)}{4Pe}\right]$$	1	$2/Pe+2/Pe^2$ $\times(1+\exp(-Pe))$
	N-CST	$N.\tau$	$E(t) = \dfrac{t^{N-1}}{(N-1)!\tau^N}\exp(-t/\tau)$	1	$1/N$
Probabilistic models:	Exponential	χ	$E(t) = \exp(-t/\chi)$	$1/\chi\tau$	$1/(\chi\tau)^2$
	Normal	$t_m.\sigma$	$E(t) = \sqrt{\dfrac{1}{2\pi\sigma^2}}\exp\left\{-\dfrac{(t-\tau)^2}{2\sigma^2}\right\}$	t_m/τ	σ^2/τ^2

N-CST, continuos stirred tank; *RTD*, residence time distribution.

Field and Davidson, 1980; Joseph et al., 1984; Kulkarni and Shah, 1984; Wachi and No-jima, 1990; Ekambara and Joshi, 2003). The most common approach to modeling the non-ideal mixing behavior in bubble columns is the one-dimensional ADM. The mixing is considered to be result from fluctuations caused by different flow velocities and molecular and turbulent dispersion. In this model, all of the mechanisms leading to axial gas mixing are lumped into a single ADC. The vast popularity of the model is due to its simplicity and ease of use. In addition, it contains only one unknown parameter: the ADC. However, the validity of the ADM to describe two-phase flows with large degrees of mixing, such as those in bubble columns, is dubitable (Myers et al., 1987; Lefebvre et al., 2004). The theoretical estimation of the dispersion coefficient is difficult because it is a complex and poorly understood function of the liquid and gas hydrodynamics. Despite the lack of a sound basis, the ADM still remains extremely popular and is used to model phase mixing by almost all of the authors working in the field. The ADM is the simplest mathematical representation of the flow system in which both convection and diffusion are important. Because of its simplicity, the ADM is often used to characterise the mixing in bubbly flow units. In some instances when hydrodynamics are not limiting, this model may give rough predictions of the reactors' performance. From Fick's law of diffusion, the model equation can be expressed for the dispersion phenomenon with the following assumptions:

1. The liquid velocity distribution is uniform.
2. The bulk flow exists only in the axial direction.
3. The ADC is independent of spatial coordinates.
4. Axis-symmetric species concentration distribution exists.

The axial dispersion of the material, which is governed by diffusion, transport by bulk flow (convection), uA_cC. The molar flow rate of inert tracer material (J) is expressed by

$$J = -D_z A_c \frac{\partial C}{\partial z} + A_c \frac{u_{sl}C}{(1-\varepsilon_g)} \tag{6.10}$$

The mole balance on the inert tracer both convective dispersion and diffusion in two-phase bubbly flow, which is not take part for reaction, yields as

$$-\frac{1}{A_c}\frac{\partial J}{\partial z} = \frac{\partial C}{\partial t} \tag{6.11}$$

But if the material takes part in reaction, a mole balance can be written on a particular component of the mixture over a short length Δz of a reactor of cross-section A_c

$$-\frac{1}{A_c}\frac{\partial J}{\partial z} + R = \frac{\partial C}{\partial t} \tag{6.12}$$

Substituting for J from equation (6.10) in equation (6.12), one can get

$$\frac{\partial C}{\partial t} = D_z \frac{\partial^2 C}{\partial z^2} - \frac{u_{sl}}{(1-\varepsilon_g)}\frac{\partial C}{\partial z} + (-R) \tag{6.13}$$

It is nonlinear when R is other than zero or first order. The parameter D_z is the ADC uniquely characterizing the intensity of mixing during the flow. R is the kinetic term of the component. The ADC depends on the definition of the velocity "u_a" used. The actual velocity of the liquid phase in the bubbly flow system is defined as $u_a = u_{sL}/(1-\varepsilon_g)$ where u_{sl} is the superficial velocity of the liquid in the column, and ε_g is the gas hold-up. Hence, for two-phase bubbly flow systems, the equation (6.13) can be expressed (in dimensionless form) as

For non reactive systems:

$$\frac{\partial \psi}{\partial \theta} = \frac{1}{Pe}\frac{\partial^2 \psi}{\partial \lambda^2} - \frac{\partial \psi}{\partial \lambda} \tag{6.14}$$

For reactive first-order systems

$$\frac{\partial \psi}{\partial \theta} = \frac{1}{Pe}\frac{\partial^2 \psi}{\partial \lambda^2} - \frac{\partial \psi}{\partial \lambda} - Da\psi \tag{6.15}$$

where

$$\theta = \frac{u_{sl}t}{(1-\varepsilon_g)L} \tag{6.16}$$

$$\psi = \frac{C(z,t)}{C_0} \tag{6.17}$$

$$\lambda = \frac{z}{L} \tag{6.18}$$

$$Pe = \frac{u_{sl}z}{(1-\varepsilon_g)D_z} = \frac{\text{Rate of transport by convection}}{\text{Rate of transport by diffusion or dispersion}} \tag{6.19}$$

$$Da = k\tau = \frac{\text{Rate of consumption of component by reaction}}{\text{Rate of transport of component by convection}} \tag{6.20}$$

The purpose of creating this dimensionless distribution function is that the flow performance inside reactors of different sizes can be compared directly. For example, if the dimensionless function ψ is used, all perfectly mixed cells in the reactor have numerically the same RTD. If the simple function C is used, numerical values of ψ can differ substantially for different cells. There are two different types of Peclet numbers in common use. One is called the reactor Peclet number (Pe_r); it uses the reactor length, z, as the characteristic length. The reactor Peclet number for mass dispersion is often referred to in reacting systems as the Bodenstein number, Bo, rather than the Peclet number (Holdsworth, 1992). The other type of Peclet number is the fluid Peclet number (Pe_f). It uses the characteristic length that determines the fluid's mechanical behavior. Generally, this length is considered as a particle diameter (d_p), tube, or column diameter (d_t or d_c). The fluid Peclet number can be easily converted to the reactor Peclet number by multiplying by z/d_p or z/d_t or z/d_c. It is important to note that as $Pe \rightarrow \infty$ (no dispersion), the behavior is that of a plug flow reactor (PFR), and as $Pe \rightarrow 0$ (maximum dispersion), it is that of a continuous stirred tank reactor (CSTR) of complete backmixing. Thus, the axially dispersed reactor can simulate all types of behaviors between the ideal limits of no backmixing and complete backmixing. The reciprocal of Peclet number (Pe) is called the vessel dispersion number, which measures the extent of axial dispersion of the phase. For small extents of dispersion or large Pe numbers ($Pe \geq 500$), this model does not depend on the boundary conditions, but for small values of Pe, it depends on the type of boundary conditions imposed.

Boundary Conditions

To solve the equation, one needs to consider boundary conditions for closed vessels and open vessels. In the case of closed-closed vessels, it is assumed that there is no dispersion or radial variation in concentration either upstream (closed) or downstream (closed) of the flow section. In this case, tracer is considered even distribution across the flowing stream. In an open vessel, dispersion occurs both upstream (open) and downstream (open) of the flow section. In this case, tracer inputs are considered proportional to the velocity of flow. A closed-open vessel boundary condition is one in which there is no dispersion in the entrance section but there is dispersion in the reaction and exit sections. These three cases are shown in Figure 6.2. The solution for the expression of the ADM and estimation of the corresponding characteristic parameter (Pe) for the various combinations of boundary conditions and input functions were reported by Laan (1958), Aris (1959), Bischoff (1960), and Bischoff and Levenspiel (1962).

Closed-Closed Boundary Condition

In this condition, it is considered plug flow or no dispersion at the left of z = 0 (z = 0⁻) plane and right of z = L plane (z = L⁺). However, in between these two planes, there is

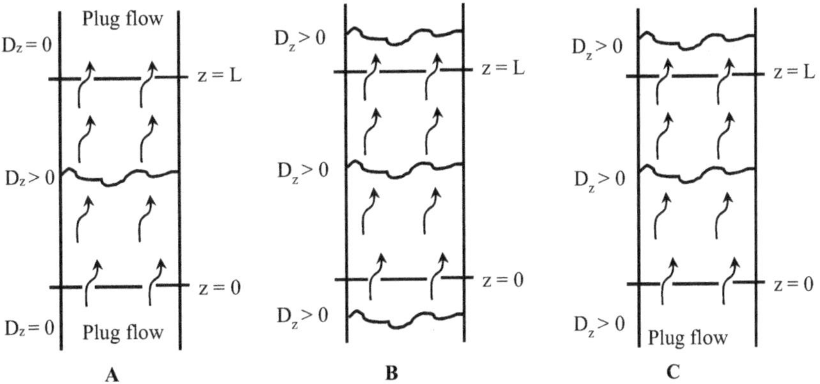

FIGURE 6.2 Schematic representation of flow for closed-closed (**A**), open-open (**B**), and closed-open (**C**) conditions.

dispersion (in $z = 0^+$ and $z = L^-$). Therefore, the entrance boundary conditions are as follows (Fogler, 1992):

$$J(0^-,t) = -D_Z A_c \left(\frac{\partial C}{\partial z} \right)_{z=0^-} + \frac{u_{sl}}{(1-\varepsilon_g)} A_c C(0^-,t) \tag{6.21}$$

$$J(0^+,t) = -D_Z A_c \left(\frac{\partial C}{\partial z} \right)_{z=0^+} + \frac{u_{sl}}{(1-\varepsilon_g)} A_c C(0^+,t) \tag{6.22}$$

$$\text{At} \quad z=0 \quad J(0^-,t) = J(0^+,t) \tag{6.23}$$

$$-D_Z A_c \left(\frac{\partial C}{\partial z} \right)_{z=0^-} + \frac{u_{sl}}{(1-\varepsilon_g)} A_c C(0^-,t) = -D_Z A_c \left(\frac{\partial C}{\partial z} \right)_{z=0^+} + \frac{u_{sl}}{(1-\varepsilon_g)} A_c C(0^+,t) \tag{6.24}$$

$$C(0^-,t) = -\frac{(1-\varepsilon_g)D_Z}{u_{sl}} \left(\frac{\partial C}{\partial z} \right)_{z=0^+} + C(0^+,t) \tag{6.25}$$

At entrance

$$C(0^-,t) = C_0 \tag{6.26}$$

which results in

$$C_0 = -\frac{(1-\varepsilon_g)D_Z}{u_{sl}} \left(\frac{\partial C}{\partial z} \right)_{z=0^+} + C(0^+,t) \tag{6.27}$$

At the exit of the reactor section, the concentration of tracer is uniform, and there is no gradient in tracer concentration. Hence,

$$at \quad z=L, \quad C(L^+,t) = C(L^-,t) \tag{6.28}$$

$$\frac{\partial C}{\partial z} = 0 \tag{6.29}$$

These two boundary conditions are known as Danckwert's (1953) boundary conditions. The discontinuity in concentration at $z = 0$ because of the concentration drops immediately on entering from C_0 to C_e for an ideal CSTR.

Open-Open Boundary Condition
For an open-open system, there is continuity of flux at the boundaries. At the plane $z = 0$, there is continuity of flux. So,

$$\text{At } z=0, t=0 \, J(0^-,t) = J(0^+,t) \tag{6.30}$$

$$J(0^-,t) = -D_Z A_c \left(\frac{\partial C}{\partial z} \right)_{z=0^-} + \frac{u_{sl}}{(1-\varepsilon_g)} A_c C(0^-,t) \tag{6.31}$$

$$J(0^+,t) = -D_Z A_c \left(\frac{\partial C}{\partial z} \right)_{z=0^+} + \frac{u_{sl}}{(1-\varepsilon_g)} A_c C(0^+,t) \tag{6.32}$$

$$-D_Z A_c \left(\frac{\partial C}{\partial z} \right)_{z=0^-} + \frac{u_{sl}}{(1-\varepsilon_g)} A_c C(0^-,t) = -D_Z A_c \left(\frac{\partial C}{\partial z} \right)_{z=0^+} + \frac{u_{sl}}{(1-\varepsilon_g)} A_c C(0^+,t) \tag{6.33}$$

Now at $z = L$, the concentration of tracer is constant, not varying with position for a long tube or $(u_{sl}L)/((1-\varepsilon_g)D_z) > 100$. Therefore,

$$\text{At } \quad z = L \qquad C(L^-,t) = C(L^+,t) \tag{6.34}$$

$$-D_Z A_c \left(\frac{\partial C}{\partial z}\right)_{z=L^-} + \frac{u_{sl}}{(1-\varepsilon_g)}A_c C(L^-,t) = -D_Z A_c \left(\frac{\partial C}{\partial z}\right)_{z=L^+} + \frac{u_{sl}}{(1-\varepsilon_g)}A_c C(L^+,t) \tag{6.35}$$

$$\frac{\partial C}{\partial z} = 0 \tag{6.36}$$

Closed-Closed System Residence Time Distribution Profile
Equation (6.5) can be solved either numerically or analytically (for a pulse injection) for exit concetration profile based on the following conditions:
The initial condition is

$$\text{At } \quad t = 0, z > 0, \qquad C(0^+,0) = 0; \quad \psi(0^+) = 0 \tag{6.37}$$

For a perfect input of tracer material (pulse input), C_0 is defined as the mass of tracer injected, m, divided by the volume, V of bubble column. The mass of tracer material injected as a pulse input is

$$m_t = \frac{u_{sl}A_c}{(1-\varepsilon_g)}\int_0^\infty C(0^-,t)\,dt \tag{6.38}$$

In dimensionless form, the Danckwerts boundary conditions are:

$$\text{At } \quad \lambda = 0: \quad \frac{C(0^-,t)}{C_0} = \left(-\frac{1}{Pe}\frac{\partial \psi}{\partial \lambda}\right)_{\lambda=0^+} + \psi(0^+) = 1 \tag{6.39}$$

$$\text{At } \quad \lambda = 1: \quad \frac{\partial \psi}{\partial \lambda} = 0 \tag{6.40}$$

The result for analytical solution under the above boundary conditions is to be in an infinite series, which can be expressed as (Froment and Bischoff, 1979)

$$E(\theta) = \exp\left[\frac{Pe}{2}\right]\sum_{i=1}^\infty (-1)^{i+1}\frac{8\alpha_i^2}{4\alpha_i^2 + 4Pe + Pe^2}\exp\left[\frac{-\theta(Pe^2 + \alpha_i^2)}{4Pe}\right] \tag{6.41}$$

$$E(\theta) = \frac{\psi(1,\theta)}{\int_0^\infty \psi(1,\theta)\,d\theta} \tag{6.42}$$

$$Tan(\alpha_i) = \frac{4\alpha_i Pe}{4\alpha_i - Pe^2} \tag{6.43}$$

Open-Open System Residence Time Distribution Profile
The analytical solution of equation (6.14) for the normalized response curves in case of nonreactive system has been proposed by Levenspiel (1972) by using open-open boundary conditions as follows

$$\psi = \frac{C(z,t)}{C_0} = \frac{1}{2\sqrt{\pi\theta/Pe}} \exp\left[-\frac{(1-\theta)^2 Pe}{4\theta}\right] \tag{6.44}$$

Profile by Any Kind of Boundary Conditions
An expression for the final concentration as a function of the kinetic rate parameter for a first-order reaction can be represented for any kind of boundary conditions as (Wehner and Wilhelm, 1956):

$$\psi = \frac{C(z,t)}{C_0} = f(Pe, Da) = \frac{4a\exp(Pe/2)}{(1+a)^2 \exp(aPe/2) - (1-a)^2 \exp(-aPe/2)} \tag{6.45}$$

$$a = \sqrt{1 + 4Da/Pe} \tag{6.46}$$

where the Damkohler number, Da, is a dimensionless form of expressing the reaction rate constant, k, which is expressed as $k\tau$. Outside the limited case of a first-order reaction, a numerical solution of the equation is required. If it is a split-boundary value problem, an iterative technique is required. The Peclet number can be obtained by solving analytically the differential equation by a mass balance to a reactant in an axially dispersed plug flow reactor.

Tank-in-Series Model

According to this model, to analyze the RTD, one should determine the number of ideal tanks, N, in series that will give approximately the same RTD as the non-ideal reactor. For the development of RTD by the tank-in-series model, N equal number of continuous stirred tank (CST) is considered in series. The output from one tank is considered as an input of immediete successive tank. Using the definition of the RTD, the fraction of material leaving the system of N reactors (i.e., leaving the Nth reactor) that has been in the system between time t and $t + \Delta t$ is (this can be looked up in any standerd textbook on reaction engineering, e.g., Fogler, 1992)

$$E(t)\Delta t = \frac{vC_N(t)\Delta t}{N_0} = \frac{C_N(t)}{\int_0^\infty C_N(t)dt}\Delta t \tag{6.47}$$

Then

$$E(t) = \frac{C_N(t)}{\int_0^\infty C_N(t)dt} \tag{6.48}$$

$C_N(t)$ is the concentration of tracer in the effluent from the Nth reactor, and v is the volumetric flow rate of fluid. From the material balance of tracer from 1 to N reactors and generalizing to a series of N number of CST reactors gives the RTD for N number of CST reactors in series (one can follow Fogler (1992) for derivation) as

$$E(t) = \frac{t^{N-1}}{(N-1)!(\tau/N)^N} e^{-tN/\tau} \tag{6.49}$$

$$E(\theta) = \tau E(t) = \frac{N(N\theta)^{N-1}}{(N-1)!} e^{-N\theta} \tag{6.50}$$

where, $\theta = t/\tau$.

Dispersion Model versus Tank-in-Series Model

One can apply both of these one-parameter models bubble column reactors using the variance of the RTD. For first-order reactions, the two models can be applied with equal ease. However, the tanks-in-series model is mathematically easier to use to obtain the solute concentration and conversion for reaction orders other than one and for multiple reactions. These two models are equivalent when the Peclet or Bodenstein number is related to the number of tanks in series, N, by the equation (Elgeti, 1996).

$$N = \frac{Pe_r}{2} + 1 \tag{6.51}$$

where $Pe_r = u_{sl}z/(1-\varepsilon_g)D_z$. For reactions other than first order, one would solve successively for the exit concentration and conversion from each tank in series in order to bound the desired values.

Theoretical Model to Account for Both the Axial and Radial Dispersion Simultaneousely

The model developed by Rubio et al. (2004) can be used to calculate the axial and radial dispersion coefficients from experimentally measured RTD data. The radial dispersion in bubble columns has generally been collocated with the ADC and almost exclusively used as an index of mixing in such reactors. The time course of a tracer's concentration at a particular measurement location in a bubble column can be presented by the following complete dispersion model:

$$\frac{\partial C}{\partial t} = D_z \frac{\partial^2 C}{\partial z^2} - \frac{u_{sl}}{1-\varepsilon_g} \frac{\partial C}{\partial z} + \frac{D_r}{r} \frac{\partial C}{\partial r} + D_r \frac{\partial^2 C}{\partial r^2} \tag{6.52}$$

where D_r is the radial dispersion coefficient and r is the radial distance measured from the center of the column. The equation can be expressed as dimensionless forms using the definitions of the variables:

$$\varphi = \frac{D_z t}{L^2} \tag{6.53}$$

$$z' = \frac{z}{L} \tag{6.54}$$

$$r' = \frac{r}{L}\sqrt{\frac{D_z}{D_r}} \tag{6.55}$$

where, z' is the dimensionless axial distance and L is the height of gas–liquid dispersion in the reactor and r' is the dimensionless radial distance. The dimensionless form of equation (6.52) is then expressed as (Rubio et al., 2004)

$$\frac{\partial C_\varphi}{\partial \varphi} = \frac{\partial^2 C_\varphi}{\partial z'^2} - \frac{u_{sl}L}{(1-\varepsilon_g)D_z}\frac{\partial C_\varphi}{\partial z'} + \frac{1}{r'}\frac{\partial C_\varphi}{\partial r'} + \frac{\partial^2 C_\varphi}{\partial r'^2} \tag{6.56}$$

where C_ϕ is the dimensionless tracer concentration defined as follows:

$$C_\varphi = \frac{C - C_0}{C_\infty - C_0} \tag{6.57}$$

and C, C_0 and C_∞ are the instantaneous, the initial, and the final equilibrium concentrations of the tracer in the liquid batch, respectively. For a batch of liquid in a bubble column, where, $u_{sl} = 0$, and considering the cylindrical symmetry of the tracer response, then one can say:

$$\text{(i)} \quad if \quad r'=0, \quad \partial C_\varphi/\partial r'=0 \tag{6.58}$$

$$\text{(ii)} \quad if \quad r'=R/L\sqrt{D_z/D_r}=\beta \quad (i.e., \quad if \quad r=R), \quad \partial C_\varphi/\partial r'=0 \tag{6.59}$$

$$\text{(iii)} \quad if \quad \varphi>0 \quad and \quad z'=0, \quad \partial C_\varphi/\partial z'=0 \quad (\text{i.e. at the surface of the dispersion}) \tag{6.60}$$

$$\text{(iv)} \quad if \quad \varphi>0 \quad and \quad z'=1, \quad \partial C_\varphi/\partial z'=0 \quad (\text{i.e. at the bottom of the reactor}) \tag{6.61}$$

Using the above initial and boundary conditions, solution of the equation (6.56) can be expressed as:

$$C_\varphi = \sum_{n=1}^{\infty} \frac{J_0(v_n r')}{J_0^2(v_n\beta)}\exp(-v_n^2\varphi)\times\left(1+2\sum_{m=1}^{\infty}\cos(m\pi z')\exp(-m^2\pi^2\varphi)\right) \tag{6.62}$$

where J_0 is the zero-order Bessel function and v_n is the nth root of the first-order Bessel function, J_1. It is to be noted that when the C_ϕ in equation (6.62) is radially invariant (i.e., $D_r = \infty$), v_n, β and r' become zero and $J_0(t\,v_n\beta) = J_0(v_n\,r') = 1$. In this case, equation (6.62) reduces to

$$C_\varphi = 1+2\sum_{m=1}^{\infty}\cos(m\pi y)\exp(-m^2\pi^2\varphi) \tag{6.63}$$

which is identical to the solution of the ADM without considering the radial component of the dispersion.

Theoretical Model to Analyze the Dispersion Coeficient by the Heat Tracer

The dispersion coefficients can be measured from the temperature distribution with time in the bubble column by the one-dimentional dispersion model. For determination of the dispersion coefficient, pulses of the same liquid, about 20 K higher or lower in temperature, should be injected at a position of inlet of the liquid stream for cocurrent continuous phase in an open-open system. The heat tracer can be injected at the top of the dispersion for batch process where liquid is still. The time courses of the local temperatures at several downstream positions should be measured with heat-detecting sensors (time constant <3 s). The records are then analyzed with the one-dimensional dispersion model (Ohki and Inoue, 1970):

$$T = \left(1 + 2\sum_{n=1}^{\infty}\left[\left(\cos\frac{n\pi}{L_d}z\right)\exp\left\{-\left(\frac{n\pi}{L_d}\right)^2 D_z t\right\}\right]\right)(T_0 - T_\infty)\exp(at) + T_\infty \qquad (6.64)$$

Heat transfer at the wall are accounted for by using the hypothetical initial temperature T_0 and the constant a as additional fitting parameter.

At steady state and adiabatic conditions of the system, heat as the tracer can be added continuously. In this case, the latent heat of vaporization is also assumed to be negligible. If the radial variations of phase fraction and temperatures are found to be absent, the one-dimensional ADM can be used for the dispersion of phases. The measured temperature distribution is then correlated by this model with the following additional assumptions:

- The thermal conductivity and the density of the gas phase are small compared with those of liquid and therefore are assumed to be neglected if a gas–liquid bubble column is considered.
- The molecular thermal conductivity of fluid is negligible compared with the ADC of heat.
- No chemical reaction is occurring.
- Mixture behaves as a homogeneous mixture.
- Gas phase backmixing is relatively small.

The last assumption is required to calculate the thermal dispersion coefficient from the energy balance, which involves the properties of all phases. Based on these assumptions, the following heat balance equation can be written:

$$\rho_l u_l C_p \frac{dT}{dz} = D_z \rho_l C_p (1 - \varepsilon_g)\frac{d^2T}{dz^2} \qquad (6.65)$$

Solution of this differential equation with the general boundary conditions:

$$\begin{aligned} \text{at} \quad z = z_c: \quad T = T_c \\ \text{at} \quad z = z_h: \quad T = T_h \end{aligned} \qquad (6.66)$$

results in

$$\frac{T - T_c}{T_h - T_c} = \frac{\exp\left\{\left(\dfrac{u_l}{D_z(1 - \varepsilon_g)}\right)(z - z_c)\right\} - 1}{\exp\left\{\left(\dfrac{u_l}{D_z(1 - \varepsilon_g)}\right)(z_h - z_c)\right\} - 1} \tag{6.67}$$

Using best fit of equation (6.67) with the experimental data by regression analysis, one can obtain the values of ADC D_z at different operating variables.

Experimental Guideline to Estimate the Mixing Parameter

The RTDs are determined experimentally by using the classical tracer response technique. The techniques involve the injection of a tracer at the inlet stream or at some point within a reactor and the observation of the corresponding response at the exit stream or at some other downstream point within the reactor. Because the tracer experiments are supposed faithfully to illustrate the RTD of actual reacting fluids, a proper choice of tracer in a given reacting system is very important. The basic requirements for tracer experiments are as follows:

- The tracer should be nonreacting with other streams.
- The tracer should not be absorbed or adsorbed by the phases.
- The tracer should be miscible and should have closely similar physical properties of the fluid stream.
- The tracer should be accurately detectable even in small concentrations.
- The tracer should not be volatile.
- The analytical device to detect the tracer should cause the least amount of disturbance in the flow pattern. In this regard, a radioactive tracer has advantage over other chemical tracers in that the detection device can be installed externally.
- Heat balance on the system should be checked properly if heat is used as a tracer.

The tracer inputs should be treated as perfect impulse inputs, although in practice, it is impossible to obtain a perfect pulse. However, a rapid, turbulent injection of the tracer may approximate a perfect pulse. Other forms of injection are the step change, oscillatory, and random forms. The tracer for both the steady and unsteady tracer injection methods can be an electrolyte, a dye, or heat. The mass dispersion method normally uses electrolyte as the tracer, and the change in electrical conductivity is measured. This method is only suitable for aqueous liquids. The thermal dispersion method uses heat as the tracer, and the temperature distribution within the column is measured. This method can also be applied to nonaqueous liquids (e.g., hydrocarbon liquids). The injection should take place over a very short period compared with the residence time in the various segments of the reactor. It is to be maintained that negligible amounts of dispersion between the point of injection and the entrance of the reactor system to be produced. When the steady-state flow condition is reached, a small amount ($\sim 1\%$ in relation to the total volume of the column; Reith et al., 1968; Levenspiel and Smith, 1957) of a tracer is injected at a certain position (top or bottom) where the flow is fully developed. The tracer is injected as quickly (fraction of second) and smoothly as possible by a high-pressure syringe pump or by other suitable device. Then the samples at a certain time interval from downstream at a/several axial distance are to be analyzed by suitable analytical instrument or by a detection probe with a data logging system. These data are noted as response data of tracer concentration as a function of time. The small increment of time Δt is to be chosen so that the concentration of tracer, $C(t)$, exiting between Δt and $t+\Delta t$ remained essentially constant. One must keep in mind that the concentrations to be taken into account for RTD analysis are mixing cup concentrations at the inlet, the outlet, and eventually at different levels of the bed. But most measurement techniques determine volume or column diameter averaged concentrations and not mixing cup concentrations (Briens et al., 1995). The measurement of the concentration may be disturbed by the presence of bubbles, and hence the actual signal has to be filtered. The type of tracer detection device depends on the nature of the tracer. So it is necessary to know the type of tracer to be selected with a suitable tracer detecting device. Three types of tracer have mainly been used to estimate the intensity of mixing in bubbling beds, which are briefed as follows:

Mixing by Tracing of Solid

Colored Tracers

Qualitative information on the mixing can be obtained by investigating the behavior of layers of particles of different colors (Fortin, 1984; Boyer et al., 2002).

Magnetic Tracers

Impregnate porous catalyst particles with a ferromagnetic substance (e.g., nickel oxide, which is reduced to nickel) can be used as a tracer (Euzen and Fortin, 1987; Wild and Poncin, 1996). The magnetic property of the material results in a response as a signal in the presence of an electric current at a level. Furthermore, the magnetic properties of the tracer particles can be used to separate them from the inert particles so they can be used again (Flaschel et al., 1987; Euzen and Fortin, 1987; Wild and Poncin, 1996). The concentration of the solid phase can also be measured by a capacitance probe if the dielectric constants of the tracer material and the remaining phase are different.

Fluorescent Tracers

A solid particle can be used as a tracer that can adsorb fluorescing substance (Flaschel et al., 1987) such as including gypsum, talc, opal, agate, quartz, and amber.

Mixing by Tracing of the Gas Phase

The tracer gas should be selected in such a way that it would not be soluble with other gases. There are no completely insoluble gases. Even with the less soluble gases (e.g., helium in water), absorption and desorption phenomena make the interpretation of gas-phase tracing experiments complex (Kunii and Levenspiel, 1991; Joseph and Shah, 1986). That is why investigation of gas-phase dispersion is more difficult than that of liquid. It is therefore necessary to consider at the same time mass transfer and mixing of phase (Mawlana et al., 1992). The mixing can also be analyzed by a stochastic distribution of the tracer gas with a cross-correlation technique of the outlet signal with the inlet signal of the tracer gas concentration. Some radioactive tracer is convenient to use for both cases in the gas phase. The benefit of radiotracer technology over the conventional methods is that the investigation can be carried out on stream and without disrupting the operating process. A series of radiotracer experiments for the measurement of the RTD, mixing time, and flow rate was conducted by Kasban et al. (2010) using 300 $\mu Ci^{99}Mo$ as the radioactive tracer. Pant et al. (2009) conducted an RTD study in a pilot and an industrialised fluid catalytic cracking unit (FCCU) using radioactive tracer, lanthanum-140, and sodium-24. The tracers were characterised using neutron activation analysis to investigate the degree of mixing.

Mixing by Tracing of the Liquid

Measuring the RTD by a liquid tracer is a well-known technique to determine the intensity of mixing of phase in a device (Levenspiel, 1999). Most of the cases salt is used as a tracer to find liquid mixing. Quite often the tracers such as KCl, H_2SO_4, NaCl, HCl, and NH_4Cl are used. The concentration change of the tracer with time can be detected by a device (probe) directly inserted into the reactor or by continuous sampling at any fixed position and measuring by an electrical conductivity cell with a data logger. Analytical procedure such as titration can be followed if the tracer concentration is to be measured by collecting the sample a discrete way from the fixed position of downstream. Colored dyes can also be used, mainly with nonaqueous liquids. Radioactive isotope tracer techniques present the advantage over other noninvasive techniques under high-temperature and high-pressure conditions (e.g., coal liquefaction process) (Tarmy et al., 1984; Blet et al., 1999; Pant et al., 2000). A nonradioactive tracer can also be used by interacting with neutrons (Sakai et al., 2000). The change of refractive index with intense mixing may tender the study of liquid mixing by a liquid tracer whose refractive index is different from that of the main liquid. Instead of using a dissolved liquid-phase tracer, a particle tracking technique (Dudukovic et al., 1991; Larachi et al., 1995; Chaouki et al., 1997) can be used.

Mixing by Tracing Heat

Heat as a tracer can be used to measure the thermal dispersion coefficient (Holcombe et al., 1983; Yang and Fan, 2003). In this case, from the axial distribution of temperature, the dispersion coefficient is measured. The method is thought to give more accurate results than the mass tracing method because the experimental operation is easily achieved for the former. The ADCs of the liquid phase is measured by the steady-state thermal dispersion method. For the heat tracer experiments, thermocouples at different axial locations can be used in the test section. All thermocouples should be connected to a single digital readout device with a data logging system. A suitable heat exchanger can be used in the outlet line of the bed to cool the heated fluid to the desired inlet temperature during the steady-state heat-tracer experiments. A steam-heated coil can be used to provide heat into the flowing stream. Generally, dispersion coefficients are independent of the amount of heat input to the system. The bubble column should be insulated with proper insulation material to prevent heat losses during the experiments with heat input. Sometimes at low fluid velocities, significant temperature fluctuations with time can persist. These fluctuations are more pronounced near the heat source. To take into account these fluctuations, 20 to 30 recorded values of temperatures can be averaged.

Moment Method

A complete analysis of RTD by the dispersion model requires an appropriate raw data treatment and the subsequent fitting of models to the experimental data. Various methods are applied to obtain the characteristic parameters of the dispersion model such as time domain analysis and Laplace domain analysis. In Table 6.3, the advantages and shortcomings of these methods are summarized. The first one, the most commonly used, is done either by using analysis of moments or curve fitting by nonlinear optimization. Frequency domain analysis has also been used, although it is not widely used in chemical processing.

Some of the methods are used in the analysis of RTD curves obtained from non-ideal input functions. The RTD by dispersion model is usually characterized by using statistical moments. The definition of various moments about the mean are shown in Table 6.4. The moments can be easily calculated by using the following definition as:

$$M_i = \int_0^\infty t^i E(t)\,dt \tag{6.68}$$

To obtain complete charaterization of a distribution function, all moments are calculated and ennuntiated; however, only moments about the mean up to the fourth level are most commonly considered relevant. The first moment is called the *mean of the distribution*. The second moment, *variance* (σ^2), is an indicator of the spread of the distribution; the third moment, called *skewness* (γ^3), is an indicator of the asymmetry of the distribution; and finally, the fourth moment, *kurtosis* (κ^4), is an indicator of how the the distribution is heavily tailed. In practice, only the first two moments of the distribution function (mean and variance) are considered to characterize the distribution (Levenspiel, 1972; Fogler, 1992). Many problems are solved with only a knowledge of these two quantities. To evaluate the integrals for moments calculation, the integration has to be done up to some finite time. If the first moment equals the mean residence time τ, the RTD curves can be normalized with respect to time; otherwise, this parameter is preferably estimated together with the parameters of the model.

Table 6.3 Advantage and Shortcomings of the Parameter Estimation Methods

Method		Advantages	Shortcomings
Time domain analysis	Specific curve regions analysis	Rapid method; good initial estimates for further curve fitting; mMay allow choice of most reliable data for analysis	May have high error associated; does not include information of whole RTD; applicable only if input is ideal
	Moments methods (ideal and non-ideal injection)	Includes information on whole RTD; may be independent of boundary conditions; may allow analysis of RTD from non-ideal input functions	Very sensitive to data errors; may overweight tailing
	Curve fitting by nonlinear optimization	Most reliable procedure; allows simultaneous estimation of τ	Same weight to all RTD information; time consuming; may overweight tailing; applicable only if input is ideal
Laplace domain curve fitting		Includes information of whole RTD; more consistent results; may be manipulated to underweight errors induced by tailing effects; allows simultaneous estimation of τ, allows analysis of RTD from non-ideal input functions	Further mathematical development required

RTD, Residence time distribution.

Table 6.4 Various Moments About the Mean of the Residence Time Distribution Function

Moment	Definition	Formula	Relative Measure
Mean	t_m or $\tau = \int_0^\infty tE(t)dt$	t_m or $\tau = M_1$	1
Variance	$\sigma^2 = \int_0^\infty (t-\tau)^2 E(t)dt$	$\sigma^2 = M_2 - \tau^2$	σ^2/τ^2
Skewness	$\gamma^3 = \int_0^\infty (t-\tau)^3 E(t)dt$	$\gamma^3 = M_3 - 3\tau M_2 + 2\tau^3$	$(\gamma^3)^2/(\sigma^2)^3$
Kurtosis	$\kappa^4 = \int_0^\infty (t-\tau)^4 E(t)dt$	$\kappa^4 = M_4 - 4\tau M_3 + 6\tau^2 M_2 - 3\tau^4$	$\kappa^4/(\sigma^2)^2$

Mean Residence Time

For treating ideal reactors, the space time (τ) or mean residence time (t_m), which is defined as the ratio of reactor volume (V) to the volumetric flow rate (uA_c), is generally used. No matter what RTD exists for a particular reactor, ideal or non-ideal, this space time is equal to the mean residence time, t_m. It is also called the *mixing time*. From the first moment of RTD function, the mean residence time is defined as:

$$M_1 = t_m = \frac{\int_0^\infty tE(t)dt}{\int_0^\infty E(t)dt} = \int_0^\infty tE(t)dt \qquad (6.69)$$

where

$$E(t) = \frac{C(t)}{\int_0^\infty C(t)\,dt} \tag{6.70}$$

$E(t)$ is called the exit-age distribution function. It represents the "age" of a species as the time it resides during reaction or in the reactor without reaction. The integration can be evaluated numerically by any one of the methods such as the trapezoidal rule, Simpson's (1/3 or 3/8) rules. Aris (1959) reported that to guarantee an error lower than 1% in the estimates, integration should be conducted up to a time period equivalent to three times the mean residence time. The $E(t)$ then concerns the age distribution of the tracer stream. It is the most often used of the distribution functions connected with reactor analysis because it characterizes the lengths of time that various tracer elements (or reactant) spend in the reactor. The mean residence time of the tracer for an open-open system as per the dispersion model is

$$t_m = \left(1 + \frac{2}{Pe}\right)\tau \tag{6.71}$$

For closed system, it is defined as

$$t_m = \tau = \frac{V}{uA_c} \tag{6.72}$$

Variance of Residence Time Distribution
The second moment is taken about the mean and is called the *variance* or *square of the standard deviation*. It is defined by

$$\sigma^2 = \int_0^\infty (t - t_m)^2 E(t)\,dt \tag{6.73}$$

The magnitude of this moment is an indication of the "spread" of the distribution. The greater the value of this moment, the greater a distribution's spread. The method of moments can be used to estimate the axial dispersion number (ADN) (=*1/Pe*) by the following equations by taking the average of repeated experiments.
For closed-closed boundary conditions:

$$\sigma_\theta^2 = \frac{\sigma^2}{\tau^2} = \frac{2}{Pe}\left[1 - \frac{1}{Pe}[1 - \exp(-Pe)]\right] \tag{6.74}$$

For open-open boundary conditions:

$$\sigma_\theta^2 = \frac{\sigma^2}{\tau^2} = \frac{2}{Pe} + \frac{8}{Pe^2} \tag{6.75}$$

The ADC (D_z) can be calculated by equation (6.76) from the values of Pe by equations (6.74) and (6.75) for closed-closed and open-open conditions, respectively, as

$$D_z = \frac{u_{sl}z}{(1-\varepsilon_g)Pe} \tag{6.76}$$

There are two steps for which one can use equations (6.74) and (6.75) to determine the system parameters:

Step 1: If the space time τ is known by measured values of V and Q, one can determine the Peclet number by determining t_m and σ^2 from the concentration–time data. The Pe again can be calculated by using equation (6.74) or (6.75). One can also calculate t_m and then use equation (6.71) as a check, but this is usually less accurate.

Step 2: If the space time τ is unknown in a situation in which there are no-flow zones exist in the reactor/column, one can calculate t_m and σ^2 from the data as in step 1. One can use equation (6.71) to eliminate τ^2 from equation (6.75) to deduce

$$\frac{\sigma^2}{t_m^2} = \frac{2Pe+8}{Pe^2+4Pe+4} \tag{6.77}$$

One can solve for the Peclet number in terms of experimentally determined variables σ^2 and t_m^2. Knowing Pe, one can solve equation (6.71) for τ and hence V. The dead volume is the difference between the measured volume and the effective volume calculated from the RTD. Sometimes perfect tracer input as a pulse is not possible during an experiment. In this case, the differences in the variances between the input and output tracer measurements can be used to calculate the Peclet number by

$$\Delta\sigma^2 = \sigma_{in}^2 - \sigma_{out}^2 \tag{6.78}$$

where σ_{in}^2 is the variance of the tracer measured at some point upstream (near the entrance) and σ_{out}^2 is the variance measured at some point downstream (near the exit). For open-open systems, Aris (1959) showed that

$$\frac{\Delta\sigma^2}{t_m^2} = \frac{2}{Pe} \tag{6.79}$$

In case of tank-in-series model, the number of tanks in series can be calculated by

$$\sigma_\theta^2 = \frac{\sigma^2}{\tau^2} = \int_0^\infty (\theta-1)^2 E(\theta)d\theta = \int_0^\infty \theta^2 E(\theta)d\theta - 1 \tag{6.80}$$

Substituting the equation for $E(\theta)$, it becomes

$$\sigma_\theta^2 = \frac{\sigma^2}{\tau^2} = \frac{(N)^N}{(N-1)!} \int_0^\infty \theta^{N+1} e^{-N\theta} \, d\theta - 1$$
$$= \frac{(N)^N}{(N-1)!} \left[\frac{(N+1)!}{N^{N+2}} \right] - 1 \qquad (6.81)$$
$$= \frac{1}{N}$$

Then the number of tanks in series is

$$N = \frac{1}{\sigma_\theta^2} = \frac{\tau^2}{\sigma^2} \qquad (6.82)$$

The equation represents the number of ideal stirred tank reactors in series is reflecting the the degree of non-ideality of a real reactor. N tends to zero means the reactor deviating to more non-ideality, but N tends to infinity, the real reactor approaching to plug flow reactor. For more details, the reader can see the textbooks by Fogler (1992) and Levenspiel (1972).

Limitation of the Analysis and Its Optimization

Tailing of the RTD may cause severe errors in calculating the parameter. Several solutions have been proposed in the literature for defeating this problem, including (1) truncation of the output curve and approximation of the tail after the truncation point (Wen and Fan, 1975), (2) estimation of parameters from other parts of the RTD curve (Levenspiel, 1972); (3) empirical corrections of RTD curves (Wen and Fan, 1975), and (4) use of Laplace transform techniques allowing linearization of the RTD curve (Ostergaard & Michelsen, 1969; Hopkins et al., 1969). The most rapid and simple method for parameter estimation in the time domain consists in analyzing specific characteristics of the RTD curve such as the peak time and height, inflection points, or defined areas under the RTD curve. As an example, Pe in the dispersion model can be estimated from the peak height of the E-curve, *or* N in the N-CST model from the peak time (Levenspiel, 1972). The calculated moments are very sensitive to small measurement errors in the tails of the RTD curves by the ADM. This difficulty can be resolved by analysis involving the Laplace transform of the dispersion model and evaluation of a linear transfer function (Ostergaard and Michelsen, 1969). This method produces much more consistent results than the method of moments, regardless of the tailing and the location of the cut-off point of the tail. The location of the cut-off point of the response curves can affect the analysis. According to this method, the transfer function $F(s)$ for the linear system obeying the axially dispersed model can be calculated from the relation:

$$F(s) = \frac{C_2(s)}{C_1(s)} = \exp\left[\frac{Pe}{2} \left\{ 1 - \left(1 + \frac{4 s t_m}{Pe} \right)^{1/2} \right\} \right] \qquad (6.83)$$

where

$$t_m = \frac{L(1-\varepsilon_g)}{u_{sl}}$$

(6.84)

By differentiation of equation (6.83) with respect to Laplace transform parameter, s, one obtains

$$-\frac{F'(s)}{F(s)} = -\left(\frac{C_2'(s)}{C_2(s)} - \frac{C_1'(s)}{C_1(s)}\right) = t_m\left(1+\frac{4st_m}{Pe}\right)^{-1/2}$$

(6.85)

$C(s)$ and $C'(s)$ can be evaluated from the recorded RTD functions:

$$C(s) = \frac{\displaystyle\int_0^\infty C(t)e^{-st}\,dt}{\displaystyle\int_0^\infty C(t)\,dt}$$

(6.86)

$$C'(s) = \frac{-\displaystyle\int_0^\infty C(t)te^{-st}\,dt}{\displaystyle\int_0^\infty C(t)\,dt}$$

(6.87)

and thus

$$M(s) = \frac{-C'(s)}{C(s)} = \frac{-\displaystyle\int_0^\infty C(t)te^{-st}\,dt}{\displaystyle\int_0^\infty C(t)e^{-st}\,dt}$$

(6.88)

From equation (6.85),

$$\Delta M_s = M_{2,s} - M_{1,s} = t_m\left(1+\frac{4st_m}{Pe}\right)^{-1/2}$$

(6.89)

From equation (6.89), one can represent

$$\left(\frac{1}{\Delta M_s}\right)^2 = \frac{1}{t_m^2} + \frac{4}{t_m Pe}s$$

(6.90)

A plot of $(1/\Delta M_s)^2$ versus s thus gives a straight line of slope $4/(t_m Pe)$ and intercept $1/t_m^2$ on the ordinate axis. Using best fit of equation (6.90) by regression analysis, one can obtain the value of the Peclet number. The moments, M_s can be evaluated for different values of s. The range of s over which the regression is carried out is very important. The recommended optimum range of s values is that for which st_m lies between 0.5 and 5

(Hopkins et al., 1969). The advantage of this method is that the increasing weighting functions t and t^2 are replaced by the decreasing weighting functions e^{-st} and te^{-st}.

Effects of Variables on Liquid Mixing Intensity

The liquid phase dispersion results from the entrainment of the surrounding liquid by the moving gas bubbles that carry the entrained liquid (Deckwer, 1992). The liquid phase mixing in bubble columns is characterized by using dispersion coefficients that arise from convective motion of fluid caused by the following the main factors: relative movement of the phases, bubble coalescence and break -p, the carry forward of fluid in wakes behind the moving gas bubbles and the consequent return flow generated for maintaining mass balance, and turbulence generated by any superimposed flow of liquid (Rubio et al., 2004). Numerous empirical correlations for axial liquid dispersion coefficient have been published on different types of bubble columns as shown in Table 6.5. However, most of the empirical correlations are available on axial liquid dispersion coefficient only for upflow bubble column with different types of distributors. Majumder et al. (2005) developed a correlation for ADC of liquid, which is also shown in Table 6.5.

From the correlations (as shown in Table 6.5), it can be seen that for the same ranges of the operating parameters, the liquid ADC of the inverse bubbly flow is higher than that of the other mentioned studies. In the inverse bubble column, higher momentum exchange of liquid jet enhances the turbulence of liquid. The dispersion coefficient of liquid phase, D_z, in the inverse system is therefore an order of magnitude larger than that of upflow. The dispersion coefficient is affected by different variables such dynamic, geometric, and physical properties. Different parameters that are related to mixing phenomena are shown schematically in Figure 6.3.

Effect of Dynamic Variables

The dependence of ADC on the superficial gas velocity (u_{sg}) usually ranges within $D_z \propto u_{sg}^{0.25-0.50}$ (Krishna et al., 2000). The relationship can be attributed to the effect of the average liquid circulation velocity (u_c), which increases with the increase of (u_{sg}), leading to a decrease in mixing time and consequently to an increase in liquid dispersion coefficient (Field and Davidson, 1980; Pandit and Joshi, 1983). Yang and Fan (2003) reported that the increase is faster in the homogeneous regime than in the heterogeneous regime. The authors also explained that this phenomenon occurs by promoting liquid recirculation velocity in the heterogeneous regime and the turbulence induced by the bubbles in the homogeneous regime. The increase of the liquid recirculation speed in heterogeneous regime does not cause as much turbulence as the one induced by the bubbles in the homogeneous regime, which implies that the increase is less important in the heterogeneous regime. The liquid Peclet number in the homogeneous regime increases with the superficial gas velocity and decreases in the heterogeneous regime, indicating that the ADC decreases in the homogeneous regime and increases in the heterogeneous regime (Zahradník et al., 1997).

Table 6.5   Correlations of Axial Dispersion Coefficient of Liquid in Bubble Columns

Flow Direction of Bubble	Systems	Equations for Axial Dispersion Coefficient	Reference
Upflow, co-, countercurrent	Air–water, ionic solutions (KCl),	$D_z = (1/3)(2u_{sg} + u_b)d_c$; $u_b \approx 30$ cm/s for $d_b = 1$ to 5 mm	Reith et al. (1968)
Upflow	Air–water	$D_z = 0.30d_c^2 u_{sg}^{1.2} + 170d_h$	Ohki and Inoue (1970)
Upflow	Air–water	$D_z = \dfrac{d_c u_{sg}[1 + 6.5(u_{sg}/\sqrt{gd_c})^{0.8}]}{13(u_{sg}/\sqrt{gd_c})}$	Kato and Nishiwaki (1971)
Upflow	Air–water	$D_z = 1.23d_c^{1.5} u_{sg}^{0.5}$	Towell and Ackerman (1972)
Upflow	Air–water	$D_z = (0.065 + 0.3u_{sg}^{0.77})d_c^{1.25}\mu_l^{-0.12}$	Hikita and Kikukawa (1974)
Upflow	Air–water, acqueous solution of Na_2SO_3, NaCl	$D_z = 0.68d_c^{1.4} u_{sg}^{0.3}$	Deckwer et al. (1974)
Upflow	Air–water, methanol solution, cane sugar sultion	$D_z = \dfrac{d_c u_{sg}[0.06 + 0.55(u_{sg}/\sqrt{gd_c})^{0.7}]}{(u_{sg}/\sqrt{gd_c})}$	Hikita and Kikukawa (1974)
Upflow	—	$D_z = 0.35g^{1/3}d_c^{4/3} u_{sg}^{1/3}$	Baird and Rice (1975)
Upflow	Air–water	$D_z = 0.30d_c(u_c + u_{sl})$ $u_c = 1.31[gd_c(u_{sg} - \varepsilon_g u_b)]^{1/3}$; $u_b = 23$ cm/s bubbles size 2.5-8 mm (Joshi and Sharma, 1979)	Joshi (1980)
Upflow	Air–water	$D_z = 0.44d_c^{1.33}[g(u_{sg} - \varepsilon_g u_s)]^{1/3}$ $u_s = u_{sg}/\varepsilon_g - u_{sl}/(1 - \varepsilon_g)$	Field and Davidson (1980)
Upflow	Air–water	$D_z = 0.068g^{3/8}d_c^{3/2} u_{sg}^{3/8}\nu^{-1/8}$	Riquarts (1981)
Upflow	Air–water	$D_z = [0.922 + 1.08(1 - 0.054u_{sg}^{1/2})^2]u_{sg}^{1/4}d_c^{3/2}$	Miyauchi et al. (1981)
Upflow	Air–water	$D_z = 0.368g^{1/3}d_c^{4/3} u_{sg}^{1/3}$	Zehner (1982)
Upflow	Air–water	$D_z = 5.04 \times 10^{-2}\left(\dfrac{d_n}{d_c}\right)^{0.9} u_{sg}^{0.32}\left(\dfrac{u_{sl}^2}{gd_c}\right)^{-0.23}$	Ogawa et al. (1982)
Upflow	Air–water	$D_z = 5.04 \times 10^{-2}(d_n/d_c)^{0.9} u_{sg}^{0.32}Fr^{-0.23}$	Kawase and Moo-Young (1986)
Upflow	Air–water	$D_z = 0.20d_c^{1.25} u_{sg}/\varepsilon_g$	Kantak et al. (1994)
Upflow	Air–water	$D_z = u_{sl}d_c$	Groen et al. (1996)
Upflow	Air–water	$D_z = 0.0651(gd_c)^{1/2}\left(\dfrac{u_{sg}}{g\upsilon_l}\right)^{1/8} d_c$	Krishna et al. (2000)
Upflow	Air–water	$D_z = 0.014u_{sg}^{0.45}\exp(-48.85u_{sl})$	Ahmed (2003)
Inverse flow flow	Air–water	$D_z = \dfrac{X}{kD_b} + D_b$; $D_b = 1.602d_r^{2.147}Re_g^{-0.671}$ $X = \dfrac{u_{sg}^{0.032}u_j^{2.532}d_n^{4.934}}{7.98 \times 10^{-4}d_c^{1.788}}$; $k = 0.115d_r^{-1.364}Re_g^{0.403}$	Majumder et al. (2005)

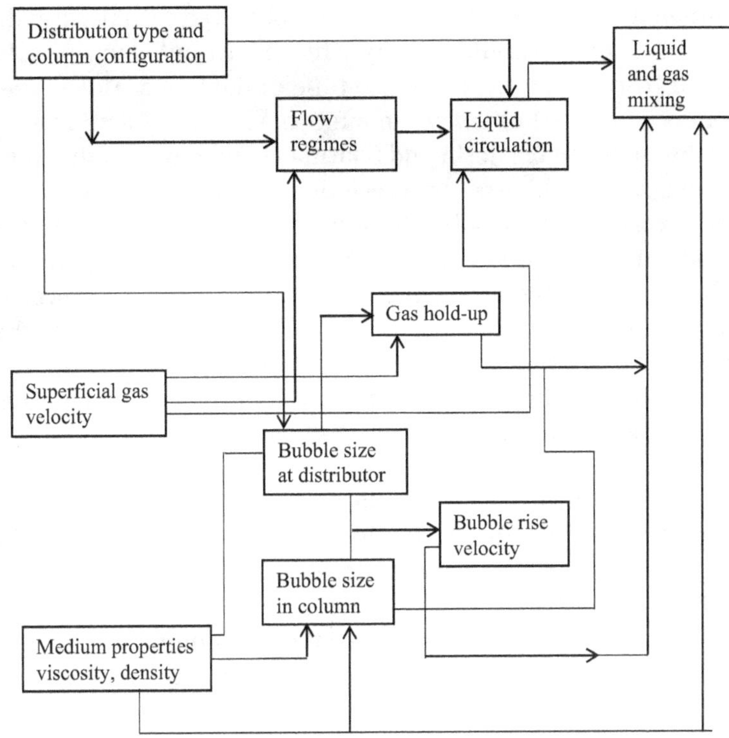

FIGURE 6.3 Interrelated parameters that affect mixing of phases.

In the homogeneous regime, the behavior of the liquid is close to plug flow. This is in contradiction with the results of other authors who conclude that the superficial gas velocity promotes the ADC even in the homogeneous regime (Ohki and Inoue, 1970; Shah et al., 2012; Smith et al., 1996; Yang and Fan, 2003). Yang and Fan (2003) found enhanced liquid dispersion to be highly dependent on superficial gas velocity; the influence of superficial liquid velocity was weaker. They also observed an increase of the ADC while increasing the superficial liquid velocity. The observed effect is higher in the case of the Paratherm National Formulary (NF) than in water. The authors explained this effect by the increase of the liquid turbulence. The authors reported that this increase is not due to the movement of bubbles but rather linked to the increased energy exchange between phases. The difference is mainly due to the effect of solvent on bubble size, which is smaller in the case of Paratherm NF of much lower surface tension despite a higher viscosity. These bubbles contribute less to the turbulence (Biń et al., 2001; Zahradník et al., 1997). In case of liquid jet–driven inverse bubbly flow, the same effect can be obtained. In this case, a high liquid flow rate also increases the turbulence and the gas hold-up, which leads to an increase in the dispersion coeffient (Majumder et al., 2005). In the slug flow regime, Shawaqfeh (2003) observed a decrease of the ADC while increasing the superficial liquid velocity, which is attributed to a predominant effect of convection compared with dispersion in this regime.

However, in countercurrent mode, Shah et al. (2012) observed a decrease of the liquid ADC while increasing the superficial liquid velocity, which is attributed to a ressult of the liquid plug-flow behavior by reducing the residence time of the liquid. Holcombe et al. (1983) and Sangnimnuan et al. (1984) observed no effect of superficial liquid velocity on liquid ADC for small column diameters. Joshi and Sharma (1979) reported that the liquid phase axial dispersion in upward bubble columns is independent of the superficial liquid velocity. However, they explained that in all of the studies, the order of magnitude of the generated liquid circulation velocity is much higher than the superficial liquid velocity (at least by a factor of 10), resulting in weak dependence of the superficial liquid velocity on Peclet number as well as ADC. Majumder et al. (2005) reported that increasing the convection transport, as characterized by the liquid superficial velocity, will increase the intensity of dispersion in case of inverse bubbly flow. The ADC increases with increasing superficial liquid velocity at a constant column diameter as shown in Figure 6.4 in the case of inverse bubbly flow systems.

The greater extent of mixing at the higher superficial liquid velocity is caused by more turbulence of liquid flow due to higher momentum transfer. The intimate contact between the phases increases momentum transfer, which creates more turbulence in the column. This may result in an increase of ADC with an increase in the superficial liquid velocity. The variation of liquid phase axial dispersion with superficial gas velocity is mainly because of the gas hold-up profile, which sets up along the column due to

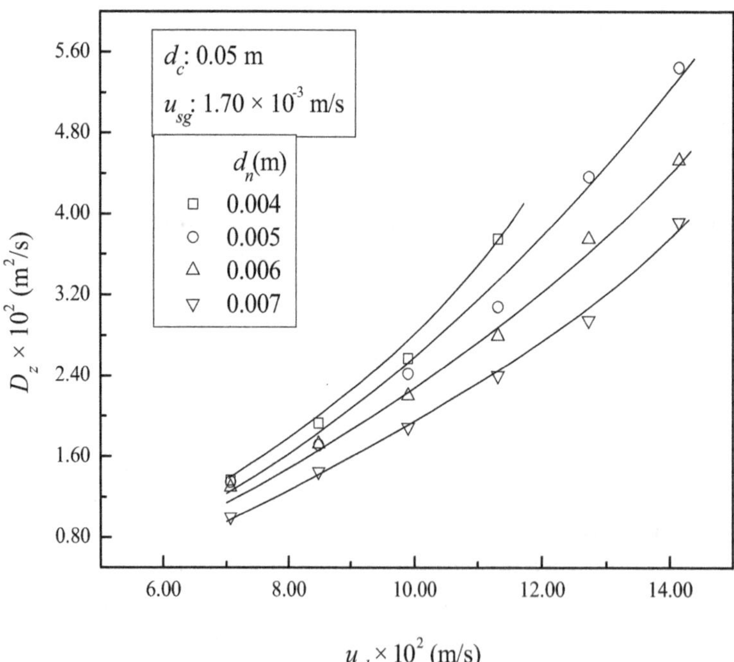

FIGURE 6.4 Variation of axial dispersion coefficient with superficial liquid velocity.

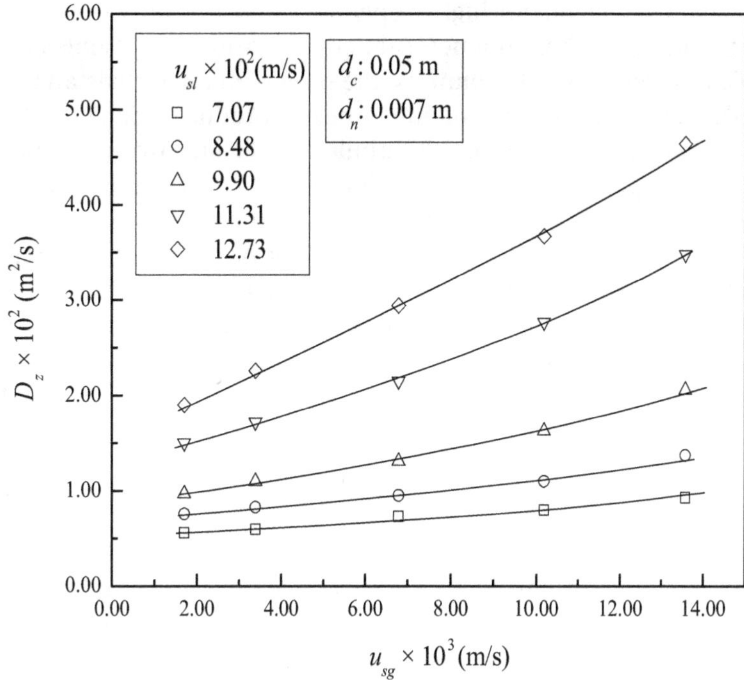

FIGURE 6.5 Variation of axial dispersion coefficient with superficial gas velocity.

inverse movement of gas bubbles through the column (Kulkarni and Shah, 1984; Kundu et al., 1995). At a higher superficial liquid velocity, gas hold-up increases because of higher entrainment of the gas. As a result, the circulation and interaction of the gas and liquid phases increase inside the column, and the flow gets more agitated. This enhances the overall mixing in the column. At a constant nozzle diameter, the ADC increases with increase in superficial gas velocity as shown in Figure 6.5. At a low superficial liquid velocity, it is nearly constant. At a low superficial liquid velocity, the momentum transfer of the liquid jet is low for which the entrainment depth of the liquid jet is reduced, and hence fluid turbulence inside the column decreases. This results in the dispersion coefficient being nearly constant at a low superficial liquid velocity.

Effect of Geometric Variables

The liquid ADC is affected by the type of sparger only in the homogeneous regime for an air/water system in atmospheric conditions. Sparger type has an immense effect on the bubble size distribution, which may result in change in ADC in the bubble column. Yang and Fan (2003) observed that the distributor design does not have a significant effect on the liquid phase mixing. The ADC is smaller for lower orifice diameters where the bubble size is lower. There is no influence of the gas sparger in the heterogeneous regime in upflow bubble columns. In the homogeneous regime, Ohki and Inoue (1970) found higher axial coefficients for higher open areas (high number of holes at constant hole diameter) and higher hole diameter (at constant number of holes), which results in higher ADC. In inverse bubble columns, the nozzle dfiameter has an immense effect on the dispersion because the small diameter nozzle produces high liquid jet, dissipates more kinetic energy, and results in more turbulence. Various investigators reported that an increase in the dispersion coefficient with an increasing column diameter d_c follows a power-law dependence $D_z \propto d_c^\phi$. The value of the coefficient φ varies between 1 and 1.5. Forret et al. (2003) showed that this dependency is linked to liquid recirculation velocity at high superficial gas velocity. The decrease in column diameter reduces the liquid circulation velocity. The reduced circulation velocity results in an increase in the mixing time, which leads to a decrease in the dispersion coefficient (Pandit and Joshi, 1982). Yang and Fan (2003) proposed a correlation to link the dispersion coefficient and the column diameter for the power φ as:

$$\phi = \phi_{atm}[1 - 0.11\ln(\rho_g / \rho_{g,atm})] \tag{6.91}$$

The index *atm* indicates atmospheric operating condition. Ohki and Inoue (1970) reported that a maximum ADC can be obtained at the transition point between the homogeneous and heterogeneous regimes for column diameters larger than 4 cm. The degree of axial dispersion is also affected by vessel internals. Generally, bubble column reactors can be equipped with two types of internals: horizontal internals, such as perforated plates (trays) or horizontal tube bundles, and vertical internals. Horizontal internals are usually used to control the flow behavior to reduce the overall liquid backmixing (Westerterp et al., 1987; Palaskar et al., 2000; Nosier, 2003). Vertical internals are preferred for higher backmixing to acheive high heat removal efficiency. The presence of internals affects the liquid and gas velocity profiles because it affects gas hold-up, bubble size, and the turbulent intensity in the system (Youssef et al., 2009). The presence of internals causes an increase in the gas hold-up, which decreases the area available for liquid flow, causing an increase in the liquid circulation. In addition, the presence of internals causes an increase in the shear stress in the bubble column, which increases the rate of bubble breakup, leading to the formation of smaller bubbles. Several researchers have shown that the presence of internals increases the centerline liquid velocity and increases the steepness of the liquid velocity profile (Forret et al., 2003), which may change the intensity of the mixing of phases in the bubble column irrespective of the flow direction of bubble whether upard or inverse direction. The ADC in the bubble column is inversely proportional to the liquid height. This may be due to a decrease in bubble diameter, which leads to a decrease in bubble velocity and consequently an increase in the liquid circulation velocity (u_c) as shown in equation (6.92) and hence an increase in ADC (Joshi and Sharma, 1979; Krishna et al., 2000). This may be due to the energy dissipipation $\propto (u_{sg} - \varepsilon_g u_b)$ in the liquid motion.

$$u_c = 1.31\{gd_c(u_{sg} - \varepsilon_g u_b)\}^{1/3} \tag{6.92}$$

This energy dissipation depends on the gas hold-up, ε_g, and and bubble velocity, u_b. It can be concluded that when u_c is lower, the mixing time is higher. Whalley and Davidson (1974) reported the relationship between the mixing time and dispersion coefficient as: $D_z \propto L_d^2 / t_m$.

Hikita and Kikukawa (1974) showed that the ADC does not depend on column length in the heterogeneous regime for L_d/d_c ratios over 7. The axial liquid dispersion coefficient decreases with increasing radial position (r/R). The radial dispersion coefficient of the liquid seems to depend on the radial component of the liquid velocity. The flow resistances to the gas phase in the direction of the flow increases, so the gas gets redistributed in the radial direction. This uniform distribution of the dispersed phase minimizes the density gradient effects, which results in the reduction in the liquid recirculation. The reduction in the liquid circulation velocities results in lower backmixing (Kawase and Moo-Young, 1989). Radial dispersion coefficient of liquid increases with the increase of column diameter. The increase in column diameter causes an decrease in the gas hold-up, which increases the liquid circulation velocity. As result, the mixing time decreases, leading to an increase in radial dispersion coefficient (Pandit and Joshi, 1983). The ADC increases with decreasing nozzle diameter at constant superficial liquid velocity and constant superficial gas velocity as shown in Figure 6.6 in case of ejector-induced inverse bubble columns (Majumder et al., 2005). As the nozzle diameter decreases, the jet velocity increases, which enhances the turbulence of the phases in the column, indicating that flow gets more agitated because of higher momentum transfer. The entrainment depth of the liquid jet increases as the nozzle diameter decreases, which causes an increase of the size of liquid circulation cells around the liquid jet in the column. This may also increase the dispersion coefficient as the nozzle diameter decreases. The column diameter has strong effect on the variation of the dispersion coefficient. At a constant superficial liquid velocity, the dispersion coefficient increases with an increase in column diameter. The increase of intensity and randomness of gas bubbles with increasing column diameter causes increased agitation of fluid mixture inside the column, which leads to greater axial mixing with column diameter.

Effect of Physical Properties

The influence of the physical properties of the liquid on the dispersion coefficient has been investigated by several authors (Hikita and Kikukawa, 1974; Riquarts, 1981; Camarasa et al., 1999; Ruzicka et al., 2001; García-Abuín et al., 2012; Anastasiou et al., 2010, 2013). The degree of axial dispersion is affected by surface-active agents that delay the coalescence of bubbles. Surfactants can produce either much more or much less backmixing than surfactant-free systems, depending on the bubble size and the gas distributor. It is known that the addition of small amounts of additives results in the extension of the homogeneous regime because it deters bubble coalescence (Camarasa et al., 1999; Ruzicka et al., 2001; Anastasiou et al., 2010). However, Hikita and Kikukawa (1974) reported that the liquid-phase dispersion coefficient decreases with increasing

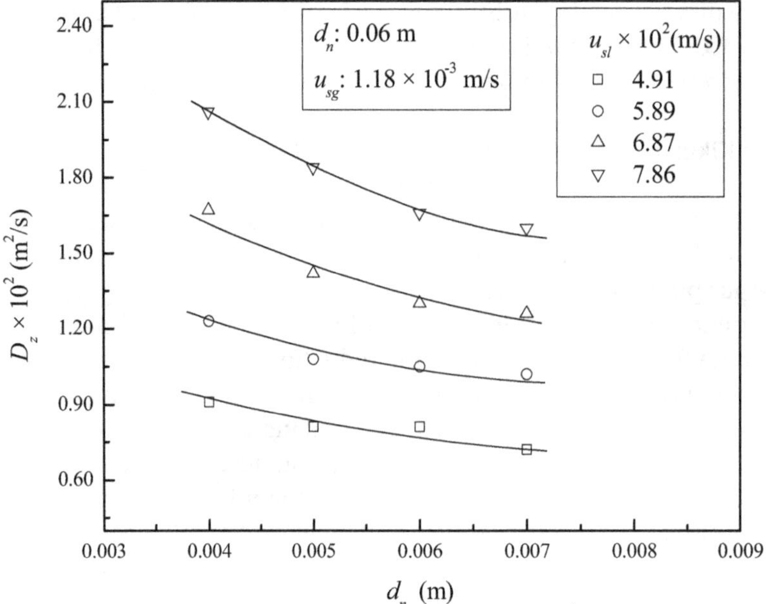

FIGURE 6.6 Variation of axial dispersion coefficient with nozzle diameter.

cane sugar concentration. The viscosity of the solution including water was ranged from 1.00 to 19.2 CP, but the surface tension of the solution changed only slightly from 72.0 to 75.7 dyne/cm. They reported that the value of liquid dispesrion coefficient is proportional to the - 0.12th power of the liquid viscosity. This effect of liquid viscosity is different from that found in the works of Aoyama et al. (1968), who investigated the effect of liquid viscosity on axial liquid dispersion coefficient using air and 61.5 vol. % aqueous glycerol solution (11.3 cP) and showed that axial liquid dispersion coefficient was not affected by the liquid viscosity. Hikita and Kikukawa (1974) also reported that the effect of liquid surface tension on axial liquid dispersion coefficient is unimportant. The same results were also stated by Aoyama et al. (1968). The viscosity reduce the gas hold-up because of less entrainment of gas in case of ejector-induced inverse bubble column due to resistance of flow of bubble in the inverse direction, which results in a reduction of the axial liquid dispersion coefficient. But in this inverse bubbly flow, the surface-active agent increases the bubble population and entrainment of the gas, which results in a wider bubble size distrubution. Continuous breakage and a coalesecence effect by the momentum exchange of the phases increase the domain of the heterogeneous flow of the phases, which drives the increase of the phase dispersion. The density of the liquid also has an effect on the dispersion of the phases. Higher densed liquid lower the dispersion coeffient, but Lorenz et al. (2005) reported that the ADC increases with an increase in the gas density caused by pressure. If a bubble moves upward inside a liquid, a wake forms behind the bubble, which entrains the liquid. The dimensions of the wake are heavily dependent on the bubble shape and size, which in turn is a function of operating parameters and liquid phase properties. A large, quickly rising bubble causes more turbulence because of its bubble wake, accelerating the liquid upward. Small, slow-moving or even temporarily stagnant bubbles, by contrast, might not induce any serious turbulence. Interactions between fast- and slow-rising bubbles occur in reactors because of high bubble density and non-uniform bubble size distributions, which results in the mixing of the phases. Lorenz et al. (2005) explained that the increase in backmixing is related to a narrower bubble size distribution. The investigators argued that small bubbles occur more often at elevated system pressures, which in turn leads to a flatter liquid radial velocity profile, which in turn reduces radial dispersion. They also argued that beyond the transition of the homogeneous and heterogeneous flow regimes, more large bubbles are formed at atmospheric pressure; therefore, the amount of axial mixing decreases (Krishna et al., 2000).

Effect of Thermodynamic Variables

The majority of the bubble columns are operated at pressure above atmospheric (Shah et al., 2012; Leonard et al., 2015). From the literature, it is found that the impact of pressure on liquid backmixing is subject to differing opinions. As in most studies, there exist three main views.

First View

The first view is that enhanced liquid backmixing is a result of pressurized conditions (Wilkinson et al., 1993; Lorenz et al., 2005; Therning and Rasmuson, 2001). Lorenz et al. (2005) explained that the increase in backmixing at elevated pressure is caused by reduced eddy diffusivity and lower liquid circulation velocities. The same opinion is shared by Wilkinson et al. (1993), who found that the liquid ADC increased with increasing pressure, especially under high gas velocity (>0.10 m/s) within a pressure range between 0.1 and 1.5 MPa. They proposed a procedure to estimate the liquid-phase dispersion coefficient at elevated pressures based on gas hold-up and the dispersion coefficient at atmospheric pressure:

$$\frac{D_z\,(at\ high\ pressure)}{D_z\,(at\ atmospheric\ pressure)} = \frac{(1-\varepsilon_g)\,(at\ atmospheric\ pressure)}{(1-\varepsilon_g)\,(at\ high\ pressure)} \tag{6.93}$$

However, their study was limited by the narrow experimental conditions (low pressure) and limited batch system (air–water), and their conclusion on the pressure effect and the proposed correlation need to be further verified. Therning and Rasmuson (2001) also found a relationship between liquid dispersion and pressure. The magnitude of the results for atmospheric pressures, 0.5 and 0.43 MPa (obtained by Therning and Rasmuson and Wilkinson et al., respectively), are the same, which is rather surprising because liquid backmixing should be lower when using packings than in the case with empty bubble columns.

Second View

The second view is that increasing pressure decreases the axial dispersion (Tarmy et al., 1984; Onozaki et al., 2000; Yang and Fan, 2003). Yang and Fan (2003) have measured liquid ADC at high pressure ($\leq$22 MPa) in different sized bubble columns. They found that the ADC decreased with an increase in pressure. The reduction in liquid mixing (especially for larger column diameters and higher superficial gas velocities) in the presence of elevated pressure (the pressure ranged up to 10.3 Mpa, and the superficial gas and liquid velocities varied upto 0.4 and 0.01m/s, respectively) is explained by the occurrence of smaller bubbles, which produce less developed bubble wakes and thus induce a lesser amount of liquid turbulence. Onozaki et al. (2000) found that the ADCs of the liquid phase under conditions (pressure = 16.8-18.8 MPa) were much smaller than those estimated from literature correlations obtained at ambient conditions. Tarmy et al. (1984) found from their experiment that the dispersion coefficients at high pressures (17 MPa) were up to 2.5 times smaller than the predictions by literature correlations, which usually were proposed for ambient conditions in air–water systems. They also reported that the values of dispersion coefficient increased with an increase in pressure.

Third View

Pressure had no influence on liquid mixing in either direction (Holcombe et al., 1983; Sangnimnuan et al., 1984; Houzelot et al., 1983). Houzelot et al.'s results showed that pressure did not have any influence nor did liquid velocity or viscosity on liquid phase dispersion because of the limited range of parameter variation in their study. They reported that axial dispersion is only dependent on gas superficial velocity. Holcombe et al. (1983) determined the axial liquid dispersion coefficient in bubble column at a pressure range of 0.3 to 0.72 MPa. They used heat as a tracer to measure the thermal dispersion coefficient, which was found to be comparable to the mass dispersion coefficient. In their study, the effect of pressure on thermal dispersion coefficients was negligible. From an industrial point of view, more studies need to be carried out under pressurized conditions in pilot-scale columns with different types of liquids. The influence of pressure on liquid ADC also depends on liquid phase properties. For the more viscous fluids, mixing may be dramatically affected by pressure because liquid recirculation turbulence is lower for these systems, and mixing is more dependent on bubble induced turbulence.

No significant studies to date reported in the literature have been conducted regarding the effect of temperature on the axial dispersion coeffient. However, Onozaki et al. (2000), in the homogeneous regime, found a lower ADC at the lowest temperature of ranges 313 to 730 K in the case of an industrial plant. This can be attributed to the increase of the liquid viscosity by reducing the temperature. As for other publications, a decrease of the liquid ADC while increasing viscosity is commonly observed (Hikita and Kikukawa, 1974; Shah et al., 2012), which is attributed to a decrease of liquid recirculation in conditions representative of the heterogeneous regime. In an ejector-induced inverse bubble column described by Majumder et al. (2005), the separator pressure has a role in reducing the backmixing. At high separator pressures, the entrainment of the gas decreases, and as a consequence, the gas hold-up in the column decreases, which leads to reduce the mixing of the phases because of resistance of the flow in inverse direction. The effect of temperature will be the same for inverse bubbly flow systems on dispersion coefficient because of the change of viscosity of liquid with temperature.

Other Models to Interpret the Mixing in Bubbly Flow

As observed from the literature, the dispersion coefficient exhibits mixing performance depending on various operating variables. Several other phenomenological models that attempted to interpret the physics of the fluid mixing have been proposed in the literature (as shown in Figure 6.7). To accurately model mixing, one needs to understand the dominant mixing mechanism of the phases. This can only be accomplished if the contribution of different mixing mechanisms can be quantified and related to the macromixing of the phases using physically based models. The different models other than plug, axial

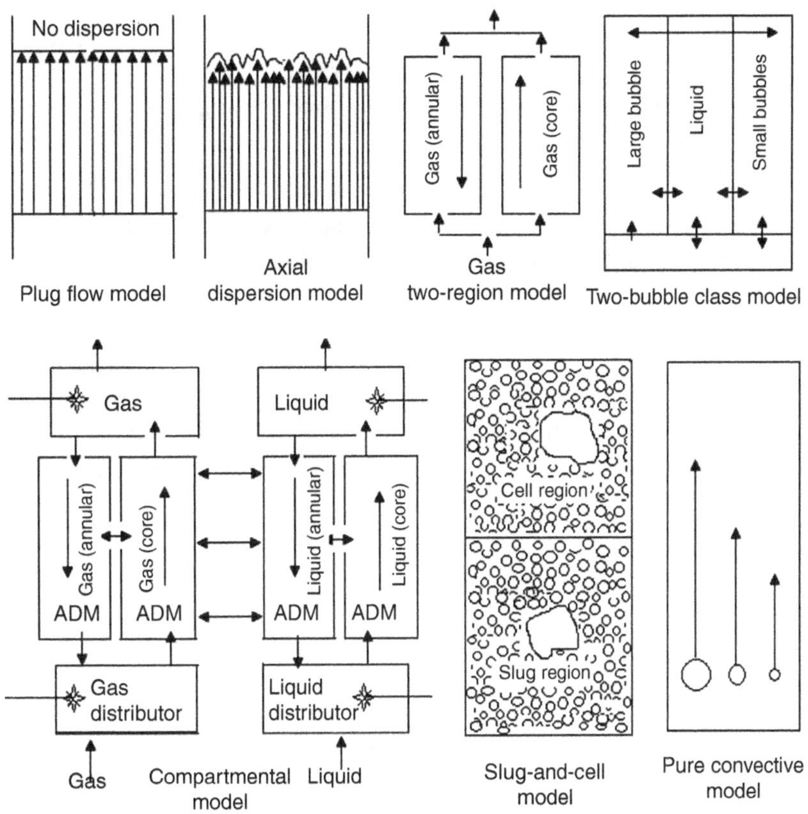

FIGURE 6.7 Different models used for analysis of phase dispersion. *ADM,* Axial dispersion model.

dispersion and tank-in-series models available to interpret the mixing behavior of gas in bubble column are discussed in the this section.

Two-Bubble Class Model

According to this model (Shah et al., 1985; Shetty et al., 1992; Kantak et al., 1995), bubbles may be separated into small bubbles driven by liquid motion and large bubbles that rise in a plug flow manner independent of liquid motion. This model is a superposition of a convective mechanism (transport at two velocities) onto a dispersive mechanism in which each bubble class is treated separately with no interaction between bubble classes. The applicability of the bimodal (two-bubble class) models to describe gas phase mixing in bubble columns is still doubtable because actual results for bubble size distribution do not follow the a bimodal distribution of the bubble size population. Furthermore, the two-bubble class model fails to explain the increase in the extent of the axial gas phase mixing with the increase in superficial gas velocity because it assumes that the increase in the fraction of large bubbles will cause the gas mixing behavior to approach plug flow, contradicting the reported experimental data.

Pure Convective Model

The model (Hyndman and Guy, 1995) is based on the assumption that a superposition of many bubbles occurs in plug flows, where bubbles rise at constant velocity along the column axis and mixing is caused solely by the differences in bubble velocities.

Compartmental Model

This model (Gupta, 2002) was developed based on the assumption that bubbles move upward in the core region and downward in the wall region with interaction between the two regions. Although the model accurately describes the physical picture of gas phase mixing emerging from numerous experimental studies, this model requires many mixing input parameters for the different reactor compartments. To overcome this problem, Gupta (2002) developed a method to estimate these parameters based on the experimental data of computer-automated radioactive particle tracking (CARPT) and computed tomography (CT) and using the radial gas hold-up profile as an input to the model.

Two-Region Model

Based on this model (Kawagoe et al., 1989), the bubble column is divided into a core region where bubbles move upward and a wall region where bubbles move in the opposite direction. Each region is modeled separately using an ADM and has its own ADC, where no interaction is assumed between different regions.

Slug-and-Cell Model

In this model (Myers et al., 1987), the bubble column is divided into gas-rich slugs, which consist of large fast-rising bubbles with small bubbles and liquid entrained in the wake, and gas-lean cells, which consist of a series of stationary pseudohomogenous dispersions of small gas bubbles in liquid. During the passage of slugs through cells, small bubbles are exchanged between slugs and cells.

Velocity Ditribution Model

The mixing of liquid occurs under the combined action of variation of velocity and the motion of gas bubbles in the liquid (Taylor, 1953; Aris, 1959).

Isotropic Turbulence Model

This model (Baird and Rice, 1975; Deckwer, 1992) is based on the the situation in which the turbulence intensity of a fluid is uniform and acts equally in any direction.

Velocity Distribution Model

In the condition of bubble flow, liquid flows in the vertical direction with wide distribution of velocities over the cross-section of the column. The mixing of liquid occurs under the combined action of variation of velocity and the motion of gas bubbles in the liquid. This is the basis of the velocity distribution model (Taylor, 1953).

Model Equations

According to Taylor (1953), the dispersion of the homogeneous system in a tube occurs when a solute is transported by a stream of non-uniform velocity. This historical say describes the evolution of dispersion in laminar single-phase flows in pipes as a function of the fluid velocity and the molecular diffusion coefficient of the tracer D_m from the following equation:

$$\frac{\partial C}{\partial t} + u\frac{\partial C}{\partial z} = D_m\left(\frac{d_c^2}{4r}\frac{\partial}{\partial r}\left(r\frac{\partial C}{\partial r}\right) + \frac{\partial^2 C}{\partial z^2}\right) \tag{6.94}$$

Aris (1959) extended Taylor's work by using method of moments and obtained a general form of dispersion coefficient as follows:

$$D_z = \frac{d_c^2 u_0^2}{kD_m} + D_m \tag{6.95}$$

where D_m and u_0 are the molecular diffusion coefficient and the velocity at column axis, respectively. The constant k determined the form of velocity distribution. When the distribution is parabolic, k is equal to 768 (Aris, 1959). Majumder et al. (2005) applied Taylor and Aris' model of homogeneous system to the ejector-induced inverse bubble colum, because of well homogenizing of two-phases in the column. To apply the above model in the inverse system, the following assumptions have been made:

1. The inverse bubble column is regarded as homogeneous.

2. The gas bubbles are dispersed uniformly. It follows a consequence that the dispersion coefficient due to bubble motion, D_b is uniform in the column, and the contribution of D_b to the overall process is much larger than that of molecular diffusion.
3. Steady flow condition is established in the axial direction.

On the basis of the above assumptions, the dispersion coefficient in the bubble column is directly derived by replacing D_m in equation (6.95) with D_b.

$$D_z = \frac{d_c^2 u_0^2}{k D_b} + D_b \tag{6.96}$$

According to Aris' (1959) analysis, any definite value of local velocity can be used as the representative velocity at column axis u_0 in equation (6.96).

Liquid Velocity at Column Axis

Whalley and Davidson (1974) proposed a model for calculation of the turbulent liquid velocity field in bubble columns. The model is based on the following energy balance:

$$E_m = E_s - E_i \tag{6.97}$$

where E_m is the energy dissipation rate in the liquid motion, E_s is rate of energy supplied, and E_i is equal to the energy dissipation rate at the bubble–liquid interface. The energy dissipation rate at the bubble–liquid interface can be neglected compared with the energy suppled to the system. The rate of energy supplied can be calculated by the following equation:

$$E_s = \frac{\pi}{4} d_c^2 L_d \rho_l g u_{sg} \tag{6.98}$$

The energy dissipation rate in the liquid motion is (Rietema and Ottengraf, 1970):

$$E_m = -\int_0^R 2\pi r L_d \tau_s \frac{du}{dr} dr \tag{6.99}$$

The shear stress (τ_s) can be expressed by two approaches. The first approach is to propose a certain form for the turbulent kinematic viscosity and use experimental data for the local liquid velocity to fit an empirical constant in this form. Based on this approach, the shear stress is expressed as:

$$\tau_s = \rho_l (v^m + v^t) \frac{du}{dr} \tag{6.100}$$

The turbulent kinematic viscosity can be expressed as (Kumar, 1994)

$$v^t = \frac{\overline{u'_{l,z} u'_{l,r}}}{du_{l,z}/dr} = l^2 \left| \frac{du_{l,z}}{dr} \right| \tag{6.101}$$

Different authors developed numerous correlations for the turbulent viscosity. Most of the correlations reported are a function only of column diameter as shown in Table 6.6.

Table 6.6   Various Correlations of Kinemaatic Viscosity

Reference	Correlations
Miyauchi and Shyu (1970)	$v^t = 0.050 d_c^{1.8} \rho_l^{-1}$
Ueyama and Miyauchi (1979)	$v^t = 0.128 d_c^{1.7} \rho_l$
Kojima et al. (1980)	$v^t = 0.053 d_c^{1.77} \rho_l^{-1}$
Miyauchi et al. (1981)	$v^t = 0.160 d_c^{3/2} \rho_l^{1/6}$
Riquarts (1981)	$v^t = 0.011 d_c^{3/2} u_{sg}^{3/8} \mu_l^{-1/8} \rho_l^{1/8} g^{3/8}$
Sekizawa et al. (1983)	$v^t = 0.265 d_c^{3/2}$
Kawase and Moo-Young (1989)	$v^t = 0.0295 d_c^{4/3} u_{sg}^{1/3} g^{1/3}$

A few other authors included the superficial gas velocity. All of these correlations imply that the kinematic viscosity is constant over the entire flow field.

The second approach is to use the Prandtl mixing length theory. Schlichting (1979) analyzed the shear stress in the turbulent flow region by using a mixing length theory, which is common for turbulent single-phase flow in a pipe. The shear stress in the turbulent flow region based on this approach can be expressed as

$$\tau_s = \rho_l \bar{l}^2 \left(\frac{du}{dr}\right)^2 \tag{6.102}$$

Serizawa et al. (1975) reported that mixing length and eddy diffusivity of momentum in two-phase flows are much larger than those in single flows. Therefore, Kawase and Moo-Young (1989) estimated the average mixing length and eddy kinematic viscosity based on the velocity profile in bubble columns by von Kerman's similarity for single-phase flow. According to Kawase and Moo-Young (1989), the average mixing length $\bar{l}$ in bubble column can be represented as

$$\bar{l}(r) = \chi \left|\frac{du/dr}{d^2u/dr^2}\right|_{at\ r=R_c/2} = \frac{\chi}{2} R_c \tag{6.103}$$

where χ is constant which is equal to $0.4n$ (Clapp, 1963). The parameter n is called the flow behavior index in the power law model; otherwise, for Newtonian liquid, it is 1. Table 6.7 lists the existing other mixing length correlations that can also be used.

The correlation for liquid velocity distribution in a bubble column at a high Reynold's number can be represented by (Kawase and Moo-Young, 1986):

$$u_l / u_{l0} = -2(r/R_c)^2 + 1 \tag{6.104}$$

Substitution of equations (6.102), (6.103), and (6.104) into equation (6.99) yields

$$E_m = 0.512 n^2 \pi \rho_l h_l d_c u_0^3 \tag{6.105}$$

Table 6.7 Mixing Length Correlations

Reference	Correlations
Schlichting (1979)	$\bar{l}(r) = (0.14 - 0.08(r/R)^2 - 0.06(r/R)^3)R$
Joshi (1980)	$\bar{l}(r) = 0.16R$
Kawase and Moo-Young (1989)	$\bar{l}(r) = 0.20R$
Geary and Rice (1992)	$\bar{l}(r) = d_b \varepsilon_g(r)/\bar{\varepsilon}_g$
Kumar (1994)	$\bar{l}(r) = a(1 - r/R)/(r/R + b)^c + d(1 - r/R)^e$

By substituting equations (6.98) and (6.105) in equation (6.97) and rearranging, one gets an expression for the liquid velocity at the column axis as

$$u_{l0} = 0.787 \left(\frac{g d_c u_{sg}}{n^2} \right)^{1/3} \tag{6.106}$$

In case of upward bubbly flow in a batch liquid phase bubble column, the axial liquid velocity is given by (Ohki and Inoue, 1970)

$$u_{l0} = 24 u_{sg}^{0.6} \tag{6.107}$$

In case of downward cocurrent bubbly flow, the axial liquid velocity is (Majumder et al., 2005)

$$u_{l0} = \left(\frac{u_{sg}^{0.082} u_j^{2.556} d_n^{4.996} d_c^{-3.534} L_d^{1.252}}{3.28 \times 10^{-4} \, \varepsilon_g^{0.082}} \right)^{1/2} \tag{6.108}$$

Estimation of Parameters D_b and k

To calculate the dispersion coefficient by equation (6.96), the value of D_b, the coefficient k, and the actual liquid velocity u_l must be known. The actual liquid velocity can be calculated as $u_l = u_{sl}/(1-\varepsilon_g)$. The actual liquid velocity (u_{la}) in the column depends on fractional gas hold-up (ε_g), which contributes to the order of k and D_b. After substituting the expression for u_o in equation (6.96), equation (6.96) can be expressed as

$$E_z = \frac{X}{kD_b} + D_b \tag{6.109}$$

where for upward bubbly flow, $X = d_c^2 u_0^2$, and for inverse bubbly flow (ejector-induced inverse bubble column),

$$X = \frac{u_{sg}^{0.082} u_j^{2.556} d_n^{4.996} L_d^{1.252}}{3.28 \times 10^{-4} \, \varepsilon_g^{0.082} d_c^{1.534}} \tag{6.110}$$

For other bubble columns, X can be accordingly represented by the operating variables. The validity of the velocity distribution model can be examined by a comparison between the dispersion coefficients calculated by equation (6.109) and the experimental data. For the calculation of dispersion coefficient (equation [6.109]), it is necessary to calculate the accurate values of k and D_b. These values are calculated from the slopes $(1/kD_b)$ and intercepts (D_b) of equation (6.109) for different experimental conditions. From the values of slope and intercept, the value of characteristic factor of velocity distribution (k) is calculated as reciprocal of the product of the slope and the intercept, that is,

$$k = \left[\text{Slope} \times \text{Intercept}\right]^{-1} = \left[\frac{1}{kD_b} \times D_b\right]^{-1} \tag{6.111}$$

As an example, at constant nozzle diameter $d_n = 0.004$ m, superficial gas velocity, $u_{sg} = 1.18 \times 10^{-3}$ m/s, in the range of superficial liquid velocity $u_{sl} = 4.91 \times 10^{-2}$ to 9.82×10^{-2} m/s and $d_c = 0.06$ m, the values of slope $(1/kD_b)$ and intercept (D_b) are 68.426 and 0.0017, respectively. The corresponding value of k as calculated by equation (6.111) is 8.59. The coefficient of velocity profile, k, is equal to 768 (Aris, 1959) for an upflow bubble column. In the case of an inverse bubbly flow system, the values of k are much less than 768. It may be due to complex velocity distribution in inverse bubble column. Because the velocity distribution in the inverse system is quite different from the up-flow system, the mixing characteristics in the inverse system is quite different. In the inverse system, the turbulence of the fluid inside the column is much more significant because of higher momentum transfer of liquid jet and variation of gas hold-up. This may cause the greater effect of nozzle diameter, column diameter, and superficial gas velocity on the dispersion coefficient of bubble motion and coefficient of velocity profile. The interstitial liquid velocity inside the column is a function of gas hold-up. As the fractional gas hold-up increases, the interstitial liquid velocity increases because of a decrease in the liquid flow area. The gas hold-up varies with the nozzle diameter, the column diameter, and the superficial gas velocity and hence causes the variation of D_b and k with nozzle diameter, column diameter, and superficial gas velocity in an ejector-induced inverse bubble column. The coefficient of velocity profile (k) decreases with an increase in nozzle diameter at a constant superficial gas velocity. This characterizes the different velocity profiles for different nozzle diameters that are liable to variation of ADC of liquid. From the experimental results, it may be seen that the values of D_b and k are different under different experimental conditions. So, it is very inconvenient to predict D_z by equations (6.109) and (6.110). Therefore, a model concerning D_b and k under various conditions u_{sg}, d_n, and d_c is necessary to predict D_z. The following correlations for D_b and k can be used to obtain the values of D_b and k (Majumder et al., 2005):

$$D_b = 1.602 D_r^{2.147} \, Re_g^{-0.671} \tag{6.111}$$

$$k = 0.115 D_r^{-1.364} \, Re_g^{0.403} \tag{6.112}$$

The above correlations are valid within the ranges of $0.08 \leq d_r \leq 0.116$ and $5.30 \leq Re_g \leq 35.38$ where $d_r = d_n/d_c$, $Re_g = (d_c \rho_g u_{sg})/\mu_g$. Therefore, predicted values of the ADC can be obtained by substituting the values of k and D_b from equations (6.111) and (6.112) in equation (6.109).

Interpretation of the Parameters Based on Different Operating Variables

Effect of Gas Distributor

The selection and design of the gas distributor is a practically important aspect of all bubble column reactors. Bubble dispersion and the construction of the mechanism of dispersion have a decisive influence on hold-up, interfacial area, and level of mass transfer in all bubble columns. The mode of dispersion may be categorized by two types: sparger and plunging jet type. In case of the sparger type mostly for upward flow, gas may be dispersed through pores or holes. These gas dispersion appliances are also known as static gas distributors in contrast to the dynamic jet, in which gas is distributed by kinetic energy generated by liquid force. In the plunging jet system mostly used in inverse bubble flow operation, a jet of liquid produced by the nozzle plunges a pool of liquid-carrying gas along with it and disperses it as a bubble caused by momentum transfer of the liquid jet (Majumder et al., 2005, 2006a, 2006b, 2007a, 2007b). The velocity pattern of the liquid and motion of gas bubbles through the liquid in the column reactor depend on the distributor hole diameter (Ohki and Inoue, 1970) or nozzle diameter (Majumder et al., 2005). It has been found from literature data that the characteristic factor of velocity distribution and the dispersion coefficient of bubble motion are strongly dependent on the hole or nozzle diameter. The value of dispersion coefficient of bubble motion (D_b) increases with the hole diameter (D_b [cm^2/s] = 170 d_h [mm]) at a constant column diameter (0.04 m) in the case of batch liquid gas upflow bubble columns (Ohki and Inoue, 1970) as shown in Table 6.8.

This is because the larger holes produce the larger bubbles, which rise with higher velocity and disperse the liquid around them more strongly. The value of characteristic factor of velocity distribution decreases with hole diameter ($k \propto 1/d_h$, where the proportionality constant is 11.76 for a column diameter of 40 mm). This may be because of inhomogeneous distribution of the bubble size, which is formed by increasing hole diameter. In gas-liquid jet reactors with two-phase nozzles such as ejectors, both gas and liquid are concentrically introduced, and the kinetic energy of the liquid jet is used to disperse gas flow into fine bubbles. In the inverse bubbly flow system, the turbulence of the fluid inside the column is much more significant because of higher momentum transfer of the plunging liquid jet and higher gas hold-up (Majumder et al., 2005). This may cause a greater effect of nozzle diameter of velocity pattern and dispersion of bubble inside the column. As the nozzle diameter increased, it has been found that the characteristic factor of velocity distribution follows a parabolic pattern ($k \propto 1/d_n^{0.68}$) (Figure 6.8) that is similar to an upflow system (Ogawa et al., 1982), but the order is different. The dispersion coefficient of bubble motion in the

Table 6.8 Correlations Developed for D_b and k as a Function of Hole/Nozzle Diameter

Range of Operating Variables	Dispersion Coefficient of Bubble Motion	Characteristic Factor of Velocity Distribution	Source of Data for D_z
$d_c = 40$ mm $u_{sg} = 3 - 25$ cm/s $d_h = 0.4$ -1.0 mm	$D_b = 170 d_h$, here D_b in cm²/s, d_h in mm $R^2 = 0.9999$	$k = 11.76 d_h^{-1.0}$ d_h in mm $R^2 = 0.998$	Ohki and Inoue (1970)
$d_c = 0.15$ m $u_{sg} = 0.018$- 0.075 m/s $d_n = 0.006$-0.016 m $d_t = 0.016$ m	$D_b = 0.0048(d_n/d_t)^{0.8918}$ at $Fr = 17.5$ $R^2 = 0.9993$	$k = 87.11(d_n/d_t)^{1.792}$ at $Fr = 17.5$ $R^2 = 0.9998$	Ogawa et al. (1982)
$d_c = 0.05$ m $u_{sg} = 0.0236$ m/s $d_n = 0.004$-0.007 m $d_t = 0.019$ m	$D_b = 0.024(d_n/d_t)^{1.68}$ $R^2 = 0.9999$	$k = 2.597(d_n/d_t)^{-0.68}$ $R^2 = 0.9999$	Majumder et al. (2005)

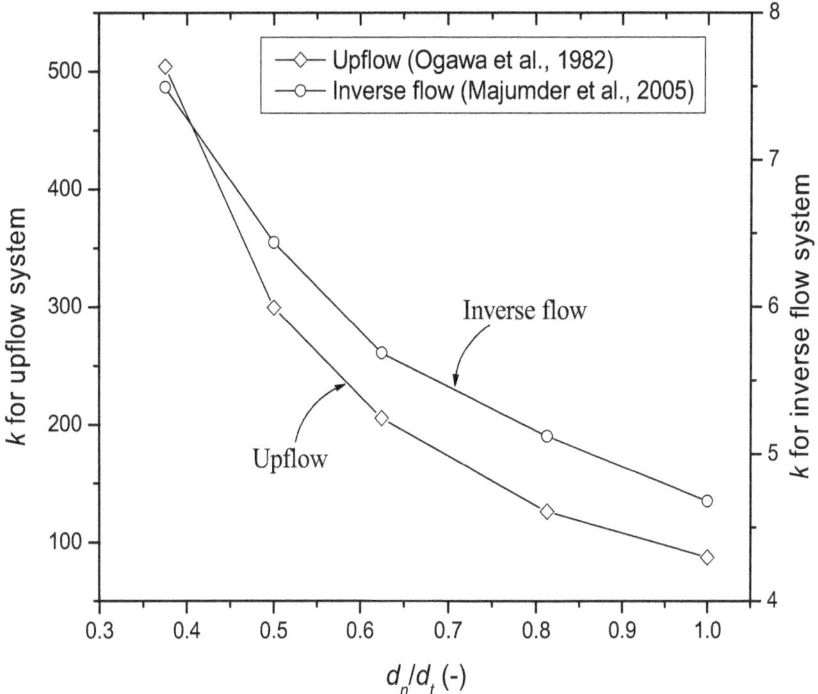

d_n/d_t (-)

FIGURE 6.8 Variation of velocity characteristic factor with nozzle-to-throat diameter.

inverse bubbly system is increased with an increase in nozzle diameter, which follows the same trend as in an upflow system ($k \propto 1/d_n^{1.792}$) (Ogawa et al., 1982). As the nozzle diameter increases, dispersion coefficient of bubble motion also increases ($D_b \propto d_n^{1.68}$) because of variation of the velocity pattern with an increase in nozzle diameter. The correlations for (k, D_b) based on nozzle diameter for upflow and inverse flow systems are shown in Table 6.8.

The order of k is much higher in upflow than inverse bubbly flow because of the intensity of turbulence of phases. It has been found that more turbulence follows less k. In case of inverse bubbly flow system, because of the opposite force, resistance of bubble buoyancy produces the more turbulence with higher momentum exchange of fluid by liquid jet (Majumder et al., 2005, 2006a, 2006b, 2007a, 2007b).

Effect of Column Diameter

The column diameter has one of the most significant influences on the fluid mixing. So far, it has been found that literature theories on liquid mixing usually account for mixing on the basis of entrainment, turbulence or liquid circulation (Wilkinson et al., 1993). The scale of liquid circulation is proportional to the column diameter. Because the degree of liquid mixing is determined by these scales, it is natural that the dispersion coefficient is strongly influenced by the column diameter (Ohki and Inoue, 1970). In the case of an upflow bubble column reactor, liquid flows upward in the center and downward near the wall region. Obviously because of turbulence, the up- and downflowing liquid interacts and therefore exchanges the liquid. This contribution is represented by the characteristic factor of velocity distribution. As the column diameter increases, the interaction and formation of large circulation will increase, which may cause the decrease in velocity characteristic factor (Figure 6.9). The decrease in characteristic factor increases the representative velocity, which causes the increase in circulation with an increase in column diameter. For the same reason, the effect of liquid circulation enhances the dispersion coefficient of

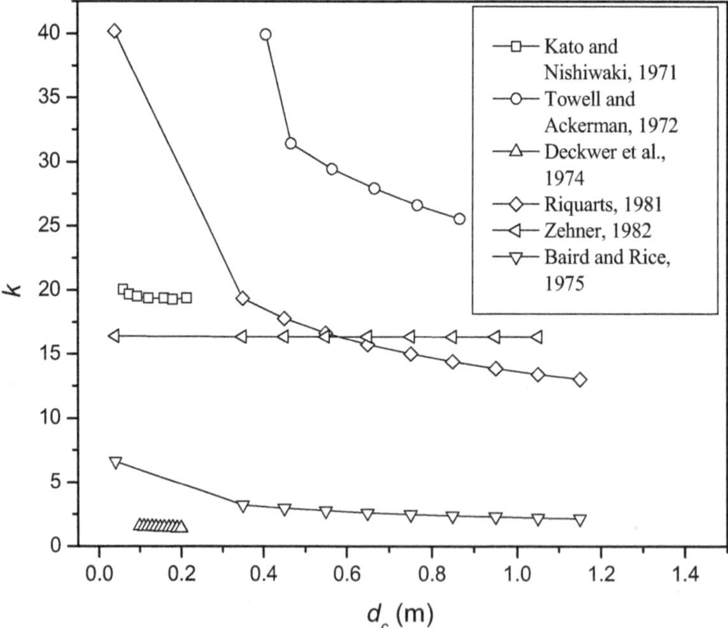

FIGURE 6.9 Variation of velocity characteristic factor with column diameter.

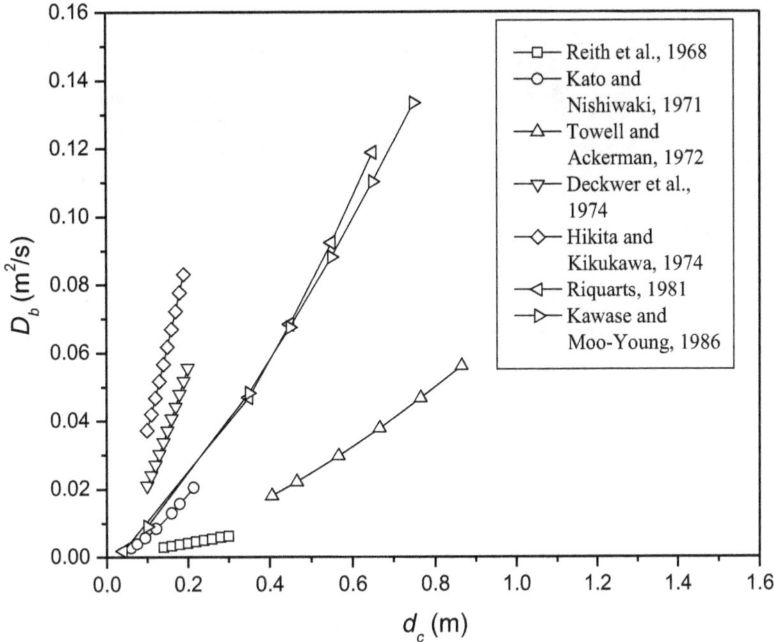

FIGURE 6.10 Variation of dispersion coefficient of bubble motion with column diameter.

bubble motion with an increase in the column diameter (Figure 6.10). The characteristic factor of velocity distribution (k) increases with column diameter (Figure 6.11).

This may be due to the mode of distribution of gas phase, which is one important criterion to influence the velocity pattern in the column. At a low gas flow rate, a larger diameter decreases the fluid circulation and turbulence and hence increases the velocity characteristic factor. The dispersion coefficient of bubble motion satisfies the relation $D_b = \alpha_1 d_c^{\beta_1}$ (Table 6.9).

Similarly, the characteristic factor of velocity distribution follows the proportionality as $k = \alpha_2 d_c^{\beta_2}$. The proportionality constants α and exponents β are shown in Table 6.10, which is estimated from investigations reported by different authors. In the case of inverse bubbly flow systems (Majumder et al., 2005), the absence of liquid circulation with respect to the column diameter results in the opposite effect on the characteristic factor of velocity and dispersion coefficient of bubble motion (Figure 6.12).

An increase in column diameter provides uniform gas distribution in inverse bubbly flow systems, which is economically viable when used, for example, in biological effluent treatment (Deckwer, 1992). In inverse bubbly flow systems, the entrainment of gas decreases with an increase of column diameter because of high resistance of the liquid pool. The high resistance of the liquid pool inside the column decreases the gas hold-up and the motion of the gas bubble. This results in a decrease in the dispersion coefficient of bubble motion with column diameter.

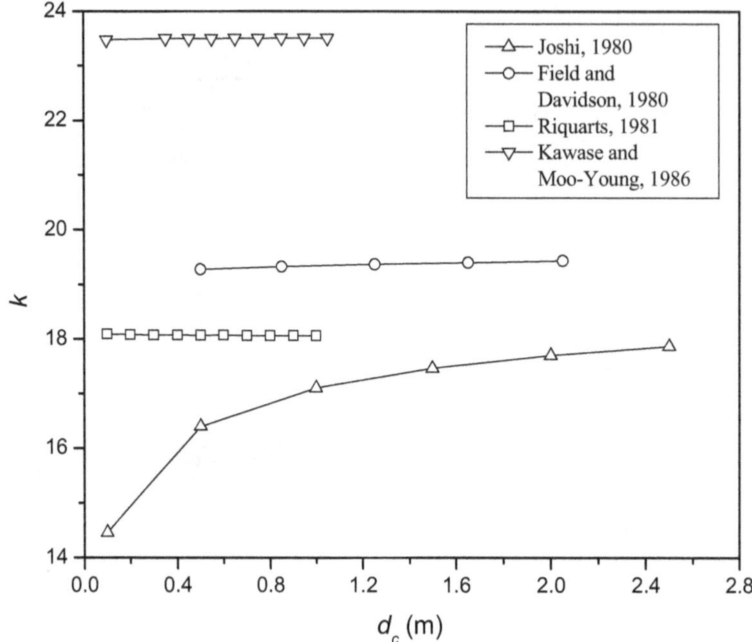

FIGURE 6.11 Variation of velocity characteristic factor with column diameter.

Effect of Liquid Velocity

In upflow bubble columns, the intensity of phase dispersion is dependent on the liquid superficial velocity and liquid circulation velocity. According to Joshi (1980), the intensity of phase dispersion is directly proportional to the summation of average liquid circulation and superficial liquid velocity in the heterogeneous bubbly flow regime. The liquid circulation is mainly because of the gas hold-up profile that sets up along the column diameter because of the tendency of the gas bubbles to move up through the central part of the column. Hence, the radial gas hold-up profile is formed, and this results in density difference across the column cross-section, which ultimately results in phase circulation and backmixing (Palaskar et al., 2000). The rate of liquid circulation increases with increasing superficial liquid velocity, which causes the increase in dispersion coefficient of bubble motion (Figure 6.13).

The velocity characteristic factor depends on the intensity of turbulence (Ohki and Inoue, 1970). As per Joshi (1980), an increase in turbulence with increase in liquid velocity results in a decrease in the velocity characteristic factor as shown in Figure 6.14. Superficial liquid velocity has an immense effect on inverse bubbly flow systems because as the liquid velocity increases, the transfer of momentum by liquid jet in the inverse bubbly flow system (Majumder et al., 2005) makes more turbulence and intense mixing of phases, which result in deviation of D_b and K from the upflow system. The functionality of the dispersion coefficient of bubble motion (D_b) and the velocity characteristic factor (k) on the superficial liquid velocity are developed as equations (6.113) and (6.114). The correlation

Table 6.9   Dispersion Coefficient of Bubble Motion (D_b) as a Function of Column Diameter and Superficial Gas Velocity

Range of u_{sg} (m/s)] and d_c (m)	Coefficients (α_1 and β_1) of Correlations $D_b = \alpha_1 d_c^{\beta_1}$	Coefficients (γ_1 and δ_1) of Correlations $D_b = \gamma_1 u_{sg}^{\delta_1}$	Source of Data for E_z
u_{sg} = 0.100-0.450 d_c = 0.14-0.30	α_1 = 0.0210, β_1 = 1.0145	γ_1 = 0.0899, δ_1 = 0.5661	Reith et al. (1968)
u_{sg} = 0.03-0.25 d_c = 0.04-0.16	α_1 = 0.0068, β_1 = 0.0 At hole diameter 0.4 mm	γ_1 = 29.41, δ_1 = 0.0 At hole diameter 0.4 mm	Ohki and Inoue (1970)
u_{sg} = 0.010-0.270 d_c = 0.06-0.214	α_1 = 0.2407, β_1 = 1.5927	γ_1 = 0.0248, δ_1 = 0.3715	Kato and Nishiwaki (1971)
u_{sg} = 0.010-0.080 d_c = 0.406-0.866	α_1 = 0.0694, β_1 = 1.4983	γ_1 = 0.3057, δ_1 = 0.5	Towell and Ackerman (1972)
u_{sg} = 0.010-0.080 d_c = 0.1-0.2	α_1 = 0.5287, β_1 = 1.3992	γ_1 = 0.0865, δ_1 = 0.3006	Deckwer et al. (1974)
u_{sg} = 0.003-0.185 d_c = 0.1-0.19	α_1 = 0.6603, β_1 = 1.2483	γ_1 = 0.1628, δ_1 = 0.3821	Hikita and Kikukawa (1974)
u_{sg} = 0.003-0.550 d_c = 0.1-1.0	α_1 = 0.4856, β_1 = 1.3339	γ_1 = 0.1368, δ_1 = 0.3334	Baird and Rice (1975)
u_{sg} = 0.040-0.450 d_c = 0.1-2.5, u_{sl} = 0.10	α_1 = 0.1853, β_1 = 1.2671	γ_1 = 1.276, δ_1 = 0.3845	Joshi (1980)
u_{sg} = 0.040-0.050 d_c = 0.5-2.05	α_1 = 0.1024, β_1 = 1.3307	γ_1 = 0.8795, δ_1 = 0.4096	Field and Davidson (1980)
u_{sg} = 0.125-0.325 d_c = 0.04-1.15	α_1 = 0.2269, β_1 = 1.5026	γ_1 = 0.2366, δ_1 = 0.3748	Riquarts (1981)
u_{sg} = 0.175-0.325 d_c = 0.04-0.65	α_1 = 0.8519, β_1 = 1.5006	γ_1 = 0.6197, δ_1 = 0.2366	Miyauchi et al. (1981)
u_{sg} = 0.100-0.400 d_c = 0.04-1.15	α_1 = 0.2348, β_1 = 1.3343	γ_1 = 0.3307, δ_1 = 0.3335	Zehner (1982)
u_{sg} = 0.175-0.325 d_c = 0.1-0.75	α_1 = 0.1958, β_1 = 1.3329	γ_1 = 0.2512, δ_1 = 0.3334	Kawase and Moo-Young (1986)
u_{sg} = 0.040-0.450 d_c = 0.1-3.5	α_1 = 0.0336, β_1 = 1.2489	γ_1 = 0.4751, δ_1 = 0.4504	Kantak et al. (1995)
u_{sg} = 0.00118-0.00943 d_c = 0.05-0.06	$\alpha_1 = 0.389\, d_n^{1.68} u_{sg}^{-0.49} v_g^{0.49}$, β_1 = -2.17	$\gamma_1 = 0.389 d_n^{1.68} d_c^{-2.17} v_g^{0.49}$, δ_1 = -0.49	Majumder et al. (2005)

developed by Joshi (1980) for E_z used to calculated D_b and k to develop the following correlations for a range of gas velocity 0.04 to 0.45 m/s. But for different gas velocities and a constant column diameter, it is found that the D_b and k follow the correlations developed as equations (6.115) and (6.116) in the range of column diameter 0.10 to 2.5 m.

$$D_b = 0.2438 d_c^{1.158} u_{sl}^{0.1224 d_c^{-0.2785}}, \qquad R^2 = 0.995 \tag{6.113}$$

$$k = 12.904\, d_c^{0.1759} u_{sl}^{-0.1224 d_c^{-0.2785}}, \qquad R^2 = 0.993 \tag{6.114}$$

$$D_b = 1.3121 u_{sg}^{0.3403} u_{sl}^{0.0222 u_{sg}^{-0.3922}}, \qquad R^2 = 0.999 \tag{6.115}$$

$$k = 21.858 u_{sg}^{-0.0527} u_{sl}^{-0.035 u_{sg}^{-0.4012}}, \qquad R^2 = 0.997 \tag{6.116}$$

Table 6.10   Characterization Factor of Velocity Distribution (k) as a Function of Column Diameter and Superficial Gas Velocity

Range of u_{sg} (m/s) & d_c (m)	Coefficients (α_2 and β_2) of Correlations, $k = \alpha_2 d_c^{\beta_2}$	Coefficients (γ_2 and δ_2) of Correlations, $k = \gamma_2 u_{sg}^{\delta_2}$ & $k = \omega_1 u_{sg}^2 + \omega_2 u_{sg} + \omega_3$	Source of Data for E_z
$u_{sg} = 0.100–0.450$ $d_c = 0.14–0.30$	$\alpha_2 = 184.47, \beta_2 = 0.6521$	$\gamma_2 = 12.268, \delta_2 = -0.4656$	Reith et al. (1968)
$u_{sg} = 0.03–0.25$ $d_c = 0.04–0.16$	$\alpha_2 = 68, \beta_2 = 0.0$ at $d_h = 0.4$ mm	$\gamma_2 = 29.41, \delta_2 = 0.0$ at hole diameter 0.4 mm	Ohki and Inoue (1970)
$u_{sg} = 0.010–0.270$ $d_c = 0.06–0.214$	$\alpha_2 = 17.458, \beta_2 = -0.048$	$\gamma_2 = 65.627, \delta_2 = 0.3078$ for $0.010 \le u_{sg} \le 0.049$, $\omega_1 = -49.23$, $\omega_2 = -14.59$, $\omega_3 = 27.031$ for $0.049 < u_{sg} \le 0.270$	Kato and Nishiwaki (1971)
$u_{sg} = 0.010–0.080$ $d_c = 0.406–0.866$	$\alpha_2 = 24.359, \beta_2 = -0.3334$	$\gamma_2 = 8.689, \delta_2 = -0.334$	Towell and Ackerman (1972)
$u_{sg} = 0.010–0.080$ $d_c = 0.1–0.2$	$\alpha_2 = 1.153, \beta_2 = -0.1325$	$\gamma_2 = 1.9057, \delta_2 = 0.067$	Deckwer et al. (1974)
$u_{sg} = 0.003–0.185$ $d_c = 0.1–0.19$	$\alpha_2 = 1.0459, \beta_2 = 0.1683$	$\gamma_2 = 2.7452, \delta_2 = 0.427$ for $0.003 \le u_{sg} \le 0.065$, $\omega_1 = -0.8054$, $\omega_2 = -0.3557$, $\omega_3 = 0.8486$; for $0.065 < u_{sg} \le 0.34$	Hikita and Kikukawa (1974)
$u_{sg} = 0.003–0.550$ $d_c = 0.1–1.0$	$\alpha_2 = 18.061, \beta_2 = -0.0007$	$\omega_1 = -0.0907, \omega_2 = 0.0289, \omega_3 = 26.577$	Baird and Rice (1975)
$u_{sg} = 0.040–0.450$ $d_c = 0.1–2.5, u_{sl} = 0.10$	$\alpha_2 = 17.16, \beta_2 = 0.0425$	$\gamma_2 = 23.151, \delta_2 = -0.1131$	Joshi (1980)
$u_{sg} = 0.040–0.050$ $d_c = 0.5–2.05$	$\alpha_2 = 19.354, \beta_2 = 0.0058$	$\gamma_2 = 16.584, \delta_2 = -0.1527$	Field and Davidson (1980)
$u_{sg} = 0.125–0.325$ $d_c = 0.04–1.15$	$\alpha_2 = 13.65, \beta_2 = -0.333$	$\gamma_2 = 15.731, \delta_2 = -0.0831$	Riquarts (1981)
$u_{sg} = 0.175–0.325$ $d_c = 0.04–0.65$	$\alpha_2 = 2.267, \beta_2 = -0.334$	$\gamma_2 = 3.2254, \delta_2 = 0.1933$	Miyauchi et al. (1981)
$u_{sg} = 0.100–0.400$ $d_c = 0.04–1.15$	$\alpha_2 = 16.355, \beta_2 = -0.0011$	$\gamma_2 = 17.072, \delta_2 = -0.0003$	Zehner (1982)
$u_{sg} = 0.175–0.325$ $d_c = 0.1–0.75$	$\alpha_2 = 23.51, \beta_2 = 0.0006$	$\gamma_2 = 25.333, \delta_2 = -0.0003$	Kawase and Moo-Young (1986)
$u_{sg} = 0.14–0.450$ $d_c = 0.1–3.5$	$\alpha_2 = 209.1, \beta_2 = 0.1678$	$\gamma_2 = 132.69, \delta_2 = -0.5128$	Kantak et al. (1994)
$u_{sg} = 0.00118–0.00943$ $d_c = 0.05–0.06$	$\alpha_2 = 0.784\, d_n^{-0.68}\, u_{sg}^{0.23}\, v_g^{-0.23}, \beta_2 = 0.91$	$\gamma_2 = 0.784 d_n^{-0.68} d_c^{0.91} v_g^{-0.23}, \delta_2 = 0.23$	Majumder et al. (2005)

Effect of Gas Velocity

Furthermore, at the same liquid velocity with an increasing superficial gas velocity results in increased phase dispersion. The intensity of liquid phase dispersion increases with 1.2th power of superficial gas velocity (Ohki and Inoue, 1970). Joshi (1980) determined that the liquid circulation velocity in upflow bubble columns decreases as superficial gas velocity approaches the product of gas hold-up and rise velocity of a

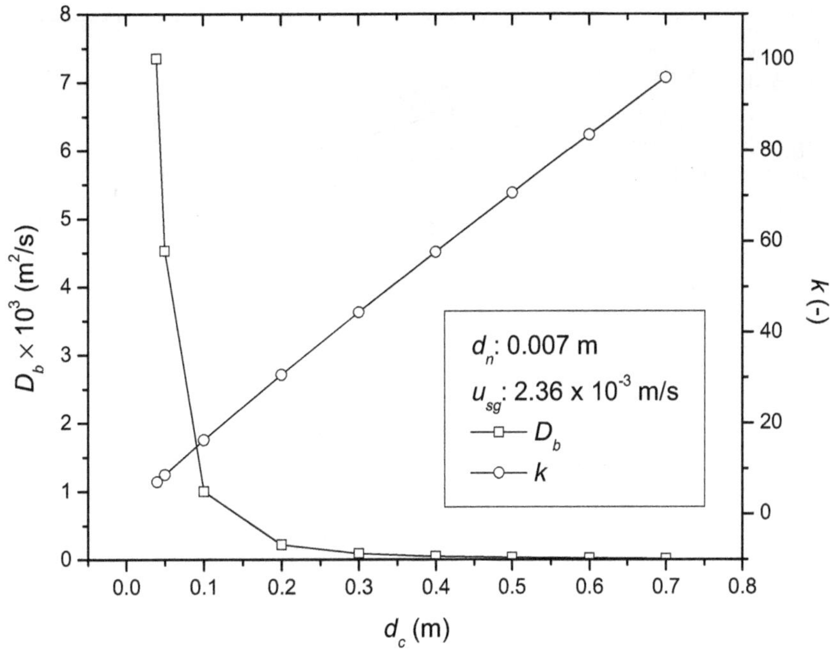

FIGURE 6.12 Variation of dispersion coefficient of bubble motion with column diameter.

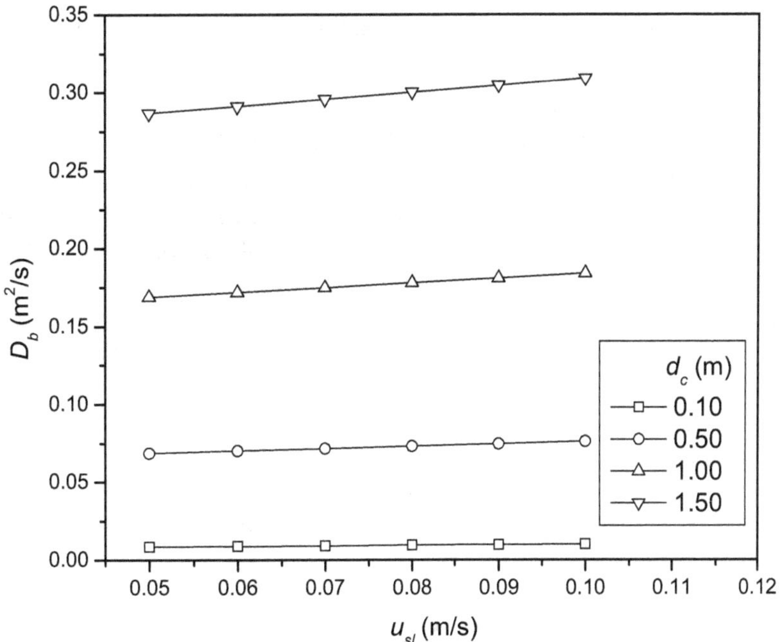

FIGURE 6.13 Variation of dispersion coefficient of bubble motion with superficial liquid velocity.

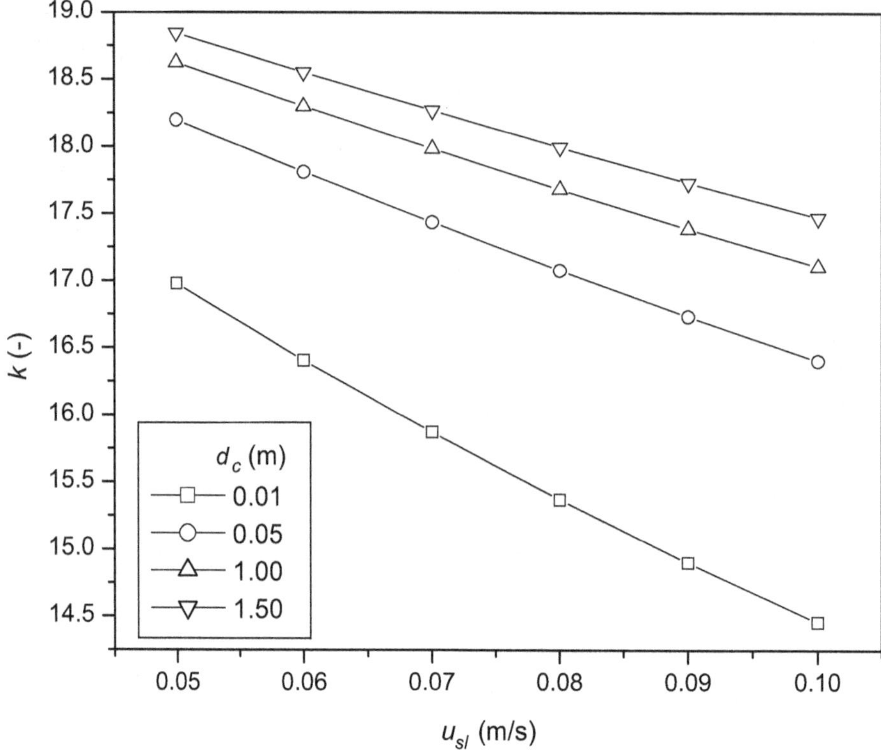

FIGURE 6.14 Variation of velocity characteristic factor with superficial liquid velocity.

single bubble. This phenomenon of phase dispersion follows the similar trend results reported by Field and Davidson (1980) and Kawase and Moo-Young (1986). That the phase circulation patterns occur in upflow bubble columns with the liquid flowing upward in the center and downward (because of continuity) near the column wall in combination with liquid turbulence may be one of the primary cause of the increase in the dispersion coefficient of bubble motion (Figures 6.15 and 6.16) and the decrease in characteristic factor (Figure 6.17) of velocity distribution with superficial gas velocity. Gas hold-up is an important factor that affects the dispersion coefficient of bubble motion and velocity pattern. The dispersion coefficient of bubble motion increases with an increase in gas hold-up, and the velocity characteristic factor decreases with an increase in gas velocity because of an increase in turbulence (Figure 6.18). The dispersion coefficient of bubble motion (D_b) and velocity characteristic factor (k) are correlated with gas hold-up as represented by equations (6.117) and (6.118) as per data of D_z by Joshi (1980).

$$D_b = (0.3337u_{sl} + 0.2433)Exp\{-(2.3654u_{sl} - 3.6259)\varepsilon_g\}, \qquad R^2 = 0.9981 \qquad (6.117)$$

$$k = (-76.923u_{sl} + 43.95)Exp\{+(3.8906u_{sl} - 1.3848)\varepsilon_g\}, \qquad R^2 = 0.9984 \qquad (6.118)$$

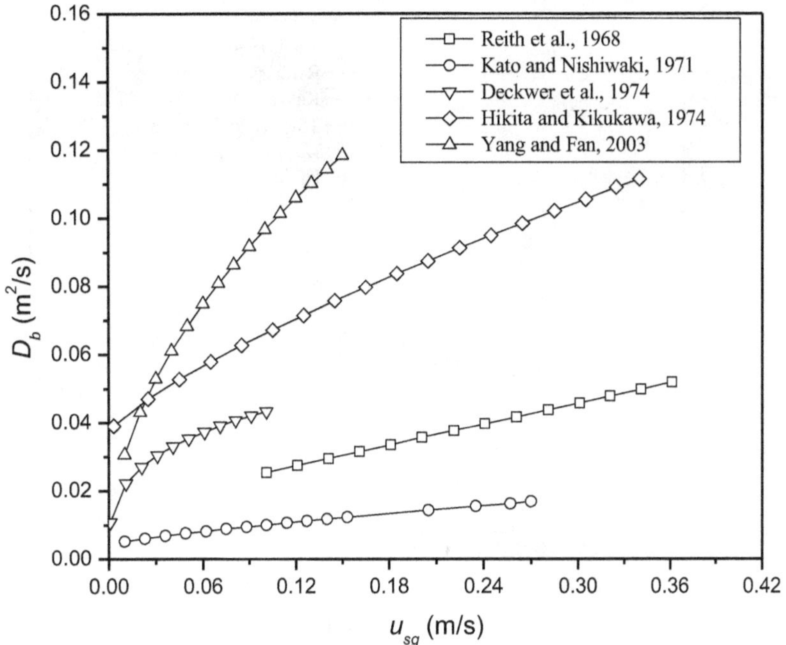

FIGURE 6.15 Variation of dispersion coefficient of bubble motion with superficial gas velocity.

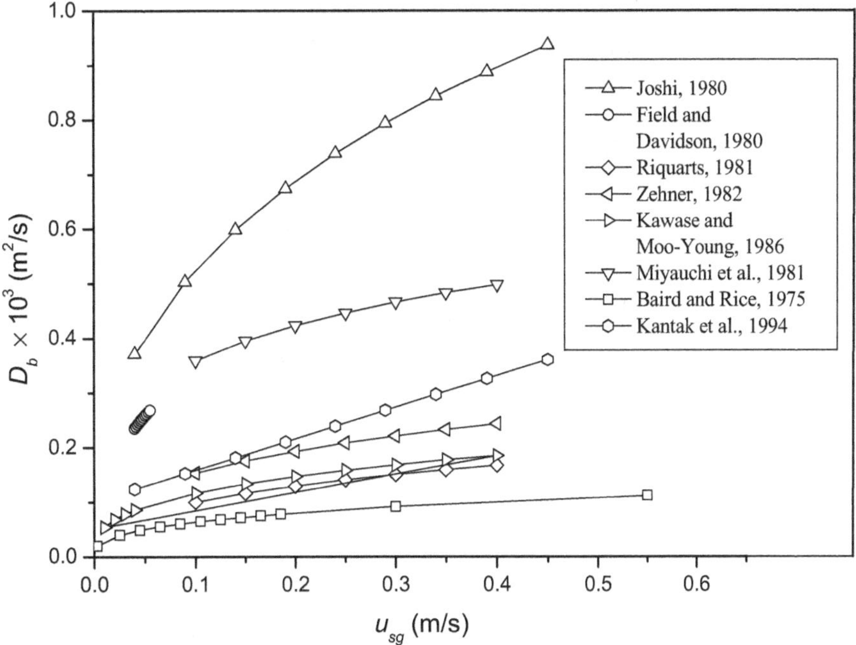

FIGURE 6.16 Variation of dispersion coefficient of bubble motion with superficial gas velocity.

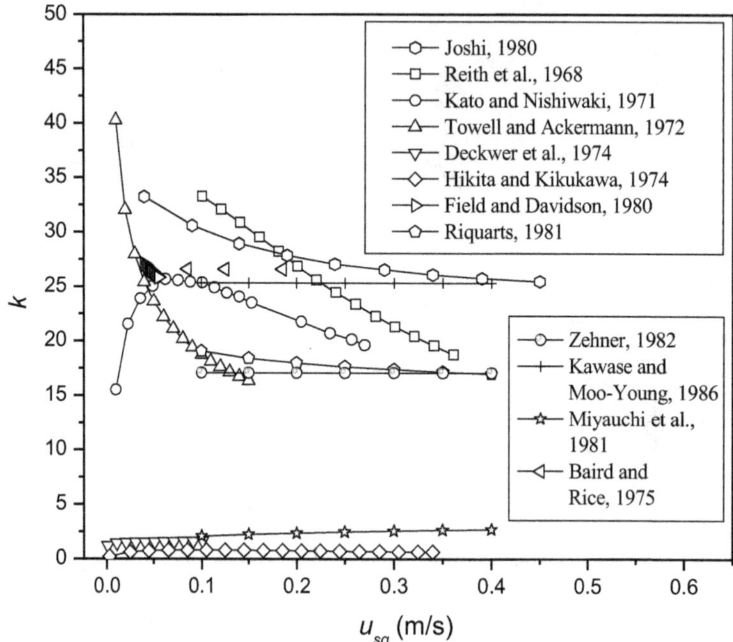

FIGURE 6.17 Variation of velocity characteristic factor with superficial gas velocity.

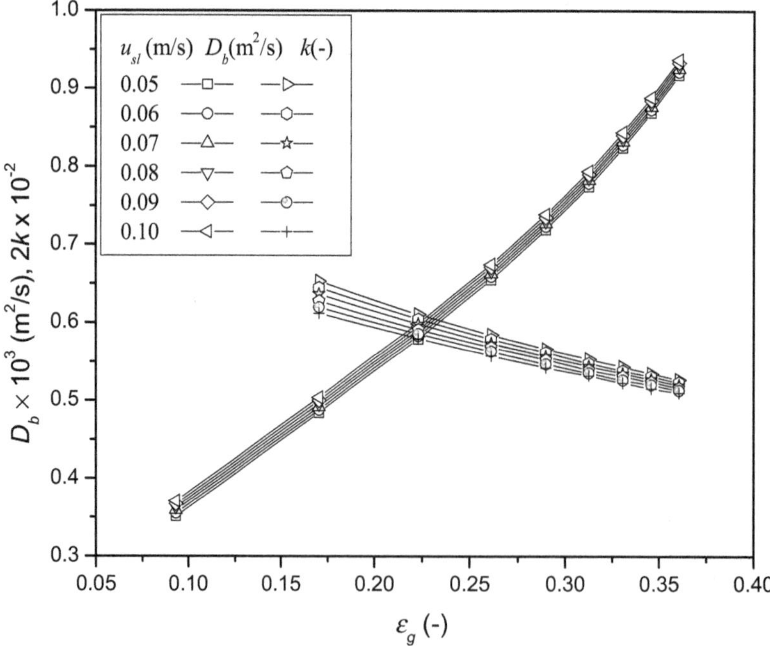

FIGURE 6.18 Variation of dispersion coefficient of bubble motion and velocity characteristic factor with gas hold-up.

For slow gas velocity, less gas hold-up decreases the turbulence of phase in a particular liquid velocity for constant geometric variables, which sometimes increases the velocity characteristic factor. It has been found that for low gas velocity ($0.003 < u_{sg} < 0.325$), the velocity characteristic factor satisfies the functionalities $k \infty u_{sg}^{+\delta_2}$ instead of $k \infty u_{sg}^{-\delta_2}$ (Kato and Nishiwaki, 1971; Deckwer et al., 1974; Hikita and Kikukawa, 1974; Miyauchi et al., 1981) as shown in Table 6.10. In accordance with the observed dependence of the dispersion coefficient of bubble motion and velocity characteristic factor on fluid velocities, these are represented as $k \infty u_{sg}^{0.967}$ and $D_b \infty u_{sg}^{0.1453}$ as per the calculation done by reported correlation of axial liquid dispersion coefficient developed by Krishna et al. (2000). It has been observed that in the inverse bubbly flow system, the values of k are much less than that of the up-flow bubble column. This may be attributable to complex velocity distribution resulting in higher momentum transfer and turbulence in inverse bubbly flow system. Because the velocity distribution in the inverse bubbly flow system is quite different from the upflow system (Majumder et al., 2005), the mixing characteristics in the inverse bubbly flow system are quite different. In the inverse bubbly flow system, the turbulence of the fluid inside the column is much more significant because of higher momentum transfer of liquid jet and variation of gas hold-up. The dispersion process in the inverse bubbly flow system also is a consequence of large bubbles whose terminal velocities exceed the liquid velocity in the column. The terminal rise velocity of the bubble is governed by the bubble size.

This may cause the greater effect of superficial gas velocity on the dispersion coefficient of bubble motion and coefficient of velocity profile (Figure 6.19). The interstitial liquid velocity inside the column is a function of gas hold-up. As the fractional gas hold-up increases, the interstitial liquid velocity increases because of a decrease in the liquid flow area. The gas hold-up varies with the superficial gas velocity and hence causes the variation of D_b and k with superficial gas velocity (Majumder et al., 2005). In the case of an inverse bubbly flow system, the entrainment of gas is a result of induction of gas by liquid jet. So the superficial gas velocity is directly proportional to the superficial liquid velocity. So the trend for variation of D_b and k with the superficial liquid velocity will be the same as the superficial gas velocity. The orders of dependence of the dispersion coefficient of bubble motion and velocity characteristic factors are shown in Tables 6.9 and 6.10.

Condition for the Dispersion Coefficient

It was found as per the discussion in an earlier section that the dispersion coefficient of bubble motion increases while characteristic factor decreases with a specific operating condition. Also from equation (6.96), the dispersion coefficient of bubble motion can be expressed as

$$D_b = \frac{D_z}{2}\left[1 \pm \sqrt{1 - \frac{4}{k}\left(\frac{d_c u_0}{D_z}\right)^2}\right] \tag{6.119}$$

From equation (6.119), it is obvious that the velocity characteristic factor is greater than $4Pe^2$. For the real value of the characteristic factor, k of velocity distribution will be greater

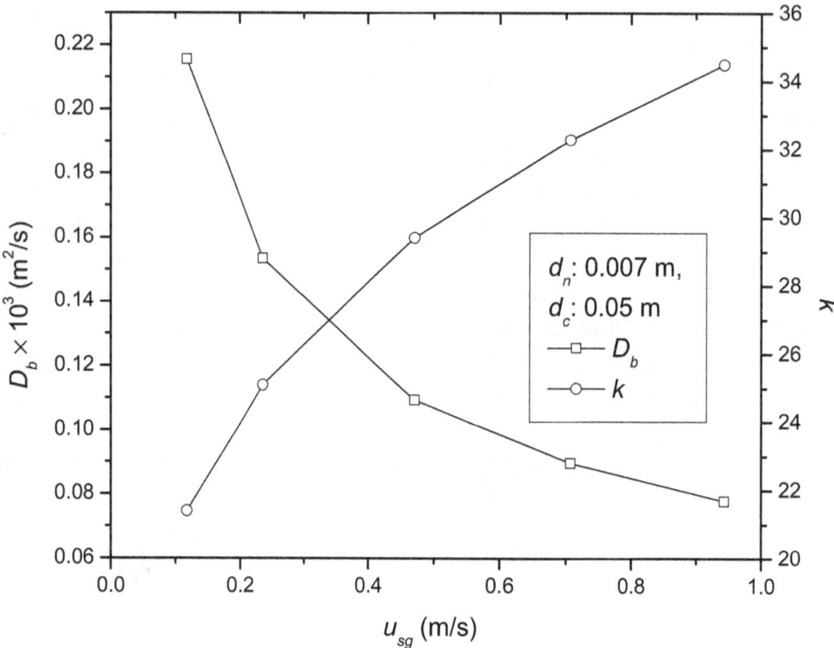

FIGURE 6.19 Variation of dispersion coefficient of bubble motion with superficial gas velocity.

than square of double of the Peclet number, $(d_c u_0/D_z)$. If the characterization factor k is equal to the square of double of the Peclet number, the coefficient of bubble motion is equal to half of the dispersion coefficient of liquid. An increase of the dispersion coefficient of bubble motion while a decrease of characteristic factor ensures that $\sqrt{1-\dfrac{4}{k}\left(\dfrac{d_c u_0}{D_z}\right)^2}$ will be in the range of 0 to 1. Therefore, the sign of the term of right side of equation (6.119) will be negative, which signifoes the decrease in k and increase in D_b.

Isotropic Turbulence Model

The isotropic turbulence model is useful in correlating data for processes such as heat and mass transfer or bubble break-up that depend on mixing of the phases. As per the model, a spectrum of eddy sizes is considered to exist. The eddies are generated continuously by an energy source, and they break down into smaller eddies that eventually are damped out by viscous effects. Fluid viscosity partly determines the minimum eddy size but plays a relatively little part in larger scale interactions. If the size of the bubble is greater than the minimum eddy size, the rate depends primarily on the specific energy dissipation (energy dissipation rate per unit mass of fluid), $\in$. Baird and Rice (1975) used Kolmogoroff's theory of isotropic turbulence and dimensional analysis to estimate dispersion coefficients. In the case of eddy diffusivity, the ADC in the bubble column depends on the specific energy dissipation rate, and they derived the following theoretical relationship in a multiphase flow system:

$$D_z = Kl^{4/3} \in^{1/3} \tag{6.120}$$

The resulting correlations can be applied to a wide range of data without experiments on the turbulence spectrum. Kolmogoroff (1941) described in some detail the situation when the turbulence intensity of a fluid is developed by the eddy diffusivity. It is to be noted that the largest eddies, which are mainly responsible for eddy diffusion, are related to a primary length parameter l and to the specific energy dissipation rate. The constant K is dimensionless proportinality constant and should have a universal value, provided the length parameter l is suitably defined for different system geometries. The energy dissipation in the gas bubbles is assumed to be negligible. At high phase flow rates, the bubble wakes interact, the liquid phase is more uniformly turbulent, and the eddy scale has a major effect on the dispersion. In bubbly flow systems in a column, the bubbles themselves can be carried about in the eddies, and the eddy scale l may be greater than the mean bubble diameter. Generally, without any machanical mixing device in the bubble column, the column diameter represents the upper value of l. Hence, from equation (6.120),

$$D_z = Kd_c^{4/3} \in^{1/3} \tag{6.121}$$

For an upflow bubble column, the specific energy dissipation rate in the liquid phase is

$$\in = \frac{\Delta P \dot{Q}_g}{m} = \frac{\rho_l g(1-\varepsilon_g)LA_c u_{sg}}{\rho_l(1-\varepsilon_g)LA_c} = u_{sg}g \tag{6.122}$$

Field and Davidson (1980) developed an equation for specific energy dissipation based on liquid circulation in the bubble column in which the liquid circulation rate is derived from an analytical solution of a differential stream function. The authors suggested that the energy dissipation rate in the bubble column should be formulated by

$$\in = (u_{sg} - \varepsilon_g u_s)g \tag{6.123}$$

where u_s is the bubble slip velocity, which has effect on the liquid circulation velocity that can be expressed as (Joshi and Sharma, 1979)

$$u_c = 13.9[d_c(u_{sg} - \varepsilon_g u_s)]^{1/3} \tag{6.124}$$

or

$$u_s = \frac{1}{\varepsilon_g}\left[u_{sg} - \frac{1}{d_c}\left(\frac{u_c}{13.9}\right)^3\right] \tag{6.125}$$

Hence, the energy dissipation rate in the bubble column is in terms of circulation velocity is

$$\in = \frac{1}{d_c}\left(\frac{u_c}{13.9}\right)^3 g \tag{6.126}$$

For ejector-induced inverse bubble columns, the rate of energy dissipation is equal to the kinetic energy used per unit time of phase dispersion in the column, which can be expressed as

$$\text{Rate of energy dissipation} = \frac{\text{Kinetic energy used } (E_u)}{\text{Phase contact time } (t_c)} \tag{6.127}$$

The whole amount of energy supplied $((1/2)m_f u_j^2)$ by the liquid jet may not be used for mixing of phase. The energy used for phase mixing can be expressed as

$$E_u = K_m \left(\frac{1}{2} m_f u_j^2 \right) \tag{6.128}$$

where K_m represents the mixing dissipation coefficient, which is equal to the fraction of jet energy used in the mixing. Majumder et al. (2007a) developed a correlation from their experimental data on an ejector-induced inverse bubble column, which can be expressed as

$$K_m = \frac{3.04 Su_n^{0.92} Mo^{0.339} A_R^{0.408}}{Re_{ln}^{0.465} H_{rn}^{0.25}} \tag{6.129}$$

They reported that there is a decrease in K_m with an increase in liquid Reynolds number, which is because of increased turbulence in the mixing zone and hence liquid velocity head. On the other hand, K_m increases with decrease in gas–liquid mixing height in the column. This is because for a particular liquid Reynolds number, gas entrainment through the secondary entrance increases with a decrease in gas–liquid mixing height, which in turn increases the bubble population in the mixing zone. They also reported that at a constant liquid Reynolds number, the energy dissipation coefficient (K_m) decreases, which results in an increase of utilization efficiency of energy for mixing. Therefore, from equations (6.127) and (6.128), the rate of energy dissipation per unit mass of fluid ($\in$) can be expressed as

$$\in = K_m \left(\frac{1}{2} u_j^2 \right) / t_c \tag{6.130}$$

Equation (6.130) can be expressed in dimensionlesss numbers when incorporating the Bodenstein number defined as follows:

$$Bo_j = u_j d_c / D_z \tag{6.131}$$

Equation (6.130) then becomes

$$Bo_j = K' \left(\frac{g t_c}{u_j} \right)^{1/3} Fr_j^{1/3} \tag{6.132}$$

where

$$K' = \frac{2^{1/3}}{KK_m^{1/3}} \tag{6.133}$$

and

$$Fr_j = \frac{u_j^2}{gd_c}$$
(6.134)

The contact time of gas and liquid phases in the column can be expressed as

$$t_c = \frac{H_p \varepsilon_g}{u_s}$$
(6.135)

where H_p is the penetration depth of bubbles, which can be calculated from equations (3.3) and (3.4) (the details are described in Chapter 3) and u_s is the relative or slip velocity between the gas and liquid phases. Attempts have been made to evaluate the value of u_s by using different relations found in the literature, and the following relation for jet mixing systems for inverse bubbly flow was found to be successful (Fujie et al., 1980).

$$u_s = u_t \left\{ 1 + 0.0167 \left(\frac{d_b^2 g \rho_l}{\sigma_l} \right)^{2.75} \right\} \left\{ 0.45 + 0.55(1 - \varepsilon_g)^{8.45} \right\}$$
(6.136)

where u_t is the terminal velocity of the bubbles in the column and can be calculated using the correlation given by Clift et al. (1978).

Model Based on Information Entropy Theory

The model is developed based on the information entropy theory. The model describes the quality of mixedness (QM) index of the phases in the bubble column (Nedeltchev et al., 1999, 2000, 2003). In information theory, the Shannon entropy or information entropy is a measure of the uncertainty associated with a random variable. A random variable is a variable in the sense that it can be used as a placeholder for a number in equations and inequalities. Its randomness is completely described by its cumulative distribution function (CDF), which can be used to determine the probability it takes on particular values. In probability theory, the CDF, also called the *probability distribution function* or *just distribution function*, completely describes the probability distribution of a real-valued random variable of X. For every real number X, the CDF of X is given by

$$x \rightarrow F_x(x) = P(X \leq x)$$
(6.137)

where the right-hand side represents the probability that the random variable X takes on a value less than or equal to x. If a < b, the probability X lies in the interval [a, b]. The information entropy of a discrete random variable X that can take the range of possible values $\{x_1, ..., x_n\}$ is defined as

$$H(X) = E(I(X)) = -\sum_{i=1}^{n} p(x_i) \log(x_i)$$
(6.138)

where $I(X)$ is the information content or self-information of X, which is itself a random variable. $p(x_i) = P(X = x_i)$ is the probability mass function of X. The information content or self-information is a measure of the information content associated with the outcome of a random variable. By definition, the amount of self-information contained in a probabilistic event depends only on the probability of that event. The smaller its probability, the larger the self-information associated with receiving the information that the event indeed occurred. Taking into account these properties, the self-information $I(w_n)$ associated with outcome w_n is represented as

$$I(w_n) = \ln\left(\frac{1}{P(w_n)}\right) = -\ln(P(w_n)) \tag{6.139}$$

where P is the probability function and w_n is the random variable. It is assumed that the mixing of phases in the bubble column happens from the standpoints of the probabilities of distribution of the fluid element in the column. To apply the information entropy theory to the mixing of tracer gas in the bubble column, the bubble column volume is devided axisymmetrically into layers of semicylindrical shells of predetermined subvolumes. In every entity semicylindrical shell, the tracer concentration is to be measured. The probability of tracer appearance in each semicylindrical shell is then defined as per probability theory as follows:

$$P_i(t) = \frac{V_0}{V_i} \cdot \frac{C_i(t).V_i}{\sum\limits_{i=1}^{N} C_i(t).V} \tag{6.140}$$

One can select V_0 to be equal to the volume of the smallest semicylindrical shell located in the vicinity of the gas distributor. The information of concentration amount is a function of the probability of tracer appearance in each region, which can be expressed as

$$I_i(t) = -\ln(P_i(t)) \tag{6.141}$$

The information entropy depends on both the probability of tracer appearance in every semicylindrical shell and the information amount. The information entropy is defined as

$$H(t) = \sum\limits_{i=1}^{N} \frac{V_i}{V_0} . P_i(t).I_i(t) \tag{6.142}$$

The information entropy takes the following minimum and maximum values (Nedeltchev et al., 1999):

$$H_{min} = 0$$
$$H_{max} = \ln \sum\limits_{i=1}^{N} \frac{V_i}{V_0} \tag{6.143}$$

The minimum value of information entropy corresponds to no mixing. The tracer will exist only in the smallest semicylindrical shell with volume V_0, and everywhere else its concentration is essentially zero; hence, the probability of tracer appearance will be unity, and in the rest of the entities, it will be essentially zero. The maximum value of information entropy corresponds to a complete mixing state. If the tracer is distributed everywhere uniformly (i.e., its concentration is the same in every semi-cylindrical shell), the probability can be represented by

$$P_i = \frac{V_0}{\sum\limits_{i=1}^{N} V_i} \tag{6.144}$$

The quality of mixedness is a measure to express the degree of asymptotical approach to the equilibrium state (Nedeltchev et al., 1999). It can be defined as follows:

$$M(t) = \frac{H(t) - H_{\min}}{H_{\max} - H_{\min}} = \frac{H(t)}{H_{\max}} \tag{6.145}$$

The quality of mixedness can vary from $0 < M(t) < 1$. The values of C_i and V_i need to be calculated experimentally. The quality of mixedness ($M(t)$) is a function of time. Nedeltchev et al. (1999) developed a correlation for quality of mixedness in case of upward bubble column which is given by

$$M(t) = 1 - 2.55 \times 10^5 \left(\frac{t}{t_c} \right)^{-2.19} \tag{6.146}$$

But in the case of inverse bubbly flow (Manish and Majumder, 2009).

$$M(t) = 1 - 0.547 \left(t_m \right)^{-2.19} \tag{6.147}$$

Gas Phase Mixing

Complex flow structure in a bubble column cause the non-idealities in the mixing behavior. Both turbulent dispersion of gas bubbles and non-uniform gas velocity profiles result in axial mixing of the gas phase. On the reactor scale, the "overall axial gas mixing" is also assessed by the dimensionless variance of the RTD curve at the reactor outlet. The overall axial gas phase mixing can also be quantified using the Peclet number (Pe), which is inversely proportional to the dimensionless variance. Any condition that causes an increase in the dimensionless variance of the RTD curve of the gas phase or a decrease in the axial gas Peclet number will cause an increase in the overall axial gas mixing. Gas mixing in bubble columns can adversely affect the reaction rates and product selectivity (Deckwer, 1992). The investigation of gas phase mixing has received significantly less attention than liquid phase mixing, partly because of the technical problems involved in determining reliable gas RTD data. The proper way to collect the RTD data at the reactor outlet from which one can quantify the extent of the overall axial gas mixing is to measure the gas concentration at different radial locations (at the reactor outlet) and calculate the mixing average from these different radial positions because the gas phase is well mixed in the radial direction. In the literature, the effect of the reactor height on the extent of the gas phase mixing (i.e., the dimensionless variance of the RTD curve) has not been addressed. Levenspiel and Fitzgerald (1983) showed that whereas the dimensionless variance of a tracer is directly proportional to the square of the reactor height if the axial mixing is dominated by dispersion, it is directly proportional to the reactor height if the mixing is mainly caused by convection. However, because the overall axial gas mixing is a result of both phenomena, the dependence of the dimensionless variance of the gas phase on the reactor height is unknown. In reactive systems, this dependence is further complicated by the fact that the contribution of the different mixing phenomena (dispersive and convective mixing) changes with the reactor height because of the shrinkage of the gas phase and the consequent change in the gas and liquid hydrodynamics. It is evident from the literature that the axial gas mixing increases with increasing the superficial gas velocity and increasing the column diameter. Kantak et al. (1995) studied the effect of liquid properties on the gas phase mixing. They found that axial gas phase mixing decreases with decreasing surface tension and increasing liquid viscosity. A decrease in turbulence in bubble columns will cause a decrease in the extent of the axial gas phase mixing even if this decrease in turbulence is coupled with an increase in gas and liquid circulation. The increase in pressure that enhances the circulation of phases and decreases turbulent intensity causes an overall decrease in the gas phase mixing. The extent of gas phase mixing is mainly controlled by turbulent dispersion rather than convective mixing within the range of superficial gas velocities ranging from 0.10 to 0.30 m/s (churn turbulent regime) and operating pressures ranging from 0.1 to 1 MPa. In simulation studies, the gas phase mixing in bubble columns was usually modeled as plug flow (Stern et al., 1983; Herbolzheimer and Iglesia, 1994). However, several authors (Joseph et al., 1984; Shetty et al., 1992; Kaštánek et al., 1993) showed that this assumption was inaccurate and that gas phase mixing cannot be neglected in bubble columns. The most common approach to modeling the non-ideal mixing behavior in bubble columns is the one-dimensional ADM. Some well-known available correlations for the gas ADC in bubble columns are shown in Table 6.11. In the majority

Table 6.11   Important Studies on Gas Phase Dispersion and Developed Correlations for the Axial Gas Dispersion Coefficients by Axial Dispersion Model

System	Column Diameter (m)	Gas Velocity (Superficial) (m/s)	Axial Gas Dispersion Coefficient (m²/s)	Correlation	Reference
Air–water	0.406-1.06	0.016-0.034	0.02-0.14	$D_g = 19.7 d_c^2 u_{sg}$	Towell and Ackerman (1972)
Air–water N₂–n-propanol Air–glycol	0.10	0.01-0.20	0.003-1	$D_g = 2.64 u_s^{3.56}$	Pilhofer et al. (1978)
Air/water	3.20	0.045 – 0.055	1-8	$D_g = 56.4 d_c^{1.33}(u_{sg}/\varepsilon_g)^{3.56}$	Field and Davidson (1980)
Air–water N₂–n-propanol Air–glycol	0.10	0.015-0.10	0.01-1	$D_g = 50.0 d_c^{3/2}(u_{sg}/\varepsilon_g)^3$	Mangartz and Pilhofer (1981)
Air–water	0.20-0.50	0.029-0.456	0.02-5	$D_g = 20 d_c^{3/2} u_{sg}$	Wachi and Nojima (1990)

of published work, the axial gas dispersion coefficient (D_g) was correlated as a function of superficial gas velocity (u_{sg}) or actual gas velocity (u_{sg}/ε_g) and of the reactor diameter. The validity of published empirical correlations is, however, limited to the experimental conditions of the particular studies.

Nomenclature

a Parameter defined in equations (6.45 and 6.64)
A_c Cross-sectional area (m²)
A_R Area of nozzle to column diameter (-)
C_A Exit concentration (kg/m³; mole/m³)
C_e Exit concentration (kg/m³; mole/m³)
C_ϕ Dimensionless tracer concentration defined in equation (6.57)
Da Damköhler numbers defined in equation (6.20)
d_h Hole diameter (m)
D_m Molecular diffusion coefficient (m²/s)
d_n Nozzle diameter (m)
d_p Particle diameter (m)
D_b Dispersion coefficient of bubble motion (m²/s)
d_c Column diameter (m)
D_g Axial gas dispersion coefficient (m²/s)
d_r Ratio of nozzle to column diameter (-)
d_t Tube diameter (m)
D_z Axial dispersion coefficient (m²/s)
E Residence time distribution function (-)
F Function to characterize the residence time distribution
$F(s)$ System transfer function
Fr Froude number (u_{sl}^2/gd_c) (-)
g Gravitational acceleration (m/s²)
$H(t)$ Information amount (-)
Hrn Ratio of nozzle diameter to liquid column height (-)
I Internal age distribution (-)
i Initial, representing number (-)
J Bassel function (-)
J_A Molar flow rate of inert tracer material (mol/s)
k Characteristic factor of velocity distribution (-)
k Rate constant (-)
l Mixing length (m)
L Length (m)
L_d Length of gas-liquid dispersion (m)
M Moment (-)
$M(t)$ Quality of mixedness (-)
m_t Mass of tracer (kg)
N Number of tanks
n Flow behavior index in power law model (-)
P Probability (-)
Pe Peclet number defined in equation (619)
Pe_f Fluid Peclet number (-)

Pe_r Reactor Peclet number (-)
r Radial distance (m)
r' Dimensionless radial distance defined in equation (6.55)
R Rate of reaction (mole/s), radius (m)
R^2 Correlation coefficient (-)
Re_g Reynolds number (= $(d_c \rho_g u_{sg})/\mu_g$) (-)
R_c Column radius (m)
s System parameter (-)
t Time (s)
t_m Mixing time or mean residence time (s)
T Temperature (°C)
T_h Hot zone temperature (°C)
T_c Cold zone temperature (°C)
u Actual velocity (m/s)
u_0 Axial liquid velocity (m/s)
u_b Bubble velocity (m/s)
u_c Circulation velocity (m/s)
u_s Slip velocity (m/s)
u_{sl} Superficial liquid velocity (m/s)
u_{sg} Superficial gas velocity (m/s)
v Volumetric flow rate (m³/s)
V Volume (m³)
X Parameter defined in equation (6.109)
z Vertical length (m), vertical direction
z' Vertical dimensionless distance defined in equation (6.54)

Greek Letters

v^m Mixture kinematic viscosity (m2/s)
v^t Turbulent kinematic viscosity (m2/s)
υ Kinematic viscosity (m²/s)
κ Kurtosis
τ Mean residence time or space time (s)
α Parameter (-)
ω Parameter (-)
ϕ Parameter defined in equation (6.53) (-)
φ Parameter (-)
δ Parameter (-)
γ Parameter (-), skewness
λ Parameter defined in equation (6.18)
ψ Concentration ratio defined in equation (6.17)
χ Constant (-)
ρ Density (kg/m³)
θ Dimensionless time (-)
σ Variance
μ Viscosity (kg/ms)
α_i Parameter defined in equation (6.43)
ε_l Liquid hold-up (-)
ε_g Gas hold-up (-)
τ_s Shear stress (N/m²)

References

Ahmed, T.S., 2003. Gas holdup and liquid axial dispersion under slug flow conditions in gas–liquid bubble column. Chem. Eng. Proc. 42 (10), 767–775.

Anastasiou, A.D., Kazakis, N.A., Mouza, A.A., Paras, S.V., 2010. Effect of organic surfactant additives on gas holdup in the pseudo-homogeneous regime in bubble columns equipped with fine pore sparger. Chem. Eng. Sci. 65, 5872–5880.

Anastasiou, A.D., Passos, A.D., Mouza, A.A., 2013. Bubble columns with fine pore sparger and non-Newtonian liquid phase: prediction of gas holdup. Chem. Eng. Sci. 98 (7), 331.

Aoyama, Y., Ogushi, K., Koide, K., Kubota, H., 1968. Liquid mixing in concurrent bubble columns. J. Chem. Eng. Jpn 1, 158–163.

Aris, R., 1959. Notes on the diffusion-type model for longitudinal mixing in flow. Chem. Eng. Sci. 9, 266–267.

Baird, M.H.I., Rice, R.G., 1975. Axial dispersion in large unbaffled columns. Chem. Eng. J. 9, 171–174.

Biń, A.K., Duczmal, B., Machniewski, P., 2001. Hydrodynamics and ozone mass transfer in a tall bubble column. Chem. Eng. Sci. 56, 6233–6240.

Bischoff, K.B., 1960. Notes on the diffusion—type model for longitudinal mixing in flow (letters to the editor). Chem. Eng. Sci. 12, 69–70.

Bischoff, K.B., Levenspiel, O., 1962. Fluid dispersion—generalization and comparison of mathematical models - I. Generalization of models. Chem. Eng. Sci. 17, 245–255.

Blet, V., Berne, P., Chaussy, C., Perrin, S., Schweich, D., 1999. Characterization of a packed column using radioactive tracers. Chem. Eng. Sci. 54, 91–101.

Bosworth, R.C.L., 1948. Distribution of reaction times for laminar flow in cylindrical reactors. Philos. Mag. 39, 846–862.

Bosworth, R.C.L., 1949. Distribution of reaction times for turbulent flow in cylindrical reactors. Philos. Mag. 40, 314–324.

Boyer, C., Duquenne, A., Wild, G., August 2002. Measuring techniques in gas–liquid and gas–liquid–solid reactors. Chem. Eng. Sci. 57 (16), 3185–3215.

Briens, C.L., Margaritis, A., Wild, G., 1995. A new stochastic model and measurement errors in residence time distributions of multiphase reactors. Chem. Eng. Sci. 50, 279–287.

Buffham, B.A., Gibilaro, 1968. A generalization of the tanks-in-series mixing model. AIChE J. 14 (5), 805–806.

Buffham, B.A., Gibilaro, L.G., Rathor, M.N., 1970. A probabilistic time delay description of flow in packed beds. AIChE J. 16 (2), 218.

Camarasa, E., Vial, C., Poncin, S., Wild, G., Midoux, N., Bouillard, J., 1999. Influence of coalescence behaviour of the liquid and of gas sparging on hydrodynamics and bubble characteristics in a bubble column. Chem. Eng. Process. 38, 329–344.

Carleton, A.J., Flair, R.J., Rennie, J., Valentin, F.H.H., 1967. Some properties of packed bubble column. Chem. Eng. Sci. 22, 1834.

Chaouki, J., Larachi, F., Dudukovic, M.P., 1997. Noninvasive tomographic and velocimetric monitoring of multiphase flows. Ind. Eng. Chem. Res. 36, 4476–4503.

Clapp, R.M., 1963. Turbulent heat transfer in pseudoplastic non-Newtonian fluids. International Developments in Heat Transfer. International Heat Transfer Conference, D211, 652–661.

Clift, R., Grace, J.R., Weber, M.E., 1978. Bubbles, drops and particles. Academic Press, New York.

Danckwerts, P.V., 1953. Continuous flow systems: distribution of residence times. Original Research Article.

Deans, H.A., Lapidus, L., 1960. A computational model for predicting and correlating the behavior of fixed-bed reactors. AICHE J. 6, 656–663.

Deans, H.A., 1963. A mathematical model for dispersion in the direction of flow in porous media. Trans. Sot. Pef. Eng. 228, 49–52.

Deckwer, W.D., 1992. Bubble Column Reactors. Wiley, New York.

Deckwer, W.D., Burckhart, R., Zoll, G., 1974. Mixing and mass transfer in tall bubble columns. Chem. Eng. Sci. 29 (11), 2177–2188.

Dudukovic, M.P., Devanathan, N., Holub, R., 1991. Multiphase reactors: models and experimental verification. Revue de l'Institut Français du Pétrole 46, 439–465.

Ekambara, K., Joshi, J.B., 2003. CFD simulation of mixing and dispersion in bubble columns. Chem. Eng. Res. Des. 81 (8), 987–1002.

Elgeti, K., 1996. A new equation for correlating a pipe flow reactor with a cascade of mixed reactors. Chem. Eng. Sci. 51 (23), 5077–5080.

Euzen, J.-P., Fortin, Y., 1987. Partikelbewegung in einem Dreiphasen4ie bett. Chemie-Ingenieur-Technik 59, 416–419.

Field, R.W., Davidson, J.F., 1980. Axial dispersion in bubble columns. Trans. Inst. Chem. Eng. 58, 228–236.

Flaschel, E., Metzdorf, C., Renken, A., 1987. Methodezur Messung der axialen Vermischung des Feststo3s in FlNussigkeit/Feststo3-Wirbelschichten. Chemie-Ingenieur-Technik 59, 494–496.

Fogler, S.H., 1992. Elements of Chemical Reaction Engineering. Prentice Hall International, New Jersey.

Forret, A., Schweitzer, J.M., Gauthier, T., Krishna, R., Schweich, D., 2003. Influence of scale on the hydrodynamics of bubblecolumn reactors: an experimental study in columns of 0.1, 0.4 and 1 m diameters. Chem. Eng. Sci. 58, 719–724.

Fortin, Y., 1984. Réacteurs á lit fluidisé triphasique: caractéristiques hydrodynamiques et mélange des particules solides. ThLese dedocteur-ingLenieur, UniversitLe de Paris VI, ENSPM, Solaize, France.

Froment, G.F., Bischoff, K.B., 1979. Chemical Reactor Analysis and Design. Wiley, New York.

Fujie, K., Takaine, M., Kubota, H., 1980. Flow and oxygen transfer in cocurrent gas-liquid downflow. J. Chem. Eng. Jpn 13, 188–193.

García-Abuín, A., Gómez-Díaz, D., Losada, M., Navaza, J.M., 2012. Bubble column gas–liquid interfacial area in a polymer+surfactant+water system. Chem. Eng. Sci. 75, 334–341, 06.

Geary, N.W., Rice, R.G., 1992. Circulation and scale-up in bubble columns. AIChE J. 38 (1), 76–81.

Grima, E.M., Fernández, F.G.A., Camacho, F.G., Rubio, F.C., Chisti, Y., 2000. Scale-up of tubular photobioreactors. J. Appl. Phycol. 12, 355–368.

Groen, J.S., Oldeman, R.G.C., Mudde, R.F., van den Akker, H.E.A., 1996. Coherent structures and axial dispersion in bubble column reactors. Chem. Eng. Sci. 51 (10), 2511–2520.

Guariguata, C., Barreiro, J.A., 8r Guariguata, G., 1979. Analysis of continuous sterilization processes for Bingham plastic fluids in laminar flow. J. Food Sci. 44, 905–910.

Gupta, P., 2002. Churn-turbulent bubble columns—experiments and modeling, D. Sc. Thesis, Washington University in St. Louis, Saint Louis, MO, USA.

Herbolzheimer, E., Iglesia, E., 1994. Slurry bubble column, US Patent 5348982.

Hikita, H., Kikukawa, H., 1974. Liquid-phase mixing in bubble columns: effect of liquid properties. Chem. Eng. J. 8, 191–197.

Holcombe, N.T., Smith, D.N., Knickle, H.N., O'Dowd, W., 1983. Thermal dispersion and heat transfer in nonisothermal bubble columns. Chem. Eng. Commun. 21, 135–150.

Holdsworth, S.D., 1992. Aseptic Processing and Packaging of Food Products. Elsevier, London.

Hoogendoorn, C.J., Lips, J., 1965. Axial mixing of liquid in gas–liquid flow through packed beds. Can. J. Chem. Eng. 43 (3), 125–131.

Hopkins, M.J., Sheppard, A.J., Eisenklam, P., 1969. The use of transfer functions in evaluating residence time distribution curves. Chem. Eng. Sci. 24, 1131–1137.

Houzelot, J.L., Thiebaut, M.F., Charpentier, J.C., Schiber, J., 1983. Contribution à l'étude hydrodynamique des colonnes à bulles. Entropie (Paris) 19, 121–126.

Hyndman, C., Guy, C., 1995. Gas phase hydrodynamics in bubble columns. Chem. Eng. Res. Des. 73, 426–434.

Joseph, S., Shah, Y.T., 1986. Errors caused by tracer solubility in the measurement of gas phase axial dispersion. Can. J. Chem. Eng. 64, 380–386.

Joseph, S., Shah, Y.T., Kelkar, B.G., 1984. A simple experimental technique to measure gas phase dispersion in bubble columns. Chem. Eng. Commun. 28 (4–6), 223–230.

Joshi, J.B., 1980. Axial mixing in multiphase contactors—a unified correlation. Trans. Instn Chem. Eng. 58, 155–165.

Joshi, J.B., Sharma, M.M., 1979. A circulation cells model for bubble column. Trans. Inst. Chem. Eng. 57, 244–251.

Kantak, M.V., Hesketh, R.P., Kelkar, B.G., 1995. Effect of gas and liquid properties on gas phase dispersion in bubble columns. Chem. Eng. J. (Lausanne) 59 (2), 91–100.

Kantak, M.V., Shetty, S.A., Kelkae, B.G., 1994. Liquid phase backmixing in bubble column reactors—a new correlation. Chem. Eng. Comm. 127, 23–34.

Kasban, H., Zahran, O., Arafa, H., El-Kordy, M., Elaraby, S.M., Abd El-Samie, F.E., 2010. Laboratory experiments and modeling for industrial radiotracer applications. Appl. Radiat. Isot. 68 (6), 1049–1056.

Kaštánek, F., Zahradnik, J., Kratochvil, J., Cermak, J., 1993. Chemical Reactors for Gas Liquid Systems. Ellis Horwood, New York.

Kato, Y., Nishiwaki, A., 1971. Longitudinal dispersion coefficient of liquid in bubble column. Kagaku Kogaku 35, 912–916.

Kawagoe, M., Otake, T., Robinson, C.W., 1989. Gas-phase mixing in bubble columns. J. Chem. Eng. Jpn 136–142.

Kawase, Y., Moo-Young, M., 1986. Liquid phase mixing in bubble columns with Newtonian and non-Newtonian fluids. Chem. Eng. Sci. 41 (8), 1969–1977.

Kawase, Y., Moo-Young, M., 1989. Turbulence intensity in bubble columns. Chem. Eng. J. 40 (1), 55–58.

Kojima, E., Unno, H., Sato, Y., Chida, T., Imai, H., Endo, K., Inoue, I., Kobayashi, J., Kaji, H., Nakanishi, H., Yamamoto, K., 1980. Liquid phase velocity in a 5.5 m diameter bubble column. J. Chem. Eng. Jpn 13 (1), 16–21.

Kolmogorov, A.N., 1941. The local structure of turbulence in incompressible viscous fluid for very large Reynolds numbers. Proceedings of the USSR Academy of Sciences (in Russian) 30: 299–303., translated into English by V. Levin: Kolmogorov, Andrey Nikolaevich, July 8, 1991. The local structure of turbulence in incompressible viscous fluid for very large Reynolds numbers (PDF). Proceedings of the Royal Society A 434 (1991): 9–13.

Krishna, R., Urseanu, M.I., Van Baten, J.M., Ellenberger, J., 2000. Liquid phase dispersion in bubble columns operating in the churn-turbulent flow regime. Chem. Eng. J. 78 (1), 43–51.

Kulkarni, A., Shah, Y.T., 1984. Gas phase dispersion in a downflow bubble column. Chem Eng. Commun. 28, 311–326.

Kumar, S.B., 1994. Computed tomographic measurements of void fraction and modeling of the flow in bubble columns, PhD. Thesis. Florida Atlantic University.

Kundu, G., Mukherjee, D., Mitra, A.K., 1995. Experimental studies on a co-current gas-liquid downflow bubble column. Int. J. Multiphase Flow 21, 893–906.

Kunii, D., Levenspiel, O., 1991. Fluidization Engineering, second ed. Butterworth-Heinemann, Stoneham.

Laan, E.T.V.D., 1958. Notes on the diffusion type model for the longitudinal mixing of flow. Chem. Eng. Sci. 7 (3), 187–191.

Larachi, F., Cassanello, M., Marie, M., Chaouki, J., Guy, C., 1995. Solids circulation patterns in three-phase fluidized beds containing binary mixtures of particles as inferred from RPT. Trans. IChemE 73 (A), 263–268.

Lefebvre, S., Chaouki, J., Guy, C., 2004. Phase modeling in multiphase reactors containing gas bubble: a review. Int. J. Chem. Reactor Eng. 2, R2.

Leonard, C., Ferrassea, J.-H., Boutin, O., Lefevre, S., Viand, A., 2015. Bubble column reactors for high pressures and high temperatures operation. Chem. Eng. Res. Des. 100, 391–421.

Levenspiel, O., 1972. Chemical Reaction Engineering. John Wiley and Sons, New York.

Levenspiel, O., 1999. Chemical Reaction Engineering, third ed. Wiley, New York.

Levenspiel, O., Bischoff, K.B., 1963. Patterns of flow in chemical process vessels. In: Drew, T.B., Hoopes, Jr., J.W., Vermeulen, T. (Eds.), Advances in Chemical Engineering, vol. 4, Academic Press, New York, pp. 95–198.

Levenspiel, O., Fitzgerald, T.J., 1983. A warning in the misuse of the dispersion model. Chem Eng. Sci. 38, 489–491.

Levenspiel, O., Smith, W.K., 1957. Notes on the diffusion-type model for the longitudinal mixing of fluids in flow. Chem. Eng. Sci. 6, 227–233.

Lin, S.H., 1980. Continuous sterilization of non-newtonian liquid foods in a tubular sterilizer. Lehensmittel- Wissenschafl und Technologie 13 (3), 135–141.

Lorenz, O., Schumpe, A., Ekambara, K., Joshi, J.B., 2005. Liquid phase axial mixing in bubble columns operated at high pressures. Chem. Eng. Sci. 60, 3573–3586.

Majumder, S.K., Kundu, G., Mukherjee, D., 2005. Mixing mechanism in a modified co-current downflow bubble column. Chem. Eng. J. 112, 45–55.

Majumder, S.K., Kundu, G., Mukherjee, D., 2006a. Efficient dispersion in a modified two-phase Non-Newtonian downflow bubble column. Chem. Eng. Sci. 61 (20), 6753–6764.

Majumder, S.K., Kundu, G., Mukherjee, D., 2006b. Bubble size distribution and interfacial phenomena in ejector induced downflow bubble column. Chem. Eng. J. 122 (1-2), 1–10.

Majumder, S.K., Kundu, G., Mukherjee, D., 2007a. Energy efficiency of two-phase mixing in a modified bubble column. Can. J. Chem. Eng. 85 (3), 380–389.

Majumder, S.K., Kundu, G., Mukherjee, D., 2007b. Pressure drop and bubble-liquid interfacial shear stress in a modified gas-non-Newtonian-liquid downflow bubble column. Chem. Eng. Sci. 62, 2482–2490.

Mangartz, K.H., Pilhofer, T., 1981. Interpretation of mass transfer measurements in bubble columns considering dispersion of both phases. Chem. Eng. Sci. 36 (6), 1069–1077.

Manish, P., Majumder, S.K., 2009. Quality of mixing in downflow bubble column based on information entropy theory. Chem. Eng. Sci. 64 (8), 1798–1805.

Mawlana, A., Huang, H., Wilhelm, A.M., Delmas, H., 1992. Transferable tracer analysis for multiphase reactor. Chem. Eng. Commun. 115, 67–82.

Mecklenburgh, J.C., Hartland, S., 1976. The Theory of Backmixing. Wiley, New York.

Michell, R.W., Furzer, I.A., 1972. Mixing in trickle flow through packed beds. Chem. Eng. J. 4, 53–63.

Miyauchi, T., Shyu, C.N., 1970. Flow of fluid in gas bubble columns. Kagaku Kogaku 34, 958–964.

Miyauchi, T., Furusaki, S., Morooka, S., Ikeda, Y., 1981. Transport phenomena and reaction in fluidized catalysts beds. Adv. Chem. Eng. 11, 275–448.

Myers, K.J., Dudukovic, M.P., Ramachandran, P.A., 1987. Modeling churn turbulent bubble columns-I, Liquid-phase mixing. Chem. Eng. Sci. 42 (10), 2301–2311.

Nedeltchev, S., Ookawara, S., Ogawa, K., 1999. A fundamental approach to bubble column scale-up based on quality of mixedness. J. Chem. Eng. Jpn 32, 431–439.

Nedeltchev, S., Ookawara, S., Ogawa, K., 2000. Time-dependent mixing behaviors in lower and upper zones of bubble column. J. Chem. Eng. Jpn 33, 761–767.

Nedeltchev, Stoyan, Kumar, Sailesh B., Dudukovic, M.P., 2003. Flow regime identification in a bubble column based on both kolmogorov entropy and quality of mixedness derived from CARPT data. Can. J. Chem. Eng. 81, 367–374.

Nosier, S.A., 2003. Solid-liquid mass transfer at gas sparged tube bundles. Chem. Eng. Technol. 26 (11), 1151–1154.

Ogawa, S., Kobayashi, M., Tone, S., Otake, T., 1982. Liquid phase mixing in the ga-liquid jet reactor with liquid jet ejector. J. Chem. Eng. Japan 15 (6), 469–473.

Ohki, Y., Inoue, H., 1970. Longitudinal mixing of the liquid phase in bubble columns. Chem. Eng. Sci. 25, 1–16.

Onozaki, M., Namiki, Y., Sakai, N., Kobayashi, M., Nakayama, Y., Yamada, T., Morooka, S., 2000. Dynamic simulation of gas–liquid dispersion behavior in coal liquefaction reactors. Chem. Eng. Sci. 55, 5099–5113.

Ostergaard, K., Michelsen, M.L., 1969. On the use of the imperfect tracer pulse method for determination of hold-up and axial mixing. Can. J. Chem. Eng. 47 (4), 107–112.

Palaskar, S.N., De, J.K., Pandit, A.B., 2000. Liquid phase RTD studies in sectionalized bubble column. Chem. Eng. Technol. 23 (1), 61–69.

Pandit, A.B., Joshi, J.B., 1983. Mixing in mechanically agitated gas-liquid contactors, bubble columns and modified bubble columns. Chem. Eng. Sci. 38, 1189–1215.

Pant, H.J., Saroha, A.K., Nigam, K.D.P., 2000. Measurement of liquid holdup and axial dispersion in trickle bed reactors using radiotracer technique. Nukleonika 45 (4), 235–241.

Pant, H.J., Sharma, V.K., Nair, A.G., Tomar, B.S., Nathaniel, T.N., Reddy, A.V., Singh, G., 2009. Application of 140La and 24Na as intrinsic radiotracers for investigating catalyst dynamics in FCCUs. Appl. Radiat. Isot. 67 (9), 1591–1599.

Pilhofer, Th., Bach, M.F., Mangartz, K.M., 1978. Determination of fluid dynamic parameters in bubble column design. ACS Symp. Ser. 65, 372.

Raghuraman, J., Varma, Y.B.G., 1974. A stochastic model for residence time and contact time distributions of the gas in multistage fluidised beds. Chem. Eng. Sci. 29 (3), 697–703.

Rao, K.S.M.S.R., Joshi, J.B., 1998. Liquid-phase mixing and power consumption in mechanically agitated solid–liquid contactors. Chem. Eng. J. 39 (2), 111–124.

Reith, T., Renken, S., Israël, B.A., 1968. Gas hold-up and axial mixing in the fluid phase of bubble columns. Chem. Eng. Sci. 23 (6), 619–629.

Rietema, K., Ottengraf, S.P.P., 1970. Laminar liquid circulation and bubble street formation in gas–liquid system. Trans. Inst. Chem. Eng. 48, 54–62.

Riquarts, H.P., 1981. A physical model for axial mixing of the liquid phase for heterogeneous flow regime in bubble columns. Ger. chem. Engng 4, 18–23.

Rubio, C., Sánchez Mirón, A., Cerón García, M.C., García Camacho, F., Molina Grima, E., Chisti, Y., 2004. Mixing in bubble columns: a new approach for characterizing dispersion coefficients. Chem. Eng. Sci. 59, 4369–4376.

Ruzicka, M.C., Drahoš, J., Fialová, M., Thomas, N.H., 2001. Effect of bubble column dimensions on flow regime transition. Chem. Eng. Sci. 56, 6117–6124.

Sakai, O., Saegusa, I., Hayashi, K., Tachikawa, I., Morooka, S., 2000. Fluid dynamics in coal liquefaction reactors using neutron absorption tracer technique. Am. Inst. Chem. Eng. J. 46, 1688–1693.

Sangnimnuan, A., Prasad, G.N., Agnew, J.B., 1984. Gas hold-up and backmixing in a bubble-column reactor undercoal-hydroliquefaction conditions. Chem. Eng. Commun. 25, 193–212.

Sawinsky, J., Simandi, B., 1982. The residence time distribution for laminar flow of a non-Newtonian fluid in a straight circular tube. Transac. Inst. Chem. Eng. 60, 188–190.

Schlichting, H., 1979. Boundary-Layer Theory, seventh ed. McGraw-Hill, New York, pp. 578–595.

Sekizawa, T., Kubota, H., Chung, W.C., 1983. Apparent slip velocity with recirculating turbulent flow in bubble columns. J. Chem. Eng. Jpn 16 (4), 327–330.

Serizawa, A., Katoka, I., Michiyoshi, I., 1975. Turbulent structure of air-water bubble flow. Intern. J. Multiphase Flow 2, 235.

Shah, M., Kiss, A.A., Zondervan, E., Van der Schaaf, J., De Haan, A.B., 2012. Gas holdup, axial dispersion, and mass transfer studies in bubble columns. Ind. Eng. Chem. Res. 51, 14268–14278.

Shah, Y.T., Joseph, S., Smith, D.N., Ruether, A.J., 1985. Two-bubble class model for churn-turbulent bubble-column reactors, Ind. Eng. Chem. Process Des. Dev. 24 (4), 1096–1104.

Shah, Y.T., Kelkar, B.G., Godbole, S.P., Deckwer, W.-D., 1982. Design parameters estimations for bubble column reactors. A. I. Ch. E. J. 28, 353–379.

Shawaqfeh, A.T., 2003. Gas holdup and liquid axial dispersion under slug flow conditions in gas–liquid bubble column. Chem. Eng. Process. 42, 767–775.

Shetty, S.A., Kantak, M.V., Kelkar, B.G., 1992. Gas-phase back-mixing in bubble column reactors. AIChE J. 38 (7), 1013–1026.

Smith, J.S., Burns, L.F., Valsaraj, K.T., Thibodeaux, L.J., 1996. Bubble column reactor for wastewater treatment. 2. The effect of sparger design on sublation column hydrodynamics in the homogeneous flow regime. Ind. Eng. Chem. Res. 35, 1700–1710.

Stern, D., Bell, A.T., Heinemann, H., 1983. Effects of mass transfer on the performance of slurry reactors used for Fischer-Tropsch synthesis. Chem. Eng. Sci. 38 (4), 597–605.

Tarmy, B., Chang, M., Coulaloglou, C., Ponzi, P., 1984. Hydrodynamic characteristics of three phase reactors. Chem. Eng., 18–23.

Taylor, G.I., 1953. Dispersion of solute matter in solvent flowing slowly through a tube. Proc. R. Soc. Lond. A 219, 186.

Therning, P., Rasmuson, A., 2001. Liquid dispersion, gas holdup and frictional pressure drop in a packed bubble column at elevated pressures. Chem. Eng. J. 81, 331–335.

Towell, G.D., Ackerman, G.H., 1972. Axial mixing of liquid and gas in large bubble reactor. Proceedings of the Fifth European Second International Symposium on Chemical Reaction Engineering B3-1.

Ueyama, K., Miyauchi, T., 1979. Properties of recirculating turbulent two-phase flow in gas bubble columns. AIChE J. 25, 258.

Van Swaaij, W.P.M., Charpentier, J.C., Villermaux, J., July 1969. Residence time distribution in the liquid phase of trickle flow in packed columns. Chem. Eng. Sci. 24 (7), 1083–1095.

Wachi, S., Nojima, Y., 1990. Gas-phase dispersion in bubble columns. Chem. Eng. Sci. 45 (4), 901–905.

Wehner, J.F., Wilhelm, R.H., 1956. Boundary conditions in a flow reactor. Chem. Eng. Sci. 6, 89–93.

Wen, C.Y., Fan, L.T., 1975. Models for flow systems and chemical reactors. Chemical Processing and Engineering Monograph Series. Marcel Dekker, Inc., New York.

Westerterp, K.R., van Swaaij, W.P.M., Beenackers, A.A.C.M., 1987. Chemical reactor design and operation. Wiley, Chicester.

Whalley, P.B., Davidson, J.F., 1974. Liquid circulation in bubble columns. Proc. Symp. Multiphase Flow Syst. 38, J5.

Wild, G., Poncin, S., 1996. Hydrodynamics of three-phase sparged reactors. In: Nigam, K.D.P., Schumpe, A. (Eds.), Three-Phase Sparged Reactors. Gordon and Breach Science Publishers, Topics in Chemical Engineering Series, New York, pp. 11–113.

Wilkinson, P.M., Haringa, H., Stokman, F.P.A., Van Dierendonck, L.L., 1993. Liquid mixing in a bubble column under pressure. Chem. Eng. Sci. 48, 1785–1791.

Yang, G.Q., Fan, L.-S., 2003. Axial liquid mixing in high-pressure bubble columns. AIChE J. 49, 1995–2008.

Youssef, A.A., Al-Dahhan, M.H., 2009. Impact of internals on the gas holdup and bubble properties of a bubble column. Ind. Eng. Chem. Res. 48 (17), 8007–8013.

Zahradník, J., Fialová, M., Ružička, M., Drahoš, J., Kaštánek, F., Thomas, N.H., 1997. Duality of the gas–liquid flow regimes in bubble column reactors. Chem. Eng. Sci. 52, 3811–3826.

Zehner, P., 1982. Impuls-, Stoff- und Warmetransport in Blasensaulen. Verfahrenstechnik 16, 347–351, 514–517.

7

Bubble Size Distribution and Gas–Liquid Interfacial Area

Introduction

Gas–liquid mass transfer is a crucial factor in the design of chemical and biochemical processes in which the dispersed phase is a reaction-limiting factor. Often the gas–liquid mass transfer is empirically determined from the time profile of concentration of dispersed phase in the system in which information about bubble size and its distribution and interfacial area are neglected. However, some evidence has emphasized the significance of bubble properties in controlling the mass transfer in gas–liquid contacting systems. For bubble columns, bubble size is usually found to decrease with increasing gas input to the system; therefore, a high gas–liquid interfacial area is obtained at high gas throughput (Akita and Yoshida, 1974; Couvert et al., 1999; Hebrard et al., 2001). In general, the gas–liquid interfacial area is a function of the unit's geometric size and operating parameters, as well as the physical and chemical properties of the phases. In the case of absorption accompanied by a slow or instantaneous reaction, the mass transfer rate is determined by the volumetric mass transfer coefficient, which is a function of the specific interfacial area. So to optimize the processes on the mass transfer phenomena, it is essential to know the bubble size distribution and interfacial phenomena in the particular system at different operating conditions.

Several authors have studied the bubble size and interfacial phenomena in different types of bubble column reactors. The hydrodynamics of bubble columns is strongly dependent on the gas distributors and the correct estimation of the influence of distributor design and process parameters is done based on bubble size distributions (Kumar and Kuloor, 1970; Miyahara et al., 1983; Miyahara and Hayashino, 1995; Miyahara and Hashimoto, 1999; Rabiger and Vogelphol, 1982). The mean diameter of bubbles is related to the gas hold-up and the gas and liquid superficial velocities (Moleurs and Kurtin, 1985). Generally, the population balance model has been used to describe the distributions over the length of the bubble column. The model adequately predicts the mean bubble size and the bubble size distribution caused by bubble growth and coalescence. The model is based on break-up and coalescence efficiency and variable relative velocities of the bubbles. The gas hold-up can be analyzed from the bubble size distributions using the theory for the conservation of the bubble size distribution function (Lage and Espósito, 1999). The evolutions of bubble characteristics can be interpreted successfully by considering the simultaneous influence of the hydrodynamic regime of the gas–liquid flow and of the operating regime of the distributor (Camarasa et al., 1999). It has also

Hydrodynamics and Transport Processes of Inverse Bubbly Flow. http://dx.doi.org/10.1016/B978-0-12-803287-9.00007-2

255

been shown that bubble frequency measurements are good tools to evaluate distributor efficiency. Bubble shape and bubble size distributions can be modified with the addition of a small amount of surface tension modifier in water. Colella et al. (1999) studied the interfacial mechanisms, focusing on the coalescence and breakage phenomena of bubbles in three different bubble columns. An elaborate database was established that describes the evolution of the bubble size distribution (BSD) in three different types of bubble columns. They investigated the influences of gas superficial velocity and different hydrodynamic configurations on bubble size distribution in the bubble columns. Pohorecki et al. (1999) reported that the bubble size distributions can be controlled along the column axis by changing the geometry of distributor. They also reported that there are considerable effects of geometry of the distributor on the interfacial area and bubble velocity. Bouaifi et al. (2001) stated that the gas flow rate, sparger type, and column diameter have unavoidable roles in the bubble size and interfacial area.

The evolution of BSD in bubble columns can be further developed by implementing a detailed physical model for predicting the initial BSD at the sparger where the model is only based on design and process parameters (Polli et al., 2002). Schäfer et al. (2002) discussed the influence of operating conditions and physical properties of gas and liquid phase on the initial and stable bubble sizes in a bubble column reactor under industrial conditions. They reported that bubbles tend to become smaller with decreasing surface tension, increasing gas density, and decreasing liquid viscosity, resulting in reduced stable bubble sizes at increased pressure and increased temperature. They also reported that impurities in aqueous and organic liquids can severely influence bubble sizes by restraining coalescence. For such systems BSDs in a column mainly depend on the size of the primary bubbles, which are determined by the sparger design. Hu et al. (2003) studied the effect of sodium carbonate concentration and processing conditions on bubble size and their break-up and coalescence in turbulent, batch air–aqueous dispersions at atmospheric and elevated pressures. They reported that under pressure, a decrease in the Sauter mean bubble diameter at the same volume fraction of air can be observed in bubble columns.

In recent years, studies on bubble characteristics in ejector- or Venturi-type bubble columns are increasing because they offer distinct advantages over other conventional bubble columns in their ability to generate fine bubbles and a high gas–liquid interfacial transfer area. The mass transfer and hydrodynamic characteristics of a loop reactor using inverse flow liquid jet ejectors fitted with straight- and Venturi-type throats are very interesting. The interfacial area and volumetric mass-transfer coefficient are reported with power input and the interfacial areas achieved with a Venturi ejector are far superior to those obtained with a straight throat ejector (Dutta and Raghavan (1987). Crammers et al. (1992) investigated the hydrodynamics and mass transfer characteristics of a loop-Venturi reactor using an inverse-flow liquid jet ejector. They determined the specific interfacial area of the ejector and the main holding vessel separately. They estimated the interfacial areas, which were in the range of 40,000 to 70,000 m^2/m^3 in the ejector and 500 to 2500 in the total system.

Evans et al. (1992) presented a model to predict the maximum bubble size generated within the mixing zone at the top of a plunging liquid jet bubble column. The model uses a critical Weber number, in which the energy dissipation rate per unit volume is derived from the theory of liquid-jet gas ejectors. They found that the bubble diameter distribution fit well by a log-normal distribution, with a Sauter-mean-to- maximum-diameter ratio of 0.61. Evans et al. (1996) investigated the flow characteristics in the pipe-flow zone of a plunging liquid jet bubble column. They analyzed the effect of the column diameter and gas and liquid flow rates on the bubble size and gas void fraction for the bubbly and churn-turbulent flow regimes for an air–water system. They also reported that the bubble size remained constant in the bubbly flow regime. In the churn-turbulent flow regime, however, an increase in bubble size is observed because of coalescence. Boyer and Lemonnier (1996) evaluated the influence of the bubble diameter on the pressure drop calculations and flow rate to determine the maximum admissible bubble diameter and then to size the mixer set upstream of a Venturi bubble column. From the literature, it was found that studies regarding the bubble characteristics and interfacial area in a jet Venturi type bubble column reactor have increased, but there is little literature available for the bubble characteristics and the interfacial area in an ejector-type inverse bubble column reactor.

Bubble Size Measurement and Its Distribution

Measurement Techniques

Several methods for determining bubble size in a gas–liquid reactor have been developed over the past few years. Examples of various methods used to measure the bubble size is shown in Table 7.1. Such methods are useful in several practical complex flows. However, there is no standard method to calibrate them. The most used methods are high-speed photography (Lunde and Perkins, 1995; Leifer et al., 2000; Wu and Gharib, 2002; Ellingsen and Risso, 2001; Alves et al., 2002), electrical impedance tomography (Kim et al., 2002; Aw et al., 2014), capillary suction probes (Barigou and Greaves, 1991), optical wave-guided sensors (Saberi et al., 1997; Rząsa and Pląskowski, 2003), endoscopic optical probes (Leifer et al., 2000; Zhu et al., 2001), capacitance probes (Leifer et al., 2000), passive acoustic detection (Manasseh, 1997; Chanson and Manasseh, 2003), and the inverted funnel (Vazquez et al., 2005). Other different methods used by different investigators are the gas disengagement technique (Schumpe and Grund, 1986; Patel et al., 1989), two-point conductivity method (Buchholz et al., 1981), light scattering method (Fujiwara, 2007), laser diffraction scattering methods (Maruyama, 2007), interferometric image method (Kawaguchi et al., 2002), Coulter count method or pore electrical resistance method (Fujiwara, 2007), and dynamic light scattering method (Takahashi, 2009).

There are some dissensions among the authors because some are rapid for detection of bubbles, although they are constrained by sensitivity of the sensing element or by the

Table 7.1   Examples of Various Methods Used to Measure the Bubble Size Distribution

Methods	Bubble Sampling Features	Methods for Bubble Generation	Bubble size Range (μm)	Reference
Imaging	Viewing box	Orifice plate	300-2000	Unno and Inoue (1979)
hHigh-speed	Fuel cell	Porous plate	75-655	Ahmed and Jameson (1985)
photographic,	Cuvette cell	DAF	10-300	De Rijk et al. (1994)
digital images,	Bubble viewer	IAF	350-1750	Chen et al. (2001)
and videos)	Viewing chamber	Porous plate	400-1200	Grau and Heiskanen (2002)
	Viewing chamber	IAF	1500-3600	Zhou et al. (1993)
		Glass capillaries	1000-2000	Wu and Gharib (2002)
Electroresistivity	Particle counters	DAF	13-96	Han et al. (2002)
Electrical impedance tomography	Online detection by probe	—	—	Kim et al. (2000)
Optical	Bubbles drawn into a capillary	Filter cloth	300-2000	Biswal et al. (1994)
Optical tomography	Online detection by sensors of the capacitance tomography	Mixing chamber	—	Rząsa and Pląskowski (2003)
Optical (optical sensors)	Particle counters, laser fiber optics	DAF	15-85	Han et al. (2002)
		EF	15-65	Han et al. (2002)
		Perforated plate	3700-4100	Saberi et al. (1995)
Porous plate	Drawing bubbles into a capillary	IAF	390-2230	Tucker et al. (1994)
		IAF	530-1450	Gorain et al. (1995)
		Porous plate	1620-3340	O'Connor et al. (1990)

DAF, Dissolved air flotation; *EF,* electroflotation; *IAF,* induced air flotation.

issues of data interpretation. Some are dependent on the purity of the liquid medium and lighting effects. The lighting may bring about several reflections and refractions. These should be considered to determine a threshold level for the bubble size estimation because some reflections on the frontal part of the bubble may reduce the apparent bubble size. The above methods may be used for determining BSD and bubble diameter in inverse bubbly flow systems.

The photographic method is very popular in view of its simplicity (Kölbel et al., 1961; Akita and Yoshida, 1974; Deckwer, 1992; Gaillard et al., 2015). As per the photographic method, photographs of the gas-in-liquid dispersion are taken alongside a suitable standard. A standard photographic arrangement with the rectangular view box is shown in Figure 7.1.

The photographs are analyzed using suitable equipment or suitable software packages to count and classify the bubbles. An online bubble picture evaluation process is also suitable for bubble analysis (Buchholz et al., 1981; Chandrasekera et al., 2015). A reliable average value must be based on a sample of at least 500 bubbles. Errors always arise when bubbles are not spherical in shape and hence a two-dimensional projection through a surface

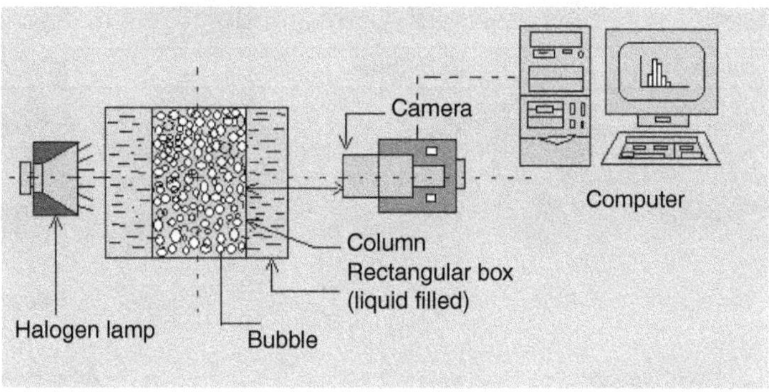

FIGURE 7.1 Photographic arrangement.

contact sphere can only give an approximate value. According to Kölbel et al. (1961), bubbles are only reliably spherical in shape up to a diameter of 1 mm; larger ones become ellipsoidal or even indented in excess of a bubble diameter of 5 mm. The photographic process is relatively tedious (especially the counting section) and covers only bubbles in the wall region, not the larger diameter ones that move up through the central area of the column.

Mean Bubble Size

After capturing the bubble image, if the size of the bubble is not spherical, and the maximum and minimum axes are known, an equivalent spherical bubble diameter can be calculated by the following equation (Couvert et al., 1999; Hebrard et al., 2001).

$$d_{be} = \sqrt[3]{l_{max}^2 l_{min}} \tag{7.1}$$

where l_{max} and l_{min} are the maximum and minimum axis length of a bubble. In fact, there may be different sizes of bubbles in the experimental test section. For this it is necessary to represent the mean size of the bubble. There are several options to represent the mean of bubble size. The various mean calculations are defined in several standard documents (ISO 9276-2:2001). A general mean diameter can be defined by (ASTM E 799)

$$d_{b,pq} = \left[\frac{\int_0^\infty N(d_b) d_b^p \, dd_b}{\int_0^\infty N(d_b) d_b^q \, dd_b} \right]^{1/(p-q)} \tag{7.2}$$

or in terms of a finite number of discrete size classes,

$$d_{b,pq} = \left[\frac{\sum_{i=1}^\infty N_i d_{bi}^p}{\sum_{i=1}^\infty N_i d_{bi}^q} \right]^{1/(p-q)} \tag{7.3}$$

Table 7.2 Different Mean Diameters with Corresponding Field of Application

Mean Diameter	Name	Definition	Field of Application
d_{10}	Arithmetic/linear mean	$d_{b,10} = \dfrac{1}{N}\sum_{i=1}^{N} N_i d_{bi}$	Evaporation
d_{20}	Surface mean	$d_{b,20} = \left[\dfrac{1}{N}\left(\sum_{i=1}^{N} N_i d_{bi}^2\right)\right]^{1/2}$	Absorption
d_{30}	Volume mean	$d_{b,30} = \left[\dfrac{1}{N}\left(\sum_{i=1}^{N} N_i d_{bi}^3\right)\right]^{1/3}$	Hydrology
d_{21}	Surface diameter mean	$d_{b,21} = \sum_{i=1}^{\infty} N_i d_{bi}^2 / \sum_{i=1}^{\infty} N_i d_{bi}$	Adsorption
d_{31}	Volume diameter mean	$d_{b,31} = \left[\sum_{i=1}^{\infty} N_i d_{bi}^3 / \sum_{i=1}^{\infty} N_i d_{bi}\right]^{1/2}$	Evaporation, molecular diffusion
d_{32}	Sauter mean	$d_{b,32} = \sum_{i=1}^{N} (N_i d_{bi}^3) / \sum_{i=1}^{N} (N_i d_{bi}^2)$	Efficiency studies, mass transfer, gas liquid reaction
d_{43}	De Brouke	$d_{b,43} = \sum_{i=1}^{N} (N_i d_{bi}^4) / \sum_{i=1}^{N} (N_i d_{bi}^3)$	Combustion equilibrium, Spray system

Definitions of the different type of expressions for "mean" are given in Table 7.2. Large bubbles contribute largly to the volume average bubble diameter and hence $d_{b,30} > d_{b,10}$. Sauter mean bubble diameter is another way to represent the mean bubble size. The mean diameter with the same ratio of volume to surface area is known as the Sauter mean diameter. For most BSDs, the Sauter mean diameter, $d_{b,32}$, is larger than the arithmetic, $d_{b,10}$; surface, $d_{b,20}$; and volume, $d_{b,30}$, mean diameters. The different mean diameters are appropriate for different purposes as illustrated in Table 7.2. The Sauter mean diameter is probably the most commonly used mean because it characterizes a number of important processes in which distribution of size of the dispersed phase is wide in range.

Error Compensation for Bubble Size Measurement

Correction of Distortion Due to Curvature

If the snap of a bubble is taken from a cylindrical bubble column ranging from 0.025 to 0.10 m, a wall curvature may cause a deviation or change of the bubble from its actual size. This can be minimized by fitting a plane parallel window or by surrounding the column with a rectangular box filled with the same liquid (Schumpe and Deckwer, 1982). The mechanism of distortion and the advantage of using rectangular operating liquid-filled viewing box to reduce the change are shown in Figure 7.2.

Pressure and Temperature Correction

The raw size of the bubble taken into account in a particular axial position is under the axial pressure and the room temperature. Therefore, corrections for the pressure and

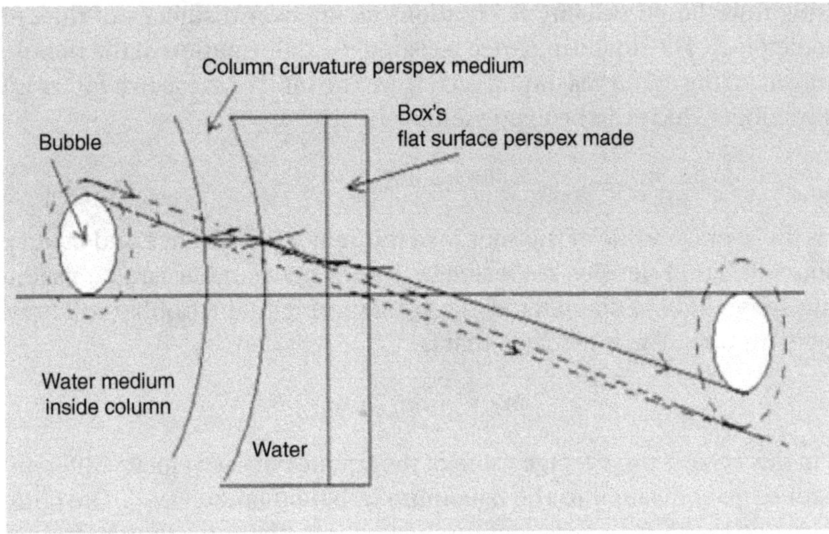

FIGURE 7.2 Mechanism of distortion by wall curvature and minimization by straight plan surface.

temperature of the raw bubble size have to be made. Bubble size can be reported at standard conditions (298 K and 1 atm) using the approximation given by following equations:

According to ideal gas law:

$$\frac{\text{Bubble volume at standard condition}}{\text{Bubble volume at experimental condition}} = \frac{P_{exp}}{P_{st}} \cdot \frac{T_{st}}{T_{exp}} \tag{7.4}$$

which implies

$$\frac{d_{b,st}^3}{d_{b,exp}^3} = \frac{P_{exp}}{P_{st}} \cdot \frac{T_{st}}{T_{exp}} = \frac{10.33 - H_i}{10.33} \cdot \frac{298}{T_i + 273} \tag{7.5}$$

which gives

$$d_{b,st} = d_{b,exp} \sqrt[3]{\left(\frac{10.33 - H_i}{10.33}\right)\left(\frac{298}{T_i + 273}\right)} \tag{7.6}$$

where $d_{b,st}$ is the bubble diameter corrected to standard conditions, $d_{b,exp}$ is the bubble diameter as taken from images, H_i is the distance between the gas–liquid mixing height and the point at which the images are collected, and T_i is the water temperature in the viewing chamber.

Maximum Bubble Size

The size of maximum stable bubbles produced in the bubble column can be determined by the forces acting on the bubble. In low-viscosity liquids, the bubbles are deformed by forces arising from liquid velocity fluctuations acting over distances of the order of the bubble diameter, d_b. The doctoring force resisting the deformation of the bubble is due to surface tension acting at the gas–liquid interface. The ratio of these two forces is known as the Weber number, which can be expressed as

$$We = \rho_l \overline{u^2} d_b / \sigma \tag{7.7}$$

where $\overline{u^2}$ is the average value of the square of the velocity difference and σ and ρ_l are surface tension and liquid density, respectively. The Weber number can be used to predict a maximum stable bubble diameter, $d_{b,max}$, by assuming that a bubble will break when a critical Weber number, We_c is reached, that is,

$$We_c = \rho_l \overline{u^2} d_{b,\mathrm{max}} / \sigma \tag{7.8}$$

where $\overline{u^2}$, in this case, is the average value of the squares of the velocity differences acting over a length scale equivalent to the maximum bubble diameter, d_{bm}. The critical Weber number is related to the average energy dissipation rate per unit volume within the liquid. The maximum stable bubble diameter can be calculated from equation (7.8) provided the liquid physical properties, critical Weber number, and average of the squares of the velocity fluctuations acting over length scales of the same order as the bubble diameter are known. If the bubble is small compared with the turbulent macroscale but large compared with the microscale, the eddies responsible for break-up are isotropic and lie within the inertial subrange such that their kinetic energy is independent of viscosity and follows the Kolmogoroff energy distribution law (Evans et al., 1992), that is,

$$\overline{u^2} = \lambda \left(\frac{\dot{E}_v d_b}{\rho_l} \right)^{2/3} \tag{7.9}$$

where ε is the average energy dissipation rate per unit volume, and according to Batchelor (1951) (Evans et al., 1992), the proportinality constant (λ) is equal to 2.0. Substituting this expression into equation (7.8), one can obtain an expression for the maximum stable bubble diameter:

$$d_{b,\mathrm{max}} = \left(\frac{We_c^3 \sigma^3}{\lambda^3 \rho_l} \right)^{1/5} \dot{E}_v^{-2/5} \tag{7.10}$$

The high energy dissipation in the gas distributor zone results in the generation of a number of very fine bubbles, which are carried inverse diection (in case of inverse bubbly flow system) by the bulk liquid motion and leave the zone. However, a number of larger bubbles are also generated, which recirculate within the distributor zone, where they are eventually broken up and carried either direction by the bulk liquid flow. A critical Weber number can be chosen from a number of theoretical and numerical studies relating to the break-up of bubbles. Some values of critical Weber number given by different investigators are shown in Table 7.3.

Table 7.3   Critical Weber Number by Various Investigators

Investigators	Critical Weber Number	Comments
Hinze (1955)	1.18	Based on the break-up of a droplet or bubble in a viscous shear flow
Sevik and Park (1973)	1.24	Based on the assumption that a bubble will oscillate violently and break up when the characteristic frequency of the turbulence is equal to one of the resonant frequencies of the bubble
Miksis et al. (1981)	3.23	For a bubble in a uniform flow, a cylindrical bubble be-
Lewis and Davidson (1982)	4.7	come unstable in an axisymmetric inviscid shear flow
Ryskin and Leal (1984)	0.95–2.76	The critical Weber number is a function of the Reynolds number for a bubble in a uniaxial extensional Newtonian flow
Evans et al. (1992)	1.2	For bubble break-up in viscous shear flow by liquid jet
Ishii et al. (2004)	6.0	Bubble break-up by impact of turbulent eddie
Majumder et al. (2006)	1.16	For bubble break-up in viscous shear flow by ejector-induced liquid jet
Rodríguez-Rodríguez et al. (2006)	2.22–2.3	based on the velocity difference between both interface points at the straining axis
Revuelta et al. (2008)	1.80	the break-up of bubbles owing to their interaction with an array of successive eddies, modeled by a train of straining flows
Revuelta (2010)	1.68	Based on framework of the classical Kolgomorov–Hinze theory on bubble break-up

Bubble Size Distribution

The distributions of bubble size can be obtained by sorting the equivalent diameters of bubbles into different uniform classes. An outline for determination of size distribution by defining the bubble class is given in Table 7.4. It has been pointed out earlier that in the inverse bubbly flow system, gas–liquid mixing occurs in the extended contactor of an ejector-induced inverse bubble column (Majumder et al., 2006). After coming out from the divergent diffuser in an inverse bubble column, the liquid jet can be arrested in the contactor by adjusting the liquid column height inside the column. It is well known that when a liquid jet of uniform initial velocity is discharged from a nozzle into a fluid, a tangential separation surface is created between the jet and the surrounding medium. The instability of this surface creates eddies that move both in the main direction of flow and across it. This causes exchange of momentum with the surrounding medium. As a result, a region of finite thickness with a velocity is formed on the border of the jet and surrounding fluid. When this jet impinges on the liquid surfaces, it entrains the surrounding gas along with it. The entrained gas undergoes subsequent breakdown and disintegrates as fine bubbles within a finite length, called intense mixing zone. Majumder et al. (2006) reported BSD in three axial locations A (of height 0.2-0.4 m), B (of height 0.4-0.6 m) and C (of height 0.8-1.0 m) from the bottom of the column (Figure 7.3) of an ejector-induced inverse bubble column.

Table 7.4 Discretization in Bubble Classes and Their Respective Frequencies

Class, i (mm)	Class Width Difference of bound of class (Δ Bin)	Bubble Number Frequency in Class i, N_i	Sum of d_i^r in Each Size Class				Cumulative Frequency in Class i, $CF_i = \sum_1^i N_i$	Relative Frequency, $N_i/\Sigma N_i$	Volume % in Class ($d_{bi}^3 / \sum d_{bi}^3$) $\times$ 100
			d_{bi}	d_{bi}^2	d_{bi}^3	d_{bi}^4			
$d_{bmin} - d_{b1}$	UB-LB	N_1	$\sum_0^{N_1} d_{bi}$	$\sum_0^{N_1} d_{bi}^2$	$\sum_0^{N_1} d_{bi}^3$	$\sum_0^{N_1} d_{bi}^4$	N_1	$N_1/\Sigma N_i$	$\sum_0^{N_1} d_{bi}^3 / \sum d_{bi}^3$
$d_{b1} - d_{b2}$	UB-LB	N_2	$\sum_0^{N_2} d_{bi}$	$\sum_0^{N_2} d_{bi}^2$	$\sum_0^{N_2} d_{bi}^3$	$\sum_0^{N_2} d_{bi}^4$	$N_1 + N_2$	$(N_1 + N_2)/\Sigma N_i$	$\sum_0^{N_2} d_{bi}^3 / \sum d_{bi}^3$
$d_{b2} - d_{b3}$	UB-LB	N_3	$\sum_0^{N_3} d_{bi}$	$\sum_0^{N_3} d_{bi}^2$	$\sum_0^{N_3} d_{bi}^3$	$\sum_0^{N_3} d_{bi}^4$	$N_1 + N_2 + N_3$	$(N_1 + N_2 + N_3)/\Sigma N_i$	$\sum_0^{N_3} d_{bi}^3 / \sum d_{bi}^3$
$d_{b3} - d_{b4}$	UB-LB	N_4	$\sum_0^{N_4} d_{bi}$	$\sum_0^{N_4} d_{bi}^2$	$\sum_0^{N_4} d_{bi}^3$	$\sum_0^{N_4} d_{bi}^4$	$N_1 + N_2 + N_3 + N_4$	$(N_1 + N_2 + N_3 + N_4)/\Sigma N_i$	$\sum_0^{N_4} d_{bi}^3 / \sum d_{bi}^3$
.	.	.	.	.	.	.	.	.	.
.	.	.	.	.	.	.	.	.	.
.	.	.	.	.	.	.	.		.
$d_{b,n-1} - d_{b,max}$	UB-LB	N_n	$\sum_0^{N_n} d_{bi}$	$\sum_0^{N_n} d_{bi}^2$	$\sum_0^{N_n} d_{bi}^3$	$\sum_0^{N_n} d_{bi}^4$	ΣN_i	1	$\sum_0^{N_n} d_{bi}^3 / \sum d_{bi}^3$
		ΣN	Σd_{bi}	Σd_{bi}^2	Σd_{bi}^3	Σd_{bi}^4			

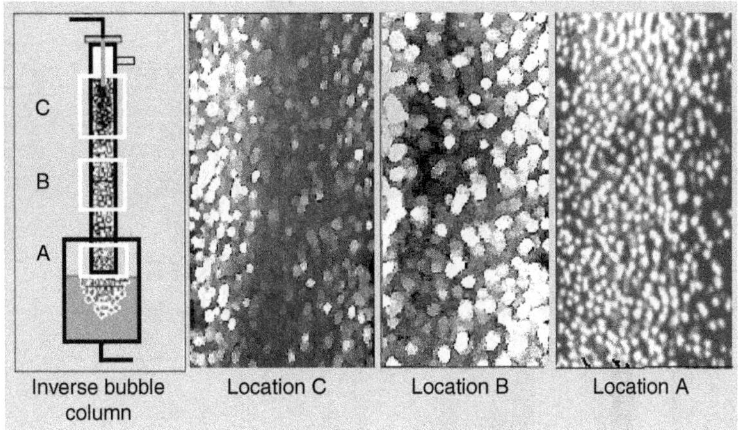

FIGURE 7.3 Photograph taken from different location of homogeneous bubbly flow zone at d_n = 0.004 m, u_{sl} = 7.07 × 10^{-2} m/s, and u_{sg} = 1.70 × 10^{-3} m/s.

Typical results for BSDs at these three locations are presented in Figures 7.4 to 7.6. It is seen that the bubble sizes in locations B and C are greater than at location A. The average bubble size in locations B and C are almost same. The coalesced bubbles of location A go up because of their buoyancy and accumulate in the location B and C. This results in the wider BSD in locations B and C compared with location A. Also, the bubble number flux varies in different locations for the same reason.

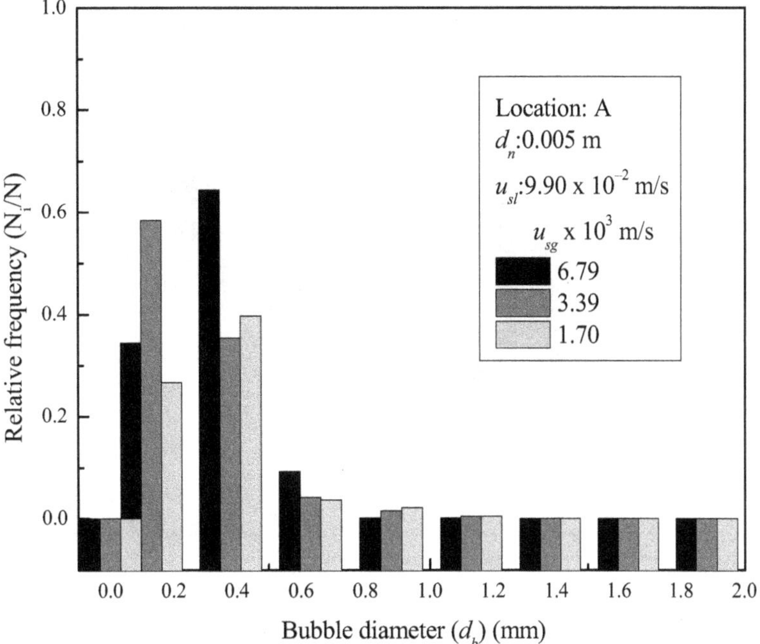

FIGURE 7.4 Bubble size distribution at different superficial gas velocities.

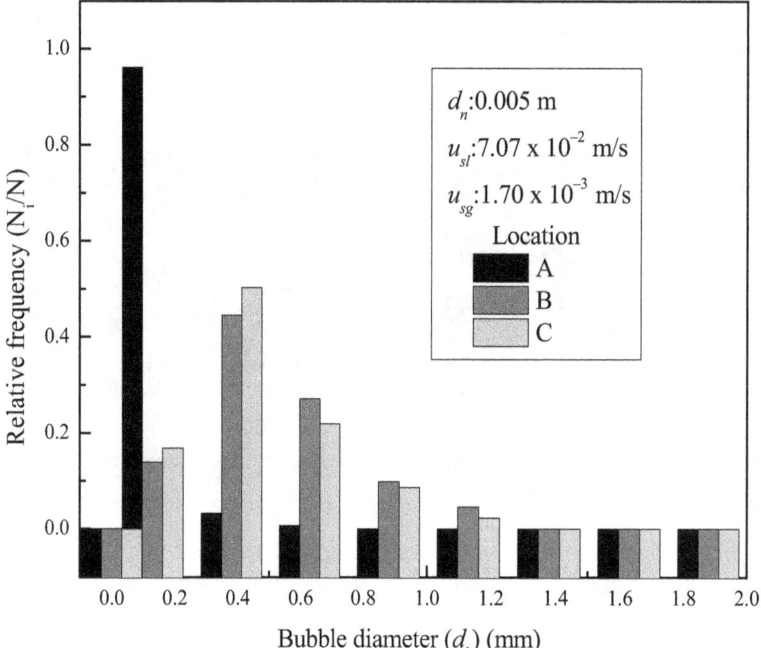

FIGURE 7.5 Bubble size distribution at different locations.

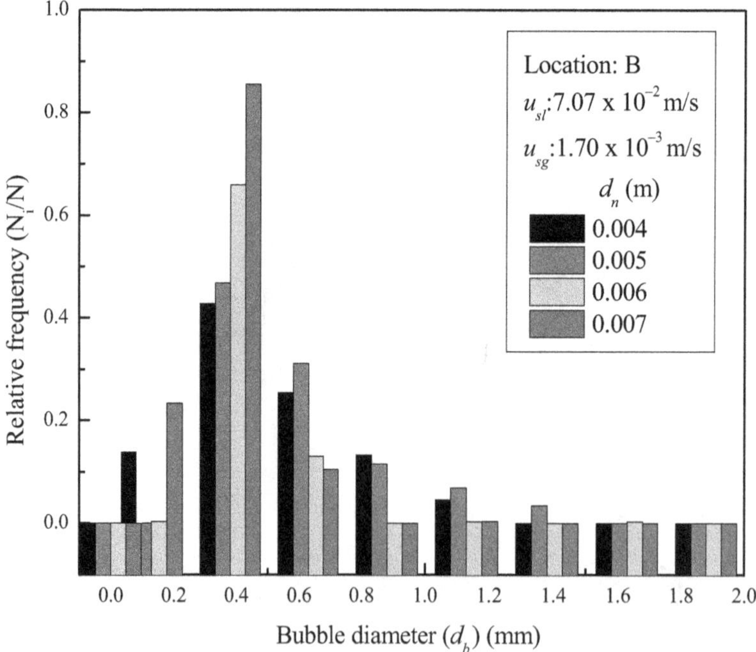

FIGURE 7.6 Bubble size distribution at different nozzle diameters.

In the intense mixing zone where bubble are broken up, the jet energy is used to form fine gas bubbles that flow inversely by the bulk liquid motion. The region below the mixing zone is referred to as the *homogeneous bubbly zone.* The homogeneous bubbly flow zone consists of a relatively slow-moving packed bed of bubbles. From the literature, it is seen that models of BSDs in swarms have been developed based on either (1) a binary coalescence process, whereby each coalescence event decreases the number of bubbles by one or (2) a nonbinary process, which is based on simultaneous coalescence of bubble clusters. Stewart et al. (1993) suggested that the binary model is adequate for predicting the BSD for systems in which coalescence does not occur. However, it is inappropriate for more routinely observed coalescing systems. In the following section, the coalescence phenomena in the homogeneous bubbly flow zone has been emphasized from the size distribution model. Majumder et al. (2006) reported that in the intense mixing zone, a high rate of energy dissipation results in the generation of bubbles. Larger bubbles generated in the mixing zone are broken up into fine bubbles by the momentum of the liquid jet and are carried inversely by the bulk liquid motion and distributed along the axis. The distribution of the bubble size occurs according to axial energy dissipation profile (Evans et al., 1992).

In all cases, the BSD is not symmetrical (normal distribution) with an extended tail for the larger bubble sizes, which is typical of a log-normal distribution. Figure 7.7 shows the typical well-fitted log-normal distribution compared with other distributions (exponential, gamma, normal) for a particular operating condition.

The details of exponential, gamma, and normal distribution are described as follows.

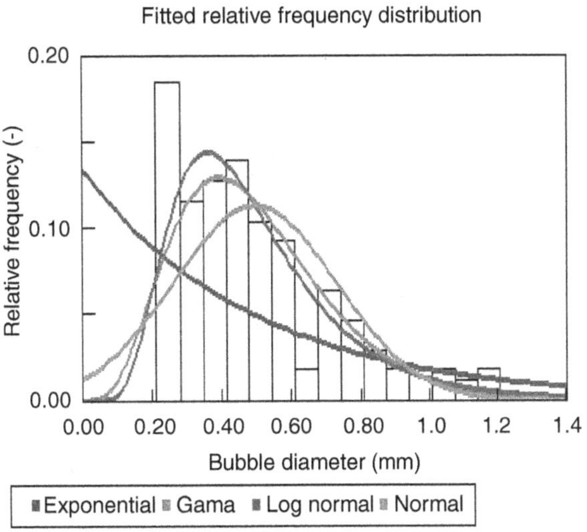

FIGURE 7.7 Fitted different distributions of bubble sizes at location B: $d_n = 0.005$ m, $u_{sl} = 7.07 \times 10^{-2}$ m/s, $u_{sg} = 1.70 \times 10^{-3}$ m/s; best fit: log-normal.

Exponential Distribution

The exponential distribution is a continuous distribution bounded on the lower side. Its shape is always the same, starting at a finite value at the minimum and continuously decreasing at larger x (Law and Kelton, 1991; Johnson et al., 1994).

$$F(x) = \frac{1}{\beta} \exp\left(-\frac{[x-\min]}{\beta}\right) \tag{7.11}$$

where, min = minimum x value, β = scale parameter = mean of x.

Normal Distribution

The normal distribution is an unbounded continuous distribution. It is defined as (Johnson et al., 1994)

$$F(x) = \frac{1}{\sqrt{2\pi\sigma^2}} \exp\left(-\frac{[x-\mu]^2}{2\sigma^2}\right) \tag{7.12}$$

where $\mu = \text{mean} = \sum_{i=1}^{N} x_i / N$, σ = standard deviation. The normal distribution is an unbounded continuous distribution. It sometimes called a Gaussian distribution.

Log-Normal Distribution

The log-normal distribution is a continuous distribution bounded on the lower side. The log-normal distribution is defined as

$$F(x) = \frac{1}{(x-\theta)\sqrt{2\pi\hat{\sigma}^2}} \exp\left(-\frac{[\ln(x-\theta)-\hat{\mu}]^2}{2\hat{\sigma}^2}\right) \tag{7.13}$$

where θ = minimum of x, $\hat{\mu} = \sum_{i=1}^{n} \ln(x_i)/N$, $\hat{\sigma} = \sqrt{(\sum_{i=1}^{n}(\ln x_i - \hat{\mu}))/N}$. $F(x)$ is always 0 at minimum x, rising to a peak that depends on both $\hat{\mu}$ and $\hat{\sigma}$ and then decreases monotonically for increasing x. By definition, the natural logarithm of a lognormal random variable is a normal random variable. The log-normal distribution can also be used to approximate the normal distribution, for small sigma, while maintaining its strictly positive values of x (Johnson et al., 1994).

Gamma Distribution

The gamma distribution is a continuous distribution bounded on the lower side. The gamma distribution is defined as (Johnson et al., 1994)

$$F(x) = \frac{(x-\min)^{\alpha-1}}{\beta^\alpha \Gamma(\alpha)} \exp\left(-\frac{[x-\min]}{\beta}\right) \tag{7.14}$$

where min = minimum x, α = shape parameter > 0, β = scale parameter > 0.

Beta Distribution

The beta distribution is a continuous distribution that has both upper and lower finite bounds. Many real situations can be bounded in this way. The beta distribution can be used empirically to estimate the actual distribution before much data is available. The distribution can be expressed as (Johnson et al., 1994)

$$f(x) = \frac{1}{B(p,q)} \frac{(x-\min)^{p-1}(\max-x)^{q-1}}{(\max-x)^{p+q-1}} \tag{7.15}$$

where $\min \leq x \leq \max$, min = minimum value of x, max = maximum value of x, p = lower shape parameter > 0, q = upper shape parameter > 0, $B\,(p, q)$ is beta function. The uniform distribution is a special case of the beta distribution with m, $q = 1$. The beta distribution can approach zero or infinity at either of its bounds, with p controlling the lower bound and q controlling the upper bound. Values of p, $q < 1$ cause the beta distribution to approach infinity at that bound. Values of p, $q > 1$ cause the beta distribution to be finite at that bound.

Weibull Distribution

The Weibull distribution is a continuous distribution bounded on the lower side (Weibull, 1951). It provides one of the limiting distributions for extreme values. The distribution is expressed as

$$F(x) = \frac{\alpha}{\beta}\left(\frac{x-\min}{\beta}\right)^{\alpha-1} \exp\left(-\left(\frac{[x-\min]}{\beta}\right)^{\alpha}\right) \tag{7.16}$$

where min = minimum x, α = shape parameter > 0, β = scale parameter > 0. Similar to the gamma distribution, it has three distinct regions. For $\alpha = 1$, the Weibull distribution is reduced to a distribution, starting at a finite value at minimum x and decreasing monotonically thereafter. For $\alpha < 1$, the Weibull distribution tends to infinity at minimum x and decreases monotonically for increasing x. for $\alpha > 1$, the Weibull distribution is 0 at minimum x, peaks at a value that depends on both α and β, decreasing monotonically thereafter. Uniquely, the Weibull distribution has negative skewness for $\alpha > 3.6$. The Weibull distribution can also be used to approximate the normal distribution for $\alpha = 3.6$ while maintaining its strictly positive values of x (actually $[x\text{-min}]$), although the kurtosis is slightly smaller than 3, the normal value.

Analysis of Axial Bubble Size Distribution

Axial Number Flux Model

The size distribution model can be written based on a bubble number flux, $\dot{N}(Z)$. The balance for the bubble number flux over a volume element inside the column can be represented as follows (Atkinson et al., 2003)

$$\dot{N}(z_2) = \dot{N}(z_1) - \int_{z_1}^{z_2} P_c \dot{f}(z)dz \tag{7.17}$$

where $\dot{f}(z)$ is the rate of collision per unit volume and P_c is the probability that a collision between two bubbles will result in a coalescence event. The collision frequency of bubbles can be found from statistical considerations analogous to the kinetic theory of gases in turbulent motions in the inertial subrange (Kamp et al., 2001). Using this approach Kuboi et al. (1972) arrived at an expression for the number of collisions of equal particles per unit time and volume. The number of collisions per unit time and volume of equal size bubbles is given by (Kuboi et al., 1972)

$$\dot{f}(z) = \left(\frac{8\pi}{3}\right)^{0.5} n^2 d_b^2 u_t \tag{7.18}$$

where n is the bubble number density, which for spherical bubbles is equal to

$$n = \frac{6_g}{\neq d_b^3} \tag{7.19}$$

and u_t is the turbulent velocity, which provides a measure of relative velocity of two bubbles in the liquid at a distance of bubble diameter, d_b (Hinze, 1955, 1975). According to Hinze (1975) the turbulent velocity is equal to

$$u_t = \left(\frac{\dot{E}_v d_b}{\rho_l}\right)^{1/3} \tag{7.20}$$

where $\dot{E}_v$ is the average energy dissipation rate per unit volume in the homogeneous bubbly zone. According to Kamp et al. (2001), the rate of energy dissipation of turbulent kinetic energy per unit volume in the bubbly flow zone can be estimated by the following equation

$$\dot{E}_v = \frac{2\rho_l u_*^3}{\kappa d_c} \tag{7.21}$$

where κ is the von Karman constant, equal to 0.41, and u_* is the friction velocity. The friction velocity can be estimated from Blasius equation (Colin et al., 1991). According to Colin et al. (1991), the friction velocity can be calculated for the bubbly pipe flow as follows

$$u_* = (u_{sl} + u_{sg})\sqrt{\frac{0.079 \, Re_m^{-1/4}}{2}} \tag{7.22}$$

where Re_m is the mixture Reynolds number, which is defined as

$$Re_m = \frac{\rho_l (u_{sl} + u_{sg}) d_c}{\mu_l} \tag{7.23}$$

Equations (7.17) to (7.23) can be used to predict the axial bubble number flux in the ejector-induced inverse flow bubble column provided the probability of coalescence, P_c, is known. According to coalescence theory (Chesters, 1991), coalescence occur because of collision of two bubbles provided that the interaction time, t_i, between bubbles exceeds the drainage time, t_d. The drainage time is defined as the time required for drainage of the liquid film between bubbles to a critical rupture thickness. Kamp et al. (2001) reported the probability of coalescence in terms of the ratio of the drainage, t_d, and interaction, t_i, times for two bubbles. The ratio t_d/t_i can provide a first indication of whether coalescence will or will not occur under a given condition (Kamp et al., 2001):

$$P_c(t_d/t_i)=0 \quad if \quad t_d/t_i>1 \qquad =1 \quad if \quad t_d/t_i<1 \tag{7.24}$$

where

$$\frac{t_d}{t_i}=k\frac{\rho_l u_0 d^2/8\sigma}{\frac{\pi}{4}\left(\rho_l C_{vm}d^3/3\sigma\right)^{1/2}} \tag{7.25}$$

In equation (7.25), u_0 is the relative velocity of the two bubbles at the onset of deformation, and C_{vm} is the virtual mass coefficient normally taken to be a constant between 0.5 and 0.8 (Chesters, 1991; Kamp et al., 2001). The virtual mass coefficient C_{vm} describes the volume of displaced fluid that contributes to the effective mass of unit volume of the dispersed phase (Cook and Harlow, 1984; Drew and Lahey, 1987). The virtual (also called "added") mass term in the momentum equations for dispersed two-phase flow represents the force required to accelerate the mass of the surrounding continuous phase in the immediate vicinity of a dispersed-phase fragment, such as a bubble or droplet, when the relative velocity of the phases changes. It is significant only if the continuous-phase density is of the same order of magnitude as or much greater than that of the dispersed phase. The variation of C_{vm} depends on the ratio of the bubble diameter. For a spherical equal-sized bubble diameter, the value of $C_{vm} = 0.785$ has been reported by Kamp et al. (2001). The parameter k in equation (7.25) is a correction factor, which is a function of bubble size and interfacial properties, to take into account hydrodynamic effects such as bubbles bouncing of each other prior to coalescence occurring. Kamp et al. (2001) developed an explicit expression involving an exponential function for the probability of coalescence proposed by Coulaloglou and Tavlarides (1977) as

$$P_c \propto \exp(-t_d/t_i) \tag{7.26}$$

But the main limitation with equation (7.26) is that the mathematical definition for t_d does not include the effect of surface tension, which can change the drainage time by orders of magnitude. To some degree, the constant K can be used to account for such effects. According to Atkinson et al. (2003), the drainage time in equation (7.26) can be replaced with a constant, K, times the bubble persistence time, t_p. The bubble persistence time is defined as the period that a bubble remains in the liquid before rupturing. It is a function of bubble size and liquid composition. Atkinson et al. (2003) developed a correlation for the persistence time with an air–water system, which is obtained as

$$t_p = 2.60 \times 10^7 d_b^{3.02} \tag{7.27}$$

Therefore, equation (7.26) can be written as

$$P_c = \lambda \exp\left(-K t_p / t_i\right) \tag{7.28}$$

where λ is the constant of proportionality. The probability of coalescence is determined by calculating the persistence and interaction times from equations (7.27) and (7.25), respectively, for the measured bubble size. λ and K are unknown parameters, to be adjusted to obtain a best fit for the bubble number flux ($\dot{N}(z)$) over a volume element inside the column calculated in accordance with the equation (7.17). The profile of the bubble flux, $\dot{N}(z)$, in the bubbly low zone can be calculated based on the gas volumetric flowrate, Q_g; column cross-sectional area, A_c; and measured Sauter mean bubble diameter, d_{32}. The profile for the bubble number flux is represented by

$$\dot{N}(z) = \frac{\text{Gas volumetric flowrate}}{\text{Bubble volume} \times \text{Cross-sectional area}} = \frac{Q_g}{\dfrac{\pi}{6} d_{32}^3 \times \dfrac{\pi}{4} d_c^2} = \frac{24 Q_g}{d_{32}^3 \times d_c^2} \tag{7.29}$$

Bubble diameter increases with increasing distance from the bottom of the column because of the coalescence of smaller bubbles. Also, the bubble number flux varies in different locations for the same reason. A typical profile of bubble number flux obtained in an ejector-induced inverse bubble column is shown in Figure 7.8. A decrease in bubble

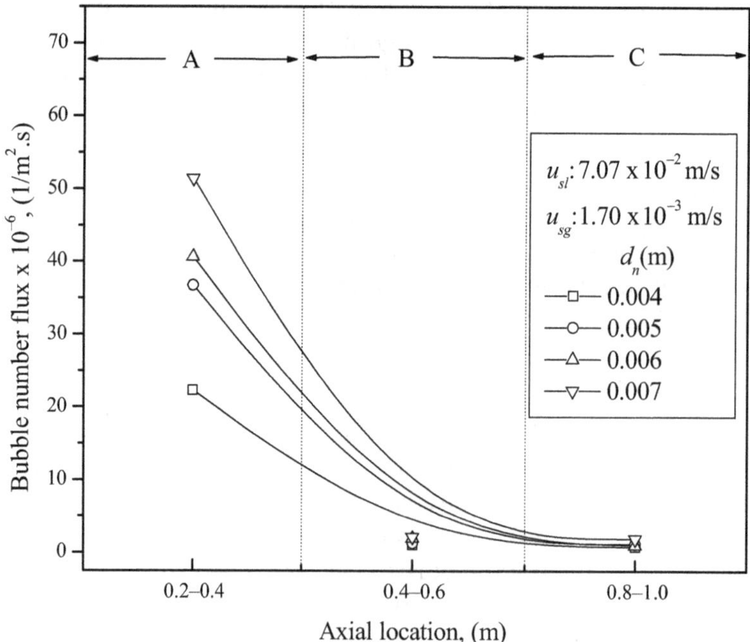

FIGURE 7.8 Variation of bubble number flux with axial location at different nozzle diameters.

number over locations is the result of an increase in bubble size caused by coalescence. The probability of coalescence (P_c) depends on bubble–bubble interaction time (t_i) and bubble persistence time (t_p). The probability of coalescence increases with the increase in liquid flow rate. This is because of increasing collision frequency between bubbles with an increase in the liquid flow rate. As the liquid flow rate increases, kinetic energy increases, which increases the turbulence intensity, bubble–bubble interaction, and velocity of bubbles (Sommerefeld et al., 2003).

That bubble number flux decreases in locations B and C over location A is the result of an increase in bubble size caused by coalescence. The probability of coalescence (P_c) depends on bubble–bubble interaction time (t_i) and bubble persistence time (t_p). The calculated value of probability of coalescence is 0.000664, which indicates that the probability of coalescence when two bubbles collide is very low. The calculated probability corresponds to a persistence and interaction time of 0.0012 and 0.00035 seconds, respectively (equation [7.28]). The probability of coalescence increases with the increase in liquid flow rate as shown in Figure 7.9. This is because of increasing collision frequency between bubbles with increase in liquid flowrate. As the liquid flow rate increases, kinetic energy increases, which increase the turbulence intensity, bubble–bubble, interaction and velocity of bubbles (Sommerefeld et al., 2003). It is also seen from Figure 7.9 that the probability of coalescence is higher in region A than other regions.

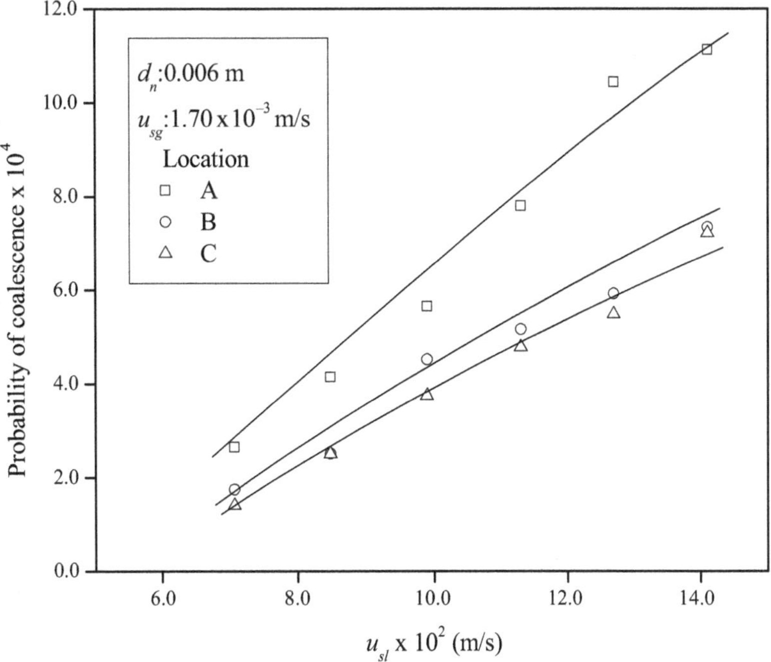

FIGURE 7.9 Variation of probability of coalescence with superficial liquid velocity.

Table 7.5 Typical Data for Bubble Number Flux, Probability of Coalescence, and Collision Frequency per Unit Volume for Different Operating Conditions

d_n, (m)	$u_{sl} \times 10^2$, (m/s)	$u_{sg} \times 10^3$, (m/s)	At location A, $\dot{f}(Z)$, $\times 10^{-10}$ (1/m³s)	At location B, $\dot{f}(Z)$, $\times 10^{-10}$ (1/m³s)	At location C, $\dot{f}(Z)$, $\times 10^{-10}$ (1/m³s)	At location A, $P_c \times 10^4$, (-)	At location B, $P_c \times 10^4$, (-)	At location C, $P_c \times 10^4$, (-)
0.004	7.07	1.7	0.46	0.03	0.03	1.63	0.91	1
0.004	8.48	1.7	1.61	0.1	0.08	3.55	1.88	1.92
0.004	9.9	1.7	2.42	0.24	0.17	4.09	2.71	2.75
0.004	11.31	1.7	9.12	0.77	0.41	6.64	3.97	3.8
0.004	12.73	1.7	50.48	1.99	0.88	9.54	4.89	4.67
0.004	14.14	1.7	63.37	5.37	2.8	9.73	5.7	5.91
0.007	7.07	1.7	1.27	0.13	0.14	3.42	2.29	2.93
0.007	8.48	1.7	1.56	0.24	0.21	3.49	2.89	3.22
0.007	9.9	1.7	6.64	0.54	0.43	6.25	3.69	4.03
0.007	11.31	1.7	57.12	1.48	1.7	9.89	4.69	5.59
0.007	12.73	1.7	461.33	3.15	2.72	12.02	5.33	5.98
0.007	14.14	1.7	841.4	21.41	16.25	12.37	6.61	7.35
0.005	7.07	1.7	16.67	1.6	1.62	8.59	5.17	5.97
0.005	7.07	3.39	20.75	1.95	1.71	8.9	5.33	6
0.005	7.07	6.79	26.81	2.14	2.23	9.22	5.38	6.21
0.005	7.07	10.18	36.85	2.47	2.85	9.61	5.47	6.4
0.005	7.07	13.58	38.87	2.57	3.21	9.67	5.47	6.48
0.005	9.9	1.7	2.63	0.31	0.18	4.25	3.02	2.83
0.005	9.9	3.39	3.38	0.35	0.26	4.72	3.17	3.31
0.005	9.9	6.79	3.89	0.41	0.34	4.9	3.3	3.61
0.005	9.9	10.18	5.38	0.55	0.49	5.5	3.62	4.08
0.005	9.9	13.58	19.23	1.07	1.18	8.07	4.37	5.18

In region A at a particular liquid flow rate, the persistence time t_p decreases as the bubble size decreases (equation [7.27]), which causes a relatively higher probability of coalescence in region A. Some typical data for bubble number flux, probability of coalescence, and collision frequency per unit volume for different operating conditions are presented in Table 7.5.

Correlation Model for Mean Bubble Size

The formation of bubbles by plunging liquid jet in an inverse flow column is a very complex phenomenon, and there is no reliable theoretical model available to predict the bubble size in the plunging liquid jet bubble column. It is very difficult to develop a theoretical model to predict the bubble size for the present system because of the complexity of the system. It has been found experimentally that bubble size depends on different operating variables, including fluid flow rates, nozzle diameter, and axial location in the column. A generalized correlation for an air–water system to express the Souter mean bubble diameter developed by Majumder et al. (2006) is recommended:

$$D_R = 1.48 \times 10^{-2} \frac{Z_R^{1.09} \, Re_{\ln}^{0.449}}{u_{SR}^{1.388} \, Fr_{gn}^{0759} \, Su_{\ln}^{0.303}} \tag{7.30}$$

The ranges of variation of the different parameters for the correlation are $42.85 < Z_R < 225.00$; $8.23 \times 10^{-3} < u_{SR} < 174.55 \times 10^{-3}$; $0.02 \times 10^4 < Fr_{gn} < 2.38 \times 10^4$; $2.83 \times 10^2 < Re_{\ln} < 17.97 \times 10^2$; $4.45 \times 10^5 < Su_{\ln} < 7.80 \times 10^5$.

Bubble Exchange Model

Consider the bubble column as shown in Figure 7.10 in which a liquid jet entrains gas in the dispersed gas of bubbles due to jet kinetic energy. When bubbles are formed and dispersed by a liquid jet, some bubbles will move inversely by momentum transfer of liquid jet, and some bubbles will move upward because of the buoyancy effect. At a certain cross-section of a height, some of the bubbles that are going upward and in an inverse direction may exchange between upward and inverse streams because of momentum variation. During the movement of dispersed phase, the up- and inverse-flowing bubbles may interact with each other. As a result, bubbles may coalescence or break up, which leads to different sizes of bubbles. Finer bubbles that are formed by break-up mechanism move inversely along with liquid, and relatively larger sizes of bubbles that exhibit more buoyancy force than momentum of liquid move upward. At the same time, some bubbles may exchange the stream cross currently because of momentum change distribution. Let us consider that the concentration (number density) of the bubbles flowing upward and inverse direction are n_u and n_d, respectively, at a cross-section of height z.

Because of concentration difference among the two streams (and inverse flowing), there always exists some bubbles being exchanged between these two streams. The

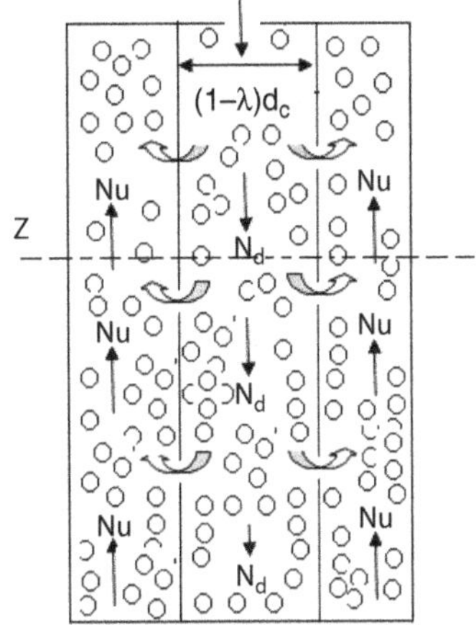

FIGURE 7.10 Schematic of bubble exchange.

exchange is quantified by the exchange factor (E). Bubble dispersion is being taking place because of velocity distribution and because of turbulence, which is accounted for by the dispersion coefficient of bubble motion. A general number balance over a control volume will then be written as

$$(Out - In)_{Bulk\,matter} + (Out - In)_{dDspersed\,matter} + Disappearance + Accumulation = 0 \qquad (7.31)$$

If the number balance is applied for upflowing bubble in a control volume that is at height z and a length of dz as shown in Figure 7.10, then

$$(Out - In)_{Bulk\,matter} = (n_{u,z+dz} - n_{u,z})u_l S_u \qquad (7.32)$$

$$(Out - In)_{Dispersed\,matter} = -S_u D_b \left(\left(\frac{\partial n_u}{\partial z} \right)_{z+dz} - \left(\frac{\partial n_u}{\partial z} \right)_z \right) \qquad (7.33)$$

Disappearance due to exchange is $-Ed_c(n_{u,z} - n_{d,z})dz$ where E is the parameter that signifies the exchange factor and accumulation is $S_u \left(\dfrac{\partial n_{u,z}}{\partial t} \right) dz$. Substituting the different terms in equation (7.31), one gets

$$S_u \left(\frac{\partial n_u}{\partial t} \right) = S_u D_b \left(\frac{\partial^2 n_u}{\partial z^2} \right) - S_u u_l \left(\frac{\partial n_u}{\partial z} \right) + Ed_c (n_{u,z} - n_{d,z}) \qquad (7.34)$$

Similarly, the number balance equation for an inverse-flowing stream becomes

$$S_d \left(\frac{\partial n_d}{\partial t} \right) = S_d D_b \left(\frac{\partial^2 n_d}{\partial z^2} \right) + S_d u_l \left(\frac{\partial n_d}{\partial z} \right) - Ed_c (n_{u,z} - n_{d,z}) \qquad (7.35)$$

If there is a much interaction between the upward- and inverse-flowing dispersed phase, the exchange factor will be large; it can be assumed that the difference in the dispersed phase concentration ($n_{u,z} - n_{d,z}$) will be relatively small. In other words, the radial concentration differences will then be small. Assume that the equivalent diameter of cross-sectional area through which bubbles move upward, $d_u i$ is equal to fraction λ of diameter of column (i.e., $d_u = \lambda d_c$). Then it can be expressed as

$$S_u = \lambda^2 S_c \qquad (7.36)$$

$$S_d = (1 - \lambda)^2 S_c \qquad (7.37)$$

Defining n as the volume average concentration at a cross-section, it can be expressed as

$$N = \frac{n_u S_u \Delta z + n_d S_d \Delta z}{S_u \Delta z + S_d \Delta z} \qquad (7.38)$$

or

$$N = \frac{n_u S_u + n_d S_d}{S_u + S_d}$$

By using equations (7.36) and (7.37), equation (7.38) can be simplified as

$$n = \frac{\lambda^2\, n_u + (1-\lambda)^2\, n_d}{1 - 2\lambda(1-\lambda)} \tag{7.39}$$

This n signifies the total number of bubbles moving at a cross-section. Defining the concentration potential difference (M_d) as equation (7.40), which signifies the difference in number of bubbles between the two passes (upflow and inverse flow) per unit volume,

$$M_d = \lambda^2\, n_u - (1-\lambda)^2\, n_d \tag{7.40}$$

Adding equations (7.34) and (7.35) and further substitution of equations (7.36) and (7.37), the resulting equation becomes

$$S_c\left(\frac{\partial n}{\partial t}\right) = S_c D_b\left(\frac{\partial^2 n}{\partial z^2}\right) - \frac{S_c u_{sl}}{1-\varepsilon_g}\left(\frac{\partial M_d}{\partial z}\right) \tag{7.41}$$

Subtracting equation (7.35) from equation (7.34) and using equations (7.36) and (7.37), one gets

$$S_c\left(\frac{\partial M_d}{\partial t}\right) = S_c D_b\left(\frac{\partial^2 M_d}{\partial z^2}\right) - \frac{S_c u_{sl}}{1-\varepsilon_g}\left(\frac{\partial n}{\partial z}\right) - 2EP(n_u - n_d) \tag{7.42}$$

From equations (7.39) and (7.40),

$$n + M_d = 2\lambda^2 n_u \tag{7.43}$$

$$n - M_d = 2(1-\lambda^2)n_d \tag{7.44}$$

Equations (7.43) and (7.44) can be solved to get

$$n_u - n_d = \frac{(1-2\lambda^2)n + M_d}{2\lambda^2(1-\lambda^2)} \tag{7.45}$$

Substituting equation (7.45) for $n_u - n_d$ in equation (7.42) and further differentiating the equation (7.42) with respect to z, one can obtain

$$S_c\left(\frac{\partial^2 M_d}{\partial t\, \partial z}\right) = S_c D_b\left(\frac{\partial^2 M_d}{\partial z^2}\right) - \frac{S_c u_{sl}}{1-\varepsilon_g}\left(\frac{\partial^2 n}{\partial z^2}\right) - \frac{EP(1-2\lambda^2)}{\lambda^2(1-\lambda^2)}\left(\frac{\partial n}{\partial z}\right) - \frac{EP}{\lambda^2(1-\lambda^2)}\left(\frac{\partial M_d}{\partial z}\right) \tag{7.46}$$

By definition, M_d can be assumed to be a small value, and hence higher order derivatives of M_d can be neglected, which modifies equation (7.46) as

$$\frac{EP}{\lambda^2(1-\lambda^2)}\left(\frac{\partial M_d}{\partial z}\right) = -\frac{S_c u_{sl}}{1-\varepsilon_g}\left(\frac{\partial^2 n}{\partial z^2}\right) - \frac{EP(1-2\lambda^2)}{\lambda^2(1-\lambda^2)}\left(\frac{\partial n}{\partial z}\right) \tag{7.47}$$

Substituting $\partial M_d/dz$ from equation (7.47) in equation (7.41) and rearranging gives

$$\frac{\partial n}{\partial t} + \frac{u_{sl}(2\lambda^2-1)}{1-\varepsilon_g}\left(\frac{\partial n}{\partial z}\right) = \left[D_b + \left(\frac{u_{sl}}{1-\varepsilon_g}\right)^2 \frac{\lambda^2(1-\lambda^2)S_c}{EP}\right]\frac{\partial^2 n}{\partial z^2} \tag{7.48}$$

Equation (7.48) forms the final model equation, which represents a modified form of the convection–dispersion model as

$$\frac{\partial n}{\partial t} + U\left(\frac{\partial n}{\partial z}\right) = \Gamma \frac{\partial^2 n}{\partial z^2} \tag{7.49}$$

where

$$U = \frac{u_{sl}(2\lambda^2 - 1)}{1 - \varepsilon_g} \tag{7.50}$$

$$\Gamma = D_b + \frac{\zeta}{E} \tag{7.51}$$

where

$$\zeta = \left(\frac{\lambda^2(1 - \lambda^2)d_c}{4}\right)\left(\frac{u_{sl}}{1 - \varepsilon_g}\right)^2 \tag{7.52}$$

Equation (7.48) represents modified dispersion model, which expresses the overall dispersion coefficient of bubbles as the dispersion resulting because of turbulence and because of bubble exchange between upward- and inverse-flowing liquid (E). At steady-state operation after integration of equation (7.48) with the following limiting conditions of equations (7.53) and (7.54), the number density profile can be expressed by equation (7.55) as follows:

At

$$z = z_f, \qquad n = n_0 \tag{7.53}$$

and

$$z = L, \qquad n = \frac{\Gamma}{U}\left(\frac{dn}{dz}\right) \tag{7.54}$$

$$n = n_0 \exp\left[\frac{U}{\Gamma}(z - z_f)\right] \tag{7.55}$$

The profiles can be obtained by plotting n versus z. The parameter Γ can be calculated from the slope of the profile gradient. z_f is the distance from the top of the column where bubbles are started to form by gas entrainment because of plunging of the liquid jet. The number density of bubbles that are formed at the formation zone at $z = z_f$ can be calculated as

$$n_{z=z_f} = n_0 = \frac{\text{Gas volume}}{\text{Bubble volume} \times \text{Bed volume}} = \frac{6\varepsilon_g}{\pi d_{b,e}^3} \tag{7.56}$$

Specific Interfacial Area

Specific Interfacial Area in Inverse Bubbly Flow

The interfacial area is one of the most important parameters for gas–liquid reactor design. Generally, the interfacial area depends on the unit's geometrical size, the operating parameters, and the physical and chemical properties of the liquid. Experimental methods for determining the interfacial area in bubble columns can be divided into two groups, one being physical, the other chemical. The optical method is also used to determine the interfacial area. The optical method involves direct refraction, diffraction, or reflection measurements on the actual bubble surface, whereby the reflected light flux penetrates through the gas–liquid dispersion (Deckwer, 1992). Photography is one of the mostly used method for determining d_{32}. Transmission methods are only suitable up to certain values for the product of specific interfacial area (a) and optical path length (Calderbank, 1958; Landau et al., 1977). Calderbank (1958) used reflected light measurements to determine the interfacial area. Physical measuring processes give local values of interfacial area, but the whole reactor volume can be assessed by using chemical methods (Deckwer, 1992). The specific interfacial area (a) can be defined by:

$$a = \frac{\text{Surface area of bubble}}{\text{Volume of dispersed bubble bed}} = \frac{S_b}{V_b/\varepsilon_g} = \frac{\pi d_{b,32}^2 \varepsilon_g}{(1/6)\pi d_{b,32}^3} = \frac{6\varepsilon_g}{d_{b,32}} \tag{7.57}$$

A review of the literature shows that considerable work has been carried out on the measurement of interfacial area in different types of equipment. The most widely used methods to measure or estimate the specific interfacial area are light attenuation, high-speed photography, dynamic gas disengagement process, and the chemical method. In the light attenuation technique proposed by Calderbank (1958), a collimated beam of light is passed through the gas–liquid dispersion. The intensity of the emerging beam is measured by a photocell, photo-multiplier tube, or other photosensitive devices. This intensity depends on the number of interphases the beam has to cross and hence is a measure of the interfacial area in the system (Vermeulen et al., 1955; Calderbank (1958)). The direct applicability of this method is limited to situations in which the effect of multiple and forward light scattering are negligible.

In the photographic method, a high-speed photograph of the system is analyzed for determining the mean bubble size of the dispersed phase. For a quantitative estimate of the BSD, it is necessary to ensure that only the bubbles in the focal plane are measured. This requirement is difficult to satisfy, especially when the bubble size is large. In case of two-phase flow in which bubble concentrations are high, the image received by the camera is the result of a series of complex optical interactions, which are difficult to analyze. The dynamic gas disengagement technique requires an accurate measurement of the rate at which the level of the gas–liquid dispersion drops when gas flow to the bubble column is shut off. The BSD, and therefore the gas–liquid interfacial area, can also be determined using appropriate correlations that relate bubble sizes to bubble rise velocities (Patel et al., 1989). The main advantage of the technique lies in its simplicity and its ability to provide a wide range of information. However, there are some limitations resulting from deficiencies in the data collection and from its dependence on correlations relating terminal bubble rise velocity to bubble size.

The chemical method of interfacial area determination is based on the theory of gas absorption followed by a chemical reaction. It can be calculated from the physicochemical rate of the reaction. Among all chemical methods, the sulfite oxidation and CO_2 absorption into caustic solutions have found the most widespread application (Schumpe and Deckwer, 1982). Because in the inverse bubbly flow system, the intense mixing between the phases generates a very high interfacial area and turbulence, the first three methods may not give accurate results; therefore, the chemical method can be used for better result. The chemical method is described in Chapter 8.

The Sauter mean bubble diameters ($d_{b,32}$) in three different axial positions were investigated with different nozzle diameters by Majumder et al. (2006) in an ejector-induced inverse-flow bubble column. They observed that the bubble size decreases with increasing superficial liquid velocity in each location of the column. This is due to bubble break-up with increasing momentum transfer of liquid jet with increasing liquid flow rate. Typical profiles of the Sauter mean diameter ($d_{b,32}$) as a function of superficial liquid velocity for different nozzle diameters and for different locations are shown in Figures 7.11 and 7.12,

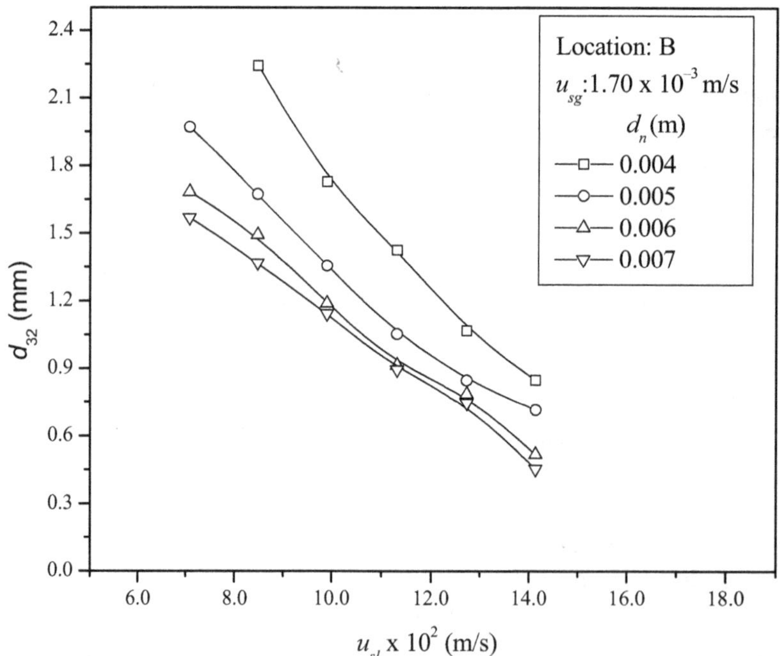

FIGURE 7.11 Effect of superficial liquid velocity on Sauter mean bubble diameter for different nozzle diameters.

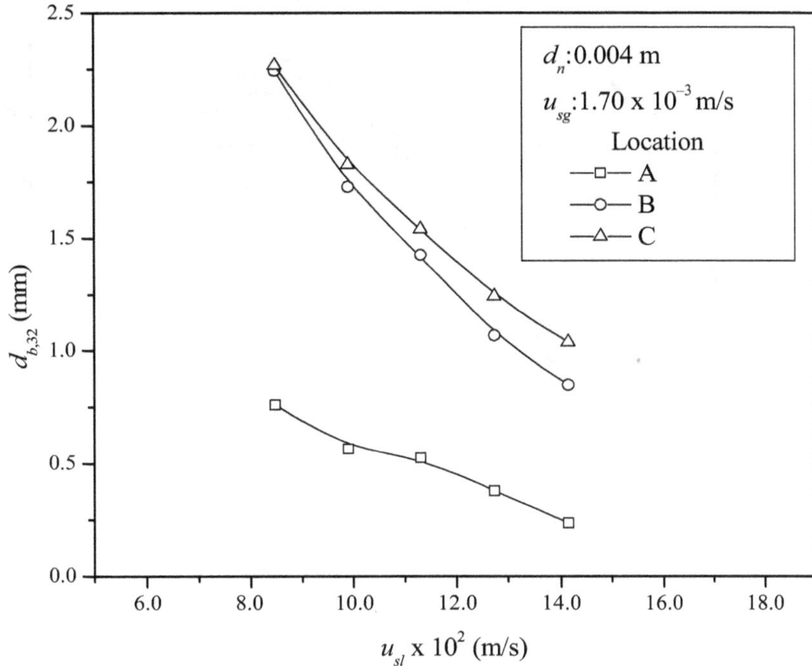

FIGURE 7.12 Effect of superficial liquid velocity on Sauter mean bubble diameter in different locations.

respectively. The bubble size decreases with increasing superficial liquid velocity in each location. This is due to bubble break-up with increasing momentum transfer of liquid jet with increasing liquid flow rate. As shown in Figure 7.12, there is no significant variation of bubble size in locations B and C, but the bubble in location A is much smaller than in location B and C. The bubble size decreases with increasing nozzle diameter as shown in Figure 7.11. This is expected because the energy input to the mixing zone is inversely proportional to the square of the nozzle diameter ($E \propto V_j^2$ or $1/d_n^2$). At a smaller nozzle diameter, relatively smaller bubbles are formed because the energy input is higher. It was observed that as the nozzle diameter decreases, the bubble size increases at locations B and C. This is due to coalescence of the smaller bubbles of location A, which move up to location B and C because of their buoyancy effect. Also, the rate of coalescence increases as the gas flow rate increases. As the gas flow rate increases, the gas hold-up increases, which enhances bubble–bubble interactions. This results in an increase in the probability of coalescence with an increase in the superficial gas velocity. For this reason, the BSD changes with the superficial gas velocity (see Figure 7.4). Also, as the superficial gas velocity increases, the Sauter mean bubble diameter increases (Figure 7.13). This may be due to less liquid resistance inside the column. The bubble size increases with the decrease of liquid resistance inside the column. The liquid resistance decreases as the superficial gas velocity increases because of an increase of gas hold-up with an increase in superficial gas velocity. This may be due to less liquid resistance inside the column. The bubble size increases with the decrease of liquid resistance inside the

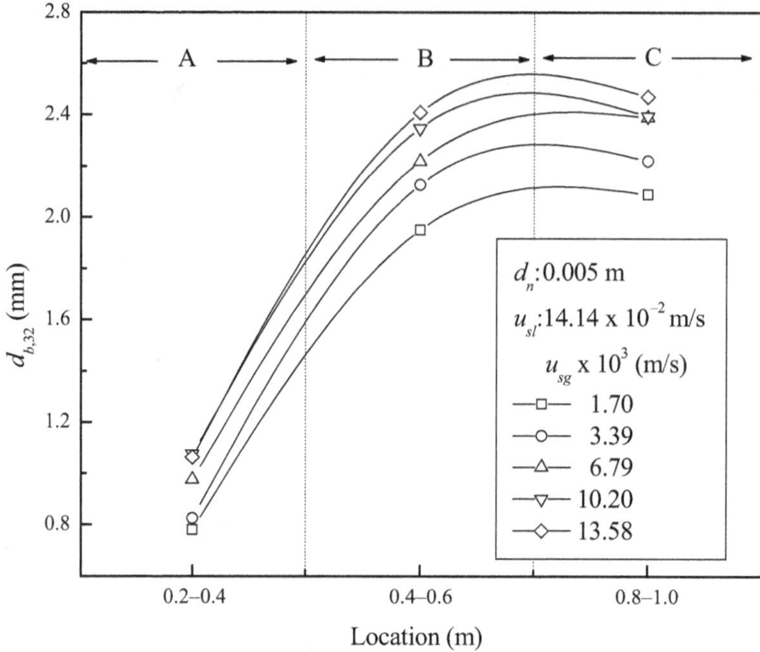

FIGURE 7.13 Variation of Sauter mean bubble diameter with axial location.

column. The liquid resistance decreases as the superficial gas velocity increases because of the increase of gas hold-up with the increase in superficial gas velocity. From the experimental results, it is seen that the specific interfacial area varies with the operating variables as well as the location of the column. The specific interfacial area is related to the Sauter mean bubble diameter and the gas hold-up. As discussed earlier, the Sauter mean bubble diameter decreases, and the gas hold-up increases with the increase of both the superficial liquid and gas velocities. A decrease of the Sauter mean bubble diameter and an increase of gas hold-up would result in an increase of the specific interfacial area with liquid velocity (Figures 7.14 and 7.15). Also, the specific interfacial area varies with the axial location.

This is because of variation of the Sauter mean bubble diameter and gas hold-up with the axial location. The Sauter mean bubble diameter varies with an axial location because of the coalescence effect, but the variation of gas hold-up is attributable to variation of bubble number flux as discussed earlier.

Correlation Model for Interfacial Area

Majumder et al. (2006) also developed a correlation to predict the specific interfacial area within a same range of operating conditions in an inverse bubble column, which is given by

$$ad_n = 0.571 \left(\frac{z}{d_n} \right)^{-0.358} \left(\frac{u_{sg}}{u_{sl}} \right)^{1.898} \left(\frac{d_n g}{u_{sg}^2} \right)^{0.997} \left(\frac{d_n u_{sl} \rho_l}{\mu_l} \right)^{-0.089} \left(\frac{d_n \rho_l \sigma_l}{\mu_l^2} \right)^{0.202} \tag{7.58}$$

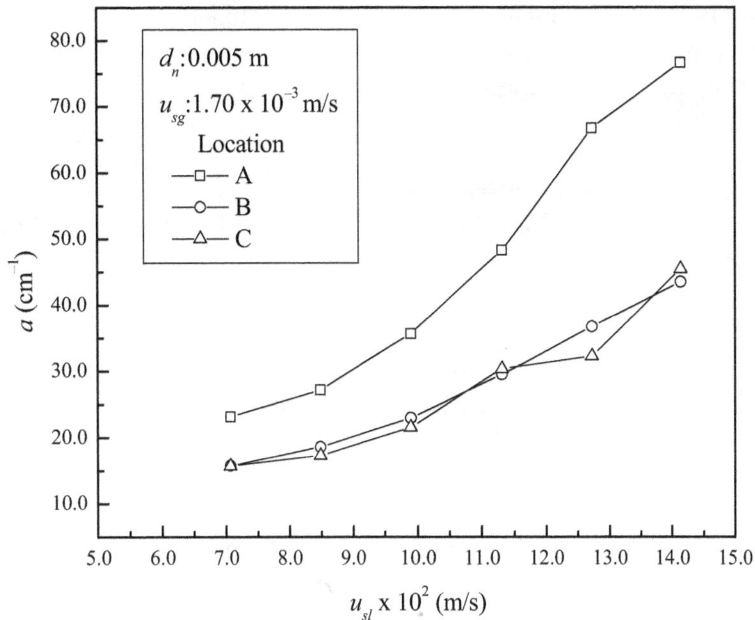

FIGURE 7.14 Variation of specific interfacial area with superficial liquid velocity for different axial locations.

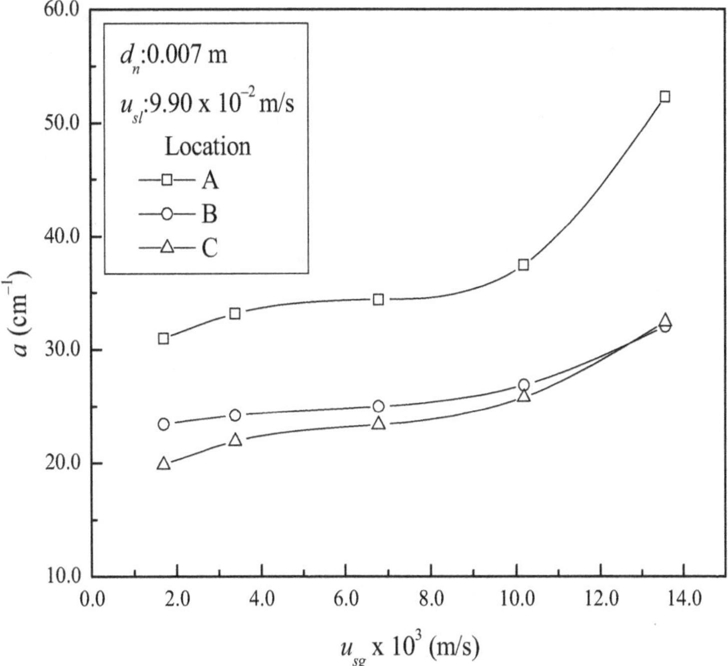

FIGURE 7.15 Variation of specific interfacial area with superficial gas velocity for different nozzle diameters.

Nagel et al. (1973) carried out extensive studies regarding the interfacial area in various gas–liquid reactors. They correlated their results with Kolmogoroff's isotropic turbulence theory, obtaining the following theoretical proportionality:

$$a \propto \dot{E}_v^{0.4} \varepsilon_g^\beta \tag{7.59}$$

in which $\dot{E}_v$ is the rate of energy dissipation per unit of dispersion volume (kWm^{-3}) and ε_g is the local gas hold-up. Equation (7.59) is valid only under isotropic conditions. In practice, the following general relation formula

$$a \propto \dot{E}_v^\alpha \varepsilon_g^\beta \tag{7.60}$$

has provided a good basis for assessing measured data from various types of equipment (Deckwer, 1992). Kastenek (1976) has also developed similar formula for volume based value of interfacial area.

$$a = 0.09942 \times 10^3 \, \dot{E}_v^{0.8322} \varepsilon_g^{0.659} \tag{7.61}$$

A comparative picture of specific interfacial area and Sauter mean bubble diameter is shown in Tables 7.6 and 7.7.

They observed that the value of specific interfacial area in jet-induced inverse bubbly flow is about an order of magnitude higher than the other systems. Mandal et al. (2003) estimated the specific interfacial area by the chemical method. They obtained values of interfacial area much lower than the present results of Majumder et al. (2006). The discrepancy is not unusual because the physical method always provides a higher interfacial area, as reported by Schumpe and Deckwer (1982). Also, Mandal et al. (2003) estimated the specific interfacial area with an air–Carboxymethyl Cellulose (CMC) system, which was strong coalescing medium. They obtained a Sauter mean bubble diameter of about 4.5 mm. But with air–water systems having relatively less coalescing medium, the Sauter mean bubble diameter was obtained in the range of 0.116 to 2.31 mm. Because the specific interfacial area is related to the bubble size, d_{vs}, a smaller bubble size would result in a larger interfacial area. Gas hold-up also has a strong effect on the interfacial area. In jet-induced inverse bubbly systems, a relatively small bubble size and higher gas hold-up resulted in an higher interfacial area compared with other works. A comparative picture of the range of interfacial area with the specific power inputs for different gas–liquid contactors are presented in Table 7.8. From the energy point of view, inverse bubbly flow offers better performance than the other systems as far as specific interfacial area and mass transfer operations are concerned.

Interfacial Area Transport Modeling

Interfacial Area Transport Equation

Interfacial area concentration can be defined as the total surface area of dispersed fluid particles per unit mixture volume. As discussed earlier, there are certain phase interaction terms in the governing equation to characterize the interfacial property transfer. These terms all contain the parameter interfacial area concentration. Therefore, it is required

Table 7.6 Comparison of Sauter Mean Bubble Diameter Between Upflow and Inverse Bubble Flow in Bubble Columns

Investigators	Dimensions (m)	Flow Direction	Phases	Superficial Gas Velocity (m/s)	Superficial Liquid Velocity (m/s)	Sauter Mean (mm)	Techniques
Fukuma et al. (1987)	$d_c = 0.15$, $Z = 1.2$, 1.7, and 3.2	Upflow	Air–water	0.0086-0.115	0.005-0.05	3.3-19.6	Probe
Patel et al. (1989)	$d_c = 0.05$ and 0.23, $Z = 3.0$	Upflow	Air–water	0.01-0.09	–	2.5-7.2	Dynamic gas disengagement
Bensler (1990)	0.040×0.040, $Z = 0.675$	Upflow	Air–water	0.035-0.253	0.001-0.059	3.05-4.83	Ultrasonic
Wolff et al. (1990)	$d_c = 0.3$, 0.2; $Z = 11.0$, 5.0	Upflow	Air–water	0.01-0.10	–	4.3-5.9	Probe
Wilkinson et al. (1994)	$d_c = 0.15$, $Z = 1.5$	Upflow	Nitrogen–water	0.02-0.18	–	2.81-7.56	Chemical
Lage and Espósito (1999)	$d_c = 0.072$, $Z = 0.5$-0.8	Upflow	Air–isopropanol	0.0065-0.0702	–	3.2-5.5	Photograph
Pohorecki et al. (1999)	$d_c = 0.304$, $Z = 3.99$	Upflow	Nitrogen–water	0.002-0.020	0.0014	6.7	Photograph
Kashinsky and Randin (1999)	$d_c = 0.0423$, $Z = 4.8$	Inverse flow	Air–water	0.01-0.095	0.5-1.25	0.96-1.73	Probe
Hibiki et al. (2001)	$d_c = 50.8$, $Z = 2.74$	Inverse flow	Air–water	0.0358-4.87	0.491-5.00	2.6-5.8	Probe
Ishii et al. (2004)	$d_c = 0.025$, 0.051, $Z = 2.74$	Inverse flow	Air–water	0.004-1.97	1.25-4.97	2.1-3.6	LDA, probe
Majumder et al. (2006)	$d_c = 0.05$, $Z = 1.65$	Inverse flow	Air–water	0.0017-0.0135	0.07-0.14	0.116-2.31	Photograph

LDA, laser Doppler anemometry.

Table 7.7 Comparison of Specific Interfacial Area Between Published and Present Results in Bubble Columns

Investigator	Geometry Dimensions (m)	Flow Direction	Phases	Length to Diameter Ratio, (-)	Superficial Gas Velocity (m/s)	Superficial Liquid Velocity (m/s)	Specific Interfacial Area (m^{-1})	Techniques
Liu (1989)	$d_c = 0.0381$	Upflow	Air–water	83.0	0.018-0.54	0.50-5.0	53.2-338	Probe
Bensler (1990)	0.040×0.040	Upflow	Air–water	65.0	0.035-0.253	0.503-3.00	67.0-391.6	Photograph
Kalkach-Navarro (1992)	$d_c = 0.0381$	Upflow	Air–water	50.0	0.056-0.332	0.30-1.25	93.4-298	Probe
Grossetete (1995)	$d_c = 0.0381$	Upflow	Air–water	155.0	0.0895-0.181	0.877-1.75	75.1-145	Probe
Hibiki et al. (1998)	$d_c = 0.0508$	Upflow	Air–water	62.0	0.0178-0.0936	0.60-1.30	25.8-135	Probe
Hibiki et al. (2001)	$d_c = 0.0508$	Upflow and inverse flow	Air–water	54.0	0.0358-4.87	0.491-5.00	101-1083	Probe
Takamasa et al. (2003)	$d_c = 0.009$	Upflow	Air–water	91.0	0.013-0.052	0.580-1.00	31.6-150	Photograph
Ishii et al. (2004)	$d_c = 0.0254$	Inverse flow	Air–water	52.5	0.004-1.97	1.25-4.97	10.0-538.59	LDA, Probe
Mandal et al. (2003)	$d_c = 0.0516$	Inverse flow	Air–CMC	36.8	0.001-0.09	0.05-0.11	450-800	Chemical
Majumder et al. (2006)	$d_c = 0.050$	Inverse flow	Air–water	33.0	0.0017-0.0135	0.07-0.14	1294.14-9181.04	Photograph

CMC, Carboxymethyl Cellulose; *LDA*, laser Doppler anemometry.

Table 7.8   Comparison of Interfacial Area for Different Gas–Liquid Contactors

Author	Type of Reactor	Superficial Liquid Velocity, Vsl, m/s	Superficial Gas Velocity, Vsg, m/s	Gas Hold-up, ε_g (-)	Specific Interfacial Area a (1/m)	Power Input E/V_R (Kw/m³)
Stein and Schafer (1984)	Inverse flow liquid jet system	—	0-0.0133	—	30-200	0.50-5.0
Ohkawa et al. (1987)	Inverse flow liquid jet bubble column	0.30-0.60	0.001-0.02	0.10-0.40	250-450	0.02-0.10
Dutta and Raghavan (1987)	Jet loop reactor	0.001-0.005	0.001-0.006	0.02-0.15	30-500	0.10-10.0
Bin (1993)	Liquid jet absorber	0.002-0.014	—	—	500-1000	0.03-4.50
Havelka et al. (2000)	Upflow ejector loop reactor	0.017-0.025	0.02-0.09	0.15-0.35	180-450	0.10-5.0
Meikap (2000)	Multistage bubble column	0.003-0.010	0.11-0.20	0.21-0.65	250-600	0.20-0.45
Ide et al. (2001)	Plunging jet absorber	0.028-0.071	—	0.09-0.35	80-800	0.40-2.0
Majumder et al. (2006)	Inverse flow ejector-induced column	0.0017-0.0135	0.07-0.14	0.38-0.61	1294.14-9181.04	0.278-20.87

to obtain the transport equation for the interfacial area concentration. The interfacial area concentration can be obtained from the number density of bubbles by the following relation.

$$n=\frac{\varepsilon_g}{(\pi/6)d_e^3}=\frac{\varepsilon_g}{\frac{\pi}{6}\left(\frac{d_e}{d_{32}}\right)^3 d_{32}^3}=\frac{\varepsilon_g}{\frac{\pi}{6}\left(\frac{d_e}{d_{32}}\right)^3\left(\frac{6\varepsilon_g}{a}\right)^3}=\frac{a^3}{36\pi\left(\frac{d_e}{d_{32}}\right)^3\varepsilon_g^2}=\psi\left(\frac{a^3}{\varepsilon_g^2}\right) \tag{7.62}$$

$$\psi=\frac{1}{36\pi}\left(\frac{d_{b,32}}{d_{b,e}}\right)^3 \tag{7.63}$$

The factor ψ depends on the shape of the bubbles. For spherical bubbles, it is $\psi=1/(36\pi)$. Substitution of equation (7.62) in equation (7.49) yields

$$\frac{\partial a}{\partial t}+\left(U-\frac{2\Gamma}{a}\frac{\partial a}{\partial z}\right)\left(\frac{\partial a}{\partial z}\right)=\Gamma\frac{\partial^2 a}{\partial z^2} \tag{7.64}$$

Equation (7.64) represents the transport equation for interfacial area concentration. From equation (7.64), it is obvious that the interfacial area concentration is a strong function of parameter Γ (defined in equation [7.51]). As discussed in earlier, to solve equation (7.64), the bubble interaction mechanisms such as coalescence and breakage should be modeled by treating different bubble groups. The possible combinations of interactions are shown in Figures 7.16 and 7.17.

A certain critical bubble volume V_c corresponding to the maximum distorted bubble limit is defined to separate the bubble groups. The parameter Γ, which is a measure of the degree of mixing and dispersion, can be analyzed based on the breakage coalescence rate and hydrodynamic parameters. The parameter D_b is called the *dispersion coefficient of bubble motion,* which is described in Chapter 6. It is observed that the dispersion coefficient of bubble motion increases with the superficial velocity. With increasing superficial liquid velocity, the rate of liquid circulation increases, which causes an increase in the dispersion coefficient of bubble motion. The exchange factor E depends on the intensity of

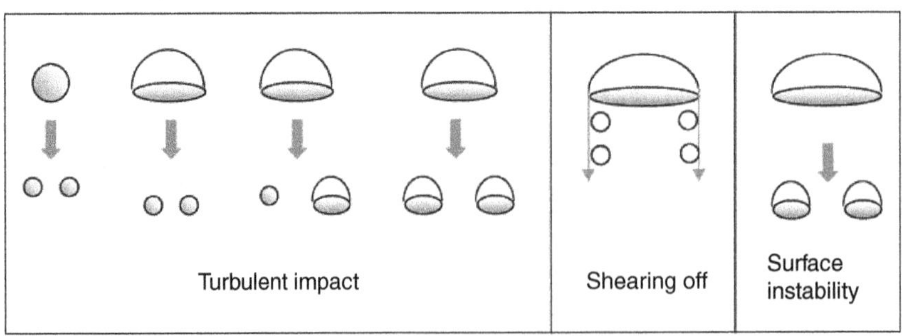

FIGURE 7.16 Break-up mechanisms.

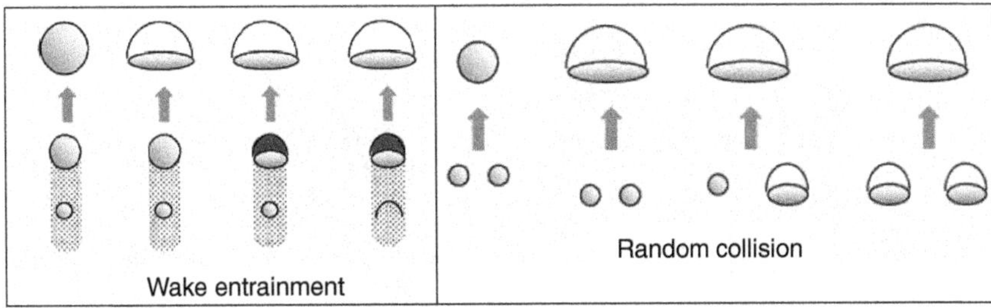

FIGURE 7.17 Coalescence mechanism.

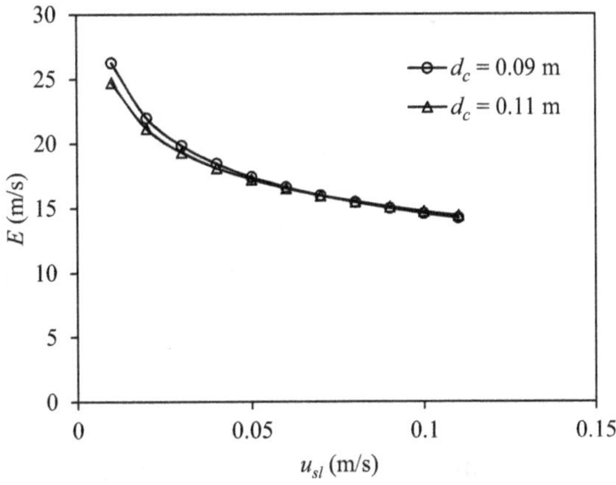

FIGURE 7.18 Variation of dispersion velocity characteristic factor with U_{sl}.

turbulence. With increasing superficial liquid velocity, the degree of turbulence increases, which reduces the exchange factor as shown in Figure 7.18.

The parameter contributes the additive effects of dispersion coefficient of bubble motion and velocity characteristic factor in which the dispersion coefficient plays major role. The trend observed in Figure 7.19 implies that an increasing superficial liquid velocity increases the parameter Γ. As the superficial liquid velocity increases, the degree of mixing increases. The superficial liquid velocity has immense effect on inverse flow system because as the liquid velocity increases, the transfer of momentum by liquid jet in the inverse flow system (Majumder et al., 2006) makes more turbulence of phases. The dispersion coefficient of bubble motion is an indication of the dispersion mode of mixing, and the exchange factor is an indication of the convective mode of mixing. Figure 7.20 shows how the mixing parameter changes with the dispersion coefficient of bubble motion. The parameter Γ also depends on the rate of change of number density. There are two ways in which rate of change of number density can change, that is, through breakage and coalescence. Bubble breakage occurs because of an imbalance in interfacial tension. The dimensionless number that accounts for breakage is the Weber number. It is the ratio of inertial forces to surface tension (Wilkinson and van Dierendonck, 1990). The breakage starts at the critical

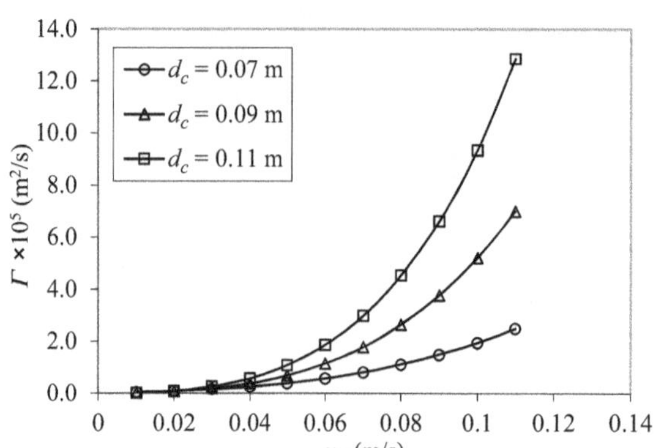

FIGURE 7.19 Variation of parameter Γ with superficial liquid velocity.

Weber number. At a low liquid flow rate with a small void fraction, bubble breakage can be neglected because of a very small value of the Weber number compared with the critical value. The correlations that are used to calculate the rate of change of number density attributable to coalescence and breakage as per different mechanisms are given in Table 7.9.

Interpretation on Parameter on Transport Model
EFFECT OF COALESCENCE RATE ON THE PHASE MIXING PARAMETER
Coalescence due to random collision is undesirable phenomena in a bubble column because it reduces the interfacial area and hence the mass transfer rate. The coalescence process is considered to be a two-stage process after the collision of the fluid particles that includes (1) the drainage of liquid films between the two contacting surfaces of the dispersed phase and (2) the rupture of the film. The coalescence process can be analyzed by examining bubble collision events and the probability of collisions resulting in coalescence. The collision should happen before the coalescence takes place. The coalescence

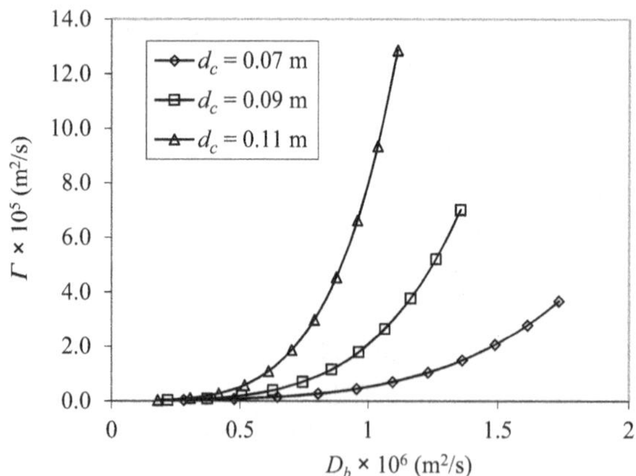

FIGURE 7.20 Variation of parameter Γ with dispersion coefficient of bubble motion.

Table 7.9   Correlations Used to Calculate Break-up or Coalescence Rate (Wu et al., 1998)

Mechanism	Correlation
Random collision	$R_{RC} = C_{RC}(u_t n^2 d_{32}^2)\left\{\dfrac{1}{\varepsilon_{g,max}^{1/3}(\varepsilon_{g,max}^{1/3} - \varepsilon_g^{1/3})}\right\}\left\{1 - \exp\left(-C\dfrac{\varepsilon_{g,max}^{1/3}\varepsilon_g^{1/3}}{\varepsilon_{g,max}^{1/3} - \varepsilon_g^{1/3}}\right)\right\}$, C and C_{RC} are adjustable parameter
Wake entrainment	$R_{WE} = C_{WE}d_{32}^2 u_r n^2;\ \ C_{WE} = \dfrac{1}{8}F\left(\dfrac{d_{32}}{L_W}\right)$
Turbulent impact	$R_{TI} = C_{TI}\exp\left(-\dfrac{We_c}{We}\right)n\dfrac{u_t}{d_{32}}\left(1 - \dfrac{We_c}{We}\right)^{1/2}$; $We > We_c$; C_{TI} is the adjustable parameter

rate can be formulated by modeling the collision rate and coalescence efficiency separately (Prince and Blanch, 1990). Prince and Blanch (1990) developed phenomenological models to determine the rates of bubble coalescence and break-up. Three different collision mechanisms are considered by them, namely, collisions due to the random motion of the bubbles driven by turbulence, collision due to different rise velocity caused by different buoyancy forces for different sizes of bubbles, and collision due to laminar shear for bubbles located in different regions of the liquid velocity field. Then the coalescence efficiency is determined by evaluating the time required for coalescence and the contact time of two bubbles in turbulent flow. Wu et al. (1998) proposed a mechanistic model for the bubble coalescence due to random collision driven by turbulent eddy in bubbly flow and modeled the bubble coalescence rate due to the wake entrainment phenomenon in the course of developing the one group. The two colliding bubbles were assumed to be the same size.

In addition to the coalescence due to random motion of fluid particles driven by turbulence in the continuous phase, the collision caused by acceleration of following bubbles in the wake of a preceding bubble is considered another major contribution to fluid particle coalescence. When a following bubble enters the wake region of another bubble (preceding bubble), the following bubble will accelerate and collide with the preceding bubble, resulting in coalescence (Ishii et al., 2004). From Figure 7.21, it is observed that mixing increases with coalescence due to random collision. Γ increases with R_{RC}, but the order of magnitude is moderate. This occurs generally in the heterogeneous regime. The turbulence causes mixing and circulating eddies. Wake entrainment occurs at the extremities of the heterogeneous flow regime. A lower magnitude of R_{WE} does not influence Γ much. As the magnitude of R_{WE} increases, Γ increases as shown in Figure 7.22. This is due to dispersion caused by the wake-entraining phase.

EFFECT OF THE BREAK-UP RATE ON THE PHASE MIXING PARAMETER

The break-up phenomenon in two-phase mixtures has also gained significant attention because of its great importance in view mass transfer. This is because the break-up of the dispersed phase greatly increases the available surface area. Similar to the coalescence mechanisms, there can be various driving mechanisms for fluid particle break-up. One of the major break-up mechanisms in two-phase systems is known as the turbulent

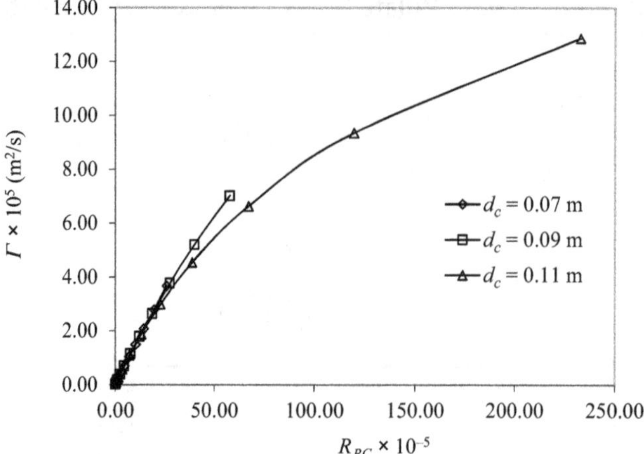

FIGURE 7.21 Variation of parameter Γ with rate of coalescence by random collision.

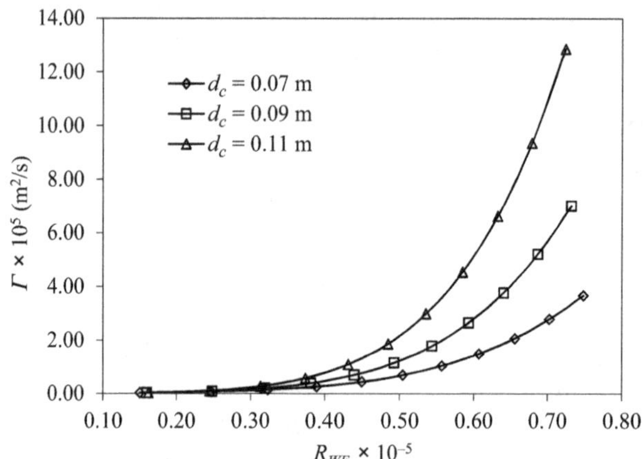

FIGURE 7.22 Variation of parameter Γ with rate of coalescence by wake entrainment.

impact of eddies on the dispersed phase. When the inertia of the turbulent impact overcomes the surface tension at the interface, the fluid particle will disintegrate (Prince and Blanch, 1990). Prince and Blanch (1990) formulated the total break-up rate for bubbles by incorporating the contribution to break-up from eddies of various sizes through summation over various sizes of bubbles and eddies. As the size of the bubbles increases, the transport mechanisms become further complicated by additional phenomena such as shearing off and surface instability. One of the interesting mechanisms in large bubble break-up is the shearing off of small bubbles at the rim of cap, slug, and churn-turbulent bubbles. The "skirt" is formed when the disruptive viscous force pulling at the rim of the large bubbles cannot be balanced by the cohesive surface tension forces. The second phenomenon (i.e., large bubble break-up due to the surface instability) becomes important

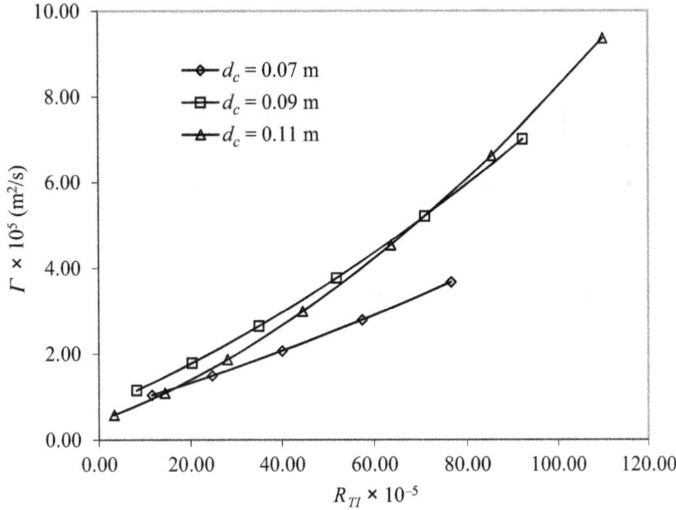

FIGURE 7.23 Variation of parameter Γ with rate of break-up by turbulent impact.

when a bubble grows bigger and reaches a certain limit such that the bubble surface can no longer be steadily sustained because of the interfacial wave instability developed at the gas–liquid interface. How the rate of break-up by turbulent impact influence on the mixing parameter Γ in an ejector-induced inverse bubbly flow system is shown in Figure 7.23.

The relation between the Γ parameter and different types of breakage or coalescence mechanisms can be represented by the general correlation $y = \alpha x^\beta$, where y denotes the Γ parameter value. The coefficients are shown in Table 7.10.

EFFECT OF GEOMETRIC PARAMETERS ON THE PHASE MIXING PARAMETER

For any bubble column model, the geometric parameters have great impact on the hydrodynamics of the bubble column. In the case of an ejector-induced inverse flow bubble column, the nozzle diameter and column diameter are of major importance. Moreover, the theoretically defined λ (the ratio of an equivalent diameter of cross-section through which bubbles move upward to the column diameter) has significant effect on the Γ parameter. The intensity of circulation depends on the column diameter. As the column diameter increases, the interaction and formation of large circulation will increase, which may cause the decrease in velocity characteristic factor. The larger the diameter of column, the greater the liquid circulation and hence the greater the mixing. Hence, the mixing parameter Γ

Table 7.10   Values of Coefficients Used in Equation $y = \alpha x^\beta$

Mode of Breakage or Coalescence	α	β	R^2
Turbulent impact	1.0 E-09	0.6589	0.9784
Random collision	3.0 E-10	0.7789	0.9799
Wake entrainment	3.0 E-20	3.0629	0.9678

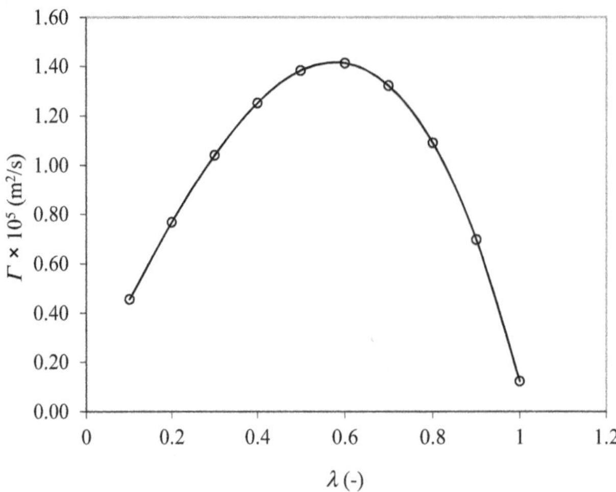

FIGURE 7.24 Variation of parameter Γ with area fraction.

increases with an increase in the column diameter. Because λ represents the ratio of the cross-sectional area available for upfow to inverse flow, maximum values of λ indicate more cross-sectional area available for upflow and less for inverse flow. An intermediate value of λ indicates an approximate equal area for both types of flow. From Figure 7.24, it can be observed that at lower values of λ viz. 0.1 and higher 0.9, the dispersion is low; this is because of incomplete mixing. But at $\lambda = 0.55$, the higher values of dispersion are obtained; this is because complete mixing occurs because of intimate contact of both directional flows. If $\lambda = 1$, the exchange of the bubble will be negligible. The functional representation can be expressed as

$$\Gamma \times 10^5 = 0.123 + 3.36\lambda + 6.0 \times 10^{-13}\lambda^2 - 3.36\lambda^3 \tag{7.65}$$

EFFECT OF THE PHASE MIXING PARAMETER ON THE INTERFACIAL AREA CONCENTRATION

In view of the mass transfer phenomena in bubble columns, the interfacial area concentration is a very important parameter. The larger the interfacial area, the greater the mass transfer rate. In general, spherical bubbles possess more interfacial area than distorted ones. Furthermore, it also depends on the size of the bubble. Finer bubbles, as occur in the homogeneous regime, have more interfacial area than larger ones. From equation (7.64), it can be seen that the rate of change of the interfacial area concentration is dependent on parameter Γ. The interfacial area concentration increases with an increase in the mixing parameter as shown in Figure 7.25. As the superficial liquid velocity is raised, the dispersion coefficient increase, and increasing the mixing and consequently the interfacial area. The following correlation can be obtained by relating the interfacial area concentration and parameter Γ.

$$a = 134.38\ln(\Gamma) + 2332.6 \tag{7.66}$$

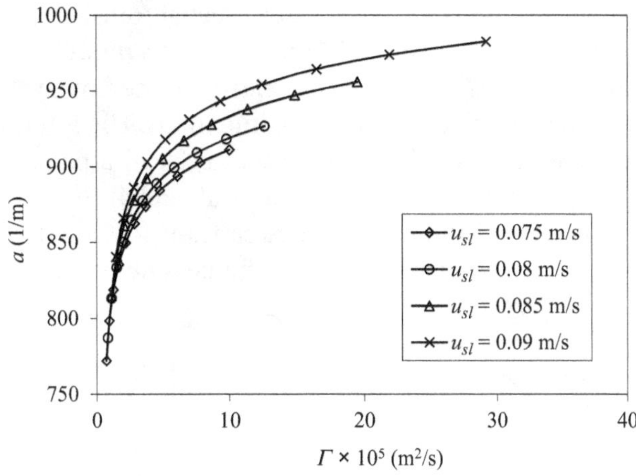

FIGURE 7.25 Variation of interfacial area concentration with Γ.

The dependence of a_i on other factors is summarized in Table 7.11.

Interfacial Area Transport Model by Ishii et al. (2004)

The transport of interfacial area in the inverse bubbly flow system was developed by Ishii et al. (2004) based on the interfacial area transport equation developed by themselves for upward flow with certain modifications in bubble interaction terms. The local profiles of different hydrodynamic parameters as well as their axial development is used to describe the nature of the interfacial structures and the bubble interaction mechanisms occurring in the flow. As per the two-fluid model for transport equation, each phase is considered separately for conservation equations (Ishii, 1975). The conservation equations for mass, momentum, and energy can be written separately for each phase by its proper averaging. Because the averaged macroscopic fields for each phase are not independent of each other, these two sets of the conservation equations are coupled through the interfacial interaction terms (Ishii et al., 2004). The phase interaction terms can be expressed in terms of the interfacial area concentration, a, and the corresponding driving

Table 7.11 Co-relation for Interfacial Area Concentration for $d_c = 0.07$ m Based on Bubble Dispersion, Velocity Characteristic Factor, and Source and Sink Terms as: $a = 134.38\ln(\alpha P^\beta) + 2332.6$

Parameter (P)	α	β
D_b	9.0×10^{10}	2.6965
I	43605.0	−7.9785
R_{TI}	1.0×10^{-9}	0.6589
R_{RC}	3.0×10^{-10}	0.7789
R_{WE}	3.0×10^{-20}	3.0629

forces (Kocamustafaogullari and Ishii, 1995) as Interfacial transfer term = Interfacial area concentration $\times$ Driving force. The interfacial area transport model is a dynamic approach for predicting the interfacial area concentration in two-phase flow systems, as opposed to the static, algebraic models based on the flow regime maps. The interfacial area transport equation dynamically predicts the change of the interfacial area concentration along a flow field from given boundary conditions (Ishii et al., 2004). In this regard, it is worth mentioning that foundations of the interfacial area transport equation were first established by Kocamustafaogullari and Ishii (1995). The general form of the interfacial area transport equation is given by

$$\frac{\partial a}{\partial t} + \nabla \cdot (a\vec{u}_i) = \frac{2a}{3\varepsilon_g}\left(\frac{\partial \varepsilon_g}{\partial t} + \nabla \cdot (\varepsilon_g \vec{u}_g) - \eta_{ph}\right) + \sum_j \varphi_j + \varphi_{ph} \tag{7.67}$$

The first term on the right-hand side of the equation represents the source in the interfacial area concentration due to the change in the volume of the dispersed phase. In this term, η_{ph} is the net rate of volume generated by nucleation and collapse of bubble due to condensation (if any) per unit mixture volume. It may be noted that $\vec{u}_i$ is the velocity of the interface, and $\vec{u}_g$ is the gas phase velocity. For bubbly flow, these two velocities are approximately the same. The terms φ_j and φ_{ph} represent the source and sink terms due to the bubble interaction mechanisms and phase change, respectively. Constitutive relations for the source and sink terms appearing on the right-hand side of the interfacial area transport equation can be achieved by mechanistically modeling the bubble interaction mechanisms in a two-phase flow. In this regard, the one-group interfacial area transport equation (Wu et al., 1998) for adiabatic air–water bubbly flows can be followed. In the one-group transport equation, three significant bubble interaction mechanisms for group-one bubbles were modeled, namely, bubble break-up due to the impact of turbulent eddies in the continuous phase (TI), bubble coalescence due to random collision of the turbulent eddies in the continuous phase (RC), and bubble coalescence due wake entrainment of following bubble into preceding bubble (WE). Finally, the one-group interfacial area transport equation for air–water two-phase flows without phase change was given by Wu et al. (1998), in one-dimensional, steady-state form as

$$\frac{d(\bar{a}\bar{u}_g)}{dz} = A + B - C - D \tag{7.68}$$

where

$$A = \left(\frac{2\bar{a}\bar{u}_g}{3p}\right)\left(\frac{dp}{dz}\right) \tag{7.69}$$

$$B = \frac{C_{TI}}{18}\left(\frac{\bar{a}^2}{\bar{\varepsilon}_g}\bar{u}_t\right)\left(1 - \frac{We_{cr}}{We}\right)^2 \exp\left(-\frac{We_{cr}}{We}\right) \tag{7.70}$$

$$C = \frac{C_{RC}}{3\pi}\frac{\bar{a}^2\bar{u}_t}{\bar{\varepsilon}_g^{1/3}(\bar{\varepsilon}_{g,\max}^{1/3} - \bar{\varepsilon}_g^{1/3})}\left(1 - \exp\left(-C\frac{\bar{\varepsilon}_{g,\max}^{1/3}\bar{\varepsilon}_g^{1/3}}{\bar{\varepsilon}_{g,\max}^{1/3} - \bar{\varepsilon}_g^{1/3}}\right)\right) \tag{7.71}$$

$$D = C_{WE} C_d^{1/3} \frac{\bar{a}^2 \bar{u}_t}{3\pi} \tag{7.72}$$

In this equation, the first term on the right-hand side represents the change in the interfacial area concentration due to compressibility of the dispersed phase, and the remaining three terms represent the change in the interfacial area concentration due to the impact of turbulent eddies in the continuous phase (TI), random collision of bubbles driven by the turbulent eddies in the continuous phase (RC), and coalescence of bubbles due to wake entrainment (WE), respectively. These interaction terms contain empirical coefficients (i.e. C_{TI}, C_{RC}, C, and C_{WE}), which have to be found from experimental data. In this equation, p, u_t, We, and C_D are the local pressure, turbulent velocity fluctuation in the liquid phase, particle Weber number, and particle drag coefficient, respectively. The interfacial area concentration along the test section is calculated by solving equation (7.68) from the boundary conditions measured experimentally. Ishii et al. (2004) determined the empirical coefficients in the bubble interaction mechanisms experimentally, and they reported that the coefficient that accounts for the rate of disintegration of the bubbles due to the impact of the turbulent eddies in inverse bubbly flow is 0.4 times of that for the upward flow. For this reason, the different turbulence structure in the inverse flow compared with upward flow can be observed. Further investigation is required to clearly understand this phenomenon. The values of the model coefficients that can be used in equation (7.68) are summarized by Ishii et al. (2004) as shown in Table 7.12. Typical results of contribution of individual mechanisms to the change in interfacial area concentration for 25.4-mm ID test section obtained by Ishii et al. (2004) is shown in Figure 7.26.

Ishii et al. (2004) concluded that it is worth studying the relative contributions of the different mechanisms to the change in the interfacial area concentration under different boundary conditions. Their study helps to understand the sensitivity of the model and that of the individual mechanisms to the different boundary conditions. The changes in pressure and gas velocity also contribute to the change in the interfacial area concentration.

Ishii et al. (2004) reported the following observations from their study on the contribution of the individual mechanisms:

- Pressure increases in the downstream direction, leading to a decrease in the void fraction and hence in the interfacial area concentration.
- The change in the void-weighted gas velocity has a significant impact on the change in the interfacial area concentration. Decreasing void-weighted gas velocity tends to

Table 7.12 Coefficients in Bubble Interaction Mechanisms

Flow Direction	Bubble Break-up by Impact of Turbulent Eddies (TI)	Bubble Coalescence by Wake Entrainment (WE)	Bubble Coalescence by Random Collision (RC)
Upward bubbly flow	$C_{TI} = 0.085$, $We_{cr} = 6.0$	$C_{WE} = 0.002$	$C_{RC} = 0.0041$, $C = 3$, $\varepsilon_{g,max} = 0.75$
Inverse bubbly flow	$C_{TI} = 0.034$, $We_{cr} = 6.0$	$C_{WE} = 0.002$	$C_{RC} = 0.0041$, $C = 3$, $\varepsilon_{g,max} = 0.75$

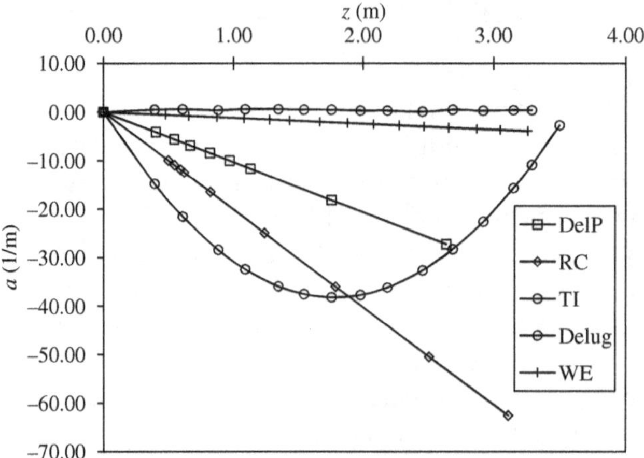

FIGURE 7.26 Contribution of individual mechanisms to the change in interfacial area concentration. *DelP*, due to pressure change; *RC*, Bubble coalescence by random collision; *TI*, Bubble breakup by impact of turbulent eddies; *Delug*, due to the divergence of the gas velocity; *WE*, Bubble coalescence by wake entrainment.

increase the interfacial area concentration. This may be due to the change in the axial derivative of the void-weighted gas velocity because of formation of bigger bubbles, which is near the bubbly-to-slug transition region.

- Low superficial liquid velocity conditions do not show any contribution from the mechanisms of disintegration of bubbles due to turbulent impact. This is due to low liquid phase turbulence in those flow conditions. On the other hand, a high superficial liquid velocity condition (>5.07 m/s) shows a significant contribution from the turbulent impact mechanism.
- Coalescence of bubbles due to random collision may be a dominant mode in the low superficial liquid velocity conditions.
- The bubble coalescence is due to wake entrainment has a small contribution to the change in the interfacial area concentration. However, a comparatively higher void fraction, compared with the two other flow conditions, showed higher contribution from wake entrainment. The larger the bubble diameter, the higher the contribution of WE results.
- In the inverse flow, the bubbles move slower than the surrounding liquid. This can be explained based on the directions of the buoyancy and drag forces acting on the bubbles.
- In most of the flow conditions, the area-averaged void fraction tends to decrease in downstream direction because of an increase in the hydrostatic pressure.
- An increase in the superficial liquid flow rate increases the turbulence and hence the bubble disintegration due to the impact of turbulent eddies. On the other hand, bubble coalescence dominates in the flow conditions with low superficial liquid velocities.

- The one-group interfacial area transport model works reasonably well in most of the bubbly flow conditions. Because turbulence in a two-phase flow system is a combination of bubble-induced and wall-induced turbulence, the center-peak void profile in inverse flow creates a different turbulent structure than that in most of the upward flows. This has an impact on the interaction of turbulent eddies with bubbles.

Nomenclature

a Specific interfacial area, (1/m)

A,B,C Axial location defined along axial length, parameter

C_{vm} Virtual mass coefficient (-)

d_{32} Sauter mean bubble diameter (m)

d_b Bubble diameter (m)

D_b Dispersion coefficient of bubble motion (m²/s)

$d_{b,\,min}$ Minimum bubble diameter (m)

$d_{b,e}$ Equivalent bubble diameter (m)

$d_{b,exp}$ Bubble diameter during experiment (m)

$d_{b,max}$ Maximum bubble diameter (m)

$d_{b,st}$ Bubble diameter at standard condition (m)

d_{bi} Bubble diameter at ith class (m)

d_c Column diameter (m)

d_d Diameter of cross-sectional area through which bubble moves inversely (m)

d_n Nozzle diameter (m)

D_R Ratio of Sauter mean bubble diameter to nozzle diameter, (d_{32}/d_n) (-)

d_u Diameter of cross-sectional area through which bubble moves upward (m)

E Exchange factor (m/s)

E Power input per unit of dispersion volume (Watt)

$\dot{f}(z)$ Rate of collisions per unit volume (1/m³s)

Fr_{gn} Gas Froude number based on nozzle diameter $(d_n g / u_{sg}^2)$ (-)

H_i Distance between gas–liquid mixing height and the point where bubble image taken (m)

n Bubble number density (1/m³)

N_i Number of bubbles in ith class (-)

$\dot{N}$ Axial bubble number flux (1/m²s)

P_c Probability that a collision between two bubbles will result in a coalescence event (-)

P_{exp} Pressure during experiment (N/m²)

P_{st} Pressure at standard condition (N/m²)

Q_g Volumetric flow rate of gas (m³/s)

Q_l Volumetric liquid flow rate (m³/s)

R^2 Correlation coefficient (-)

Re_{ln} Liquid Reynolds number based on nozzle diameter $(d_n u_{sl} \rho_l / \mu_l)$ (-)

Re_m Gas–liquid mixture Reynolds number (-)

S_d Cross-sectional area through which bubble moves inversely (m²)

S_u Cross-sectional area through which bubble moves upward (m²)

Su_{ln} Liquid Suratman number based on nozzle diameter $(d_n \sigma_l \rho_l / \mu_l^2)$ (-)

t_d Drainage time (s)

T_{exp} Temperature during experiment (K)

t_i Interaction time (s)

t_p Persistence time (s)
u_0 Relative velocity of the two bubbles at the onset of deformation (m/s)
u_j Jet velocity (m/s)
u_{sg} Superficial gas velocity (m/s)
u_{sl} Superficial liquid velocity (m/s)
u_{SR} Ratio of superficial gas velocity to superficial liquid velocity (-)
u_t Turbulent velocity (m/s)
u_* Friction velocity (m/s)
u^2 Average value of the squares of the velocity differences (m^2)
V_R Dispersed volume of reactor (m^3)
We Weber number (-)
We_c Critical Weber number (-)
z Axial distance (m)
Z_R Ratio of axial length to nozzle diameter (z/d_n) (-)

Greek Letters

κ Von Karman constant (-)
λ Constant of proportionality (-)
α Parameter (-)
β Parameter (-)
ε Rate of energy dissipation of turbulent kinetic energy per unit volume (kg/ms^3)
ε_g Fractional gas hold-up (-)
ρ_l Density of liquid (kg/m^3)
σ_l Surface tension of liquid (N/m)
μ_l Viscosity of liquid (kg/m.s)

Abbreviations

ASTM American Society for Testing and Materials
BSD Bubble size distribution
DAF Dissolved air flotation
EF Electroflotation
IAF Induced air flotation
RC Random collision
TI Turbulent impact
WE Wake entrainment

References

Ahmed, N., Jameson, G.J., 1985. The effect of bubble size on the rate of flotation of fine particles. Int. J. Miner. Process. 14 (3), 195–215.

Akita, K., Yoshida, F., 1974. Bubble size, interfacial area and liquid-phase mass transfer coefficient in bubble columns. Ind. Eng. Process Des. Dev. 13, 84–91.

Alves, S., Maia, C., Vasconcelos, J., Serralheiro, A., 2002. Bubble size in aerated stirred tanks. Chem. Eng. J. 89, 109–111.

Atkinson, B.W., Gameson, G.J., Nguyen, Anh V., Evans, G.M., Machniewski, Piotr, M., 2003. Bubble breakup and Coalescence in a plunging liquid jet bubble column. Can. J. Chem. Eng. 81, 519–527.

Aw, S.R., Rahim, R.A., Rahiman, M.H.F., Yunus, F.R.M., Fadzil, N.S., Zawahir, M.Z., Jumaah, M.F., Pusppanathan, M.J., Ayub, N.M.N., Wahab, Y.A., Bunyamin, S., 2014. Application study on bubble detection in a metallic bubble column using electrical resistance tomography. Journal Technol. (Sciences & Engineering) 69 (8), 19–25.

Barigou, M., Greaves, M., 1991. A capillary suction probe for bubble size measurement. Meas. Sci. Technol. 2, 318–326.

Batchelor, G.K., 1951. Pressure fluctuations in isotropic turbulence. Proc. Cambridge Phil. Soc. 47, 359, refereed by Evans et al., 1992.

Bensler, P.H., 1990. Ultrasonic determination of interfacial area, void fraction, and Sauter mean diameter in bubbly flow, PhD Thesis. Institut National Polytechnique de Grenoble, France.

Bin, A.K., 1993. Gas entrainment by plunging liquid jet. Chem. Eng. Sci. 48, 3585–3630.

Biswal, S.K., Reddy, P.S.R., Bhaumik, S.K., 1994. Bubble size distribution in a flotation column. Can. J. Chem. Eng. 72 (1), 148–152.

Bouaifi, M., Hebrard, G., Bastoul, D., Roustan, M., 2001. A comparative study of gas hold-up, bubble size, interfacial area and mass transfer coefficients in stirred gas–liquid reactors and bubble columns. Chem. Engg. Proc. 40 (2), 97–111.

Boyer, C., Lemonnier, H., 1996. Design of a flow metering process for two-phase dispersed flows. Int. J. Multiphase Flow 22 (4), 713–732.

Buchholz, R., Zakrzewski, W., Schugerl, K., 1981. Techniques for determining the properties of bubbles in bubble columns. Int. Chem. Eng. 21, 180–187.

Calderbank, P.H., 1958. Physical rate processes in industrial fermentation. Part 1: The interfacial area in gas–liquid contacting with mechanical agitation. Trans. Inst. Chem. Eng. 36, 443–463.

Camarasa, E., Vial, C., Poncin, S., Wild, G., Midoux, N., Bouillard, J., 1999. Influence of coalescence behavior of the liquid and of gas sparging on hydrodynamics and bubble characteristics in a bubble column. Chem. Eng. Proc. 38, 329–344.

Chandrasekera, T.C., Li, Y., Moody, D., Schnellmann, M.A., Dennis, J.S., Holland, D.J., 2015. Measurement of bubble sizes in fluidised beds using electrical capacitance tomography. Chem. Eng. Sci. 126, 679–687.

Chanson, H., Manasseh, R., 2003. Air entrainment processes in a circular plunging jet: void-fraction and acoustic measurements. J. Fluids Eng. Trans. ASME 125, 910–921.

Chen, F., Gomez, C.O., Finch, J.A., 2001. Bubble size measurement in flotation machines. Miner. Eng. 14 (4), 427–432.

Chesters, A.K., 1991. The modelling of coalescence process in fluid–fluid dispersions—a review of current understanding. Trans I. Chem. E 69 (part A), 353–361.

Colella, D., Vinci, D., Bagatin, R., Masi, M., Abu Bakr, E., 1999. A study on coalescence and breakage mechanisms in three different bubble columns. Chem. Eng. Sci. 54, 2331.

Colin, C., Fabre, J., Dukler, A., 1991. Gas liquid flow at microgravity conditions – I (Dispersed bubble and slug flow). Int. J. Multiphase Flow 17, 533–544.

Cook, T.L., Harlow, F.H., 1984. Virtual mass in multiphase flow. Int. J. Multiphase Flow 10 (5), 691.

Coulaloglou, C.A., Tavlarides, L.L., 1977. Description of interaction processes in agitated liquid–liquid dispersions. Chem. Eng. Sci. 32, 1289–1297.

Couvert, A., Roustan, M., Chatellier, P., 1999. Two-phase hydrodynamic study of a rectangular airlift loop reactor with an internal baffle. Chem. Eng. Sci. 54, 5245–5252.

Crammers, P.H.M.R., Beenackers, A.A.C.M., Van Dierendonck, L.L., 1992. Hydrodynamics and mass transfer characteristics of a loop-venturi reactor. Chem. Eng. Sci. 47, 3557–3564.

De Rijk, S.E., Van der Graaf, J.H.J.M., den Blanken, J.G., 1994. Bubble size in flotation thickening. Water Res. 28 (2), 465–473.

Deckwer, W.D., 1992. Bubble Column Reactors. Wiley, New York.

Drew, D.A., Lahey, T.J., 1987. The virtual mass and lift force on a sphere in rotating and straining inviscid flow. Int. J. Multiphase Flow 13 (1), 113.

Dutta, N.N., Raghavan, K.V., 1987. Mass transfer and hydrodynamics of loop reactors with downflow liquid jet ejector. Chem. Eng. J. 36, 111–121.

Ellingsen, K., Risso, F., 2001. On the rise of an ellipsoidal bubble in water: oscillatory paths and liquid-induced velocity. J. Fluid Mech. 440, 235–268.

Evans, G.M., Jameson, G.J., Atkinson, B.W., 1992. Prediction of the bubble size generated by a plunging liquid jet bubble column. Chem. Eng. Sci. 47, 3265–3272.

Evans, G.M., Jameson, G.J., Rielly, C.D., 1996. Free jet expansion and gas entrainment characteristics of plunging liquid jet. Exp. Ther. Fluid Sci. 12, 142–149.

Fujiwara, H., 2007. Spectroscopic Ellipsometry: Principles and Applications. Wiley, New York.

Fukuma, M., Muroyama, K., Yasunishi, A., 1987. Properties of bubble swarm in a slurry bubble column. J. Chem. Eng. Jpn 20, 28–33.

Gaillard, T., Honorez, C., Jumeau, M., Elias, F., Drenckhan, W., 2015. A simple technique for the automation of bubble size measurements. Colloid. Surf. A: 473, 68–74.

Gorain, B.K., Franzidis, J.-P., Manlapig, E.V., 1995. Studies on impeller type, impeller speed and air flow rate in industrial scale flotation cell—part 1: effect on bubble size distribution. Miner. Eng. 8 (6), 615–635.

Grau, R.A., Heiskanen, Kari, 2002. Kari heiskanen visual technique for measuring bubble size in flotation machines. Miner. Eng. 15 (7), 507–513.

Grossetete, C., 1995. Caracterization experimentale et simulations de l'evolution d'un ecoulement diphasique a bulles ascendant dans une conduite verticale, PhD Thesis. Ecole Centrale Paris, France.

Han, M.Y., Park, Y.H., Yu, T.J., 2002. Development of a new method of measuring bubble size. Water Supply 2 (2), 77–83.

Havelka, P., Linek, V., Sinkule, J., Zahradnik, J., Fialova, M., 2000. Hydrodynamic and mass transfer characteristics of ejector loop reactors. Chem. Eng. Sci. 55, 535–549.

Hebrard, G., Bouaifi, M., Bastoul, D., Roustan, M., 2001. A comparative study of gas hold-up, bubble size, interfacial area and mass transfer coefficients in stirred gas–liquid reactors and bubble columns. Chem. Eng. Proc. 40 (2), 97–111.

Hibiki, T., Hogsett, S., Ishii, M., 1998. Local measurement of interfacial area, interfacial velocity and liquid turbulence in two-phase flow. Nucl. Eng. Des. 184, 287–304.

Hibiki, T., Ishii, M., Xiao, Z., 2001. Axial interfacial area transport of vertical bubbly flows. Int. J. Heat Mass Transfer 44, 1869–1888.

Hinze, J.O., 1955. Fundamentals of the hydrodynamic mechanism of splitting in dispersion processes. AIChE J. 1, 289–295.

Hinze, J.O., 1975. Turbulence—An introduction to its Mechanism and Theory. McGraw-Hill, New York, NY.

Hu, B., Nienow, A.W., Pacek, A.W., 2003. The effect of sodium caseinate concentration and processing conditions on bubble sizes and their break-up and coalescence in turbulent, batch air/aqueous dispersions at atmospheric and elevated pressures. Colloids Surf. B 31 (1–4), 3–11.

Ide, M., Ucheiyama, H., Ishikura, T., 2001. Mass transfer characteristics in gas bubble dispersed phase generated by plunging jet containing small solute bubbles. Chem. Eng. Sci. 56, 6225–6231.

Ishii, M., 1975. Thermo-fluid Dynamic Theory of Two-phase Flow, Eyrolles Paris: Scientific and Medical Publication of France, Paris.

Ishii, M., Paranjape, S.S., Kim, S., Sun, X., 2004. Interfacial structures and interfacial area transport in downward two-phase bubbly flow. Int. J. Multiphase Flow 30, 779–801.

ISO 9276-2, 2001. Representation of results of particle size analysis—Part 2: Calculation of average particle sizes/diameters and moments from particle size distributions.

Johnson, N.L., Kotz, J.S., Balakrishnan, N., 1994. Continuous Univariate Distributions. 1, John Wiley & Sons, p. 207.

Kalkach-Navarro, S., 1992. The mathematical modeling of flow regime transition in bubbly two-phase flow, PhD Thesis. Rensselaer Polytechnic Institute, USA.

Kamp, A.M., Chesters, A.K., Colin, C., Fabre, J., 2001. Bubble coalescence in turbulent flows: a mechanistic model for turbulence-induced coalescence applied to microgravity bubbly pipe flow. Int. J. Multiphase Flow 27 (8), 1363–1396.

Kashinsky, O.N., Randin, V.V., 1999. Downward bubbly gas–liquid flow in a vertical pipe. Int. J. Multiphase Flow 25, 109–138.

Kastenek, F., 1976. The relation between interfacial area and the rate of energy dissipation in bubble column. Collect. Czech. Chem. Commun. 41, 3709–3714.

Kawaguchi, Akasaka, Y., Maeda, M., 2002. Size measurements of droplets and bubbles by advanced interferometric laser imaging technique. Meas. Sci. Technol. 13, 308–316.

Kim, M.C., Kim, S., Lee, H.J., Lee, Y.J., Kim, K.Y., 2002. An experimental study of electrical impedance tomography for the two-phase flow visualization. Int. Commun. Heat Mass Transfer 29, 193–202.

Kim, S., Fu, X.Y., Wang, X., Ishii, M., 2000. Development of the miniaturized four-sensor conductivity probe and the signal processing scheme. Int. J. Heat Mass Transfer 43, 4101–4118.

Kocamustafaogullari, G., Ishii, M., 1995. Foundations of interfacial area transport equation and its closure relations. Int. J. Heat Mass Transfer 38, 481–493.

Kölbel, H., Borchers, E., Langemann, 1961. Grobenverteilung der Gasblasen in Blasensaulen (Size distribution of gas bubbles in bubble columns). Chem. Eng. Tech. 33, 668.

Kuboi, R., Komosawa, I., Otake, T., 1972. Collision and coalescence of dispersed drops in turbulent liquid flow. J. Chem. Eng. Jpn 5, 423–424.

Kumar, R., Kuloor, R., 1970. The formation of the bubbles and drops. Adv. Chem. Eng. 8, 255.

Lage, P.L.C., Espósito, R.O., 1999. Experimental determination of bubble size distributions in bubble columns: prediction of mean bubble diameter and gas hold up. Powder Technol. 101 (2), 142–150.

Landau, J., Boyle, J., Gomaa, H.G., Al Toweel, A.M., 1977. Comparison of methods for measuring interfacial areas in gas-liquid dispersions. Can. J. Chem. Eng. 55 (1), 13–18.

Law, A.M., Kelton, W.D., 1991. Simulation Modeling & Analysis. McGraw-Hill, New York, p. 330.

Leifer, I., Patro, R., Bowyer, P., 2000. A study on the temperature variation of rise velocity for large clean bubbles. J. Atm. Ocean. Tech. 17 (10), 1392–1402.

Lewis, D.A., Davidson, J.F., 1982. Bubble splitting in shear flows. Trans. Inst. Chem. Eng. 60, 283–291.

Liu, T.J., 1989. Experimental investigation of turbulence structure in two-phase bubbly flow, PhD Thesis. Northwestern University, USA.

Lunde, K., Perkins, R.T., 1995. In: Serizawa, A., Yukano, T., Bataille, J. (Eds.), A method for the detailed Study of bubble motion and deformation, Advances in Multiphase Flows 95. Elsevier Pub. Co, Amsterdam, pp. 395–405.

Majumder, S.K., Kundu, G., Mukherjee, D., 2006. Bubble size distribution and interfacial phenomena in ejector induced downflow bubble column. Chem. Eng. J. 122 (1–2), 1–10.

Manasseh, R., 1997. Acoustic sizing of bubbles at moderate to high bubbling rates. In: Giot, M., Mayinger, F., Celata, G.P. (Eds.), Experimental Heat Transfer, Fluid Mechanics and Thermodynamics. Edizioni ETS, Pisa, pp. 943–947.

Mandal, A., Kundu, G., Mukerjee, D., 2003. Interfacial area and liquid-side mass transfer coefficient in downflow bubble column. Can. J. Chem. Eng. 81, 212–219.

Maruyama, M., 2007. Laser Diffraction Scattering Method, The Latest Technology of Microbubbles and Nanobubbles. CMC Book Co. Japan, 31–41.

Meikap, B.C., 2000. Abatement of particulate laden sulphur in a modified multi-stage bubble column scrubber, PhD Dissertation. Chemical Engineering Department, IIT, Kharagpur, India.

Miksis, M., Vanden-Broeck, J.M., Keller, J.B., 1981. Axisymmetric bubble or drop in a uniform flow. J. Fluid Mech. 108, 89–100.

Miyahara, T., Hayashino, T., 1995. Size of bubbles generated from perforated plates in a non-Newtonian liquids. J. Chem. Eng. Jpn. 28, 596–601.

Miyahara, T., Hashimoto, S., 1999. Bubble splitting by jet emitted form orifice. J. Chem. Eng. Japan 32, 91–96.

Miyahara, T., Matsuha, Y., Takahashi, T., 1983. The size of bubbles generated from perforated plates. Int. Chem. Eng. 23 (3), 517.

Moleurs, O., Kurtin, M., 1985. Hydrodynamics of bubble columns in the uniform bubbling regime. Chem. Eng. Sci. 40 (4), 647–652.

Nagel, O., Kurten, H., Hegner, B., 1973. Criteria for the selection and design of gas liquid reactors. Int. Chem. Eng. 21 (981), 161–172.

O'Connor, C.T., Randall, E.W., Goodall, C.M., 1990. Measurement of the effect of physical and chemical variables on bubble size. Int. J. Miner. Proc. 28, 139–149.

Ohkawa, A., Kawai, Y., Ksabiraki, D., Saki, N., Endoh, K., 1987. Bubble size, interfacial area and volumetric liquid-phase mass transfer coefficient in downflow bubble columns with gas entrainment by a liquid jet. J. Chem. Eng. Jpn 20, 99–101.

Patel, S.A., Daly, J.G., Bukur, D.B., 1989. Holdup and interfacial area measurements using dynamic gas disengagement. A. I. Ch. E. J. 35, 931–942.

Pohorecki, R., Moniuk, W., Zdrójkowski, A., 1999. Hydrodynamics of a bubble column under elevated pressure. Chem. Eng. Sci. 54, 5187–5519.

Polli, M., Stanislao, M.D., Bagatin, R., Abu Bakr, E., Masi, M., 2002. Bubble size distribution in the sparger region of bubble columns. Chem. Eng. Sci. 57 (1), 197–205.

Prince, M., Blanch, H., 1990. Bubble coalescence and break-up in air-sparged bubble columns. AIChE J. 36 (10), 1485–1499.

Rabiger, N., Vogelphol, A., 1982. Bubble formation in stagnant flowing newtonian liquids. German Chem. Eng. 5, 314.

Revuelta, A., 2010. On the interaction of a bubble and a vortex ring at high Reynolds numbers. Eur. J. Mech. B/Fluids 29 (2), 119–126.

Revuelta, A., Rodríguez-Rodríguez, Javier, Martínez-Bazán, Carlos, 2008. On the breakup of bubbles at high Reynolds numbers and subcritical Weber numbers. Eur. J. Mech. B/Fluids 27 (5), 591–608.

Rodríguez-Rodríguez, J., Gordillo, J.M., Martínez-Bazán, C., 2006. Break-up time and morphology of drops and bubbles in a high Reynolds number flow. J. Fluid Mech. 548, 69–86.

Ryskin, G., Leal, L.G., 1984. Numerical solution of free boundary problems in fluid mechanics part 3. Bubble deformation in an axysymmetric straining flow. J. Fluid Mech. 148, 37–43.

Rząsa, M.R., Pląskowski, A., 2003. Application of optical tomography for measurement of aeration parameters in large water tanks. Meas. Sci. Technol. 14, 199–204.

Saberi, S., Shakourzadeh, K., Bastoul, D., Militzer, J., 1995. Bubble size and velocity measurement in gas—liquid systems: Application of fiber optic technique to pilot plant scale. The Canadian J. Chem. Eng. 73 (2), 253–257.

Saberi, B., Shakourzadeh, K., Militzer, J., 1997. Application of fiber optic probe to the hydrodynamic study of an industrial fluidized bed furnace. Proc. Int. Conf. Fluid Bed Combust. 14, 595–601.

Schäfer, R., Merten, C., Eigenberger, G., 2002. Bubble size distributions in a bubble column reactor under industrial conditions. Exp. Ther. Fluid Sci. 26 (6–7), 595–604.

Schumpe, A., Deckwer, W.D., 1982. Gas holdups, specific interfacial areas, and mass transfer coefficients of aerated carboxymethyl cellulose solutions in a bubble column. Ind. Eng. Chem. Process Des. Dev. 21, 706–711.

Schumpe, A., Grund, G., 1986. The gas disengagement technique for studying gas holdup structure in bubble column. Con J. Chem Eng. 64 (89), 1–896.

Sevik, M., Park, S.H., 1973. The splitting of drops and bubbles by turbulent fluid flow. J. Fluid Eng. Trans. ASME 95, 53–60.

Sommerefeld, M., Bourloutski, E., Bröder, D., 2003. Euler/Lagrange calculations of bubbly flows with consideration of bubble coalescence. Can. J. Chem. Eng. 81, 508–518.

Stein, W.A., Schafer, H., 1984. Aeration of viscous liquids with a downward directed two-fluid nozzle. Ger. Chem. Eng. 7, 115–125.

Stewart, C.W., Crowe, C.T., Saunders, S.C., 1993. A model for simultaneous coalescence of bubble clusters. Chem. Eng. Sci. 48, 3347–3354.

Takahashi, M., 2009. Fundamentals and engineering applications of microbubbles and nanobubbles. Mater. Integr. 22, 2–9.

Takamasa, T., Goto, T., Hibiki, T., Ishii, M., 2003. Experimental study of interfacial area transport of bubbly flow in small diameter tube. Int. J. Multiphase Flow 29 (3), 395–409.

Tucker, J.P., Deglon, D.A., Franzidis, J.P., Harris, M.C., O'connor, C.T., 1994. An evaluation of a direct method of bubble size distribution measurement in a laboratory batch flotation cell. Miner. Eng. 7 (5–6), 667–680.

Unno, H., Inoue, I., 1979. Size reduction of bubbles by orifice mixer. Chem. Eng. Sci. 35, 1571–1579.

Vazquez, A., Sanchez, R.M., Salinas-Rodríguez, E., Soria, A., Manasseh, R., 2005. A look at three measurement techniques for bubble size determination. Exp. Ther. Fluid Sci. 30, 49–57.

Vermeulen, T., Williams, G.M., Langlois, G.E., 1955. Interfacial area in liquid–liquid and gas–liquid agitation. Chem Eng Prog 51 (2), 85F–94F.

Weibull, W., 1951. A statistical distribution function of wide applicability (PDF). J. Appl. Mech. Trans. ASME 18 (3), 293–297.

Wilkinson, P., van Dierendonck, L., 1990. Pressure and gas density effects on bubble break up and gas hold-up in bubble columns. Chem. Engg. Sci. 42, 2309–2315.

Wilkinson, P.M., Haringa, H., Van Dierendonck, L.L., 1994. Mass transfer and bubble size in a bubble column under pressure. Chem. Eng. Sci. 49, 1417–1427.

Wolff, C., Briegleb, F.U., Bader, J., Hektor, K., Hanmer, H., 1990. Measurements with multi-point microprobes: Effect of suspended solids on the hydrodynamics of bubble columns for application in chemical and biotechnological processes. Chem. Eng. Tech. 13, 172–184.

Wu, M. & Gharib, M. 2002. Experimental studies on the shape and path of small air bubbles rising in clean water. Phys. Fluids 14, L49.

Wu, Q., Kim, S., Ishii, M., Beus, B.G., 1998. One-group interfacial area transport in vertical bubbly flow. Int J. Heat Mass Transfer 41, 1103–1112.

Zhou, Z.A., Egiebor, N.O., Plitt, L.R., 1993. Frother effects on bubble size estimation in a flotation column. Miner. Eng. 61, 55–67.

Zhu Y., Wu J., Manasseh R., 2001. Rapid measurement of bubble size in gas–liquid flows using a bubble detection technique. Proceedings of fourteenth Australasian Fluid Mechanics Conference, Adelaide University, Adelaide, Australia, December 10–14, 2001.

8

Mass Transfer Characteristics

Introduction

Mass transfer by bubble dispersion is widely used in biological, chemical, and environmental applications. There is a great value in trying to intensify mass transfer using bubbles moving inversely with minimal energy consumption. Before any consideration of the effects of liquid flow, a simple analysis suggests the use of bubbles first because the great amount of contact surface area between the gas and liquid phases that intensifies mass transfer, and second, longer residence times may be expected because of inverse flow against the buoyancy and hence slow relative velocity. However, because the residence time of bubbles is determined not only by the bubbles' relative velocity but also by the velocity of the surrounding liquid, the consideration of liquid flow becomes the more important for inverse flow bubbles. It is expected that the inverse flow bubbles increase the efficiency of gas–liquid contact devices, including bubble columns, chemical reactors, gas absorbents, and fermentors.

As one of the key determinants of reactor performance, the gas–liquid mass transfer in bubble columns is directly affected by hydrodynamics, phase mixing, and physical properties. Bubble diameter and gas hold-up are two significant parameters with respect to mass transfer area available in bubbly flow devices. Recent studies show that a high mass transfer rate is obtained when the bubbles are small and the residence time is high. Bubbles with larger diameters promote a high degree of turbulence and shorter residence times and have high rise velocity. On the other hand, bubbles with small diameters and inverse flow offer low relative velocity, long residence times, and uniform mixing. The mass transfer coefficient is one of the important parameters in designing a reactor. The product of the mass transfer coefficient k_l and specific interfacial area a (based on dispersion volume) is known as the volumetric mass transfer coefficient ($k_l a$) of liquid. When the gas-side mass transfer resistance is insignificant, then the mass transfer properties can be characterized by liquid-side mass transfer alone.

Gas–Liquid Mass Transport Process

When gas is blown through the liquid as a stream of bubbles, whether upflow or inverse flow in a bubble column, diffusion, convection, and reaction proceed simultaneously. Mass transfer accompanied by a chemical reaction not only can enhance the mass transfer rate but also can reduce the height of bubbly flow unit (Levenspiel, 1998). Therefore, in mass transfer accompanied by a chemical reaction system, chemical reaction kinetics play an important role in the overall rate of mass transfer process. To quantitatively evaluate

Hydrodynamics and Transport Processes of Inverse Bubbly Flow. http://dx.doi.org/10.1016/B978-0-12-803287-9.00008-4

the mass transfer rate, its analysis by a mass transfer model is required. In the literature (Danckwerts, 1970; Sherwood et al., 1975), four models are generally incorporated to interpret the mass transfer rate; these are the film model, still surface model, surface-renewal model, and penetration model.

Theories on Mass Transfer

Film Theory

The film model assumes that the mass transfer resistance depends on the velocity of the solute diffusive transport in each contacting phases and that it is localized near the interface between two stagnant liquid and gas film with finite thickness, δl and δg. There is a thermodynamic equilibrium between interfacial phase concentrations; thus, the interface itself does not represent a mass transfer resistance. The mass transfer through the stagnant film is governed by molecular diffusion in steady state in which the local flux across each element of area is constant. A two-film model is a more practical model for applications in bubbly flow systems, as shown in Figure 8.1.

For a two-phase bubbly flow mass transfer, three steps, as shown in Figure 8.1, must be followed by the reactants if both transport and kinetic resistances are important. The mass transfer between bubble and phases can then be represented by Fick's Law with the following expression:

$$\frac{\partial C}{\partial t} = -D_{AB}\frac{\partial^2 C}{\partial z^2} \tag{8.1}$$

where D_{AB} is the diffusivity of phase A into B. The products of the chemical reaction will follow the inverse direction back to the gas phase in the case of volatile gaseous products. Using the two-film model based on a modified Fick's law and initially developed by Lewis and Whitman (1924), rectant transport from gas film to gas–liquid interface (step 1), gas–liquid interface to liquid film (step 2), and then liquid film to liquid bulk (step 3) can be described by a steady-state mass transfer flux, J, across a stagnant gas–liquid interface by the following expressions:

$$J_g = k_g a(P_g - P_e) \tag{8.2}$$

$$J_l = k_l a(C_e - C_l) \tag{8.3}$$

Pg is the partial pressure of the gas phase and P_e is the pressure at equilibrium defined as:

$$P_e = HC_e \tag{8.4}$$

where C_e is the concentration at equilibrium (solubility) and H is Henry's law constant. The concentration profile of two-phase flow is shown in Figure 8.1. The gas solute in bulk phase overcomes the resistance of gas film and moves to the gas–liquid interface; later, solute moves across the interface and overcomes the liquid film resistance and goes through the bulk liquid. In general, the reactant concentration in the gas film is high enough to

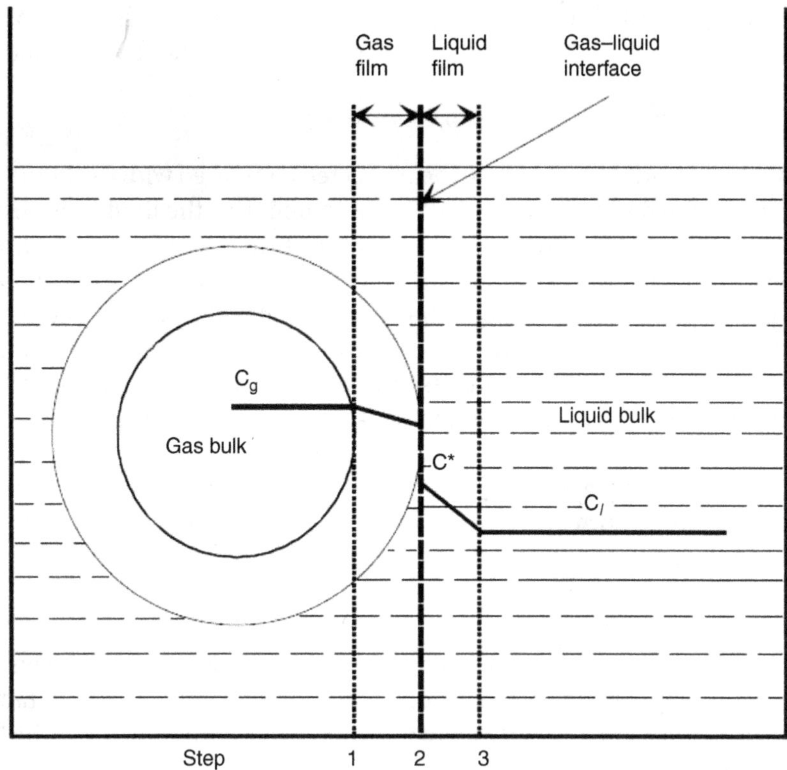

FIGURE 8.1 Concentration profile of two-phase flow.

prevent the partial pressure of the liquid in the gas phase from imposing any resistance to transport. Therefore, the resistance due to the transport of the reactant through the gas film can be neglected. Consequently, the main resistance occurs in transport from the gas–liquid interface to the liquid film. As a result, equation (8.3) becomes the overall rate of mass flux, where k_l, the liquid-side mass transfer coefficient, is related to the gas–liquid diffusivity and the liquid film thickness, δ, as shown by the following equation:

$$k_l = \frac{D_{AB}}{\delta} \tag{8.5}$$

The overall rate of mass transfer should include the transport rates of mass flux from the gas phase to the bulk liquid and the kinetic term, which can be represented as

$$J_A = K_l a(C_g - C_l) \tag{8.6}$$

where $K_l a$ is the overall mass transfer coefficient, which is given by

$$\frac{1}{K_l a} = \frac{H}{k_g a} + \frac{1}{k_l a} \tag{8.7}$$

At 25° C, $D_{O2\text{-}air} = 0.1937$ cm^2/s, $D_{O2\text{-}water} = 2.10 \times 10^{-5}$ cm^2/s (Cussler, 1984; Sherwood et al., 1975), and $H_{O2} = 3.1 \times 10^{-2}$ (Perry and Green, 1984). As a result, it becomes

$$1/(k_l a) = 3098\, H/(k_g a) \tag{8.8}$$

Therefore, the gas-side resistance can be neglected compared with the liquid-side resistance. If the gas phase resistance is negligible compared with the liquid phase, the overall mass transfer coefficient can be denoted by the liquid-side mass transfer coefficient. Thus, proper design of bubble column reactors for a specific commercial process and knowledge of both the transport and kinetic parameters are essential in the determination of the rate-limiting step. Because the overall rate of mass transport flux during the reaction process is reduced to equation (8.3), it is therefore essential to determine the liquid-side mass transfer coefficient, k_l, and the gas liquid interfacial area per unit volume, a. The estmation of the liquid-side mass transfer coefficient, based on equation (8.5), requires knowledge of the gas–liquid diffusivity and the liquid film thickness. The diffusivity can be obtained using available literature correlations; however, the measurement of the liquid film thickness can be a challenging task.

Penetration Theory (Higbie, 1935)
Higbie (1935) emphasized that in many situations, the contact times between phases are too short for the steady state to be achieved. It is assumed that if t_c is the contact time that a liquid particle is subject to unsteady-state diffusion (or penetration), then the liquid-side mass transfer coefficient is given by $k_l = (2\sqrt{1/\pi t_c})\sqrt{D_{AB}}$.

Surface-Renewal Theory (Danckwerts, 1970)
The surface-renewal theory is an extension of the penetration theory that allows eddies of fluid to be exposed at the surface for varying lengths of time. On the assumption that the change of a surface element being replaced is independent of its age, the liquid-side mass transfer coefficient is given by $k_l = (\sqrt{s})\sqrt{D_{AB}}$ where s is the fractional rate of surface renewal (Treybal, 1968; Perry and Green, 1984).

In the stagnant-film model, the liquid-side mass transfer coefficient (k_l) is proportional to the diffusion coefficient (D_{AB}), and the other models considered here predict a square-root dependency on D_{AB}.

Estimation Methods of Mass Transfer Coefficient

Several methods are available to estimate the liquid-side mass transfer coefficeint. Table 8.1 summarizes available literature studies on methods to measure k_l by experiments. The volumetric mass transfer coefficient can be measured by dynamic methods (Deckwer, 1992; Kabdasli and Arslan-Alaton, 2010).

In bubbly flow systems, the driving force of gas–liquid mass transfer is usually produced by dynamic change in the gas or liquid input (pulse or step), by pressurizing the

Table 8.1   List of Important Studies of the Gas–Liquid Mass Transfer Coefficient

Reference	System	Operating Conditions	Measurement Method
Akita and Yoshida (1973)	O_2–N_2–water, glycerol solution, methanol, Na_2SO_3 solution	$d_c = 0.15$ m, $L_c = 4.0$ m $u_{sg} \leq 0.33$ m/s $T = 293$ K, $P = 0.1$ MPa	Oxygen absorption
Deckwer et al. (1974)	Air–tap water and solutions of salts and molasses	$d_c = 0.2$ m, $L_c = 7.23$ m and $d_c = 0.15$ m, $L_c = 4.4$ m $u_{sg} \leq 0.15$ m/s $T = 298$ K, $P = 0.1$ MPa	Oxygen absorption and desorption
Hikita et al. (1981)	Air, O_2, H_2, CH_4, CO_2–water, sucrose, n-butanol, methanol	$d_c = 0.10$ m, $L_c = 1.5$ m and $d_c = 0.19$ m, $L_c = 2.4$ m $u_{sg} \leq 0.39$ m/s $T = 283$-303 K, $P = 0.1$ MPa	Oxygen absorption and desorption
Sada et al. (1983)	CO_2–$Ca(OH)_2$ solution, aluminum oxide, calcium carbonate	$d_c = 0.078$ m, $L_c = 1.5$ m $u_{sg} \leq 0.2$ m/s $T = 293$ K, $P = 0.1$ MPa	Oxygen absorption
Lewis and Davidson (1985)	Air–water	$d_c = 0.45$ m, $L_c = 4.5$ m $u_{sg} \leq 0.23$ m/s, $u_{sl} \leq 0.34$-0.68 m/s $T = 293$ K, $P = 0.1$ MPa	Oxygen absorption
Kawase et al. (1987)	CO_2–water, aqueous solutions of CMC	$d_c = 0.23, 0.76$ m, $L_c = 1.22, 3.21$ m $u_{sg} = 0.6$ m/s $T = 298$ K, P = 0.1 MPa	Dynamic CO_2 gas analysis method
Ozturk et al. (1987)	Air, N_2, CO_2, He, H_2–organic liquids	$d_c = 0.095$ m $u_{sg} \leq 0.1$ m/s $T = 293$ K, $P = 0.1$ MPa	Oxygen absorption
Schumpe et al. (1989)	Oxygen–glycerol solution, CMC solution	$d_c = 0.14$ m, $L_c = 2.65$ m $u_{sl} \leq 12$ cm/s, $u_{sg} \leq 12$ cm/s $T = 298$ K, $P = 0.1$ MPa	Oxygen absorption, oxygen electrode
Kim and Kim (1990)	Glass beads-air–water	$d_c = 0.142$ m, $L_c = 2.0$ m $u_{sl} = 0.02$-0.1 m/s, $u_{sg} = 0.02$-0.2 m/s $T = 298$ K, $P = 0.1$ MPa,	Oxygen absorption
Tang and Fan (1990)	Polystyrene, acrylic, acetate, nylon-air–water	$d_c = 0.0762$ m, $u_{sg} \leq 4.14$ cm/s $T = 298$ K, $P = 0.1$ MPa	Oxygen absorption
Linek et al. (1989)	O_2/air–distilled water and aqueous Na_2SO_4	$d_c = 0.29$ m, $u_{sg} = 0.004$ m/s $T = 293$ K, $P = 0.1$ MPa	Oxygen absorption
Zheng et al. (1995)	Air–water	$d_c = 0.285$ m, $L_c = 4.1$ m, $u_{sg} \leq 11$ cm/s, $T = 298$ K, $P = 0.1$ MPa	Oxygen absorption
Terasaka et al. (1998)	N_2/O_2–tap water, solutions of microbial polysaccharides xanthan or gellan	$d_c = 0.06, L_c = 0.114$ m $u_{sg} \leq 0.15$ m/s $T = 292$ K, $P = 0.1$ MPa	Oxygen absorption
Peeva et al. (2001)	Silicon oil–n-decane	$d_c = 0.087$ m, $L_c = 0.54$ m, $u_{sg} = \leq 0.018$ m/s $T = 298$ K, $P = 0.1$ MPa	Decane absorption into emulsion
Chaumat et al. (2005)	N_2, CO_2, organic (cyclohexane), aqueous (tap water)	$d_c = 0.2$ m; $L_c = 1.6$ m, $u_{sg} = \leq 0.18$ m/s $T = 298$ K, $P = 0.1$ MPa	IR spectrometer

(Continued)

Table 8.1 List of Important Studies of the Gas–Liquid Mass Transfer Coefficient (*cont.*)

Reference	System	Operating Conditions	Measurement Method
Therning and Rasmuson (2006)	O_2–dDeionized water, sulphite solution	d_c = 0.05, 0.2 m, L_c =1, 2 m u_{sg} = ≤ 0.25 m/s T = 288-303 K, 308-318 K; P = 0.1 MPa	Oxygen absorption
Maceiras et al. (2007)	CO_2–amine solution	6 cm × 6 cm and L_c = 1.03 m, Q_g = 10-25 l/h T = 295 K, P = 0.1 MPa	Absorbed CO_2 (determined by titration
Gómez-Díaz et al. (2009)	O_2–aqueous and organic liquid phase forming an emulsion stabilized by Tween 80	d_c = 0.04 m, L_c =0. 65 m. Q_g = 0.25-0.7 l/m T = 298 K, P = 0.1 MPa	Polarographic-membrane probe, static gassing-out technique
Mandal (2010)	CO_2–$Ca(OH)_2$ solution, aluminum oxide, calcium carbonate	d_c = 0.0516 m, L_c = 1.5 m u_{sg} = 0.01-0.09 m/s, u_{sl} = 0.05-0.11 m/s T= 298 K, P = 0.1 MPa	Dynamic CO_2 gas analysis method
Muroyama et al. (2013)	O_2–N_2 mixture–water	d_c = 0.15 m, L_c = 1.0-2.0 m u_{sg} = 0.0944-0.6610 m/s, 0.00472-0.01415 m/s T = 298 K, P = (0.028-2.8)	Oxygen absorption
Jin et al. (2014)	H_2, CO, CO_2–paraffin	d_c = 0.10 m, L_c = 1.25 m u_{sg} = 0.03-0.10 m/s T = 293-473 K, P = 1.0 to 3.0 MPa	Gas chromatograph (H_2) and infrared analyzer (CO, CO_2).
McClure et al. (2015a, 2015b)	Air–water, sulphite containing solution in the presence of a cobalt catalyst	d_c = 0.39 m, L_c = 2 m u_{sg} = 0.14-0.28 m/s T = 298 K, P = 0.1 MPa	Sulphite oxidation method

CMC, Critical micelle concentration; *IR*, infrared.

gas phase (called physical method), or by the presence of chemical reactions (called the chemical method). Mass transfer studies in the literature are commonly conducted with oxygen absorption or the desorption method using either air or oxygen as the gas phase under atmospheric conditions, especially for air–water systems.

Physical Method

From the very beginning of the gas–liquid reactor study of Bartholomew et al. (1950), the dynamic gas absorption or desorption method has been popularly used in bubble columns for its simplicity and reliability. The application of the method is typically based on switching the gas flow between air and nitrogen (or air and oxygen for some rare cases). However, this creates difficulties in bubble columns at high operating pressure and high gas velocity. The measured transient concentration profiles of the introduced gas represent not only the extent of gas–liquid mass transfer rate but also the phase mixing. Hence, a reactor model is needed to estimate the $k_l a$ values by fitting the experimental data. The experiment can be performed by dissolving oxygen in a liquid in a pressurized bubble column (if pressure is required) as it is done in majority of the studies because its partial pressure in the liquid phase is easy to determine electrochemically using polarographic electrodes (Deckwer, 1992; Van de Sande and Smith, 1975; Bin and Smith, 1982; Tojo and Miyanami, 1982; Wachsmann et al., 1985). In this case, compressed and filtered oxygen-enriched air can be used as the continuous gas flow, and filtered tap water can be used as the liquid phase. Initially, nitrogen is passed through the liquid phase to remove the initial oxygen content in the liquid. The dissolved oxygen (DO) concentration can be measured by a suitable sensor (fast-response, stable, long-life optical oxygen probes have been recently used in bubble column mass transfer studies by Terasaka et al., 1998; Jordan et al., 2002; Lau et al., 2004; Han et al., 2007). Because of the sensor delay, the measured DO response may not exactly represent the actual DO concentration and hence needs to be deperverted. The delay depends on the liquid phase viscosity. A period of between 5 to 10 s is generally acceptable for most commercial electrodes (Deckwer, 1992). As shown in equation (8.9), the measured DO response follows the actual DO concentration with a certain extent of delay described by a probe constant (K_s) (Han et al., 2007).

$$\frac{dC_s(t)}{dt} = K_s\{C_l(t) - C_s(t)\} \tag{8.9}$$

The K_s of the oxygen probe should be estimated by a calibration process using liquid stripped of oxygen by a nitrogen flow and liquid saturated with oxygen in the air flow. By quickly switching water with zero-DO concentration (liquid) to water with air-saturated DO concentration (liquid), the DO response, $C_s(t)$, can be obtained. For this calibration process, equation (8.9) can be used for the DO response as (Han and et al., 2007)

$$C_s(t) = C_{l,e,0}\{1 - \exp(-K_s.t)\} \tag{8.10}$$

According to equation (8.9), the obtained DO response should be deconvoluted to obtain the actual DO concentration profiles by (Han et al., 2007)

$$C_l(t) = C_s(t) + \frac{1}{K_s}\frac{dC_s(t)}{dt} \tag{8.11}$$

The curves (as per equation [8.11]) can be considered responses of identical gas dynamic changes, except with random noises. The absorption data can be used for the $k_l a$ measurements.

Complete Mixing Model

For unsteady state operation with the complete mixing assumptions, the DO concentration in the bubble column can be written for the absorption method as

$$\frac{dC_{l,t}}{dt} = k_l a\left(C_{l,e} - C_{l,t}\right) \tag{8.12}$$

After integration of equation (8.12) with boundary condition at $t = 0$, $C_{l,t} = C_0$ and t $t = t$, $C_{l,t} = C_{l,t}$, one can get

$$\frac{C_{l,t} - C_{l,0}}{C_{l,e} - C_{l,0}} = 1 - e^{-k_l a t} \tag{8.13}$$

where $C_{l,e}$ is the concentration of gas in liquid at equilibrium, $C_{l,0}$ is initial concentration of gas in liquid, and $C_{l,t}$ is concentration of gas in liquid at time t. The coefficient k_l is the liquid-side mass transfer coefficient, and a is the specific interfacial area with respect to the dispersed volume. In the case of steady-state operation, the liquid phase is passed through the reactor, and the concentration of gas is measured at the inlet and outlet of the reactor. If it is assumed that complete liquid mixing takes place and saturated concentration of gas in liquid does not depend on height, then the volumetric mass transfer coefficient can be calculated as

$$k_l a = \frac{u_{sl}}{(1 - \varepsilon_g) L_c} \left(\frac{C_l - C_0}{C_{l,e} - C_l} \right) \tag{8.14}$$

where u_l is interstitial or actual velocity of liquid in column, L_c is the length of the column, and C_0 and C_l are concentration of gas in liquid at the inlet and outlet of the reactor. The unknown term $k_l a$ can be determined by a minimum square error fit between equation (8.13) and the concentration data. Equations (8.13) and (8.14) have been used by several investigators many times to determine $k_l a$ values from inlet and outlet concentration by assuming $C_{l,e}$ is constant throughout the whole column, but in real cases, hydrostatic pressure along the column may change the $C_{l,e}$ in the tall bubble column. In this case, consider the bubble column the effective volume of the bubble column reactor for mass transfer having a gas–liquid mixture height of z_m is divided axially into a series of segments at z_1, z_2, z_3,...., z_i from the distribution source of gas. Assume that gas leaving from each axial segment h_i is in equilibrium with liquid with respect to both heat and mass transfer. Consider the liquid flows at the rate Q_l m^3/s. The interfacial surface area between gas and liquid is a m^2 per unit volume of gas–liquid mixture. As the gas–liquid mixture raises a differential height dz, the area of contact is adz per unit active area of the column. If the gas solute undergoes a concentration change dC_l in this gas–liquid mixture height (dz_i), the solute balance can be represented as

$$Q_l dC_{l,z_i} = A_c k_l a dz_i (C_{l,e,z_i} - C_{l,z_i}) \tag{8.15}$$

Then

$$\int_{C_{l,z_{i-1}}}^{C_{l,z_i}} \frac{dC_{l,z_i}}{C_{l,e,z_i} - C_{l,z_i}} = \frac{A_c k_l a}{Q_l} \int_{z_{i-1}}^{z_i} dz_i \tag{8.16}$$

This gives

$$\frac{C_{l,z_i} - C_{l,z_{i-1}}}{C_{l,e,z_i} - C_{l,z_{i-1}}} = 1 - \exp\left(-\frac{k_l a (z_i - z_{i-1})(1 - \varepsilon_g)}{u_{sl}} \right) \tag{8.17}$$

The equilibrium concentration in liquid phase, $C_{l,e}$, a function of height in tall bubble columns:

$$C_{l,e,z_i} = \frac{P_{z_{Ri}} y_{z_{Ri}}}{H} \tag{8.18}$$

in which

$$P_{z_{Ri}} = P_t \{1 + \alpha(1 - z_{R,i})\} \tag{8.19}$$

where α represents the ratio between hydrostatic pressure (at the gas distributor) and the total internal column pressure P_t.

$$\alpha = \frac{\rho_l g (1 - \varepsilon_g) z_m}{P_t} = \frac{(1 - \varepsilon_g)}{P_R + z_{R,i} + P_{b,R}} \tag{8.20}$$

The total internal column pressure P_t at any distance z_i below the surface is

$$P_t = P_{atm} + \rho_l g z_i + 4\sigma/d_b \tag{8.21}$$

where P_{atm} is the atmospheric pressure above the liquid and σ is the surface tension. The dimensionless distance, $z_{R,i}$ represents z_i/z_m.
where $P_R = P_{atm}/(\rho_l g z_m)$, $P_{b,R} = 4\sigma/(\rho_l g z_m d_b)$. If the gas phase molar fraction is constant, then equation (8.18) becomes

$$C_{l,e,z_i} = \frac{y P_t \{1 + \alpha(1 - z_{R,i})\}}{H} \tag{8.22}$$

Then the local solute concentration in liquid at various axial location of the column can be represented as

$$C_{l,z_i} = C_{l,z_{i-1}} e^{-St_{z_i}} + \frac{y_{z_{Ri}} P_t \{1 + \alpha(1 - z_{R,i})\}}{H}\left(1 - e^{-St_{z_i}}\right) \tag{8.23}$$

The Stanton number is defined as $St_{z_i} = k_l a (z_i - z_{i-1})(1 - \varepsilon_g)/u_{sl}$. The equation can be used to evaluate the concentration profile with respect to column height. If the bulk average concentration of liquid at the inlet and outlet are $C_{l,in}$ and $C_{l,out}$, respectively, the concentration profile can be expressed as

$$\frac{C_{l,z_i=z_m} - C_{l,in}}{C_{l,e,z_i=z_m} - C_{l,in}} = 1 - \exp\left(-\frac{k_l a z_m (1 - \varepsilon_g)}{u_{sl}}\right) \tag{8.24}$$

where C_{l,e,z_i} is in equilibrium with C_{g,z_i} and assumed to be a constant average value. The overall mass transfer rate per unit volume of the dispersion in a bubble column is governed by the liquid side mass transfer coefficient, $k_l a$, assuming that the gas-side resistance is negligible.

Axial Dispersion Model

The one-dimensional axial dispersion model for the individual phase can also be used to analyze the liquid-side volumetric mass transfer coefficient. The model can be expressed for the liquid phase and gas phases, respectively, as

$$\frac{\partial C_l}{\partial t} = D_{z,l}\frac{\partial^2 C_l}{\partial z^2} + k_l a(HC_g - C_l) \tag{8.25}$$

$$\frac{\partial C_g}{\partial t} = D_{z,g}\frac{\partial^2 C_g}{\partial z^2} - \frac{u_{sg}}{\varepsilon_g}\frac{\partial C_g}{\partial z} - k_l a(HC_g - C_l) \tag{8.26}$$

In this case, the empirical correlations are required to calculate the dispersion coefficient of gas and liquid phase, which are discussed in Chapter 7. The local gas hold-up should be considered as equal to the overall gas hold-up. Although the local gas hold-up may vary with the axial distance, it can be neglected by assuming constant axial gas hold-up. Such an assumption would be reasonable with a small error with large bubble column $(>L_c/d_c)$ and with its fully developed flow region much larger compared with the distributor and disengagement zones. The gas hold-up and $k_l a$ values have small variance along the fully developed flow region (Han et al., 2007). The $k_l a$ values can then be obtained with minimum square error fits between the model prediction $C_l(t,z_i)$ and the DO concentration data from experiment at axial distance $z = z_i$.

Chemical Method

Generally, two chemical methods, such as Danckwerts' plot and absorption rate methods, are used for determination of volumetric mass transfer coefficient.

Absorption Rate Method

The method is based on gas absorption followed by chemical reaction. In general, when physical absorption is followed by a chemical reaction, the overall rate will be governed by both the physical rate of absorption and the kinetics of the reaction (Sherwood and Pigford, 1952). The application of the chemical method for determination of interfacial area and mass transfer coefficient has been reviewed by Deckwer (1992). The chemical method actually is used by applying fast chemical reaction absorption range because the result is not affected by hydrodynamic conditions. In other words, it can be said that the absorption rate is independent of the mass transfer coefficient, k_l. Brian (1964) has discussed the theory of absorption followed by a chemical reaction of the general order. For the irreversible reaction,

$$A + B \rightarrow P \tag{8.27}$$

which is m^{th} order in A and n^{th} in B, the local rate of reaction is given by equation (8.28). If the concentration of B is same everywhere and is equal to the bulk concentration, $C_{B0,}$ then

$$-r_A = k'_{m,n} C_A^m C_{B0}^n \tag{8.28}$$

Then the local absorption rate from the absorption–reaction theory can be given as

$$R_A a = a C_{A,e} \left(\frac{2}{m+1} D_A k'_{m,n} C_{B0}^n C_A^{m-1} \right)^{0.5} \tag{8.29}$$

Because $k'_{m,n} C_{B0}^n$ is a constant, the rate depends only on the concentration, C_A. If the reduction of B is not much more at the interfacial area than in bulk (i.e., the concentration of B at the interface in practically same as in bulk liquid), then the reaction is of the pseudo-m^{th} order:

$$k'_m = k'_{m,n} C_{B0}^n \tag{8.30}$$

where C_{B0} is the concentration of B in the bulk liquid phase. Substituting value of k'_m in equation (8.29), one can get

$$R_A a = (Ha) k_l a C_{A,e} \tag{8.31}$$

where Ha is the Hatta number, which is defined as

$$Ha = \sqrt{\left(\frac{2}{m+1} D_{A,e} k_m C_{A,e}^{m-1} \right) / k_l} \tag{8.32}$$

The necessary conditions are as follows if there is no significant reduction in the concentration of B at the phase interface.

$$Ha \ll 1 + \frac{D_B C_{B0}}{D_A C_{A,e}} \tag{8.33}$$

and

$$k_l a C_{A,e} \ll \varepsilon_l k_{m,n} C_{A,e} C_{B0}^n \tag{8.34}$$

Under these conditions, the rate of absorption is given by

$$R_A a = k_l a C_{A,e} \tag{8.35}$$

If the reaction is to be proceed irrespective of hydrodynamic conditions present at the phase interface, Ha should be greater than 3. If both the conditions are satisfied, then the volumetric mass transfer coefficient can be calculated from the absorption rate. For slow and instantaneous absorption, the mass transfer rate is determined by the volumetric mass transfer coefficient ($k_l a$), which is a function of the specific interfacial area (a). As an example, the liquid phase reaction between dissolved carbon dioxide and sodium hydroxide satisfies the condition for pseudo first-order reaction at certain alkali concentrations. This reaction takes place as

$$CO_2 + OH^- \rightarrow HCO_3^- \tag{8.36}$$

$$HCO_3^- + OH^- \rightarrow CO_3^{2-} \tag{8.37}$$

The second reaction is ionic. It takes place at a much faster rate than the first one; therefore, the first reaction step is the rate-controlling step. The overall reaction can then be represented as

$$CO_2 + 2OH^- \rightarrow CO_3^{2-} + H_2O \tag{8.38}$$

The equation becomes pseudo first order if the concentration of OH^- remains essentially constant in the bulk of the solution for which the satisfied condition can be written as

$$Ha = \sqrt{\left(D_A k_2 C_{B0}^{m-1}\right)}/k_l \ll 1 + \frac{C_{B0}}{2C_{A,e}} \tag{8.39}$$

D_A, the diffusivity of carbon dioxide in aqueous solutions, can be obtained from the equation reported by Nijsing et al. (1959). The specific rate of absorption per unit area for a second-order reaction satisfying the condition of equation (8.39) is (Danckwerts, 1970; Mandal, 2010)

$$R = C_{A,e}\sqrt{D_A k_2 C_{B0} + k_l^2} \tag{8.40}$$

and

$$Ha = \sqrt{D_A k_2 C_{B0}}/k_l > 3 \tag{8.41}$$

The second term within the square root of equation (8.40) becomes small compared with the first term and, therefore, the simplified equation for the specific rate is

$$R = C_{A,e}\sqrt{D_A k_2 C_{B0}} = k_{app} C_{A,e} \tag{8.42}$$

where

$$k_{app} = \sqrt{D_A k_2 C_{B0}} \tag{8.43}$$

The rate of absorption becomes a function of the physicochemical factors without depending on the hydrodynamic conditions in the system if the condition given by equations (8.39) and (8.41) are satisfied. The concentration of carbon dioxide at the interphase, $C_{A,e}$ can be calculated from its partial pressure in the gas phase by Henry's law as per equation (8.4). In mole fraction, it can be expressed as $y = P_A/P_t$. The gas phase resistance does not exceed about 10% of total resistance and can be neglected (Porter et al., 1966; Voyer and Miller, 1968; Nagel et al., 1970; Mandal, 2010). The following equation can be derived from the material balance based on the method suggested by Voyer and Miller (1968) for the rate of absorption in a differential volume dv as

$$Radv = aC_{A,e} k_{app} dv \tag{8.44}$$

From a material balance equation on a solute-free basis, one can write

$$G' = G(1-y) \tag{8.45}$$

Therefore, the rate of absorption in a differential volume dv is

$$-d(Gy) = -d\left(\frac{G'y}{1-y}\right) = -\frac{G'dy}{(1-y)^2} \tag{8.46}$$

From equations (8.44) and (8.46), it can be written as

$$aC_{A,e}k_{lR}dv = -\frac{G'dy}{(1-y)^2} \tag{8.47}$$

Substituting the value of $C_{A,e}$ from equation (8.4), one can get

$$\frac{ak_{lR}yP_t}{H}dv = -\frac{G'dy}{(1-y)^2} \tag{8.48}$$

Therefore,

$$\frac{ak_{lR}P_t}{G'H}dv = -\frac{dy}{y(1-y)^2} \tag{8.49}$$

Integrating equation (8.49) over the total system volume, one can obtain,

$$\frac{ak_{LR}P_t}{G'H}\int_0^V dv = -\int_{y_i}^{y_o}\frac{dy}{y(1-y)^2} \tag{8.50}$$

which, on integration and rearrangement, gives the following expression for interfacial area:

$$a = \frac{G'H}{k_{lR}P_tV}\ln\left[\frac{1-y_o}{y_o}\frac{y_i}{1-y_i}\right] + \left[\frac{1}{1-y_i} - \frac{1}{1-y_o}\right] \tag{8.51}$$

Equation (8.51) can be used to determine the specific interfacial area, provided the conditions for the pseudo first-order reaction (equation [8.39]) and fast reaction (equation [8.41]) are satisfied. However, under certain conditions, if the reaction is essentially in the controlled diffusional regime, the gas component concentration in the center of the liquid phase must be practically zero, yet there should be no significant reaction in the film. At that condition, the volumetric mass transfer coefficients can be easily calculated by chemical methods. The sulphite oxidation and CO_2 absorption in carbonate buffer solution are suitable in this regard. The absorption of CO_2 into a carbonate–bicarbonate buffer solution is a convenient system and is represented by the second-order reaction (equation [8.36]). Under certain conditions, the above reaction is sufficiently fast, so that the concentration of CO_2 in the bulk of the liquid phase is equal to zero. The condition to be satisfied for this is

$$k_la \ll (1-\varepsilon_g)k_2C_{B0} \tag{8.52}$$

Furthermore, the reaction rate is such that no appreciable amount of reaction takes place in the diffusion films, and the condition to be satisfied for this is given by

$$\frac{D_A k_2 C_{B0}}{k_l^2} \ll 1 \tag{8.53}$$

When the condition given by equations (8.52) and (8.53) are satisfied, the rate of mass transfer is given by

$$Ra = k_l a C_{A,e} \tag{8.54}$$

By measuring, Ra, the rate of mass transfer per unit volume, the value of volumetric mass transfer coefficient, $k_l a$, can be determined. For the calculation of the interfacial area and the liquid-side mass transfer coefficient, values of diffusivity of CO_2 in the solutions, D_A; the Henry's law constant, H; and the second-order reaction velocity constant, k_2 are required. Furthermore, for calculating the concentration, C_{B0}, of the OH^- ions in carbonate bicarbonate buffer solution, the equilibrium constant of water, K_W, and the stoichiometric second ionization constant, $K_{I,2}$, are required. A considerable amount of information is available on the diffusivities of dissolved carbon dioxide in water, organic solvent, and electrolyte solution at various temperatures. The diffusivity of carbon dioxide, D_A, in aqueous solutions was obtained by the equation

$$D_{A,solution} = D_{A,water} \left[\frac{\mu_{water}}{\mu_{solution}} \right]^{0.85} \tag{8.55}$$

which was reported by Nijsing et al. (1959). The diffusivities in water at various temperatures can be obtained from the data of Nijsing et al. (1959) and were used with the Nernst-Einstein equation,

$$\frac{D_A \mu}{T} = C_1 = \text{Constant} \tag{8.56}$$

The solubility of the gas in the electrolyte solution can be determined by the following equations (Nijsing et al., 1959; Sharma and Danckwerts, 1970), taking the ionic strength of the solution into consideration. The solubility of carbon dioxide in water, S_W, is given by

$$\ln S_W = \frac{1140}{T} - 5.30 \tag{8.57}$$

where S_W is in gmole/(atm lit) and T in Kelvin. Solubility in electrolyte solution, S, can be obtained by the following equation:

$$\ln \frac{S}{S_W} = -hI \tag{8.58}$$

I is the ionic strength of the solution defined by

$$I = \tfrac{1}{2} \sum c_i z_i^2 \tag{8.59}$$

c_i being the concentration of ions of valency z_i. The parameter h is the sum of contributions referring to the species of positive and of negative ions present and to the species of gas:

$$h = h_+ + h_- + h_g \tag{8.60}$$

The values of the contribution due to cation, anion, and gas, h_+, h_- and h_g, can be obtained from the data as reported by Danckwerts (1970). The Henry's law constant, H, is directly can be calculated from the solubility. The second-order reaction velocity constant, k_2, can be determined by the equation suggested by Porter et al. (1966)

$$k_2 = \left[\frac{C_{H^+} C_{HCO_3^-}}{C_{CO_3^{-2}}} \right] \tag{8.61}$$

$$\ln k_2 = 16.635 - \frac{2895}{T} + 0.132 I \tag{8.62}$$

where k_2 is in cm^3/gmoles, T in °K, and the ionic strength of solution I is in gmoles/cm^3. The concentration of hydroxyl ion in a carbonate–bicarbonate mixture can be obtained by the following expression:

$$C_{B0} = \frac{K_W}{K_{1,2}} \frac{C_{CO_3^{2-}}}{C_{HCO_3^-}} \tag{8.63}$$

where

$$K_W = C_{H^+} C_{CO_3^{2-}} \tag{8.64}$$

and

$$K_{1,2} = \frac{C_{H^+} C_{CO_3^{2-}}}{C_{HCO_3^-}} \tag{8.65}$$

The stoichiometric second ionization constant, $K_{1,2}$, can be calculated by the following equation (Harned and Owen, 1958).

$$\ln K_{1,2} = -\frac{2902.4}{T} - 0.238T + 6.498 \tag{8.66}$$

The ionic product of water, K_W, is available from standard references (Danckwerts, 1970; Mandal, 2010). Many researchers have used the chemical method for determining the interfacial area and consequently mass transfer coefficient in different types of equipment. Schumpe and Deckwer (1980) measured the interfacial area by two different chemical methods, sulfite oxidation and CO_2 absorption into alkali solution in the same contactor. They observed a lower effective interfacial area by carbon dioxide absorption method compared with sulfite oxidation. The interfacial area in horizontal tubes has been measured by Biswas and Mitra (1981), vertical column by Kasturi and Stephanek (1974), and in a sieve tray by Eben and Pigford (1965). Studies on ejector or Venturi-type bubble columns showed a high gas–liquid interfacial transfer area. Cramers et al. (1992) determined mass transfer characteristics in an inverse flow liquid jet ejector by cobalt catalyzed sulfite oxidation method. Dutta and Raghavan (1987) measured the interfacial area and mass transfer coefficient in an ejector loop reactor by absorbing carbon dioxide in an aqueous sodium hydroxide solution and in an aqueous buffer solution of sodium carbonate–bicarbonate, respectively. Raghuram et al. (1992) developed a model for a fast pseudo first-order reaction with an ejector-contactor system and assume that the contactor comprises a series of stirred vessels followed by a plug-flow reactor. They adopted the process of absorption of carbon dioxide in sodium hydroxide solution to test the model and proved the acceptability of the model with the experimental results. Evans et al. (2001) analyzed the performance of confined plunging liquid jet bubble column by measuring the volumetric mass transfer coefficient in chemical method in the presence of a hypochlorite catalyst. The characterization of volumetric mass transfer coefficient of different plunging liquid jet systems is also described by several authors (Van de Sande, 1975; Evans and Machniewski, 1999; Ide et al., 2001; Mandal, 2010; Sivaiah and Majumder, 2013).

Danckwerts Plot Method

The Danckwerts plot technique is a recognized method for simultaneous determination of the gas–liquid mass transfer coefficient (k_l) and the specific interfacial area (a). From the experimentally obtained gas absorption rate, R_A (mole/s), at different apparent first-order reaction rate constants, the values of the liquid side mass transfer coefficient and specific interfacial area can be determined simultaneously using the Danckwerts surface renewal model (Danckwerts, 1970) with a (pseudo first-order reaction:

$$R_A/V_l = aC_{A,e}\left(D_A k_{l,app} + k_l^2\right)^{0.5} \tag{8.67}$$

or

$$\left(\frac{R_A}{S_w C_A V_l}\right)^2 = k_{l,app} D_A a^2 + (k_l a)^2 \tag{8.68}$$

where $C_A = S_w C_{A,e}$. Plot of $\left(R_A/(S_w C_A V_l)\right)^2$ versus $k_{l,app} D_A$ gives a straight line with slope of a^2 and an intercept $(k_l a)^2$. This is known as "Danckwerts plot." Both the interfacial area and liquid side mass transfer coefficient can be calculated from the slope and intercept provided that D_A and $C_{A,e}$ are known. The reactivity of the solution (the apparent first-order rate constant) is changed by variation of the catalyst concentration in a catalyzed pseudo first-order reaction or by changing the bulk concentration of the reactant that is used in excess in a pseudo first-order reaction. A reaction system that is very suitable for this technique is the absorption of CO_2 in carbonate–bicarbonate buffer solutions. The dissolved CO_2 can react with water as well as with hydroxyl ions, which are formed from the equilibrium between carbonate and bicarbonate ions. The overall reaction that occurs is

$$CO_2 + CO_3^{2-} + H_2O \rightarrow 2HCO_3^- \tag{8.69}$$

The reaction of CO_2 and water can be catalyzed by hypochlorite, arsenite, carbonic anhydrase, and different sugars. The apparent first-order rate constant for this system is given by (Cents, 2005a)

$$k_{1,app} = k_{H_2O} + k_{OH} - C_{OH^-} + k_{cat}C_{cat} \tag{8.70}$$

For a convenient and accurate description of the absorption process, it is important that the reaction can be assumed to be pseudo first- in the hydroxyl or catalyst concentration. To consider the reaction to be pseudo first order, it is important that the concentrations of all the ionic species (CO_3^{2-}, HCO_3^-, H^+ and OH^-) up to the interface are uniform and identical to the bulk concentrations, that is, there is no depletion of ionic species within the mass transfer zone (Cents et al., 2005a). The criterion that determines whether the concentrations of all ions are uniform throughout the mass transfer zone was derived by Danckwerts (1970):

$$S_w C_{CO_2} \left(\frac{1}{C_{CO_3^{2-}}} + \frac{2}{C_{HCO_3^-}} \right) \times \left\{ \left(1 + \frac{D_{CO_2} k_{l,app}}{k_l^2} \right)^{0.5} - 1 \right\} << 1 \tag{8.71}$$

The catalyst ion is assumed to be fully dissociated for the validity of the equation, which is generally the case when sodium hypochlorite is used as a catalyst (Danckwerts, 1970). Cents et al. (2005b) reported that although the left-hand side of equation (8.71) is much smaller than 1, large concentration gradients can still exist, and thus considerable deviations from the pseudo first-order region may occur. They also stated that the criterion is based on the fact that the transport of carbonate from the bulk can become the limiting step in the mass transfer process. Because of the influence of both the carbonate and bicarbonate ions on the hydroxyl concentration in the mass transfer zone, a possible reduction in flux is difficult to predict on beforehand. They suggested that information regarding this criterion is very important, both for research activities (the above-mentioned Danckwerts-plot technique), as well as for industrial purposes (e.g., the Benfield process; Benson et al., 1954) in which CO_2 is frequently removed from gas streams at high pressure using chemical absorption in a liquid. For more details of the reaction system, readers can follow the published article by Cents et al. (2005b).

Design Criteria for Physical Mass Transfer Experiments

To accurate estimation of $k_l a$ by the physical method with the nonreactive gas, the experiments should be well designed. A continuous operation of the liquid phase is more appropriate compared with batchwise use of liquid (Linek et al., 1989). Two choices must be made to arrive at the most accurate method to determine the mass transfer rate: (1) the outlet concentration can be measured in the gas phase or in the liquid phase or (2) the direction of mass transfer of the gas phase component such as absorption or desorption. For the first selection, a criterion is derived by Cents et al. (2005a) based on the relative change in outlet concentration with a variation in $k_l a$. According to Cents et al. (2005a), in a system with a continuous and well-mixed liquid phase and a continuous well-mixed gas phase, the steady-state mass balances for the component to be transferred in the gas and the liquid phases, respectively, can be expressed as

$$Q_g(C_{g,0}-C_g)-k_l a(mC_g-C_l)V_l=0 \tag{8.72}$$

$$Q_l(C_{l,0}-C_l)+k_l a(mC_g-C_l)V_l=0 \tag{8.73}$$

The volumetric mass transfer coefficient can be determined from these two balances by making use of the steady-state concentration in the gas or in the liquid phase outlet stream, which is measured experimentally. The $k_l a$ based on gas phase and on the liquid phase are, respectively,

$$k_l a_g = \frac{C_{g,0}-C_g}{(V_l/Q_g)(mC_g-C_{l,0})+(V_l/Q_l)(C_g-C_{g,0})} \tag{8.74}$$

$$k_l a_l = \frac{C_l-C_{l,0}}{m(V_l/Q_g)(C_{l,0}-C_l)+(V_l/Q_l)(mC_{g,0}-C_l)} \tag{8.75}$$

Cents et al. (2005a) reported that for both of these cases, it is possible to calculate the change in the outlet concentration in the steady state with a small variation in $k_l a$; thus, $dc/d(k_l a)$ with different values of the distribution coefficient, m. A small variation in $k_l a$ results in a large value for $dc/d(k_l a)$, which may cause a relatively large change in the experimentally measured outlet concentration. This can improve the accuracy of the measurement. In the case of oxygen desorption, measurement of the concentration in the liquid phase is more accurate, and in the case of desorption of CO_2, the concentration is preferably determined in the gas phase because of the variation of distribution coefficient of the gases in the liquid. Cents et al. (2005a) derived a relation from the ratio of this parameter using both gas and liquid phase analysis. This ratio can be determined, and is identical, for both absorption ($C_{l,i}=0$) and desorption ($C_{g,i}=0$) experiments, which can be expressed as

$$\frac{dC_g/d(k_l a_g)}{dC_l/d(k_l a_l)}=\frac{mQ_l}{Q_g} \tag{8.76}$$

For $mQ_l > Q_g$, concentration measurement in the gas phase is favorable and vice versa. They also reported that the second selection (absorption or desorption) depends on the first one because desorption is favored in case of liquid concentration measurement and absorption in case of gas phase measurement. They suggested that the dimensionless concentration ratio in the reactor (C/C_0) can be used and absorption should be used for gas phase analysis and desorption for liquid phase experiments.

Other Methods of Estimation of Gas–Liquid Mass Transfer Coefficient in a Bubble Column

Three other methods were developed by Inga (1996) for the calculation of the gas–liquid volumetric mass transfer coefficient in the bubble column reactor, namely, the integral method, the differential method, and the multiple linear regression method. These three methods gave similar results; however, the selection of one of them should be based on the mathematical stability of the final function. In the following, a summary of these methods is given as follows:

The Integral Method

The calculated solubility values can be modeled by Henry's law as

$$C_e = \frac{P_t - P_e}{H} \tag{8.77}$$

The total number of moles transfer, N_0, can be written as

$$N_0 = N_g + N_l = \frac{(P_t - P_e)V_g}{RT} + C_l V_l \tag{8.78}$$

Using equations (8.77) and (8.78), the rate of mass transfer between phases can be written as

$$\frac{V_g}{RT}\frac{dP_t}{dt} = -(V_l k_l a)\left(\frac{P_t - P_e}{H} - \frac{N_0 RT - (P_t - P_e)V_g}{V_l RT} \right) \tag{8.79}$$

Separating the variables and integration of equation (8.79) gives

$$\left(\frac{V_g H}{V_l RT + HV_g} \right)\ln\left\{ \left(\frac{(P_t - P_e)(V_l RT + HV_g)}{HV_l RT} \right) - \frac{N_0}{V_l} \right\} = -k_l a.t + C_I \tag{8.80}$$

If the left-hand side of the equation (8.80) is plotted versus time t, it gives a straight line. The slope of this line represents the value of $k_l a$.

The Differential Method

Equation (8.80) can be solved by approximating dP as ΔP and dt as Δt if the experimental reading is taken with a fraction of second ($\Delta t < 0.05$ s). Equation (8.79) can be expressed as

$$\lambda \Delta P = -V_l k_l a.\Delta t$$

where

$$\lambda = \left(\frac{V_g}{RT}\right) \bigg/ \left(\frac{P_t - P_s}{He} - \frac{N_0 RT - (P_t - P_s)V_g}{V_l RT}\right) \tag{8.81}$$

If the ratio between $(\lambda \Delta P)$ and $(-V_l \Delta t)$ is constant, the resulting value will correspond to $k_l a$.

The Multiple Linear Regression Method

This method can be expressed based on the linearization of equation (8.79). The function can be rewritten as a linear expression as

$$A_1 \Delta P = A_2 P_t \Delta t + A_3 \Delta t + A_4 \tag{8.82}$$

where

$$A_1 = -\frac{V_g}{V_l RT k_l a} \tag{8.83}$$

$$A_2 = \frac{V_l RT + V_g H}{V_l RTH} \tag{8.84}$$

$$A_3 = -\frac{N_0}{V_l} - P_e \left(\frac{V_l RT + V_g H}{V_l RTH}\right) \tag{8.85}$$

$$A_4 = \text{Constant} \tag{8.86}$$

Using multiple linear regression, the coefficients A_1, A_2, A_3, and A_4 can be found. A_4 should be zero without forcing the data to go through the origin. The equilibrium concentration can be expressed as

$$C_e = \left(\frac{V_l RT + V_g H}{V_l RTH}\right) P_e - \frac{V_g}{V_l RT} \tag{8.87}$$

Mass Transfer Coefficient in Inverse Bubbly Flow Columns

Gas–Liquid Interfacial Mass Transfer

The mass transfer characteristics in an ejector-induced inverse bubbly flow column was investigated by Mandal (2010) with the liquid phase reaction between dissolved carbon dioxide and sodium hydroxide, sodium carbonate–bicarbonate solutions. The schematic of experimental setup is shown in Figure 8.2. The values of both interfacial area and liquid side mass transfer coefficient $k_l a$ were obtained by chemical methods. A 50-L capacity vessel was installed in the line to provide an undisturbed supply of constant composition air–CO_2 mixture. Before the mixture entered the column, a small portion was collected

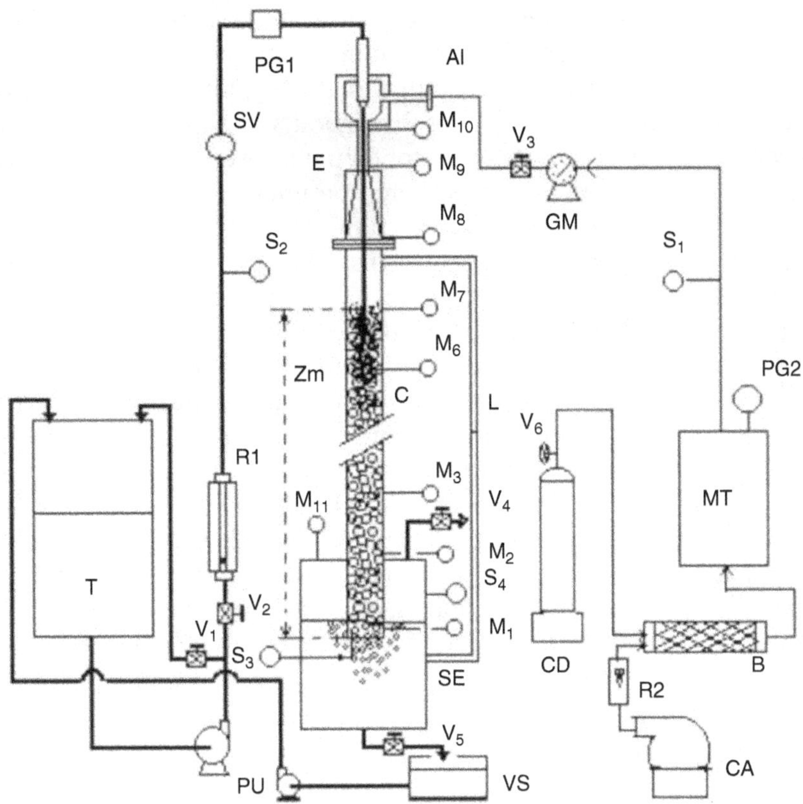

FIGURE 8.2 Schematic diagram of the experimental setup of mass transfer operation in an ejector-induced inverse bubble column (Mandal, 2010). *AI*, Air inlet; *B*, gas mixing column; *C*, contactor; *CA*, compressed air; *CD*, CO_2 gas cylinder; *E*, ejector assembly; *GM*, gas flowmeter; *L*, clear liquid column; M_1-M_{11}, manometers; *MT*, mixture tank; *PG1-2*, pressure gauge; *PU*, pump; *R1-2*, rotameter; S_1-S_4, sampling points; *SE*, separator; *SV*, solenoid valve; *T*, circulating tank; V_1-V_6, valves; *VS*, collector vessel.

for analysis. This air–CO_2 mixture was used as the secondary fluid. A sodium hydroxide solutions of desired composition were used as the primary fluid for interfacial area measurement, and sodium carbonate–bicarbonate solutions were used for the measurement of the liquid-side volumetric mass transfer coefficient. The ratio of CO_3^{2-} to HCO_3^- in the buffer solution was maintained within 1.6 to 4.3 for the measurement of the mass transfer coefficient. The inlet liquid sample was collected at a particular point. Outlet gas and liquid samples were also collected as sampling for further analysis. The gas samples were analyzed by Orsat apparatus, and liquid samples were analyzed by standard methods of titration. The experiments were conducted by pumping the primary fluid through the nozzle. The secondary fluid was dispersed by the jet momentum and two-phase mixture moved through the contactor, with transfer of the solute gas into the liquid phase.

The inlet fluid flow and CO_2 concentration were maintained at their desired values. After the system had attained steady state, the flow rate of the gas and liquid, system pressure, and system temperature were noted, and the gas and liquid samples were collected at the sampling points for subsequent analysis. The authors reported that their inverse flow bubble column with ejector-type gas distributor provided a large interfacial area and volumetric mass transfer coefficient. The gas flow rates throughout the experiments were maintained close to the self-entrainment rate of the ejector for the measurement of the interfacial area and volumetric mass transfer coefficient. Their experimental results were found to satisfy the conditions for "fast pseudo first-order reaction" (Danckwerts, 1970) as given by equations (8.39) and (8.41). A specific interfacial area can be obtained by using equation (8.51) with knowledge of D_A, k_2, and H together with the data of inlet and outlet concentration of carbon dioxide, volume of gas liquid mixture in the column, and gas flow rate. The Henry's law constant, H, in ionic solution can be directly obtained from solubility, S. D_A (i.e., diffusivity of carbon dioxide in aqueous solution) was calculated from an equation suggested by Nijshing et al. (1959), and k_2 was obtained from equation (8.62) proposed by Porter et al. (1966). They reported that increase in secondary gas flow rate is attributed to an increase in the bubble population and hence the gas hold-up. Therefore, a higher interfacial area can be observed at higher liquid and gas flow rates. They showed that the specific interfacial area increases with an increase in the rate of energy dissipation for mixing of the phases per unit volume of phase dispersion as shown in Figure 8.3, which are almost similar in trend reported by many authors (Stein and Schafer, 1984; Dutta and Raghavan, 1987; Ide et al., 2001; Bouaifi et al., 2001).

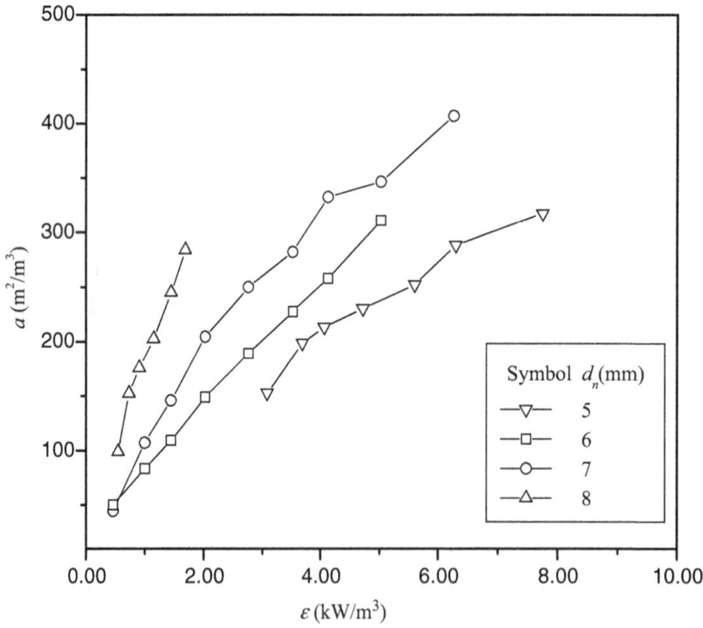

FIGURE 8.3 Variations of the interfacial area with specific rate of energy input by chemical method.

The interfacial area is strongly dependent on the superficial gas velocity because of the increase in population of gas bubbles and hence gas hold-up. Mandal et al. (2003) developed a correlation with superficial gas velocity from their experimental data, which is given by

$$a = 0.38 \times 10^4 u_{sg} \tag{8.88}$$

They also reported that the interfacial area obtained by the CO_2 absorption method from their experiment were 100 to 400 m^2/m^3. Thus, it was found that the interfacial area measured by the physical method is much higher than that of the chemical method. These discrepancies are not unusual because the physical method always provides a higher interfacial area, as reported by Schumpe and Deckwer (1980). They showed that among chemical methods, sulfite oxidation gives four times the interfacial area compared with the CO_2 absorption method. Kuerten and Zehner (1978) reported that that sulfite oxidation gives an interfacial area that is very close to data obtained by the physical method. The rate of mass transfer per unit volume, Ra, can be obtained from the gas flow rate and the composition of CO_2 at the inlet and outlet. From the knowledge of Ra and C_{Ae}, liquid-side mass transfer coefficients, k_la can be calculated from equation (8.54). Mandal et al. (2003) showed that the k_la increases with an increase in gas flow rate at a constant liquid flow rate because of increased gas bubble population, and nozzle size has little effect on k_la. For a constant height of gas–liquid mixture in the column, simultaneous effects of superficial liquid and gas velocity on k_la are shown in Figure 8.4 at a constant nozzle diameter ($d_n = 0.004$ m). Similar to the interfacial area, the volumetric mass transfer coefficient was

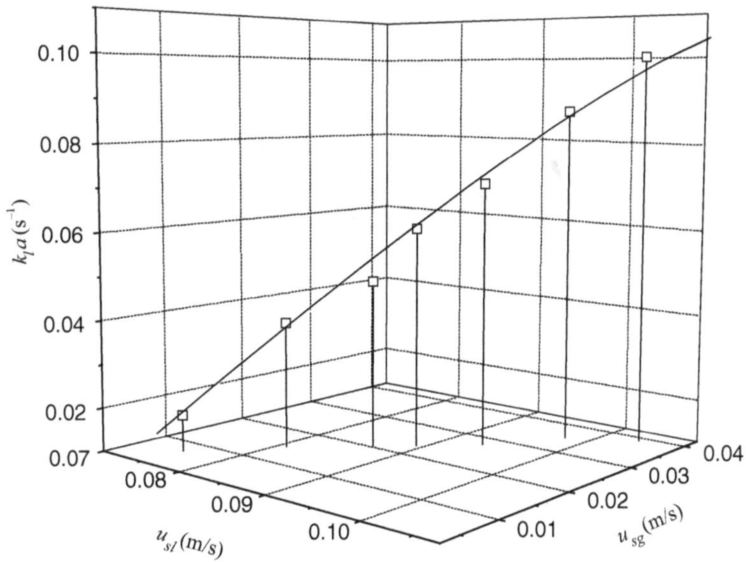

FIGURE 8.4 Variation of k_la with superficial liquid and gas velocity (Mandal, 2010).

also reported (Mandal et al., 2003) to be a strong function of superficial gas velocity, and it can be expressed as

$$k_l a = 1.80 u_{sg}^{0.92} \tag{8.89}$$

Kawase et al. (1987) showed that the exponential dependency of $k_l a$ on superficial gas velocity is 0.92, but the coefficient reported by Mandal et al. (2003) is somewhat greater, thus having higher values of $k_l a$. Shah et al. (1982) proposed the following correlation:

$$k_l a = 0.467 u_{sg}^{0.82} \tag{8.90}$$

Equation (8.90) also predicts lower values of $k_l a$ compared with the data reported by Mandal et al. (2003), which may be attributed to efficient dispersion of the gas in the inverse bubbly flow system by the liquid jet. Mandal et al. (2003) also reported the performance of the inverse bubbly flow system based on rate of energy dissipation per unit volume of phase dispersion compared with other systems, which is shown in Figure 8.5. Evans et al. (2001), with a confined liquid jet plunging bubble column, obtained significantly higher values of $k_l a$. But the values of $k_l a$ obtained in the inverse bubbly flow system as reported by Mandal (2010) are comparable with Tojo and Miyanami (1982), Dutta and Raghavan (1987), and Ide et al. (2001). Detailed comparative pictures of different gas–liquid contactors under different operating conditions are presented Table 8.2. If the values of volumetric mass transfer coefficient, $k_l a$, and

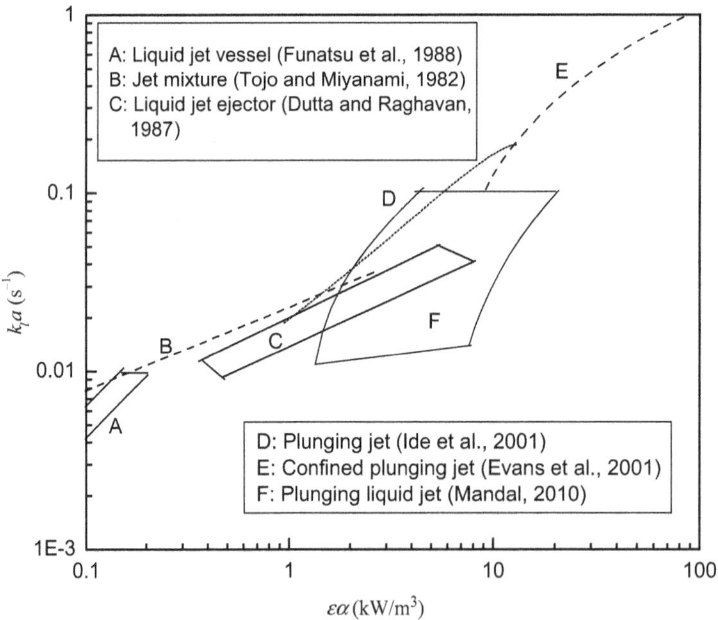

FIGURE 8.5 Comparison of $k_l a$ for different gas–liquid contacting devices.

Table 8.2 Comparison of Liquid-Side Volumetric Mass Transfer Coefficient with Specific Power Input for Different Gas–Liquid Contactors

Type of Reactor	u_{sl} (m/s)	u_{sg} (m/s)	ε_g (-)	ε (Kw/m³)	$k_l a$ (s⁻¹)	Reference
Inverse bubbly flow (jet system)	—	0–0.0133	—	0.50–5.0	0.01–0.03	Stein and Schafer (1984)
Inverse bubbly flow column	0.24–0.33	0.002–0.013	0.02–0.15	0.003–0.004	0.011–0.075	Kulkarni and Shah (1984)
Jet loop reactor	0.01–0.10	0–0.03	0.05–0.40	1.0–8.0	0.01–0.02	Wachsmann et al. (1985)
Jet loop reactor	0.001–0.005	0.001–0.006	0.02–0.15	0.4–10.0	0.01–0.08	Dutta and Raghavan (1987)
Inverse bubbly flow column	0.30–0.60	0.001–.02	0.10–0.40	0.02–0.10	0.02–0.10	Ohkawa et al. (1987)
Water jet vessel	—	—	—	0.015–0.70	0.003–0.01	Funatsu et al. (1988)
Liquid jet absorber	—	—	—	0.03–4.50	0.001–0.08	Bin (1993)
Aero-ejector contactor	0–0.80	3.40–22.5	0.04–0.25	0.20–10.0	0.015–0.10	de Billerbeck et al. (1999)
Upflow ejector loop reactor	0.017–0.025	0.02–0.09	0.15–0.35	0.10–5.0	0.03–0.20	Havelka et al. (2000)
Bubble column	—	0.0025–0.04	0.01–.30	0.50–1.00	0.005–0.10	Bouaifi et al. (2001)
Plunging jet bubble column	0.10–0.183	—	0.15–0.25	8.0–80.0	0.10–0.60	Evans et al. (2001)
Upward Venturi bubble column	0.175–0.353	0.05–0.25	0.05–0.30	0.50–6.0	0.10–0.45	Huynh et al. (1991)
Plunging jet absorber	0.028–0.071	—	0.09–0.35	1.0–15.0	0.02–0.20	Ide et al. (2001)
Concentric-tube airlift reactors	—	0.001–0.10	0.01–0.16	0.81–12.93	0.01–0.07	Cerri et al. (2009)
Ejector-induced Inverse bubbly flow	0.07–0.12	0.01–0.08	0.40–0.60	0.50–5.0	0.01–0.10	Mandal (2010)
Ejector-induced Inverse bubbly flow	0–0.14	0–0.034	0–0.35	0.278–20.87	$k_l = (0.46$–$0.83) \times 10^{-4}$, m/s	Sivaiah and Majumder (2013)

interfacial area, a, are known, the true liquid-side mass transfer coefficient, k_l, can be easily calculated as follows:

$$k_l = \frac{k_l a}{a} \qquad (8.91)$$

A series of empirical and semiempirical correlations has been proposed for liquid-side mass transfer coefficients in gas–liquid dispersions as shown in Table 8.3. The k_l values are a function of bubble size and liquid flow pattern around the bubbles. Calderbank and Moo-Young (1961) carried out detailed and extensive research on mass transfer coefficient k_l as a function of bubble diameter in stirred tank reactors and bubble layers. A rule of thumb for low-viscosity media is that k_l values range between 1.5×10^{-4} and 3.5×10^{-4} m/s. The k_l values reported by Mandal (2010) ranged between 2.5×10^{-4} and 5.0×10^{-4} m/s,

Table 8.3 Correlations for Liquid Side Mass Transfer Coefficient (kl)

Correlations	System	Conditions	Reference
$k_l = 0.5 g^{5/8} D_{AB}^{1/2} \rho_l^{3/8} \sigma^{-3/8} d_{32}^{1/2}$	Air, O_2–H_2O, glycol, methanol	$u_{sg} = 0.003\text{-}0.4$ m/s $u_{sl} = 0\text{-}0.044$ m/s $d_c = 0.152\text{-}0.6$ m $L_c = 1.26\text{-}3.5$ m $P = 0.1$ MPa $T = 298$ K	Akita and Yoshida (1974)
$k_{l(CMC)} = 1.12 \times 10^{-4} u_{sg}^{-0.03} \mu_{eff}^{-1/2}$ $\mu_{eff} = k(2800 u_{sg})$ k = flow consistency index ($Pa\ s^n$); $k = 1.97$, $\quad n = 0.951$	Air–H_2O, CMC, sodium sulfate	$P = 0.1$ MPa $T = 298$ K $u_{sg} = 0\text{-}0.24$ m/s $d_c = 0.305$ m $L_c = 3.4$ m	Godbole et al. (1984)
$k_l = Ku_{sg}^{-0.05} \mu_{eff}^{0.15}$ $K = 9.7 \times 10^{-5}$ (H_2O/salt solution), $K = 6.45 \times 10^{-5}$ (H_2O, 0.8M Na_2SO_4) $\mu_{eff} = k(2800 u_{sg})$ k = flow consistency index (Pa s^n); $k = 1.97$, $\quad n = 0.951$	N_2, O_2–H_2O, 0.8M Na_2SO_4–carbon, Kiselguhr, aluminum oxide	$u_{sg} = 0\text{-}0.07$ m/s $d_c = 0.095$ m $L_c = 0.85$ m $P = 0.1$ MPa $T = 298$ K	Schumpe et al. (1987)
$\dfrac{k_l d_{32}}{D_{AB}} = 2 \times 10^{-4} Sc^{0.5}(Re\,Mo^{0.15})^3$ $Sc = \mu_l/(\rho_l D_{AB})$; $Re = \rho_l u_{sg} d_c/\mu_l$; $Mo = (g\mu_l^4)/(\rho_l \sigma^3)$	Air–H_2O, glycerol, ethanol solutions, CMC–polystyrene	$L_c = 0.1, 1.5$ m $d_c = 0.8, 1.48$ m $u_{sl} = 0.02\text{-}0.06$ m/s $P = 0.1$ MPa $T = 298$ K	Miyahara et al. (1997)
$\dfrac{k_l}{u_{sg}} = 0.105\left(Re\,Fr Sc^2\right)^{-0.268}$ $Fr = u_{sg}/(gd_c)^{0.5}$ *for gas;* $= u_{sl}/(gd_c)^{0.5}$ *for liquid*	N_2/Fe(CN), NaOH, CMC, $HNaCO_3$, Na_2CO_3–glass, diatomite, silicon carbide, alumina	—	Neme et al. (1997)
$k_l = K\sigma^{1.35} u_{sg}^{0.5}$ $K = 0.17587$ (for $d_h = 150\text{-}200$ µm), $K = 0.18233$ $\quad$ (for $d_h = 90\text{-}50$ µm), $K = 0.18689$ $\quad$ (for $d_h = 40\text{-}90$ µm)	CO_2–sodium carbonate, bicarbonate, surfactant and sodium arsenite	$L_c = 1.086$ m $d_c = 11.3$ m $u_{sg} = 0.00087\text{-}0.00181$ m/s $P = 0.1$ MPa $T = 298$ K	Vázquez et al. (2000)

Correlation	System	Conditions	Reference
$k_l = 1.13 D_{AB}^{1/2} u_{sp}^{1/2} d_{32}^{-1/2}$	Air–H_2O	d_c = 0.15 and 0.20 m L_c = 2 m u_{sg} = 0.002-0.04 m/s. P = 0.1 MPa T = 298 K	Bouaifi et al. (2001)
$\dfrac{k_l d_{32}}{D_{AB}} = 1.546 \times 10^2 Eu^{0.052} Re^{0.076} Sc^{-0.231}\ for\ H_2$ $\dfrac{k_l d_{32}}{D_{AB}} = 8.748 \times 10^2 Eu^{-0.012} Re^{0.024} Sc^{-0.133}\ for\ CO$ $Eu = P/(\rho_l u_b d_b^2)$	H_2, CO–paraffin oil–silica gel	d_c = 0.037 m, L_c = 0.480 m T = 573 K P = 0.1-6 MPa u_{sg} = 0.003- 0.020 m/s	Yang et al. (2003)
$\left(\dfrac{k_L}{u_{sl}}\right) = 10.83 \left(\dfrac{d_c u_{sl} \rho_{sl}}{\mu_{sl}}\right)^{-0.698} \left(\dfrac{u_g^2}{d_c g}\right)^{-0.777} \left(\dfrac{\mu_{sl}}{\mu_g}\right)^{0.085}$	Air–electrochemical red-ox system in the presence of an excess indifferent electrolyte (sodium hydroxide solution).	u_{sl} = 0-0.14 m/s u_{sg} = 0-0.034 m/s d_c = 0.05 m L_c = 1.6 m P = 0.12 MPa	Sivaiah and Majumder (2013)

CMC, Critical micelle concentration.

which are in good agreement with those found in the literature (Calderbank and Moo-Young, 1961; Kasturi and Stephanek, 1974; Ohkawa et al., 1987; Bouaifi et al., 2001). The overall mass transfer rate per unit volume of the dispersion in a bubble column is governed by the liquid-side mass transfer coefficient, $k_l a$, assuming that the gas-side resistance is negligible. In a bubble column reactor, the variation in $k_l a$ is primarily due to variations in the interfacial area. Ozturk et al. (1987) studied the mass transfer coefficient in alcohols, glycol solution, and various organic liquids in an upflow bubble column reactor. They compared their results with the correlations developed by Akita and Yoshida (1974) and Hikita et al. (1981) at superficial gas velocity more than 0.01 m/s. They reported that using the original form of the correlations gives better fit if a modification factor is allowed for the correlations with a slightly smaller error.

Even if modified by an optimized factor, the literature correlations describe the data with mean errors exceeding 20%. Ozturk et al. (1987) modified the Akita and Yoshida (1974) correlation by considering the density ratio of the phases as

$$Sh = 0.62 Sc^{0.50} Bo^{0.33} Ga^{0.29} Fr^{0.68} (\rho_g/\rho_l)^{0.04}. \tag{8.92}$$

The dimensionless numbers were varied in the following ranges: $1.6 \times 10^1 \leq Sh \leq 9.7 \times 10^2$; $3.2 \times 10^1 \leq Sc \leq 1.5 \times 10^5$; $1.6 \times 10^0 \leq Bo \leq 5.4 \times 10^0$; $8.3 \times 10^2 \leq Ga \leq 1.5 \times 10^6$; $4.3 \times 10^{-2} \leq Fr \leq 6.0 \times 10^{-1}$; $9.3 \times 10^{-5} \leq \rho_G/\rho_L \leq 2.0 \times 10^{-3}$. Some other important correlations for overall volumetric mass transfer coefficient are shown in Table 8.4.

Wall–Liquid Mass Transfer in Inverse Bubbly Flow Columns

Mass transfer on the column wall at forced two-phase flow is governed by ion transfer carried out by the gas–liquid flow between the test electrode (cathode) and reference electrodes (anode) in an electrochemical cell. The method is called electrodiffusion. In the diffusion limitation regime, the diffusion current depends only on the rate of ion supply to the test electrode surface, and therefore, it is the quantitative characteristic of mass transfer on a surface. The advantage of this method is the fact that it can be used with the change in probe configuration. To determine the mass transfer coefficient, it is necessary to measure current in redox reaction (equation [8.93]) on the surface of electrode installed on the wall as shown in Figure 8.6.

$$[Fe(CN)_6]^{3-} + e \rightarrow [Fe(CN)_6]^{4-} \tag{8.93}$$

The current in a measurement cell (cathode–solution–anode) is proportional to the mass transfer coefficient. The diffusion coefficients of reacting ions in the red-ox reaction stand for Schmidt number $Sc \approx 1500$. Thickness of diffusion boundary layer δ_d, where the main change in concentration of reacting ions occurs, is importantly less than thickness of hydrodynamic boundary layer δ_h, that is, $\delta_d/\delta_h \sim Sc^{-1/3}$. The use of electrochemical method for mass transfer measurement has an advantage over other conventional methods (Kottke and Blenke, 1970). It is appropriate to measure mass transfer and wall shear stress in one experiment. It is practically useful for determination of the interconnection between heat

Table 8.4   Correlations for Overall Volumetric Mass Transfer Coefficient

Correlations	System	Conditions	Reference
$\dfrac{k_l a d_{32}^2}{D_z \varepsilon_g} = 3.31\left(\dfrac{\mu_l}{\rho_l D_{AB}}\right)^{1/3}\left(\dfrac{d_{32}\rho_l u_{sg}}{\mu_l \varepsilon_g}\right)^{1/2}$	Air–H_2O	Quiscent regime	Fair et al. (1962)
$\dfrac{k_l a d_c^2}{D_{AB}} = 0.6\left(\dfrac{\mu_l}{\rho_l D_{AB}}\right)^{0.5}\left(\dfrac{gd_c^2\rho_l}{\sigma}\right)^{0.62}\left(\dfrac{gd_c^3\rho_l^2}{\mu_l^2}\right)^{0.31}\varepsilon_g^{1.1}$	Air, O_2–H_2O, glycol, methanol	u_{sg} = 0.003-0.4 m/s u_{sl} = 0-0.044 m/s d_c = 0.152-0.6 m L_c = 1.26-3.5 m P = 0.1 MPa T = 298 K	Akita and Yoshida (1974)
$\dfrac{k_l a u_{sg}}{g} = 14.9\left(\dfrac{\mu_l}{\rho_l D_{AB}}\right)^{-0.604}\left(\dfrac{u_{sg}\mu_l}{\sigma}\right)^{1.76}\left(\dfrac{\mu_l^4 g}{\rho_l\sigma^3}\right)^{-0.248}\left(\dfrac{\mu_g}{\mu_l}\right)^{0.243}$	Air, H_2, CO_2, CH_4, C_3H_8/H_2O, 30, 50wt% sucrose, methanol, n-butanol	P = 0.1 MPa T = 298 K u_{sg} = 0.042-0.38 m/s d_c = 0.1m L_c = 1.5 m	Hikita et al. (1981)
$k_l a = 8.35\times10^{-4} u_{sg}^{0.44}\mu_{eff}^{-1.01}$ $\mu_{eff} = k(2800u_{sg})$ k = flow consistency index (Pa s^n); k = 1.97, n = 0.951	Air–H_2O, CMC, sodium sulfate	P = 0.1 MPa T = 298 K u_{sg} = 0-0.24 m/s d_c = 0.305 m L_c = 3.4 m	Godbole et al. (1984)
$\dfrac{k_l a\sigma}{\rho_l D_{AB} g} = \dfrac{M}{N}$; $M = 2.11\left(\dfrac{\mu_l}{\rho_l D_{AB}}\right)^{0.5}\left(\dfrac{\mu_l^4 g}{\rho_l\sigma^3}\right)^{-0.159}\varepsilon_g^{1.18}$ $N = 1+1.47\times10^4 C_{s,V}^{0.612}\left(\dfrac{u_{t\infty}}{\sqrt{d_c g}}\right)^{0.486}\left(\dfrac{gd_c^2\rho_l}{\sigma}\right)^{-0.477}\left(\dfrac{u_{sg}d_c\rho_l}{\mu_l}\right)^{-0.345}$	N_2/H_2O, glycerol, glycol, barium chloride, sodium sulphate–glass and bronze	P = 0.1 MPa T = 298 K u_{sg} = 0.03-0.15 m/s C_{SV} = 0-200 kg/m^3 d_c = 0.1-0.3 m L_c = 2.3-3 m	Koide et al. (1984)
$\dfrac{k_l a d_b^2}{D_{AB}} = 0.62Sc^{-0.5}Bo^{0.33}Ga^{0.29}Fr^{0.68}\left(\dfrac{\rho_g}{\rho_l}\right)^{0.04}$ * *All dimensionless numbers in terms of d_b (rather than d_c)	Air, N_2, He, CO_2, H_2–17 pure organic liquid, 5 inherently mixed liquid, 17 adj. mixtures	P = 0.1 MPa T = 298 K u_{sg} = 0.008-0.1 m/s d_c = 0.095 m L_c = 0.85 m	Ozturk et al. (1987)
$k_l a = Ku_{sg}^{0.82}\mu_{eff}^{-0.39}$; K = 0.063 ($H_2O$/salt solution), K = 0.042 (H_2O, 0.8M Na_2SO_4); $\mu_{eff} = k(2800u_{sg})$ k = flow consistency index (Pa s^n); k = 1.97, n = 0.951	N_2, O_2–H_2O, 0.8 M Na_2SO_4–carbon, Kiselguhr, aluminum oxide	P = 0.1 MPa T = 298 K u_{sg} = 0-0.07 m/s d_c = 0.095 m L_c = 0.85 m	Schumpe et al. (1987)

(Continued)

Table 8.4 Correlations for Overall Volumetric Mass Transfer Coefficient (*cont.*)

Correlations	System	Conditions	Reference
$\dfrac{k_l a d_c^2}{D_{AB}} = 1.428 n^{11/6} Re^{(2+n)/(2+2n)} Bo^{3/5} Fr^{(11n-4)/(39(1+n))} Sc^{0.5}$	Air–H_2O, carbopol, CMC	Semitheoretical $P = 0.1$ MPa $T = 298$ K	Kawase et al. (1987)
$\dfrac{k_l a \sigma}{\rho_l D_{AB} g} = 12.9 Sc^{0.5} Mo^{-0.159} Bo^{-0.184} \varepsilon_g^{1.3}$ $\times \left[0.47 + 0.53 \exp\left(-41.4 \dfrac{\Pi_\infty k_1}{\mu_l u_p} Re_b^{-1/2} \right) \right] \times \left(\dfrac{1}{1 + 0.62 C_{s,V}} \right)$ $\Pi_\infty = -C_B \left(\dfrac{d\sigma}{dC_B} \right); \quad k_1 = -\left(\dfrac{d\sigma}{dC_B} \right) \left(\dfrac{3 u_p r_B}{2 D_B} \right)^{1/2} / (r_B RT)$ C_B = concentration of alcohol, mol/m^3; D_B = diffusivity of alcohol in the liquid, m^2/s	Air, N_2–H_2O, alcohol solutions– calcium alginate gel, polystyrene	$P = 0.1$ MPa $T = 298$ K $u_{sg} = 0$-0.15 m/s $CV = 20$ vol.% $d_c = 0.14, 0.218,$ 0.3 m $L_c = 1.5$ m	Salvacion et al. (1995)
$k_l a = u_{sg}^{0.9} \mu_{eff}^{-0.55} \rho_g^{0.46}$ $\mu_{eff} = k(2800 u_{sg})$ k = flow consistency index (Pa s^n); $k = 1.97$, $n = 0.951$	He, N_2, air, sulfur hexafluoride–0.8 M sodium sulfate + xantham gum–Kieselghur, alumina	$P = 1$-10 bar $T = 298$ K $u_{sg} = 0.01$-0.08 m/s $C_{sv} = 0$-18 vol.% $d_c = 0.115$ m $L_c = 1.37$ m	Dewes and Schumpe (1997)
$k_l a = C \varepsilon_g^D \left(\dfrac{\rho_l Q^2}{\sigma d_h^{-3}} \right)^E \left(\dfrac{P}{P_{atm}} \right)^F$ The coefficients C, D, E, and F depend on the enzyme concentration	N_2, O_2–H_2O, enzyme solutions (CE)	$P = 0.1$-1.1 Mpa $T = 298$ K $u_{sg} = 0.005$-0.15 m/s $C_E = 3$-163 mg/dm3 $d_c = 0.055$ m $L_c = 0.9$-1.2 m	Kojima et al. (1997)
$k_l a = K \times 10^{-3.08} \left(\dfrac{d_c u_{sg} \rho_g}{\mu_l} \right)^{0.245}$ K is the correlation dimension	Air–CMC	$P = 0.1$-0.6 Mpa $T = 298$ K $u_{sg} = 0.02$-0.2 m/s $\mu_l = 1$-38 mPa s $d_c = 0.152$ m $L_c = 2$ m	Kang et al. (1999)
$k_l a = K_h \left(\dfrac{u_{sg}^{2/3} \sigma^{3/4} \rho_l^{3/2}}{\mu_l^{3/4}} \right)$; K_h depends on the distributor hole diameter. $K_h = 1.924 \times 10^{-7}$ if $d_h = 150$-200 µm; $K_h = 1.969 \times 10^{-7}$ if $d_h = 90$-150 µm; $K_h = 2.079 \times 10^{-7}$ if $d_h = 40$-90 µm	CO_2–aqueous solution of sucrose and surfactants	$P = 0.1$ MPa $T = 298$ K $u_{sg} = \leq 0.0016$ m/s $d_c = 0.06$ m $L_c = 0.6, 0.9$ m	Alvarez et al. (2000)

Correlation	System	Conditions	Reference
$\dfrac{k_l a d_b^2}{D_{AB}} = A Sc^{0.5} Bo^{0.34} Ga^{0.27} Fr^{0.72}\left(1+13.2 Fr^{0.37}\left(\dfrac{\rho_g}{\rho_l}\right)^{0.49}\right)$ The coefficient A depends on the perforated plate dimensions. $A = 0.522$ for plate dimension of 19×1 mm; $A = 0.599$ for plate dimension of 1×1 mm; $A = 0.488$ for plate dimension of 1×4.3 mm	N_2, He–ethanol, 1-butaol, toluene, decalin	$P = 1$- 40 bar $T = 298$ K $u_{sg} = 0.01$-0.21 m/s $d_c = 0.1$ m $L_c = 2.4$ m	Jordan and Schumpe (2001)
$k_l a = 0.40 u_{sg}^{0.625} u_{sl}^{0.26} \exp(1.477 \times 10^{-5} H_{amf})$ H_{amf} is the applied magnetic field, A/m	Air–H_2O–nickel magnetized SBCR	$P = 0.1$ MPa $T = 298$ K $u_{sg} = 0$-0.04 m/s $H_{amf} = 0$-25000 A/m $d_c = 0.05$ m $L_c = 0.5$ m	Chen and Leu (2001)
$k_l a = 0.18 Sc^{-0.6}\left(\dfrac{\rho_l \upsilon_A}{M_B}\right)^{-2.84}\left(\rho_g u_{sg}\right)^{0.49} e^{-2.66 C_{s,v}}$	H_2, CO, N_2, CH_4–isopar-M, hexanes– glass beads, iron oxide	$P = 1.7$-7.9 bar $T = 298$ K $u_{sg} = 0.05$-0.25 m/s $C_{sV} = 0$-36 vol.% $d_c = 0.316$ m	Behkish et al. (2002)
$k_l a = 1.80 u_{sg}^{0.92}$	CO_2–water–NaOH–$NaHCO_3$, Na_2CO_3	$d_c = 0.051$ m $L_c = 1.5$ m $u_{sg} = 0.01$-0.09 m/s $u_{sl} = 0.05$-0.11 m/s	Mandal et al. (2004)

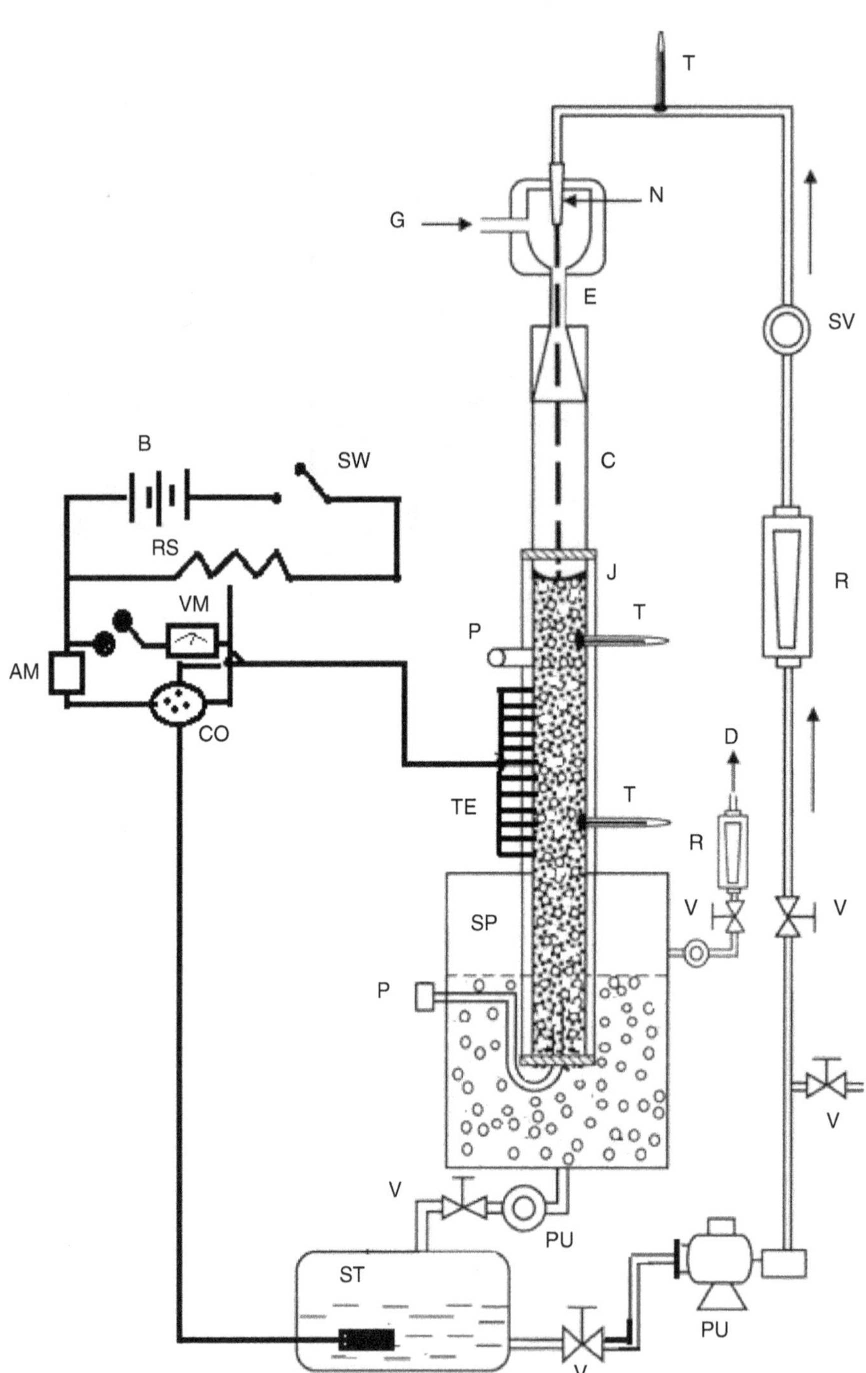

FIGURE 8.6 Schematic diagram of study on wall–liquid mass transfer in an inverse bubble column.

and mass transfer and hydrodynamics in the two-phase flows. In addition, application of the electrochemical method for mass transfer measurement significantly expands the range of physical properties of the liquids toward the higher Prandtl numbers. A relatively thin near-wall liquid layer becomes the most important zone of the flow. Experimental results show that the wall–liquid mass transfer coefficient in the inverse flow bubble column is highly influenced by the superficial phase velocity (Sivaiah and Majumder, 2013). A schematic diagram of the experimental setup is shown in Figure 8.6. The setup consists of a storage tank, centrifugal pumps for circulating the electrolyte, and two rotameters for measuring the flow rates of the electrolyte and gas. Two open manometers are used to measure the pressure at the mixing zone and the pressure inside the column, respectively. The valves are used to control the flow rates of liquid and the gas in the column. The inner wall of the test section is provided with copper point electrodes of diameter 3.024×10^{-3} m. One end of these electrodes is fixed with the surface of the inner wall of the test section, and the other end is projected outward, serving as terminal for connecting the electrodes to the external circuit. The lower end of the column is extended to about 0.5 m inside the separator to enable uniform movement of the phases and easy separation of the bubbles from the main stream. Before the assembly of the test section, the surfaces of the point electrodes were polished and cleaned thoroughly. The electrical connectivity of the point electrode can be checked using a digital multimeter. The current-potential measuring circuit is consisted of a voltmeter (V), ammeter (A), rheostat (R), commutator, and 6-volt DC source for connecting any desired electrode to the electrical circuit. The measurements were carried out for the case of an electrochemical red-ox system in the presence of an excess indifferent electrolyte (sodium hydroxide). This system is chosen because (1) the chemical polarization involved is negligible; (2) the reacting surface remains smooth and unaffected, unlike in the cases of solids dissolution or sublimation processes; and (3) the measurements are relatively fast, accurate, and reproducible.

An equimolar amount of solution of potassium ferrocyanide and potassium ferricyanide of about $0.01 \, \text{kmol/m}^3$ and $0.5 \, \text{kmol/m}^3$ of sodium hydroxide are prepared using distilled water; these are used as the electrolyte. During the experiment, the reacting ion concentration is obtained by volumetric analysis. Ferrocyanide ion concentration is estimated by the permanganometric titration method, and the ferricyanide ion concentration is obtained by the iodometric titration method (Kolthoff and Tomsicek, 1935). When the system attains steady state, the flow rates of electrolyte and gas are noted. When the liquid and gas flow rates are stabilized, the limiting currents are measured by applying an electric potential in small increments between the test electrode and the wall electrode. The measurement of limiting current can be made in the lines similar to those reported earlier in the studies on ionic mass transfer (Lin et al., 1951). From the measured value of limiting current at any given electrode area, the mass transfer coefficients are evaluated. In any electrochemical cell, the mass transfer between the bulk solution and the polarized surface takes place either by diffusion of ions caused by concentration gradient or migration of ions caused by potential difference. By adding an excess indifferent electrolyte, the ionic migration of the reacting species can be eliminated, and the electrode reaction is made diffusion controlled. The excess indifferent electrolyte can be chosen as 0.5 N of sodium hydroxide. The emphasis is thus laid on the diffusion of the reacting material toward the electrode surface and the reacted species to the bulk solution. This process is of the type steady-state diffusion of solute in a dilute solution through a stagnant solvent. The transfer rate of diffusion of particular ionic species can be written by using Fick's first law of diffusion, which is given by the following expression:

$$\frac{ds}{dt} = \frac{A_e D_i (C_i - C_s)}{\delta(1 - t_i^+)} \tag{8.94}$$

where ds/dt is the transfer rate of ionic species (equivalents/s), A_e is the exposed area of the electrode surface, D_i is the diffusion coefficient of the ions, δ is the thickness of the hypothetical diffusion layer, C_i is the concentration of the ionic species on the solution side of the diffusion layer, C_s is the concentration of the ionic species at electrode surface, and t_i^+ is the transport number of the reacting species. If the concentration of indifferent electrolyte (0.5 kmol/m³) is very high compared with the concentration (0.01 kmol/m³) of the reacting ionic species, the transport number of the discharge ions can be taken as zero ($t_i^+ = 0$). Therefore, equation (8.94) reduces to

$$\frac{ds}{dt} = \frac{A_e D_i (C_i - C_s)}{\delta} \tag{8.95}$$

The rate of discharge of ions can be expressed as

$$\frac{i_d}{zF} = \frac{D_i (C_i - C_s)}{\delta} \tag{8.96}$$

where i_d is the limiting current density (i_L/A_e), z is the number of electrons involved in the discharge process, and F is the Faraday constant (96,500 C/equivalent). The limiting current density represents the maximum rate at which the particular ion can be discharged under the given experimental conditions. At this current density, rapid increase of potential is observed. The value of the limiting current is obtained from the plots drawn between current and applied potential. The attainment of the limiting current is indicated by a sharp increase in potential for a marginal increase in the current. Figure 8.7 illustrates the graphical method of obtaining limiting current from voltage (E) versus current (i) data. When the limiting current is reached on a polarized electrode, the concentration of the ionic species at the electrode surface (C_s) becomes zero. Then equation (8.96) can be written as

$$\frac{i_d}{zF} = \frac{D_i C_i}{\delta} \tag{8.97}$$

The mass transfer coefficient (k_l) is defined as the ratio of diffusion coefficient to the thickness of the hypothetical diffusion layer (D_i/δ), which on substitution into equation (8.97) yields

$$k_L = \frac{i_d}{zFC_i} \tag{8.98}$$

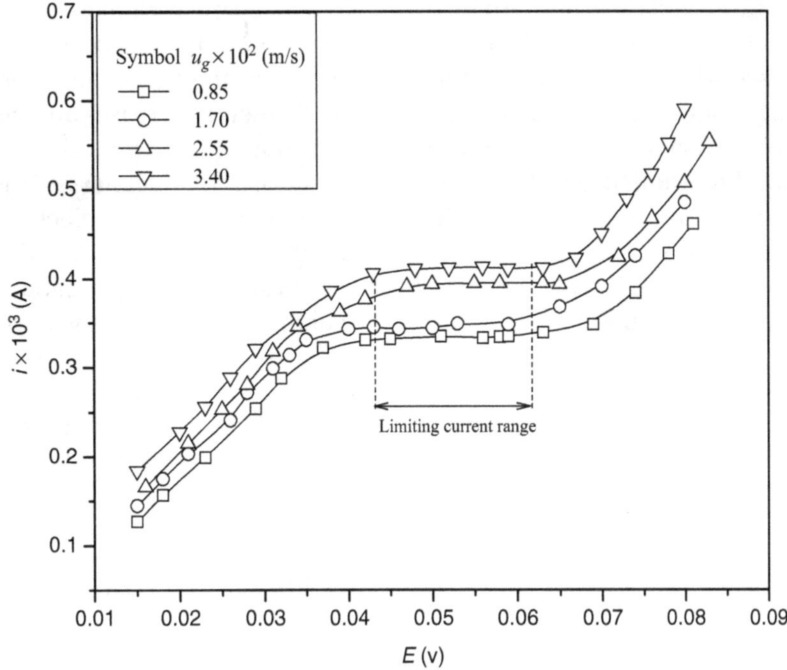

FIGURE 8.7 Graphical representation of obtaining limiting current from voltage (Sivaiah and Majumder, 2013).

The gas phase acts to favor the hydrodynamic conditions at the reacting surface. A point electrode fixed at the inside wall of a column reactor is chosen as a reacting surface. The reacting ion in the continuous phase reaches the reacting surface under an electric potential by diffusion, which depends on the polarity induced at the electrode surface.

Reduction of ferricyanide ion takes place at the electrode surface. It is observed from the experiment that the mass transfer coefficient is increased with an increase in gas and liquid velocities because of an increase of the momentum of the secondary air to be dispersed into discrete bubbles, which in turn increase the turbulence and intensity of mixing in the column. Also, at higher superficial liquid velocity, gas hold-up increases because of an higher entrainment of the gas. As a result, the flow gets more agitated. With an increase in superficial gas velocity, the coarser bubbles immediately get released from the jet because of an their higher buoyancy, and these bubbles face more resistance to move in the inverse direction and spend more time in the column, which results in more mixing because of an the interaction of phases and hence increase the mass transfer coefficient. The population of the bubble is higher because of an formation of more uniform finer bubbles with the interaction of the phases. The intimate contact between the phases increases as the gas hold-up increases with the superficial gas velocity, which creates more turbulence in the column. This may result in an increase in the mass transfer coefficient with an increase in the superficial gas velocity. The mass transfer coefficients are increased with an increase in the gas hold-up the same as those observed in case of gas velocity. In the inverse bubbly flow system, the average gas hold-up increases almost linearly with increasing superficial gas and liquid velocity. As the gas hold-up increases, more bubbles occupy the column, which decreases the flow area of the liquid and increases the true velocity of the liquid. Also, with increasing gas hold-up, the mean bubble diameter decreases and increases the specific interfacial area. An increase in the interfacial area and true velocity of the liquid increases the turbulence in the column and decrease the thickness of the hydrodynamic boundary and diffusion layers. This leads to an increase in the electron migration toward the electrode surface and hence an increase the mass transfer rate. The change of surface properties of the liquid affects the bubble coalescence and generation and hence the interfacial area and mass transfer rate in the column. The mass transfer coefficient is decreased with an increase in liquid viscosity. An increase in viscosity results in a decrease the degree of turbulence in the liquid, thereby increasing the thickness of the laminar film at the boundary layer between the electrode surface and the liquid. An increase in thickness of the laminar film decreases the electron deposition on the electrode surface, which results in a decrease in the mass transfer coefficient.

Correlation Model for the Wall–Liquid Mass Transfer Coefficient

The rate of mass transfer is related to the concentration and velocity distributions by the mass balance

$$\frac{d}{dz}\left[\int_{R}^{R+\delta} ru(C - C_s)\,dr\right] = Rk_{L,z}(C_i - C_s) \tag{8.99}$$

Near the wall of the vertical column, a logarithmic concentration profile in the radial direction is assumed as (Ravoo, 1971)

$$\theta = \frac{C - C_s}{C_i - C_s} = 1 - \frac{\ln(r/R)}{\ln\{(R+\delta)/R\}} \tag{8.100}$$

It is assumed that the concentration distribution across a concentration boundary layer with z-dependent boundary thickness, δ, is approximately governed by the diffusion equation

$$\frac{\partial}{\partial r}\left(r\frac{\partial \theta}{\partial r}\right) = 0 \tag{8.101}$$

with the boundary conditions $\theta = 1$ for $r = R$ and $\theta = 0$ for $r = R + \delta$. By assuming mass transfer flux at the wall, introducing a Sherwood number, Sh_R, related to the radius of the column as

$$Sh_R = \frac{k_{L,z}R}{D_L} \tag{8.102}$$

It can be easily shown that

$$1/Sh_R = \ln\{(R+\delta)/R\} = \psi \, (\text{say}) \tag{8.103}$$

The inverse of the Sherwood number denotes the logarithmic radial concentration distribution. This concentration distribution affects the velocity profile in the column. The velocity profile can be found from the momentum equation. For the infinite Schmidt number when incompressibility is assumed, the momentum equation can be written as

$$\frac{v}{r}\frac{\partial}{\partial r}\left(r\frac{\partial u}{\partial r}\right) + g\beta(C_i - C_s) = 0 \tag{8.104}$$

Substitution of the concentration profile in the momentum equation gives

$$\frac{\partial^2 u}{\partial r^2} + \frac{1}{r}\frac{\partial u}{\partial r} + \frac{g\beta(C_i - C_s)}{v}\left(1 - \frac{\ln(r/R)}{1/Sh_R}\right) = 0 \tag{8.105}$$

The boundary condition for the velocity distribution is $u = 0$ at $r = R$ and $\partial u/\partial r = 0$ at $r = R + \delta$. Solution of the equation (8.105) with these boundary conditions can be expressed in a dimensionless velocity profile as

$$U = \frac{Sh_R}{2}\left(\exp(2/Sh_R) + \frac{r^2}{R^2}\right)\ln\left(\frac{r}{R}\right) + \frac{1}{2}\left(1 - \frac{r^2}{R^2}\right)(1 + Sh_R) \tag{8.106}$$

where

$$U = \frac{2vu}{g\beta(C_i - C_s)R^2} \tag{8.107}$$

Equation (8.99) can be expressed in dimensionless form as

$$\frac{d}{d(z/R)}\left[\int_1^{\exp(\psi)} \frac{r}{R}U\theta d\left(\frac{r}{R}\right)\right] = \frac{2Sh_R v D_L}{g\beta(C_i - C_s)R^3} \tag{8.108}$$

The above definite integral being a function of ψ is denoted by $I(\psi)$. Separating variables and integrating with the boundary conditions $\psi = 0$ at $z = 0$, one can express the equation as

$$\frac{z^4}{R^4 Ra_z} = \frac{1}{2}\left[\psi I(\psi) - \int_0^\psi I(\psi)d\psi\right] \tag{8.109}$$

For $I(\psi)$, after substitution of U and θ with equations (8.106) and (8.100) and integration, it can be shown that (Ravoo, 1971)

$$I(\psi) = \frac{\psi - 11/8}{8\psi^2}\exp(4\psi) + \frac{\psi + 1}{4\psi^2}\exp(2\psi) - \frac{5}{64\psi^2} - \frac{3}{16\psi} - \frac{1}{8} \tag{8.110}$$

With a series expansion of equation (8.110) and taking its first four terms, one can express

$$\frac{z^4}{R^4 Ra_z} = \frac{3}{80}\psi^4 + \frac{16}{225}\psi^5 + \frac{19}{252}\psi^6 + \frac{17}{294}\psi^7 + \ldots\ldots\ldots \tag{8.111}$$

For practical applications and for comparison of theory and experimental results, it is convenient to know the average of Sherwood number as

$$<Sh>_R = \frac{1}{Z}\int_0^z Sh_R\, dz \tag{8.112}$$

With equations (8.112), (8.109), and (8.103), one can find taking ψ as the integrating variables as

$$<Sh>_R = \frac{R^4 Ra_z}{z^4}\int_0^{\psi_z}\frac{1}{\psi}\frac{d}{d\psi}\left(\frac{z^4}{R^4 Ra_z}\right)d\psi = \frac{R^4 Ra_z I(\psi)}{2z^4} \tag{8.113}$$

Hence, $<Sh>_R$ can be calculated from $I(\psi_z)$ values from equation (8.110). For calculating Sh_R, the mass transfer coefficient $k_{l,z}$ can be obtained from the developed correlation (equation [8.114]) based on the experimental data (Sivaiah and Majumder, 2013).

$$\left(\frac{k_L}{u_l}\right) = 10.83\left(\frac{d_c u_l \rho_l}{\mu_l}\right)^{-0.698}\left(\frac{u_g^2}{d_c g}\right)^{-0.777}\left(\frac{\mu_l}{\mu_g}\right)^{0.085} \tag{8.114}$$

The ranges of the variables of equation (8.114) are $0.4 \times 10^{-3} > k_L/u_l < 1.5 \times 10^{-3}$, $11.45 \times 10^2 > d_c u_l \rho_l/\mu_l < 44.12 \times 10^2$, $0.15 \times 10^{-3} > u_g^2/gd_c < 2.35 \times 10^{-3}$ and $0.79 \times 10^2 > \mu_l/\mu_g < 1.02 \times 10^2$.

Mass Transfer Efficiency in a Bubble Column

The influences of different physical, dynamic, and geometric variables cause efficiency characterization of the equipment that is involved in gas–liquid mass transfer processes. This characterization has great importance to optimize the process plant design. If the bulk average concentration of liquid at inlet and outlet are $C_{l,in}$ and $C_{l,out}$, respectively, the mass transfer efficiency of entire column can be defined as

$$\eta_{BC} = \frac{C_{l,h_i=h_m} - C_{l,in}}{C_{le,h_i=h_m} - C_{l,in}} = 1 - \exp\left[-\frac{k_l a h_m}{u_{sl}}\right] \tag{8.115}$$

where C_{le,h_i} is in equilibrium with C_{g,h_i}. The relationship between $\eta_{h_{i-1}}^{h_i}$ and η_{BC} can then be derived by integrating the local $\eta_{h_{i-1}}^{h_i}$ for the entire column of gas–liquid mixing height h_m. The mass transfer efficiency of entire column depends on the degree of mixing of gas and liquid: (1) if the liquid in column is completely mixed, everywhere of uniform concentration, C_{l,h_i} and the gas entering the segment h_i is perfectly mixed, then $\eta_{h_{i-1}}^{h_i} = \eta_{BC}$; and (2) if the movement of the liquid is in plug flow with no mixing, while gas entering the segment h_i is perfectly mixed, then according to Kister's concept (Kister, 1992) the mass transfer efficiency of entire column can be represented as (Majumder, 2008a)

$$\eta_{BC} = \frac{C_{l,h_i=h_m} - C_{l,in}}{C_{l,h_i=h_m}^* - C_{l,in}} = \frac{P_{h_{Ri}}}{HQ_R}\left[\exp\left(\frac{HQ_R\eta_{h_{i-1}}^{h_i}}{P_{h_{Ri}}}\right) - 1\right] \tag{8.116}$$

(3) If the entering gas is well mixed but the liquid is only partially mixed, then as per the concept of Gerster et al. (1958), the ratio of η_{BC} to $\eta_{h_{i-1}}^{h_i}$ is a function of dispersion number of liquid, the gas to liquid flow ratio, and the axial equilibrium distribution. (4) If both gas and liquid are only partially mixed, as per Bennett et al. (1997) concept of gas dispersion in liquid, the mass transfer efficiency of the entire column may be a function of the mixing characteristic, which depends on geometric, dynamic, and physical variables of the system. The mass transfer efficiency cannot be analyzed completely by a specific theoretical model because of the simultaneous effect of various variables on mass transfer efficiency. From the literature, it is found that the mass transfer efficiency depends on different dynamic, geometric, and physical variables of the system. The effect of all these variables independently on the mass transfer efficiency is very complicated; thus, a correlation can be developed by dimensional analysis to incorporate the mass transfer efficiency in terms of the dynamic, geometric, and physical variables of the system. Combining the relevant dimensionless groups obtained by the dimensional analysis and regression analysis, the following functional relationship was developed by Majumder (2008a):

$$\eta_{BC} = 0.5364 Ln(X) + 0.9106; \quad R^2 = 0.9906 \tag{8.117}$$

where

$$X = 9.71 \times 10^{-2} \, Re^{-1.204} \, Sc^{-0.269} \, Fr^{0.686} \, Ga^{0.672} \, Mo^{0.077} \, Pe^{0.609} \tag{8.118}$$

The correlation is valid within the range of the different variables as $20 < Re < 6.70 \times 10^4$; $70 < Sc < 4.44 \times 10^3$; $0.03 < Fr < 0.60$; $5.29 \times 10^5 < Ga < 1.44 \times 10^{13}$; $1.14 \times 10^{-11} < Mo < 6.08 \times 10^{-8}$; $0.044 < Pe < 373.88$. The correlation equation (8.117) is valid for $0.19 < X \leq 1.0$ and $0.25 \leq \eta_{BC} < 1.0$.

Effect of Different Variables on Mass Transfer Coefficient in a Bubbly Flow System

Mass transfer coefficient in bubbly flow device depends on the different operating variables. The effects of different variables on the mass transfer coefficient, either upflow or inverse flow of phases in a bubble column, are discussed in the following section.

Effect of the Distributor

Several authors reported that the gas distributor design significantly affects the mass transfer coefficient in the bubbly flow regime. Majumder (2008a) analyzed the results in detail. Their work attributes the variations of the volumetric mass transfer coefficient with superficial gas velocities are shown in Table 8.5.

By comparison, the perforated distributor with small holes has higher $k_l a$ values, and the cross-distributor with large holes has lower $k_l a$ values. This result shows the effect of the two different types of distributor (perforated plate and cross distributor) and the effect of hole size of the perforated plates on the gas–liquid mass transfer coefficient. Han et al. (2007) reported that the distributor effect on $k_l a$ at low gas velocities (<0.15 m/s) is higher because of the different bubble sizes formed at these distributors, which affect the bubble sizes and interfacial area in the whole reactor. The distributor effect becomes smaller in the high gas velocity range (>0.20 m/s). This can be explained by the nature of the coalescent system used in which the bubbles further enlarge in size from the distributor. Although the distributor design greatly affects the initial bubble formation (Terasaka et al., 1999; Hsu et al., 2000), the three distributors resulted in similar interfacial area values with only small differences at 0.30 m/s. On the other hand, an increase in the gas flow rate (superficial gas velocity) produces a clear increase on interfacial area because of the higher gas volume fed to the contactor. Also, the use of a gas distributor with a smaller pore diameter generates a higher number of bubbles, and then an increase in interfacial area results in an increase in the mass transfer coefficient with a smaller pore diameter of the sparger. Deckwer et al. (1974) found that different gas sparger designs can affect $k_l a$ values up to a factor of 2.

Effect of Column Size

Mass transfer strongly depends on the liquid-phase turbulence and large-scale liquid internal circulation, both of which depend on the column size. The gas hold-up is higher at smaller size column. The difference is more prominent in the churn-turbulent flow regime (Lau et al., 2004). The higher gas hold-up in the smaller column is mainly due to wall effects on the bubble characteristics. In a small column, the bubble size is limited by the column size; therefore, a higher gas hold-up is observed in smaller columns

Table 8.5   Gas Hold-up and Overall Mass Transfer Coefficient as a Function of Sparger Design

Type of Sparger	Values of α_2 and β_2 of Correlation: $k_l a = \alpha_2 u_{sg}^{\beta_2}$
Perforated plate with 163 holes (0.50 mm) with open area of 0.156%	$\alpha_2 = 0.273$, $\beta_2 = 0.482$, $R^2 = 0.953$
Perforated plate with 163 larger holes (1.32 mm) with open area of 1.09%	$\alpha_2 = 0.348$, $\beta_2 = 0.676$, $R^2 = 0.966$
Cross sparger with four holes (2.54 mm), each 25.4 mm off center and facing down and with open area of 0.1%	$\alpha_2 = 0.341$, $\beta_2 = 0.742$, $R^2 = 0.998$

(Lau et al., 2004). When the column size is larger than 0.1 to 0.15 m, the wall effect on gas hold-up becomes negligible (Akita and Yoshida, 1974). Thus, the bubble size becomes independent of the column dimension and is governed by the rates of bubble coalescence and break-up (Lau et al., 2004). Vandu and Krishna (2004) observed that $k_l a / \varepsilon_g$ showed a slight increase with column diameter. Computational fluid dynamics (CFD) simulations showed that $k_l a$ decrease with column diameter (Krishna and Van Baten, 2003). Verma and Rai (2003) reported that the mass transfer coefficient is independent of the initial bed height.

Effect of Operating Pressure

The performance of a bubble column based on mass transfer depends on the operating pressure because the overall gas hold-up and the volumetric gas–liquid mass transfer coefficient, $k_l a$, have strong dependency on operating pressures without varying the distributor type as shown in Table 8.6 (Majumder, 2008a).

The correlations are made based on the experimental data reported by Han et al. (2007). The values of volumetric mass transfer coefficient increased with the superficial gas velocity and the operating pressure. The volumetric mass transfer coefficient increased greatly because of the smaller bubble size generated at high pressures (Lau et al., 2004). Vafopulos et al. (1975) investigated the mass transfer in an air–water bubble column at pressures from 0.1 to 1 MPa. They reported that pressure has no significant effect on gas hold-up and the volumetric liquid phase mass transfer coefficient. However, many studies report a significant effect of pressure on mass transfer rates. The liquid side mass transfer coefficient, k_l, unlike $k_l a$, decreased with the increase of pressure, especially from ambient pressure to 0.4 MPa (Han et al., 2007). This verdict of the pressure effect on the k_l values was corroborated by Lemoine et al. (2004). The apparent decrease of k_l is attributed to the change in bubble dynamics and hydrodynamics at 0.4 MPa (Gupta et al., 2001). At high pressure, smaller bubble sizes have a lower slip velocity (Xue et al., 2003). Therefore, smaller bubbles have longer exposure times in the liquid interface renewal and cause a decrease in k_l values at high pressure. However, as pressure increases from 0.4 to 1.0 MPa, the k_l values do not seem to change much, which is due to the unnoticeable differences in bubble sizes at 0.4 and 1.0 M Pa. Calderbank and Moo-Young (1961) found that k_l is lower for smaller bubble sizes. Because an increase in pressure decreases the bubble size, these findings suggest that the liquid-phase mass transfer coefficient decreases with increasing pressure. Wilkinson and Haringa (1994) reported that both the interfacial area and the volumetric

Table 8.6 Gas Hold-up and Overall Mass Transfer Coefficient as a Function of Operating Pressure with Perforated Plate with 163 Larger Holes (1.32 mm)

Operating Pressures	Values of α_3 and β_3 of Correlation $\varepsilon_g = \alpha_3 Ln(u_{sg}) + \beta_3$	Values of α_4 and β_4 of Correlation $k_l a = \alpha_4 Ln(u_{sg}) + \beta_4$
0.1 MPa	$\alpha_3 = 0.089, \beta_3 = 0.389; R^2 = 0.985$	$\alpha_4 = 0.044, \beta_4 = 0.188; R^2 = 0.947$
0.2 MPa	$\alpha_3 = 0.145, \beta_3 = 0.654; R^2 = 0.975$	$\alpha_4 = 0.062, \beta_4 = 0.285; R^2 = 0.942$
1.0 MPa	$\alpha_3 = 0.215, \beta_3 = 0.897; R^2 = 0.992$	$\alpha_4 = 0.084, \beta_4 = 0.366; R^2 = 0.985$

mass transfer coefficient increase with pressure. They also argued that the increase in k_la is partly limited to the higher gas-phase mass transfer resistance and a decrease in the liquid-phase volume. Linek et al. (1989) found that an increase in k_la with increasing pressure is due to the increase in gas hold-up. Lau et al. (2004) reported that k_la increases significantly when the pressure is increased from ambient pressure to 4.24 MPa. The pressure effect is more noticeable at high gas velocities. For example, at a gas velocity of 10 cm/s and a liquid velocity of 0.17 cm/s, when the pressure increases from 0.1 to 2.86 MPa, k_la increases from 0.01 to 0.023 s-1 (a 130% increase), but at a gas velocity of 20 cm/s and the same liquid velocity, k_la increases from 0.015 to 0.043 1/s (a 187% increase). Furthermore, the effect of pressure on k_la became significant at higher superficial gas velocities for single-nozzle gas distributors (Kojima et al., 1997).

Effect of System Temperature

The system temperature has a significant effect on the mass transfer coefficient because the temperature directly affects the liquid diffusivity and thereby k_l. Temperature also affects the physical properties of the liquid phase. Both the viscosity and surface tension of the liquid phase are reduced by an increase in the temperature. A lower viscosity and smaller surface tension result in the formation of smaller bubbles. Therefore, the mass transfer interfacial area increases with increasing temperature. Moreover, because k_l is inversely proportional to the liquid viscosity (Calderbank and Moo-Young,1961), the decrease in the viscosity would therefore increase k_l. Also, according to Higbie, the mass transfer coefficient is inversely proportional to the square root of the contact time, which may increase by the smaller surface tension at higher temperatures. As a result, k_l would be smaller at higher temperatures because of the surface tension effect (Lau et al., 2004). The effects of the liquid surface tension and viscosity have competing effects on k_l. Lau et al. (2004) reported that an increase in the system temperature increases k_la significantly. The increase in the interfacial area, a, and k_l is much stronger than the decrease of k_l at high temperatures. Therefore, the effects of the liquid surface tension and viscosity have competing effects on k_l. They also suggested that to have a better understanding of the effect of the temperature on mass transfer, it is necessary to compare gas hold-up at elevated temperatures. They stated from their experimental data that at a superficial gas velocity of 20 cm/s, k_la increases from 0.03 to 0.17 1/s (a 470% increase) when the temperature increases from 25 to 92°C. The rate of increase of k_la is much higher than the rate of increase of gas hold-up. They have concluded that the favorable effect of temperature on k_l due to the higher liquid diffusivity may play an important role in determining the mass transfer behavior at high temperatures.

Effect of Gas and Liquid Velocity

The volumetric mass transfer coefficient increases with the superficial gas velocity, which in turn systematically induces an increase in specific interfacial area (Majumder, 2008a). At a low liquid phase flow rate, an increase in superficial gas velocity causes an increase in the mass transfer coefficient due to a higher turbulence in the liquid phase. At high values of liquid flow rate, the effect of superficial gas velocity is negligible because the influence of the liquid flow rate is higher than that produced by the gas flow rate. The ratio of $k_l a/u_{sl}$ decreases with an increase in the superficial liquid velocity, which results in a decrease in the mass transfer efficiency of the bubble column. Sivaiah and Majumder (2013) studied the effect of slurry velocity on the mass transfer coefficient in an inverse bubbly flow column. They explained that the mass transfer coefficient is increased with increase in slurry velocity as well as gas velocity as shown in Figure 8.8. As the superficial slurry velocity increased, the jet momentum causes the secondary air to be dispersed into discrete bubbles, in turn increasing the turbulence and intensity of mixing in the column. This leads to further augmentation in the mass transfer rate. Also, at a higher superficial slurry velocity, gas hold-up increases because of higher entrainment of the gas. As a result, the circulation and interaction of the gas and liquid phases increase inside the column, and the flow gets more agitated. With an increase in superficial gas velocity, the coarser bubbles immediately get released from the jet because of their higher buoyancy, and these bubbles face more resistance to move in an inverse direction and spend more time in the column, which results in more mixing caused by the interaction of phases and hence an increase in the mass transfer coefficient. The population of the bubble is higher because of formation of more uniform, finer bubbles with the interaction of the phases. The intimate contact between the phases increases

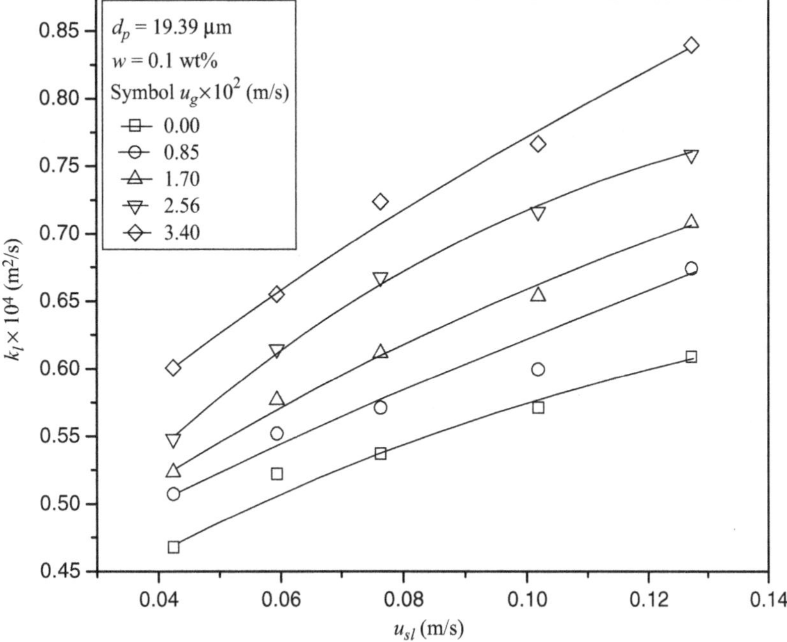

FIGURE 8.8 Variation of mass transfer coefficient with superficial slurry and gas velocity (Sivaiah and Majumder, 2013).

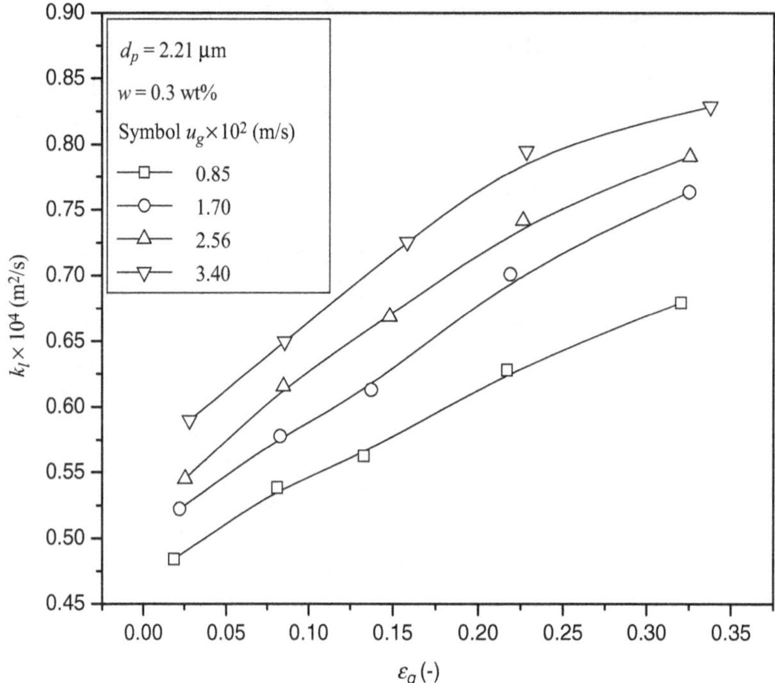

FIGURE 8.9 Effect of gas hold-up on mass transfer coefficient in an inverse bubbly flow column (Sivaiah and Majumder, 2013).

as the gas hold-up increases with the superficial gas velocity, which creates more turbulence in the column.

This may result in an increase in the mass transfer coefficient with an increase in the superficial gas velocity. The variation of the mass transfer coefficient with gas hold-up is shown in Figure 8.9. It is seen that the mass transfer coefficients are increased with an increase in gas hold-up the same as those observed in the case of gas velocity. As the gas hold-up increases, more bubbles occupy the column, which decreases the flow area of the liquid and increases the true velocity of the liquid–solid slurry. Also, with increasing gas hold-up, the mean bubble diameter decreases and increases the specific interfacial area. Increases in the interfacial area and true velocity of the slurry increase the turbulence in the column and decrease the thickness of the hydrodynamic boundary and diffusion layers. This leads to increase the electron migration toward the electrode surface and hence an increase in the mass transfer rate.

Effect of Bubble Size

The bubble size distribution has significant effects on the mass transfer coefficient because it determines the interfacial area per volume of the gas phase. The bubble size distribution is determined by bubble coalescence and break-up. In a given system, bubble coalescence and break-up rates are mainly affected by the local gas hold-up and turbulent energy dissipation rate. Because of the non-uniform radial profiles of the gas hold-up and dissipation rate, especially in the heterogeneous regime, the bubble size distribution varies not only with the superficial gas velocity but also with the radial position (Wang and Wang, 2007). According to Wang and Wang (2007), at low superficial gas velocities, the bubble size distributions are very similar at different radial positions. With an increase in the superficial gas velocity, the difference in the bubble size distributions at different radial positions becomes pronounced, especially after the flow enters the heterogeneous regime.

The bubble size distribution is quite homogeneous, for instance, around 0.0035 m at distribution level. This distribution spreads with increasing gas velocity; the mean bubble diameter becomes so large that bubbles become unstable and break up into a new population of tiny bubbles and increase in interfacial area. An increase in interfacial area increases the mass transfer coefficient. At a lower gas velocity, the bubble size tends to increase with height of the column (depending on the type of bubble column), and the smallest bubbles form near the distributor because of a higher dissipation rate of energy. The smallest bubbles may coalesce and induce an increase of the mean bubble size with the column height. At a high gas velocity in the sparger area, the small bubbles population fraction increases with an increase in height from the source of generation because of distribution of energy dissipation. These small bubbles probably result from the break-up of the large unstable bubbles and rise slowly. At low gas velocity the smaller bubbles formed may coalesce with other bubbles. When the gas velocity increases, the bubble diameter tends to increase, which may reduce the surface area and mass transfer coefficient. Lewis and Davidson measured the volumetric mass transfer coefficient in a 0.45-m-diameter column and found that the mass transfer rates were enhanced by bubble break-up, but the effect was small because of the rapid coalescence of small bubbles (Lewis and Davidson, 1985).

Effect of Surface Tension on Mass Transfer Coefficient

In bubble columns, the surface tension effect is similar to trend for single bubbles: a decrease in surface tension decreases the bubble size and bubble velocity, resulting in higher gas hold-up and a higher mass transfer coefficient (Majumder, 2008a). Consequently, the decrease in surface tension increases the coefficient of mass transfer. The trend is confirmed by the classical correlations of Shah et al. (1982). The surface tension effect is particularly effective in the homogeneous and transition regimes and less in the heterogeneous regime in which the reduction of coalescence is overshadowed by the predominant effect of macro-scale turbulence (Camarasa et al., 1999). Also, with an increase in surface tension, the transition of bubbling regimes is delayed to higher gas velocity, but the heterogeneous regime appears almost at the same gas velocity, and the transition regime tends to disappear (Dargar and Macchi, 2006). Jia et al. (2015) reported that the variation of interfacial area is insensitive to the surfactant type and its concentration. Different surfactants decrease k_l differently, which is related to the value of the critical

Table 8.7   Critical Micelle Concentration (CMC), Adsorption Equilibrium Constant (K), Surface Tension at CMC (σ_{CMC}) and C/C_M of Different Surfactants (Jia et al., 2015)

Surfactant Type	C_M/mol·L^{-1}	K/L·mol^{-1}	σ_{CMC}/mN·mol^{-1}	C/C_M
n-Octyltrimethylammonium bromide (OTABr)	0.26	34.84	35.1	1.92×10^{-4}– 3.85×10^{-3}
Sodium dodecyl benzene sulfonate (SDBS)	1.2×10^{-3}	9.874×10^3	34.2	4.17×10^{-2}–1.17
Tween 80	1.1×10^{-5}	7.407×10^8	37.9	4.55-109

micelle concentration (CMC) as shown in Table 8.7. According to Rubia et al. (2010), the presence of surfactant molecules in the liquid phase tends to stabilize the liquid surface and to reduce the liquid renewal at the interface, and hence k_l decreases. From their work, it is seen that k_l has a significant relationship with the CMC of surfactants.

They concluded that the effect of any surfactant on mass transfer of CO_2 into water can be rated by a specific concentration of the surfactant, which is defined as the ratio of the concentration to the critical micelle concentration (CMC) of the surfactant, namely C/C_M as shown in Table 8.7. They also suggested that to ensure effective mass transfer of CO_2 into water, the value of C/C_M of a surfactant should be as low as possible and must not exceed 1. For gas–liquid systems, a model has been developed by Sardeing et al. (2006) for the liquid phase mass transfer coefficient k_l in the presence of surfactants, which can be represented as:

$$k_l = S_e k_{l1} + (1 - S_e) k_{l0} \tag{8.119}$$

where k_{l0} is the liquid phase mass transfer coefficient without surfactant, which is calculated using the model developed by Higbie (1935) and k_{l1} is the liquid phase mass transfer coefficient where the surface is saturated with the surfactant. Sardeing et al. (2006) proposed an equation to calculate k_{l1} that incorporates the influence of surfactant characteristics:

$$k_{l1} = 1.744 K^{-0.0837} k_{lF} \tag{8.120}$$

where K is the surfactant adsorption equilibrium constant as also shown in Table 8.7. The surfactants used by Jia et al. (2015) differ in molecular weight and size compared with those used in Sardeing et al. (2006), and the concentration range was much lower, hence Jia et al. (2015) got different coefficient of equation (8.120) as 1.8455 instead of 1.744. The parameter k_{lF} is called the Frössling coefficient and can be expressed as

$$k_{lF} = 0.6 (u_s/d_b)^{1/2} D^{2/3} (\mu/\rho)^{-1/6} \tag{8.121}$$

Effect of Viscosity of Liquid

The efficiency of mass transfer of a bubble column is quietly controlled by the transport mechanism of species in the liquid viscous boundary layer around the bubbles. The contact time between liquid particles active for the transfer and the bubble, as well as the interfacial area active for the transfer, result in variation of mass transfer efficiency with viscosity of liquid. According to Haut and Cartage (2005), for 1 mm $< d_b <$ 1 cm, the bubbles moves with the superposition of two perfect spherical caps, and only the upper spherical cap is active for the transfer of solute. The contact time between a liquid element and the lower spherical cap is significantly larger than the contact time between a liquid element and the upper spherical cap (Haut et al., 2004). Therefore, the main part of the gas–liquid mass transfer is ensured by liquid elements in contact with the upper spherical cap. The expressions of volumetric mass transfer coefficient and specific interfacial area in terms of contact time can be expressed (Majumder, 2008a) as

$$k_l a = 2 \sqrt{\frac{D_l}{\pi}} \left[\sum_{j=1}^{n} \frac{\varepsilon_{g,j} a(d_{b,j})}{\sqrt{t_{c,j}}} \right] \tag{8.122}$$

$$a(d_{b,j}) = \frac{3(1+W_j^2)2^{1/3}}{W_j^{2/3}(3+W_j^2)^{2/3} d_{b,j}} \tag{8.123}$$

The specific interfacial area depends on the wake angle of the bubble. A wake angle of 1 radian gives $a = 5.26/d_{b,j}$ (Haut and Cartage, 2005). For the bubble group, j the aspect ratio (ratio of major and minor axis) is defined as W_j, which can be calculated by the correlation of Wellek (Clift et al., 1978) as

$$W_j = [1 + 0.163 Eo^{0.757}]^{-1} \tag{8.124}$$

where $Eo = g\rho_l d_{b,j}^2/\sigma$ is the Eotvos number of the bubble of diameter $d_{b,j}$. Also, within the liquid viscous boundary layer and if the angular position of bubble is less than the wake angle, the transport of solute in a direction normal to the interface is purely diffusional, and its transport in a direction tangential to the interface is purely convective (Haut and Cartage, 2005). A liquid element active for the transfer of solute enters the liquid boundary layer near the point of incidence and leaves it near the extreme point of the bubble. The contact time between a bubble and a liquid element active for the transfer of solute is inversely proportional to the gas–liquid relative velocity of the bubble. The mass transfer rate between the bubble and the surrounding liquid is increased with the interfacial area and is decreased with an increase in contact time (Haut and Cartage, 2005). More viscous liquid resumes the more contact time between a bubble and a liquid element active for the transfer of solute, which may cause the decrease in mass transfer coefficient with an increase in viscosity (Majumder, 2008a). Majumder and Sivaiah (2013) studied the effect of slurry viscosity on the liquid side mass transfer coefficient in an inverse bubbly flow system. They reported that the change of surface properties of the liquid affects the bubble coalescence and generation and hence the interfacial area and mass transfer rate in the column. The viscosity of the slurry varies with the slurry concentrations. The variation of mass transfer coefficient with slurry

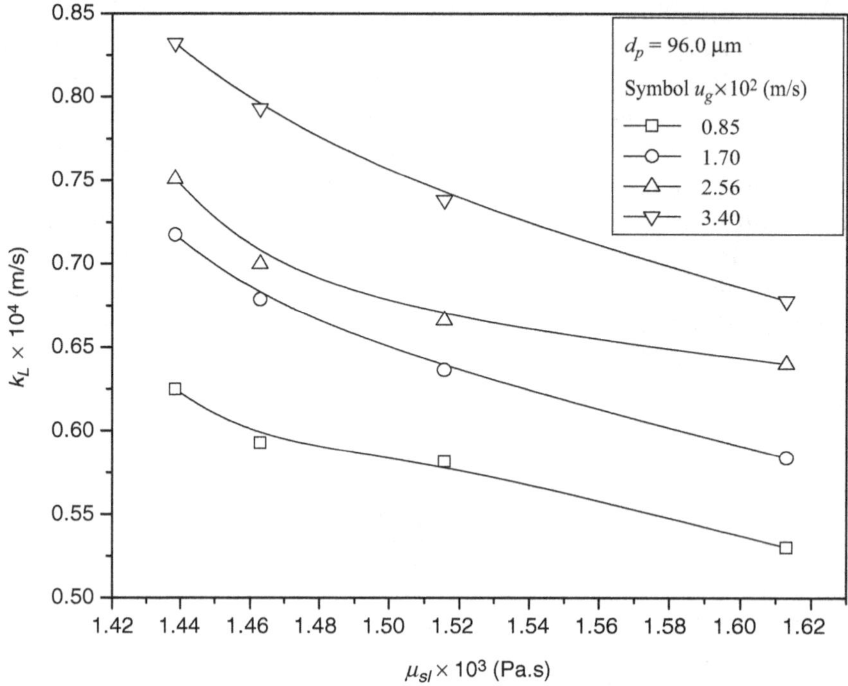

FIGURE 8.10 Variation of k_l with slurry viscosity in an inverse bubbly flow column (Sivaiah and Majumder, 2013).

viscosity at different superficial gas velocities is shown Figure 8.10. It is observed that the mass transfer coefficient is decreased with increase in slurry viscosity. An increase in slurry viscosity results in a decrease in the degree of turbulence in the liquid, thereby increasing the thickness of the laminar film at the boundary layer between the electrode surface and the slurry. An increase in the thickness of the laminar film decreases the electron deposition on the electrode surface, which results in a decrease in the mass transfer coefficient.

Effect of Diffusivity on the Mass Transfer Coefficient

Mass transfer theories are mainly developed for the process of absorption of a gas into a liquid even though their application might be extended to the cases that mass transfer occurs between any two immiscible fluid phases. The conventional and simplest picture of mass transfer between two fluid phases is that there exists a stagnant film at the interface (Lewis and Whitman, 1924). The general dependence of liquid-side film mass transfer coefficient is $k_l \propto D_l^n$ where n is dependent on the fluid–fluid system under given conditions (Kozinski and King, 1966). Higbie (1935) took a major step forward to analyze the mass transfer mechanism by introducing the "penetration theory" on the dependency of diffusivity on mass transfer.

Mass Transfer Coefficient as a Function of Energy Dissipation

Numerous mass transfer measurements have been carried out on various types of modified bubble columns. The mass transfer coefficient can be considerably increased by incorporating rigid or mobile components or by additional energy input. The mass transfer–enhancing variables are generally incorporated with the volume-based energy input to the bubble column. The mass transfer coefficient is exponentially dependent on the volume-based energy input to the bubble column as $k_l a \propto \varepsilon^n$ (Deckwer, 1992) in which $n = 1.2$ for a bubble column with perforated plates of 1 and 3 mm holes and $1.2 \leq n \leq 1.85$ for a hole diameter of 0.5 mm. This implies that energy input has a usually strong influence on mass transfer coefficient. The variations of mass transfer coefficient in different modified gas liquid contactor are shown in Table 8.2.

Interfacial Mass Transport Model

Several authors have described the different models on the transfer processes in bubble column reactors in different ways. Some of the important models are described as follows.

Model by Lau et al. (2004)

Lau et al. (2004) developed a mass transfer model to determine the transitent behavior of the mass transfer phenomena by saturating oxygen at elevated pressure. They assumed that the liquid is perfectly mixed in the radial direction. Therefore, the DO concentration is described by the equation

$$V(1-\varepsilon_g)\frac{dC}{dt} = k_l a V((1-\varepsilon_g)(C_e - C) + Q(C_0 - C)$$

(8.125)

where V is the reactor volume, ε_g is the gas hold-up, $k_l a$ is the gas–liquid mass transfer coefficient, C is the liquid-phase oxygen concentration, C_0 is the inlet liquid oxygen concentration, and C_e is the equilibrium oxygen concentration. Assuming that C_e is negligible for all pressure conditions and the inlet liquid has no oxygen content, equation (8.125) can be integrated with the boundary condition that at $t = t_0$, $C = C_i$. The solution of the above equation becomes

$$C(t) = C_i \exp(-\frac{k_l a V(1-\varepsilon_g) + Q}{V(1-\varepsilon_g)}t)$$

(8.126)

Therefore,

$$\ln C(t) = -\frac{k_l a V(1-\varepsilon_g) + Q}{V(1-\varepsilon_g)}t + \text{Constant}$$

(8.127)

The oxygen concentration in the liquid phase can be normalized by the air-saturated and nitrogen-equilibrated points as

$$C_{norm}(t) = \frac{C(t) - C_{min}}{C_{max} - C_{min}} \tag{8.128}$$

where C_{max} is the measured signal when the liquid is saturated by air and C_{min} is the signal at which oxygen is completely desorbed to the nitrogen stream. To account for the effect of axial dispersion of the liquid, equation (8.125) can be expressed as

$$\frac{\partial C}{\partial t} = D_z \frac{\partial^2 C}{\partial z^2} - \frac{u_{sl}}{(1 - \varepsilon_g)} \frac{\partial C}{\partial z} + k_l a (C_e - C) \tag{8.129}$$

where D_z is the axial dispersion coefficient and u_{sl} is the superficial liquid velocity. The initial condition for equation (8.129) is

$$C(z, t = 0) = C_i \tag{8.130}$$

The boundary conditions can be used as

$$\left(-D_z \frac{\partial C}{\partial z} + \frac{u_{sl}}{(1 - \varepsilon_g)} C \right)_{z=0} = \frac{u_{sl}}{(1 - \varepsilon_g)} C_0 (z = 0, t) \tag{8.131}$$

$$\frac{\partial C}{\partial z}(z = 0, t) = 0 \tag{8.132}$$

The above partial differential equation can be solved analytically under the boundary conditions by a Laplace transform, and the solution can be expressed as follows (Lau et al., 2004):

$$C(z, t) = C_0 A(z, t) + B(z, t) \tag{8.133}$$

where

$$
\begin{aligned}
A(z, t) = &\frac{u_{sl}}{u_{sl} + u'(1 - \varepsilon_g)} \exp\left[\frac{\{u_{sl} - u'(1 - \varepsilon_g)\}z}{2D_z((1 - \varepsilon_g)}\right] erfc\left[\frac{z - u't}{2(D_z t)^{0.5}}\right] \\
&+ \frac{u_{sl}}{u_{sl} - u'(1 - \varepsilon_g)} \exp\left[\frac{\{u_{sl} + u'(1 - \varepsilon_g)\}z}{2D_z((1 - \varepsilon_g)}\right] erfc\left[\frac{z + u't}{2(D_z t)^{0.5}}\right] \\
&+ \frac{u_{sl}^2}{2k_l a D_z (1 - \varepsilon_g)^2} \exp\left[\frac{u_{sl} z}{D_z (1 - \varepsilon_g)} - k_l a t\right] erfc\left[\frac{z(1 - \varepsilon_g) + u_{sl} t}{2(1 - \varepsilon_g)(D_z t)^{0.5}}\right]
\end{aligned} \tag{8.134}
$$

$$
B(z, t) =
\begin{bmatrix}
\dfrac{1}{2} erfc\left\{\dfrac{z(1 - \varepsilon_g) - u_{sl} t}{2(1 - \varepsilon_g)(D_z t)^{0.5}}\right\} + \dfrac{u_{sl}}{(1 - \varepsilon_g)(\pi D_z)^{0.5}} \times \exp\left\{-\dfrac{(z(1 - \varepsilon_g) - u_{sl} t)^2}{4(1 - \varepsilon_g)^2 D_z t}\right\} - \\
\dfrac{1}{2}\left(1 + \dfrac{u_{sl} z}{D_z (1 - \varepsilon_g)} + \dfrac{u_{sl}^2 t}{D_z (1 - \varepsilon_g)^2}\right) \times \exp\left(\dfrac{u_{sl} z}{D_z (1 - \varepsilon_g)}\right) erfc\left\{\dfrac{z(1 - \varepsilon_g) + u_{sl} t}{2(1 - \varepsilon_g)(D_z t)^{0.5}}\right\}
\end{bmatrix} \\
[-C_i \exp(-k_l a t)] + C_i \exp(-k_l a t) \tag{8.135}
$$

$$u' = \frac{u_{sl}}{(1 - \varepsilon_g)}\left(1 + \frac{4 k_l a D_z (1 - \varepsilon_g)^2}{u_{sl}^2}\right) \tag{8.136}$$

The mass transfer coefficient, $k_l a$, can then be solved by optimizing the objective function:

$$error = (C_{\exp} - C_{cal})^2 \tag{8.137}$$

Lau et al. (2004) reported that the dispersion model is more sensitive to the variation of the gas flow rate than the continuous stirred tank reactor model. Therefore, a larger section of the initial data point needs to be truncated to ensure the constant gas flow condition.

Cents et al. (2005a) Model

In an earlier section, it is mentioned that a commonly used reaction system for determination of mass transfer coefficient by the Dankwarts method is a carbonate–bicarbonate buffer solution, in which the reaction between CO_2 and water can be catalyzed by a number of agents. The criterion that determines whether or not the reaction of CO_2 in a carbonate–bicarbonate buffer solution can be regarded as pseudo first order is given by Danckwerts (1970). The reaction can be regarded as pseudo first order when the concentrations of all the ions are uniform throughout the mass transfer zone. However, Cents et al. (2005b) reported that considerable deviations from the pseudo first-order region may occur. The criterion is based on the fact that the transport of carbonate from the bulk can become the limiting step in the mass transfer process. However, because of the influence of both the carbonate and bicarbonate ions on the hydroxyl concentration in the mass transfer zone, a possible reduction in flux is difficult to predict beforehand. Information regarding this criterion is therefore very important, in which CO_2 is frequently removed from gas streams at high pressure using chemical absorption in a liquid. At these elevated CO_2 pressures, the Danckwerts criterion is violated more easily, and accurate knowledge regarding the solution of the complete reaction system is therefore required. To study the validity of the above criterion, they developed a model to determine the flux of CO_2 from the gas to the liquid phase. According to their model, the concentration profiles of all species in the mass transfer zone can be calculated as a function of time using the following equation:

$$\frac{\partial C_A(z,t)}{\partial t} = D_A \frac{\partial^2 C_A(z,t)}{\partial z^2} - I_A D_A \frac{F}{RT} \frac{\partial(\phi(z,t)C_A(z,t))}{\partial z} + r_A(z,t) \tag{8.138}$$

The parameter ϕ is called *electrostatic potential gradient*, which can be calculated by the use of the Nernst–Einstein equation (Newman, 1973) assuming dynamic electroneutrality

$$\phi(z,t) = \frac{RT}{F} \frac{\displaystyle\sum_{q=1}^{NC} I_q D_q \frac{\partial C_q(z,t)}{\partial z}}{\displaystyle\sum_{q=1}^{NC} I_q^2 D_q C_q(z,t)} \tag{8.139}$$

The following boundary conditions can be used:

$$t = 0, \quad z > 0 : C_A(z,t) = C_{A,bulk} \tag{8.140}$$

$$t > 0, \quad z = 0 : C_{CO_2}(z,t) = m_{CO_2} C_{CO_2(g)} \tag{8.141}$$

$$t > 0, \quad z = 0 : D_{ion} \frac{\partial C_{ion}(z,t)}{\partial x} - I_{ion} D_{ion} \frac{F}{RT} \phi(z,t) C_{ion}(z,t) = 0 \tag{8.142}$$

$$t > 0, \quad z = \infty : C_A(z,t) = C_{A,bulk} \tag{8.143}$$

The flux of CO_2 through the gas–liquid interface is calculated using the Higbie penetration model (Higbie, 1935). The complete system of equations can be solved numerically to obtain the concentration profile of solute during mass transfer.

Model by Singh and Majumder (2011)

There is a lack of comparative studies on mode of operation of the bubble column, although the mode of operation influences the efficiency of the mass transfer of the bubble column. Singh and Majumder (2011) developed a mechanistic model to interpret the mass transfer efficiency of the column and its dependency on various physical parameters, operating condition, and column geometry in the case of both cocurrent and countercurrent operations in a bubble column reactor. The concentration variation of the phases obtained by simulation of model can be used to obtain the mass transfer efficiency of bubble column reactor.

Model Description
Consider a two-phase (dispersed and continuous) system consist of i components participating in mass transfer process without chemical reaction in a bubble column reactor. In the bubble column, gas is dispersed as a dispersed phase of bubble in a continuous liquid phase. Let the concentration in the dispersed and continuous phases are respectively C_{1i} and C_{2i} where $i = 1, 2, 3, \ldots\ldots, i$. The material balance of these components can be written in both phases as

For dispersed phase:

$$\frac{\partial(\varepsilon_g \rho_g)}{\partial t} + \frac{\partial(u_g \rho_g \varepsilon_g)}{\partial r} = -\sum K_i (m_{ei} C_{1i} - C_{2i}) \tag{8.144}$$

For continuous phase:

$$\frac{\partial(\varepsilon_l \rho_l)}{\partial t} + \frac{\partial(u_l \rho_l \varepsilon_l)}{\partial r} = \sum K_i (m_{ei} C_{1i} - C_{2i}) \tag{8.145}$$

where

$$\varepsilon_g + \varepsilon_l = 1 \tag{8.146}$$

$$K_i = n K_{bi} \tag{8.147}$$

$$n = \frac{\varepsilon_g}{\frac{4}{3}\pi r_b^3} \tag{8.148}$$

In the continuous medium, in general, it is necessary to consider convective diffusion of component within the bubble and surrounding this bubble; otherwise, it is impossible to formulate the boundary condition on the surface of the bubble. This complicates the problem. Therefore, one usually takes the concentration of the component i in the dispersed phase equal to the concentration of the component in the surface of the gas bubble. In the following section, a simplified equation for both cocurrent and countercurrent flow of the gas–liquid two phases is analyzed to interpret the mass transfer process in the bubble column.

Model for Countercurrent Operation

Consider a one-dimensional flow in a countercurrent bubble column as shown in Figure 8.11. The z-axis is considered in the direction of movement of the gas phase, and the origin is selected at the place of its entrance into the column. To investigate the mass transfer from gas to liquid at steady state, equations (8.144) and (8.145) can be simplified. Introducing dimensionless quantities, the following equations can be represented:

$$(1-U)\frac{dC_1}{dZ}+k(m_e'C_1-C_2)=0 \tag{8.149}$$

$$U\frac{dC_2}{dZ}+k(m_e'C_1-C_2)=0 \tag{8.150}$$

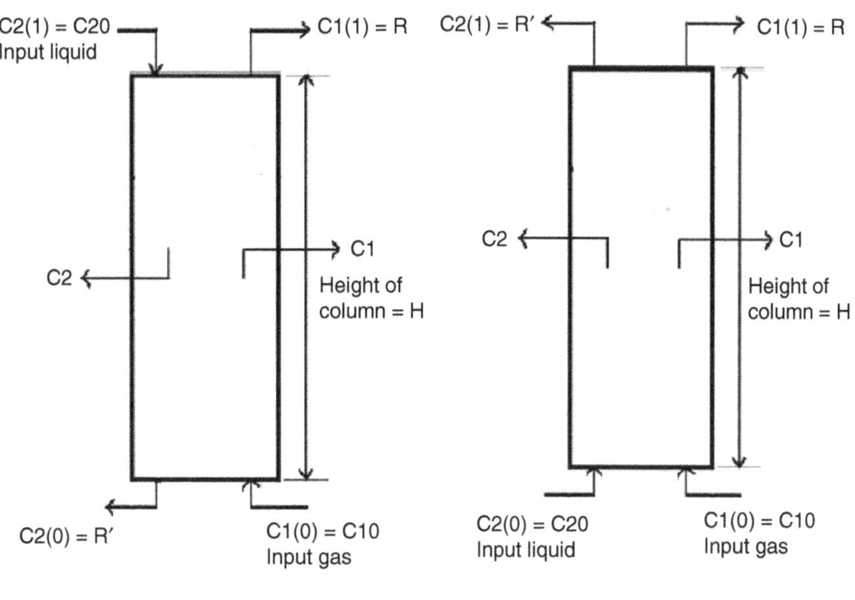

FIGURE 8.11 A, Diagram of countercurrent bubble column reactor (BCR). **B,** Diagram of cocurrent BCR.

where $U = u_l/u$, $m_e' = m_e \varepsilon_l/\varepsilon_g$, $k = (Kh_m)/(\varepsilon_l u)$, $0 \le U \le 1$, $k > 0$. Z = Dimensionless distance along the vertical z-axis ($Z = z/h_m$), $u_l = v_{sl}/\varepsilon_l$ is the actual velocity of liquid, u_{sl} = superficial liquid velocity. $u = u_g \pm u_l = u_{sg}/\varepsilon_g \pm u_{sl}/\varepsilon_l$, is the slip velocity or relative velocity of gas with respect to liquid (+ ve for countercurrent and – ve for co-current), h_m is the height of gas–liquid mixture in the column, $K = nK_b$ is volumetric mass transfer coefficient, [s^{-1}], ε_l is the fractional liquid hold-up, and m_e is the equilibrium distribution coefficient.

Here for small and moderate Reynolds numbers based on bubble diameter up to approximately equal to 700, the following expression can be expressed (Levich, 1959) as

$$u_g - u_l = \alpha_1 \frac{\rho_l r_b^2}{\mu_l} f_1(\varepsilon_g) \tag{8.151}$$

$$K_b = \alpha_2 \left(\frac{D\rho_l \varepsilon_g\, g f_1(\varepsilon_g)}{\mu_l} \right)^{0.5} f_2(\varepsilon_g) r_b^{2.5} \tag{8.152}$$

where the functions $f_1(\varepsilon_g)$ and $f_2(\varepsilon_g)$ take into account the hindered nature of the bubble flow and the effects of diffusion from their surfaces. As per Levich (1959), at a small Reynolds number, $\alpha_1 = 2/9$, $\alpha_2 = 8/3(\pi/3)^{0.5}$, and at large Reynolds number, $\alpha_1 = 1/9$, $\alpha_2 = 8/3(\pi/2)^{0.5}$. From equations (8.149) and (8.150), one can get,

$$(1-U)C_1 - UC_2 = P \tag{8.153}$$

where P is a constant. At $Z = 1$, $C_1(1) = R$ and $C_2(1) = C_{20}$. Therefore,

$$(1-U)R - UC_{20} = P \tag{8.154}$$

Similarly, between $Z = 1$ and $Z = Z$,

$$C_2 = \frac{1-U}{U}(C_1 - R) + C_{20} \tag{8.155}$$

Substituting the C_2 in terms of C_1 by equation (8.155) into the differential equation (8.149) and after rearranging, one gets

$$\frac{dC_1}{(m_e'U - (1-U))C_1 + (1-U)R - UC_{20}} = \frac{-kdZ}{U(1-U)} \tag{8.156}$$

Integrating equation (8.156) with boundary condition, at $Z = 0$, $C_1 = C_{10}$, one can obtain

$$\frac{1}{m_e'U - (1-U)} \ln\left(\frac{(m_e'U - (1-U))C_1 - (1-U)R - UC_{20}}{(m_e'U - (1-U))C_{10} - (1-U)R - UC_{20}} \right) = \frac{-kZ}{(1-U)U} \tag{8.157}$$

Defining $A = \dfrac{1-U}{m_e'U - (1-U)}$ and rearranging equation (8.157), one gets

$$C_1 = \frac{C_{20}UA}{1-U} + RA + \left(C_{10} - RA - \frac{C_{20}UA}{1-U} \right) \exp\left(\frac{-kZ}{UA} \right) \tag{8.158}$$

Putting $Z = 1$ and $C_1(1) = R$ in the above equation (8.158), one can get

$$R = \frac{C_{10}}{A + (1-A)\exp\left(\dfrac{k}{UA}\right)} + \frac{C_{20}UA\left(\exp\left(\dfrac{k}{UA}\right)-1\right)}{(1-U)\left(A + (1-A)\exp\left(\dfrac{k}{UA}\right)\right)} \tag{8.159}$$

If the liquid is having no gaseous component that is to be absorbed at the entry, $C_{20} = 0$, then

$$R = C_1(1) = \frac{C_{10}}{A + (1-A)\exp\left(\dfrac{k}{UA}\right)} \tag{8.160}$$

Similarly, substituting $C_1 = \dfrac{UC_2}{1-U} + R - \dfrac{UC_{20}}{1-U}$ from equation (8.155) in equation (8.150), integrating with boundary condition at $Z = 0, C_2 = C_2(0) = R'$, and rearranging, one can represent it as

$$\frac{1-U}{m'_e U - (1-U)}\ln\left(\frac{(m'_e U - (1-U))C_2 + m'_e R(1-U) - m'_e UC_{20}}{(m'_e U - (1-U))R' + m'_e R(1-U) - m'_e UC_{20}}\right) = \frac{-kZ}{U} \tag{8.161}$$

Now substituting $\dfrac{1-U}{m'_e U - (1-U)} = A$ and rearranging the terms, one can get

$$R' = C_{20}U\left(\left(\frac{m'_e A}{1-U}\right) + \exp\left(\frac{k}{UA}\right)\right) - m'_e RA\left(1 - \exp\left(\frac{k}{UA}\right)\right) \tag{8.162}$$

Mass transfer efficiency for countercurrent operation is defined as

$$\eta_{count} = 1 - \frac{C_1(1)}{C_1(0)} \tag{8.163}$$

Putting the values of $C_1(1)$ from equation (8.159) and $C_1(0) = C_{10}$, in equation (8.163), one gets mass transfer efficiency as

$$\eta_{count} = 1 - \frac{1}{A + (1-A)\exp\left(\dfrac{K}{UA}\right)} - \frac{C_{20}UA\left(\exp\left(\dfrac{k}{UA}\right)-1\right)}{C_{10}(1-U)\left(A + (1-A)\exp\left(\dfrac{k}{UA}\right)\right)} \tag{8.164}$$

General observation under limiting conditions of U as

$$\begin{aligned} U \to \infty, \eta_{count} \to 0, A \to 0 \\ U \to 0, \eta_{count} \to 1, A \to 1 \end{aligned} \tag{8.165}$$

Model for Cocurrent Operation

Again substituting the boundary conditions, $Z = 0, C_1(0) = C_{10}, C_2(0) = C_{20}$ in equation (8.153) it can be written as

$$(1-U)C_1 + UC_2 = (1-U)C_{10} + UC_{20} \tag{8.166}$$

Substituting the profile of C_2 in terms of C_1 from equation (8.166) into governing equation (8.149) for cocurrent operation, one can represent the equation as

$$\frac{dC_1}{(m'_e U + (1-U))C_1 - (1-U)C_{10} - UC_{20}} = \frac{-kdZ}{(1-U)U} \tag{8.167}$$

Integrating equation (8.167) with boundary condition at $Z = 0, C_1(0) = C_{10}$, one can get

$$\frac{1}{m'_e U + (1-U)} \ln\left(\frac{(m'_e U + (1-U))C_1 - (1-U)C_{10} - UC_{20}}{(m'_e U + (1-U))C_{10} - (1-U)C_{10} - UC_{20})}\right) = \frac{-kZ}{(1-U)U} \tag{8.168}$$

Substituting $Z = 1, C_1 = C_1(1)$, into equation (8.168), it can be expressed as

$$C_1(1) = C_{10}\left(A + \exp\left(\frac{-k}{UA}\right)(1-A)\right) + C_{20}\frac{UA}{1-U}\left(1 - \exp\left(\frac{-k}{UA}\right)\right) \tag{8.169}$$

Similarly, putting C_1 in terms of C_2 in equation (8.150), equation (8.150) can be represented as

$$\frac{(1-U)dC_2}{-(m'_e U + (1-U))C_2 + m'_e((1-U)C_{10} + m'_e UC_{20})} = \frac{kdZ}{U} \tag{8.170}$$

With the boundary conditions, at $Z = 0, C_2(0) = C_{20}$ integrating the equation (8.170), one gets

$$\frac{-(1-U)}{m'_e U + (1-U)} \ln\left(\frac{-(m'_e U + (1-U))C_2 + m'_e((1-U)C_{10} + m'_e UC_{20})}{-(m'_e U + (1-U))C_{20} + m'_e((1-U)C_{10} + m'_e UC_{20})}\right) = \frac{kZ}{U} \tag{8.171}$$

Defining $A' = \dfrac{1-U}{m'_e U + (1-U)}$ and substituting into the equation (8.171), one gets

$$C_2 = \left(C_{20} - m'_e\left(A'C_{10} + \frac{A'C_{20}m'_e U}{1-U}\right)\right)\exp\left(-\frac{kZ}{UA'}\right) + m'_e\left(A'C_{10} + \frac{A'C_{20}m'_e U}{1-U}\right) \tag{8.172}$$

At $Z = 1, C_2 = C_2(1) = R'$, equation (8.172) becomes

$$R' = (C_{20} - m'_e(A'C_{10} + \frac{A'C_{20}m'_e U}{1-U}))\exp\left(\frac{-k}{UA'}\right) + m'_e(A'C_{10} + \frac{A'C_{20}m'_e U}{1-U}) \tag{8.173}$$

In this operational mode with the help of equation (8.169), the mass transfer efficiency can then be expressed as

$$\eta_{co} = 1 - \frac{(C_{10}(1-A)(1-U) - UAC_{20})\exp\left(\dfrac{-k}{UA}\right)}{1-U} - A - \frac{UAC_{20}}{C_{10}(1-U)} \tag{8.174}$$

Observation under limiting condition of U implies that

$$\begin{aligned} U \to \infty, A \to 0, \eta_{co} \to 1 \\ U \to 0, A \to 1, \eta_{co} \to 0 \end{aligned} \tag{8.175}$$

For detailed discussion, readers can follow our publication (Singh and Majumder, 2011), where we reported that the countercurrent operation shows better efficiency than the cocurrent operation at different operating variables. In countercurrent operations, the intensity of mixing of the phases is more prominent than the cocurrent operation. The mixing phenomena may change the mass transfer efficiency. Majumder (2008a) and Manish and Majumder (2009) explained the mass transfer efficiency based on mixing phenomena. As the quality of mixedness increases, the mass transfer efficiency of the column increases. Also, the efficiency increases with the increase in inverse of modified effective Sherwood number $(D_z \varepsilon_g)/(k_l d_{32})$ (Manish and Majumder, 2009). For the countercurrent operation, the opposite flow directions of the phases increase the momentum transfer between the phases, which may increase the number of more circulation cell inside the column. More circulation of the phases results in more mixing of the phases. But in the case of cocurrent operation, the momentum exchange between phases is relatively less than the countercurrent operation because of the decrease in the relative velocity of the phases. This may cause the decrease in mass transfer efficiency in case of the cocurrent operation. The authors concluded that the countercurrent operation of bubbly flow may be suitable for the design and installation of the bubble column in industry for their specific application. The enhance factor (E) of mass transfer efficiency by the countercurrent operation has been correlated by them with the superficial gas velocity within the range of superficial gas velocity 0.02 to 0.08 m/s, which can be expressed as:

$$E = (\eta_{count} - \eta_{co})/\eta_{co} = 0.1353 \ln(u_{sg}) + 0.543 \tag{8.176}$$

Information Entropy Theory to Interpret Mass Transfer Efficiency

The mixing phenomenon can be explained in terms of information entropy theory. Details of the probality theory are described in the Chapter 6. Earlier Nedeltchev et al. (1999) and Manish and Majumder (2009) analyzed the quality of mixedness and the degree of mass transfer based on the entropy theory for upflow and inverse bubble columns. According to the theory put forward by Higbie (1935), a bubble of gas rises through a liquid, which absorbs the gas. The contact time characterizes the residence time of micro eddies that are responsible for mass transfer in the liquid film at the bubble–liquid interface. The area of the bubble–liquid interface is controlled by the behavior of micro-scale eddies behind the bubbles. Although several other different mechanisms have been proposed to describe the transport process near the free surface, Higbie's (1935) model and the surface-renewal model have received the greatest acceptance. The contact time (t_c) can be obtained from equation (8.177) as (Manish and Majumder, 2009)

$$t_c = \frac{4D_z}{\pi k_l^2} \tag{8.177}$$

The volumetric mass transfer coefficient ($k_l a$) can be calculated from the correlation (equation [8.89]) developed by Mandal et al. (2004) for an inverse flow bubble column.

The liquid phase axial dispersion coefficient (D_z) for an inverse flow bubble column can be calculated from the correlation developed by Majumder et al. (2005), which is shown in equation (6.110). The final form of equation for quality of mixedness ($M(t)$) as a function of dimensionless groups of Sherwood number and intrinsic mass transfer number can be represented by equation (8.179) as (Manish and Majumder, 2009)

$$1 - M(t) = f(ShM_i) \tag{8.178}$$

where the intrinsic mass transfer number can be defined as

$$M_i = \frac{k_l a t}{\varepsilon_g} \tag{8.179}$$

The Sherwood number is defined as $k_l d_{32}/D_z$. The intrinsic mass transfer number signifies the effective mass transfer from gas to liquid for a given expose time of contact. The intrinsic mass transfer number increases with the increase in both specific interfacial area and the interfacial contact time. The correlation of intrinsic mass transfer number related to the quality of maximum mixedness can be expressed as

$$M_i = \frac{5.57 \times 10^{-4}}{\{1 - M(t)\}^{1.703} Sh} \tag{8.180}$$

According to penetration theory, the main process occurring inside the column is the mass transfer between bubble and liquid. This physical process is often the major criterion for design and scale up of bubble column. The variation of intrinsic mass transfer number with the quality of mixedness is shown in Figure 8.12.

The intrinsic mass transfer number increases with an increase in the superficial liquid velocity and hence the quality of mixedness. As the superficial liquid velocity increases, the exchange of momentum increases for which the finer bubbles are formed because of the break-up of bubbles. Consequently, more interfacial area between gas and liquid results. More interfacial area gives a larger volumetric mass transfer coefficient. As a result, the intrinsic mass transfer number increases with an increase in the superficial liquid velocity. The quality of mixedness in the bubble column depends on how the flow characteristics behave in the column because of variability of the axial liquid velocity and dispersion coefficient of bubble motion in the bubble column. For a plunging liquid jet inverse flow bubble column reactor, the variations of dispersion coefficient of bubble motion and velocity characteristic factors for different operating range were reported by Majumder (2008b).

A Multiple-Column Plug-Flow Model (Chen, 2012)

Consider a bubble column in which a binary gas mixture ($A + B$) (A is soluble, and B is insoluble to liquid) is dispersed in a continuous liquid medium flowing into the column. The column is divided into N number of columns of the same height through which gas or liquid flows in the plug flow condition. The gas and liquid streams come into contact within the column adversely. The mass transfer rate of gas component A can be determined by

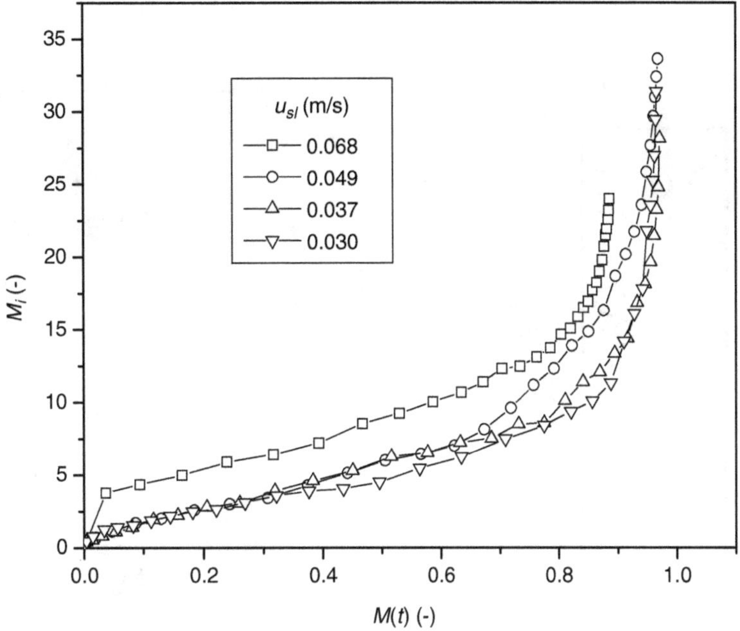

FIGURE 8.12 Variation of intrinsic mass transfer number with quality of mixing (Manish and Majumder, 2009).

using the material balance under steady-state operation for a multiple-column plug-flow model as shown in Figure 8.13.

If plug flow is assumed for the gas phase through the column, the material balance for component A in the ith column at steady state can be written as

$$\left(u_{gi} C_{Ai}\big|_z - u_{gi} C_{Ai}\big|_{z+\Delta z}\right) A_{ci} - r_{Ai}\pi d_{ci}\Delta z = 0 \tag{8.181}$$

where u_{gi} is the actual velocity in each ith column, A_{ci} is the cross-section of the ith column, r_{Ai} is the mass transfer rate per unit area of the ith column, and d_{ci} is the diameter of the ith column. For the total number of columns, the material balance can be written as

$$\sum_{i=1}^{N}\left(u_{gi} C_{Ai}\big|_z - u_{gi} C_{Ai}\big|_{z+\Delta z}\right) A_{ci} - \sum_{i=1}^{N} r_{Ai}\pi d_{ci}\Delta z = 0 \tag{8.182}$$

Divideding equation (8.182) by Δz and taken to the limit (i.e., $\Delta z \rightarrow 0$), it yields

$$-\frac{d\left(\sum u_{gi} A_{ci} C_{Ai}\right)}{dz} - \sum_{i=1}^{N} r_{Ai}(\pi d_{ci}) = 0 \tag{8.183}$$

or

$$-\frac{dF_A}{dz} - \sum_{i=1}^{N} r_{Ai}(\pi d_{ci}) = 0 \tag{8.184}$$

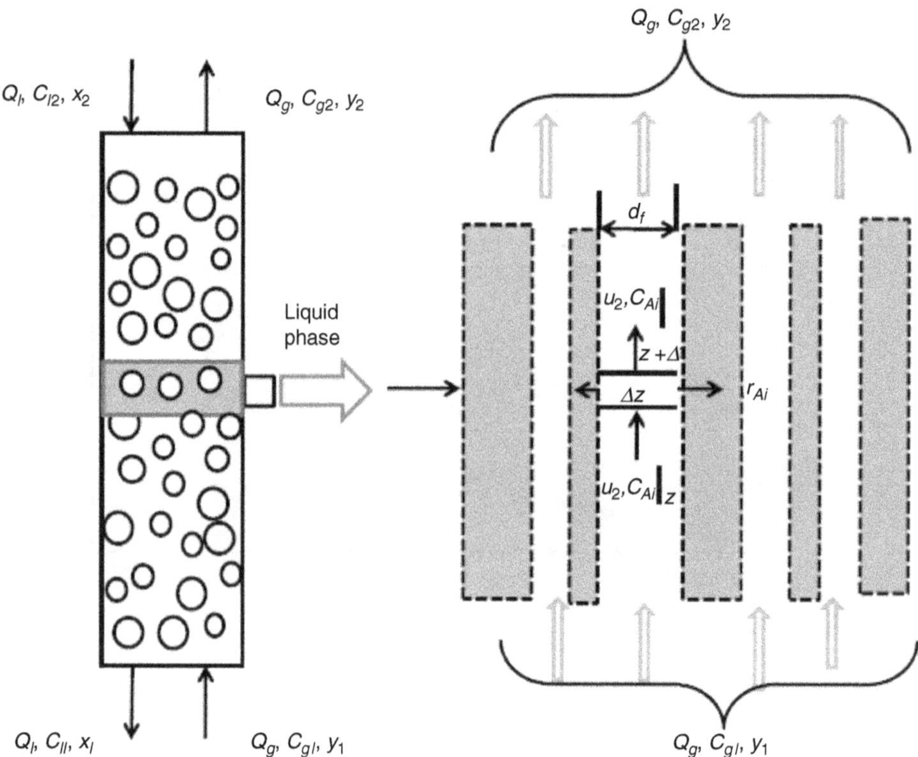

FIGURE 8.13 Schematic of a multiple-column plug-flow model (as per Chen, 2012).

where F_A is the overall molar flow rate of the gas phase, which is equal to $\sum u_{gi} A_{ci} C_{Ai}$. Integrating equation (8.184), one can get

$$F_{A1} - F_{A2} - \int_1^2 r_{Ai}\, dS_i = 0 \tag{8.185}$$

where S_i is the lateral surface area of the ith column. These equations can be rewritten as

$$F_{A1} - F_{A2} - \bar{r}_A S = 0 \tag{8.186}$$

where

$$\bar{r}_A = \frac{\sum \int_1^2 r_{Ai}\, dS_i}{\sum \int_1^2 dS_i} = \frac{\sum \int_1^2 r_{Ai}\, dS_i}{S} \tag{8.187}$$

where $\bar{r}_A$ is the mean mass transfer rate and S is the total surface area of the gas phase in the bubble column, which cannot be obtained directly. Equation (8.186) shows that the mass transfer rate, $\bar{r}_A S$, is equal to the consumption of component A through the bubble column. This value can be determined by measuring the concentration of A and the gas flow rate:

$$R_A = \frac{F_{A1} - F_{A2}}{V_l} = \frac{F_{A1} - F_{A2}}{\varepsilon_l V_b} \tag{8.188}$$

and

$$R_A = \frac{\bar{r}_A S}{V_l} \tag{8.189}$$

where V_l is the volume of the liquid phase, V_b the volume of the bubble column, and ε_g is the hold-up of the liquid phase. Because the molar flow rate of inert gas is equal to $F_{A1}(1 - y_1)/y_1$, the molar flow rate of the component A at the outlet F_{A2} is $F_{A1}[(1 - y_1)/y_1][y_2/(1 - y_2)]$. Thus, equation (188) can be rewritten as

$$R_A = \frac{F_{A1}}{V_l}\left[1 - \left(\frac{(1 - y_1)}{y_1}\right)\left(\frac{y_2}{1 - y_2}\right)\right] \tag{8.190}$$

Therefore, the overall absorption rate, R_A, can be obtained with measurable quantities. According to a two-film model, the relationship between local mass transfer rate, r_A, and local individual mass-transfer coefficients based on both the gas-side and liquid-side can be written as

$$r_A = k_g a (C_g - C_{gi}) \tag{8.191}$$

$$r_A = k_l a (C_{li} - C_l) \tag{8.192}$$

at the interface $C_{gi} = HC_{li}$, where H is Henry's law constant. Combining equations (191) and (192) in terms of the overall mass transfer coefficient, the equation becomes

$$r_A = (k_g a)_{ov,z}(C_A - HC_{lA}) \tag{8.193}$$

where C_A and C_{lA} are concentrations of A in the gas phase and liquid phase, respectively. According to equation (8.182), it is assumed that plug flow for gas phase and well-mixed flow for the liquid phase, the material balance equation between Δz at steady state can then be rewritten as

$$\left(u_g C_A\big|_z - u_g C_A\big|_{z+\Delta z}\right) A_c - r_A \varepsilon_l \pi S \Delta z = 0 \tag{8.194}$$

where u_g is the mean gas-superficial velocity based on the column cross-sectional area. Taking limit for equation (8.194) and substituting equation (8.193) into equation (8.194), one can write

$$u_g \frac{dC_A}{dz} + (k_g a)_{ov,z}(C_A - HC_{Al})\varepsilon_l = 0 \tag{8.195}$$

If $C_A >> HC_{lA}$, equation (8.195) can be written as

$$u_g \frac{dC_A}{dz} + (k_g a)_{ov,z} C_{Al} \varepsilon_l = 0 \tag{8.196}$$

If liquid hold-up, ε_l, is kept constant through the column integration of equation (8.196), the result is

$$u_g \int_{C_{A1}}^{C_{A2}} \frac{dC_A}{C_A} + \varepsilon_l \int_0^L (k_g a)_{ov,z} \, dz = 0 \tag{8.197}$$

and

$$C_{A2} = C_{A1} e^{-\frac{(k_g a)_{ov,L} L \varepsilon_l}{u_g}} \tag{8.198}$$

where

$$(k_g a)_{ov,L} = \frac{1}{L} \int_0^L (k_g a)_{ov,z} \, dz \tag{8.199}$$

By multiplying S with equation (8.198), it becomes

$$(k_g a)_{ov,L} = \frac{Q_g}{V_l} \ln \frac{C_{A1}}{C_{A2}} \tag{8.200}$$

Equation (8.200) can be used to evaluate the average overall mass transfer coefficient in terms of measurable quantities. Moreover, the overall mass transfer coefficient is correlated with the individual mass transfer coefficients as

$$\frac{1}{(k_g a)_{ov,L}} = \frac{1}{k_g a} + \frac{H}{k_l a} \tag{8.201}$$

$1/(k_g a)_{ov,L}$ is called overall resistance, and $1/(k_g a)$ and $H/(k_l a)$ are the gas and the liquid-side resistances, respetively. For a second-order reaction, equation (8.201) is written as

$$\frac{1}{(k_g a)_{ov,L}} = \frac{1}{k_g a} + \frac{H}{(D_A k_2 C_{B0})^{1/2} a} \tag{8.202}$$

where k_2 is the second-order reaction constant and D_A is the diffusion coefficient for component A. Equation (8.202) can be plotted to obtain $k_g a$ and k_g at different experimental conditions provided the other variables at the experimental conditions. Chen (2012) reported the absorption rate and overall mass transfer coefficient obtained based on the model in a bubble column scrubber for five absorption systems, that is, $NaOH/BaCl_2/H_2O$, $NaOH/Zn(NO_3)_2/H_2O$, $MEA/CO_2/H_2O$, $MEA/CaCl_2/CO_2$, and $NH_3/CO_2/H_2O$. The process variables are working temperature, gas flow rate, pH value, and gas concentration. The absorption rates and overall mass transfer coefficients are listed in Table 8.8.

Table 8.8 Absorption Rates and Overall Mass Transfer Coefficients in a Bubble Column Scrubber for Different Systems

Systems	Operating Conditions	$R_A \times 10^4$ (mol/L.s)	$(k_g a)_{ov,L}$ (1/s)
MEA–CO$_2$–H$_2$O	T = 25-45 °C, y_1 = 15%, Q_g = 4-9.5 L/min, C_{MEA} = 4 M	0.0135-6.22	0.027-1.120
MEA–CaCl$_2$–CO$_2$ (with precipitation)	T = 25 °C, y_1 =10%-30%, pH = 9–11, C_l = 0.2 M, C_{MEA} = 4 M, Q_g = 2 L/min-9.5 L/min	0.0626-3.69	0.018-0.249
NH$_3$–CO$_2$–H$_2$O (with precipitation)	T = 25-60 °C, y_1 = 15%-60%, pH = 9.5-11.5, C_{NH3} = 7.7 M, Q_g = 2-5 L/min	2.17-10.9	0.0136-0.567
NaOH–BaCl$_2$–H$_2$O (with precipitation)	T = 25 °C, y_1 = 10%-30%, pH = 12-13, C_l = 0.01-0.2 M, C_{NaOH} = 1-2 M, Q_l = 50-320 ml/min, Q_g = 2-10.7 L/min	0.44-7.69	0.0277-0.196
NaOH–BaCl$_2$–H$_2$O (without precipitation)	T = 25 °C, y_1 = 20%, pH = 12-13, C_{NaOH} = 1-2 M, Q_l = 50 mL/ min, Q_g = 2-10.7 L/min	1.03-11.3	0.0651-0.339
NaOH–Zn(NO$_3$)$_2$–H$_2$O (with precipitation)	T = 25-45 °C, y_1 = 20%, pH = 10-13, C_l = 0.2 M, C_{NaOH} = 2 M, Q_l = 50 ml/min, Q_g = 3-8 L/min	1.34-28.8	0.0145-0.590

Computational Fluid Dynamics Modelling

Computational fluid dynamics are initiated on the basis of closure of the differential equations of momentum and transport processes by various researchers. The fundamental problem of establishing the computational mass transfer is to find a method of closure for the mass transfer differential equation such as those for closing the momentum and the heat transfer differential equations. The basic equations are described as follows.

Basic Equations
The mass transfer equation of a component species with instantaneous concentration $\tilde{c}$ in incompressible fluids can be expressed as (Bird et al., 2002):

$$\frac{\partial \tilde{c}}{\partial t} + \tilde{u}_i \frac{\partial \tilde{c}}{\partial x_i} = D \frac{\partial^2 \tilde{c}}{\partial x_i \partial x_i} + \tilde{S}_c \tag{8.203}$$

where $\tilde{u}$ is instantaneous velocity, D is the molecular diffusivity, and $\tilde{S}_c$ is the source term. For the turbulent flow specially jet-induced inverse bubbly blow, if substituting $\tilde{c} = C + c$ and $\tilde{u} = U + u$ into equation (8.204), in which C and U are the time-average values, then the time-average transport equation for the concentration scalar can be expressed as

$$\frac{\partial C}{\partial t} + U_j \frac{\partial C}{\partial x_j} = \frac{\partial}{\partial x_j}\left(D \frac{\partial C}{\partial x_j} - \overline{u_j c} \right) + S_c \tag{8.204}$$

The term $\overline{u_j c}$, the second-order covariance of the velocity and concentration, appears and may be regarded as the Reynolds mass flux, analogous to Reynolds stress in the CFD.

For isotropic fluid dynamics similar to the Boussinesq postulation for turbulence, the term can be defined as proportional to the gradient of average concentration (Rosén and Trägårdh, 1995):

$$\overline{u_j c} = -D_{t,i} \frac{\partial C_i}{\partial x_j} \tag{8.205}$$

where the parameter $D_{t,i}$ is called the turbulent diffusivity of component i. The parameter is related to the velocity, temperature, component concentration, and structure of the flow field. The turbulent diffusivity plays an important role in solving the equations. The turbulent diffusivity can be estimated empirically by simple analogy to the turbulent viscosity based on the assumption that the turbulent mass transfer is simply related to the turbulent momentum transfer. This empirical analogy conduces to a simple expression for the turbulent diffusivity as

$$D_{t,i} = \mu_t / Sc_t \tag{8.206}$$

where μ_t is a turbulent viscosity that can be acquired by using a CFD model such as the k-ε model and Sct is the Schmidt number, usually assuming to be a constant taken between 0.7 and 1.0. Another empirical approach can be followed to estimate $D_{t,i}$ from the dispersion number, which is equal to the turbulent diffusivity divided by a characteristic velocity and characteristic length. The dispersion number can be applied to interpret the intensity of backmixing of the fluid in the column. The experimental techniques are described Chapter 6. The main disadvantage of the consideration of the $D_{t,i}$ by these two approach is that it not only related to the fluid velocity fluctuation but also dependent on the concentration fluctuation, which cannot be explained by these two approaches. Sun et al. (2005) suggested that $D_{t,i}$ is proportional to the product of characteristic velocity and characteristic length, or mathematically

$$D_t = C_t k^{1/2} L_m \tag{8.207}$$

where k is the turbulent kinetic energy, equal to $\overline{u_i' u_i'}/2$ and its square root represents the characteristic velocity and L_m is the characteristic length represented by $k^{1/2}/\tau_m$ where the mixed time scale τ_m is taken as geometric average $(\tau_\mu \tau_c)^{1/2}$ in which τ_μ and τ_c are the dissipation time scale of the velocity and concentration fluctuations. In CFD, $\tau_\mu = k/\varepsilon$, where ε is the dissipation rate of concentration fluctuation. Similarly, one can consider $\tau_c = c^2/\varepsilon_c$ (Xigang and Guocong 2008) where c^2 is the variance of concentration fluctuation and ε_c is the dissipation rate of concentration fluctuation. Then equation (8.207) becomes

$$D_t = C_t k \left(\frac{k \overline{c^2}}{\varepsilon \varepsilon_c} \right)^{1/2} \tag{8.208}$$

The parameters k and ε in equation (8.208) can be obtained by the k-ε model in CFD. The other two variables $\overline{c^2}$ and ε_c can be obtained by solving corresponding equations. The equations for $\overline{c^2}$ and ε_c are derived and simplified by Sun et al. (2005) as follows

$$\frac{\partial \overline{c^2}}{\partial t} + U_i \frac{\partial \overline{c^2}}{\partial x_i} = \frac{\partial}{\partial x_i}\left(\left(\frac{D_t}{\sigma_c} + D\right)\frac{\partial \overline{c^2}}{\partial x_i}\right) - 2D_t \frac{\partial C}{\partial x_i}\frac{\partial C}{\partial x_i} - 2\varepsilon_c \tag{8.209}$$

$$\frac{\partial \varepsilon_c}{\partial t} + U_i \frac{\partial \varepsilon_c}{\partial x_i} = \frac{\partial}{\partial x_i}\left(\left(\frac{D_t}{\sigma_{\varepsilon c}} + D\right)\frac{\partial \varepsilon_c}{\partial x_i}\right) - C_{c1}\frac{\varepsilon_c}{\overline{c^2}}\overline{cu_i'}\frac{\partial C}{\partial x_i} - C_{c2}\frac{\varepsilon}{k}\varepsilon_c \tag{8.210}$$

where D and D_t are the molecular and turbulent diffusivities, respectively. The parameters $\sigma_c \sigma_{\varepsilon c}$, C_1, and C_2 are constants.

Equations (8.209), (8.210), and (8.204) constitute the $\overline{c^2}$-ε_c model for the closure of equation (8.205) and are the fundamental part of the computational mass transfer. The variables k and ε can be solved by the CFD method for turbulent flow consisting of five equations, namely the continuity equation, the moment equation, the Boussinesq's D_t equation, the k equation, and the ε equation. If the process involves heat effect, such as the chemical absorption or exothermic catalytic chemical reaction, the effect of temperature distribution should be incorporated. It is to be noted that if the model involves the heat transport equation, the Boussinesq's thermal diffusivity D_h equation, the $\overline{t^2}$ equation, and ε_t equation should be considered. The article on the different case studies for the modeling of chemical processes by computational mass transfer by Xigang and Guocong (2008) is suggested for further reading.

McClure et al. (2014a, 2014b) Model
The mass transfer phenomena for gas-to-liquid oxygen transfer can be accounted by incorporating the transport equations (McClure et al., 2014a, 2014b) for the mass fraction of DO ($m_{f,DO2}$) in the liquid phase. The governing transport equation is

$$\frac{\partial(\rho_l \varepsilon_l m_{f,DO_2})}{\partial t} + \nabla.(\varepsilon_l(\rho_l u_l m_{f,DO_2} - \rho_l D_{DO_2}(\nabla m_{f,DO_2}))) = S_l + \Gamma_{gl} \tag{8.211}$$

where S_l denotes any sources of DO in the liquid phase, Γ_{gl} is a term that accounts for interphase mass transfer of oxygen from the gas to the liquid, and D_{DO2} is the kinematic diffusivity of the DO species, which includes both laminar and turbulent contributions. It is also necessary to include a transport equation for the oxygen mass fraction ($m_{f,O2}$) in the gas phase:

$$\frac{\partial(\rho_g \varepsilon_g m_{f,O_2})}{\partial t} + \nabla.(\varepsilon_g(\rho_g u_g m_{f,O_2} - \rho_g D_{O_2}(\nabla m_{f,O_2}))) = S_g - \Gamma_{gl} \tag{8.212}$$

where S_g denotes any sources of oxygen in the gas phase and D_{O2} is the kinematic diffusivity of oxygen in the gas phase. The term S_g can be set to zero as is physically correct, and the source term for DO (S_l) given a sufficiently large negative value such that the DO concentration is zero at all times. The latter condition depends on the experimental situation in which any DO immediately reacts with the other specis present. One can set the values of the laminar kinematic diffusivity in both the gas and liquid phases to be zero, and the turbulent Schmidt numbers take their default value of 0.9. The rate of interphase mass transfer is calculated according to

$$\Gamma_{gl} = k_l a (C_e - C) \tag{8.213}$$

The interfacial area density per unit volume (a) is calculated by the CFD model using $a = 6\varepsilon_g/d_b$. A single bubble size (corresponding to the experimentally measured mean size across all operating conditions) can be used in the CFD model. McClure et al. (2014a, 2014b) reported that the use of a single bubble size in the CFD model has considerable advantages from the perspective of computational efficiency. They pointed out that a single bubble size can be an excellent approximation if the experimentally measured BSD is relatively narrow. The interphase momentum transfer used in the CFD modeling arise from two mechanisms, turbulent bubble dispersion and bubble drag. The former can be described using the Favre-averaged model (Burns et al., 2004). The bubble drag can be modeled using the correlation for a single bubble developed by Clift et al. (1978) together with a modified form of the volume fraction correction term developed by Simonnet et al. (2007). The use of a liquid types of chemical compund will impact on the liquid phase properties, which needs to be accounted for in the CFD model. As an example, the presence of salts (e.g., sodium sulphite) tends to reduce the bubble terminal velocity or change the slip velocity (Jamialahmadi and Müller-Steinhagen, 1992) that should be accounted for in CFD model through the introduction of an empirical constant having a value of 2 into the bubble drag calculations. Readers can follow further details about this approach in the reference by McClure et al. (2015a). Measured values of the liquid film mass transfer coefficient (k_l) should be wisely used in the CFD model. The saturation oxygen concentration (C^*) in the CFD model can be calculated using Henry's law:

$$C_e = \frac{yP}{H} M_{w,O_2} \tag{8.214}$$

where P is the pressure, y is the mole fraction of oxygen in the gas phase, $M_{W,O2}$ is the molecular mass of oxygen, and H is the Henry's law constant (77,942 Pa m^3/mol). The oxygen transfer rate can be calculated from the CFD results by taking the average value of the interphase transfer term over the last 30 s of the simulation (McClure, 2015b), dividing this value by the liquid volume, which can be expressed as

$$OTR = \frac{1}{V_l} \frac{\int_0^{t_{ave}} \Gamma_{gl,t}\, dt}{t_{ave}} \tag{8.215}$$

McClure et al. (2015b) suggested that the t_{ave} can be taken as 30 s. A commercial CFD code can be used to simulate this. Simulations can be performed at the same superficial velocities as those that are experimentally investigated. Further details about the numerical setup, including the computational mesh and validation of the column hydro-dynamics, can be obtained from the literature (McClure et al., 2014a).

Nomenclature

a Interfacial area per volume (1/m)

C Concentration (mol/m^3)

d Diameter (m)

D Dispesrion coefficient/Diffusivity (m^2/s)

d_{32} Sauter mean bubble diameter (m)

d_o Gas distributor hole diameter (m)

d_b Bubble diameter (m)

G Molar flowrate of the reacting diluent gas (mol/s)

G' Molar flowrate of inert gas (mol/s)

H Henry's law constant (N-m/.mol)

h Sum of contribution of ions to solubility factors (l/gmole)

h_+, h_-, h_g Contributions of positive ion, negative ion, and gas to solubility factor h (l/gmol)

I Ionic charge (-)

J Mass transfer flux (mol/m^2.s)

K Constant (-)

k Dimensionless mass transfer coefficient ($k = (Kh_m)/(\varepsilon_l u)$) (-)

k'_{LR} Coefficient of chemical absorption, $(D_A k_2 C_B^o)^{1/2}$ (m/s)

$k'_{m,n}$ Rate constant for mth, nth order reaction ((m^3/mol)$^{m+n-1}$/s)

K_i Coefficient of mass transfer of the ith substance (m/s)

k_l Liquid-side mass transfer coefficient (m/s)

$K_l a$ Overall mass transfer coefficient (1/s)

$k_l a$ Volumetric mass transfer coefficient of liquid (1/s)

K_{bi} Coefficient of mass transfer of the ith substance from single bubble (m/s)

$k_g a$ Volumetric mass transfer coefficient of gas (1/s)

L Length (m)

m Coefficient (-), equilibrium constant (-)

m'_e Equilibrium distribution coefficient defined as $m_e \varepsilon_l / \varepsilon_g$ (-)

$m_{e,i}$ Equilibrium coefficient of distribution of the substance between phases (-)

m_f Mass fraction (-)

M_i Intrinsic mass transfer number (-)

N Number of moles (-)

n Number density of bubbles (-)

n Coefficient (-)

P Pressure (N/m^2)

Q Volumetric flow rate (m^3/s)

R Ideal gas constant

R Overall rate of mass transfer (mol/m^3.s)

R Unknown concentration of gas at outlet (moles/m^3)

R' Unknown concentration of liquid at outlet (moles/m^3)

r_b Radius of bubble (m)

S Solubility in solution (gmole/[atm lit])

St Stanton number (-)

S_W Solubility of in water (gmole/[atm lit])

T Temperature (K)

t Time (s)
U Ratio of liquid to slip velocity ($= u_l/u$) (-)
u Velocity (m/s)
u' Velocity fluctuation (m/s)
V Volume of the reactor (m^3)
y mole fraction (-)
y_i Composition of CO_2 in gas phase at outlet i (mol%)
y_o Composition of CO_2 in gas phase at inlet (mol%)
Z Dimensionless distance along the vertical z-axis ($= z/z_m$) (-)
z Vertical distance (m)
z_R Dimensionless distance (-)

Greek Letters

ϕ Electrostatic potential gradient (J/C.m)
α Ratio of hydrostatic to total pressure (-)
ε Energy dissipation rate per unit volume (N/m^2.s)
δ Liquid film thickness (m)
η_{co} Mass transfer efficiency for concurrent operation
η_{count} Mass transfer operation for countercurrent operation
ε_g Gas hold-up (-)
ρ_l Density of liquid (kg/m^3)
ε_l Liquid hold-up (-)
σ_l Surface tension of liquid (N/m)
μ_l Viscosity of liquid (kg/m.s)
Γ Rate of interphase mass transfer (mol/s)

Subscripts

0 Initial
A Component A
ave Average
b bubble
B Component B
c Column
cal Calculated
cat Catalyst
e Equilibrium
exp Experimeental
f Fraction
g Gas
i Indices
l Liquid
m mixture
n nozzle
s Sensor
sg Superficial gas
sl Superficial liquid
t Turbulent

z axial
z_i At distance z_i

Dimensionless Group

Eo Eotvos number of the bubble of diameter $d_{b,j}$, $(g\rho_l d_{b,j}{}^2/\sigma)$
Fr Froude number $(u_{sg}/\sqrt{gd_b})$
Ga Gallilei number $((gd_c^3\rho_l^2)/\mu_l^2)$
Ha Hatta number (-)
Mo Morton number $((g\mu_l^4)/(\rho_l\sigma^3))$
P_R Pressure ratio defined as $P_{atm}/(\rho_l gz_m)$
$P_{b,R}$ Pressure ratio defined as $4\sigma/(\rho_l gz_m d_b)$
Pe Peclet number $((u_{sl}z_i)/D_z)$
Re Reynolds number $((d_c u_{sl}\rho_l)/\mu_l)$
Sc Schmidt number $(\mu_l/(D_l\rho_l))$
W_j Aspect ratio

References

Akita, K., Yoshida, F., 1973. Gas holdup and volumetric mass transfer coefficient in bubble columns. Ind. Eng. Process Des. Dev. 12, 76–80.

Akita, K., Yoshida, F., 1974. Bubble size, interfacial area and liquid-phase mass transfer coefficient in bubble columns. Ind. Eng. Chem. Process. Des. Dev. 12, 76–80.

Alvarez, E., Sanjuro, B., Cancela, A., Navaza, J.M., 2000. Mass transfer and influence of physical properties of solutions in a bubble column. Trans IChemE 78, 889–893.

Bartholomew, W.H., Karow, E.O., Sfat, M.R., Wilhelm, R.H., 1950. Mass transfer of oxygen in submerged fermentation of *Streptomyces griseus*. Ind. Eng. Chem. 42, 1801–1809.

Behkish, A., Men, Z., Inga, J.R., Morsi, B.I., 2002. Mass transfer characteristics in a large- scale slurry bubble column reactor with organic liquid mixtures. Chem. Eng. Sci. 57, 3307–3324.

Bennett, D.L., Watson, D.N., Wiescinski, M.A., 1997. New correlation for sieve-tray point efficiency, entrainment and section efficiency. A.I. Ch. E. J. 43 (6), 1611–1626.

Benson, H.E., Field, J.H., Jimeson, R.M., 1954. CO_2 absorption employing hot carbonate solutions. Chem. Eng. Progress 50, 356–363.

Bin, A.K., 1993. Gas entrainment by plunging liquid jet. Chem. Eng. Sci. 48, 3585–3630.

Bin, A.K., Smith, J.M., 1982. Mass transfer in a plunging liquid jet absorber. Chem. Eng. Commn. 15, 367–383.

Bird, R.B., Steward, W.E., Lightfoot, E.N., 2002. Transport Phenomena, second ed. John Wiley & Sons, Hoboken, NJ.

Biswas, M.N., Mitra, A.K., 1981. Momentum transfer in horizontal multijet liquid–gas ejector. Can. J. Chem. Eng. 59, 634–637.

Bouaifi, M., Hebrard, G., Bastoul, D., Roustan, M., 2001. A comparative study of gas holdup, bubble size, interfacial area and mass transfer coefficients in stirred gas–liquid reactors and bubble columns. Chem. Eng. Proc. 40, 97–111.

Brian, P.L.T., 1964. Gas absorption accompanied by an irreversible reaction of general order. AIChE J. 10, 5–10.

Burns, A.D., T. Frank, I. Hamill, J.-M. Shi, 2004. The Favre averaged drag model for turbulent dispersion in Eulerian multi-phase flows. Fifth International Conference on Multiphase Flow, Yokohama, Japan.

Calderbank, P.H., Moo-Young, M.B., 1961. The continuous phase heat and mass transfer properties of dispersions. Chem. Eng. Sci. 16, 39–54.

Camarasa, E., Vial, C., Poncin, S., Wild, G., Midoux, N., Bouillard, J., 1999. Influence of coalescence behaviour of the liquid and of gas sparging on hydrodynamics and bubble characteristics in a bubble column. Chem. Eng. Proc. 38, 329–344.

Cents, A.H.G., de Bruijn, F.T., Brilman, D.W.F., Versteeg, G.F., 2005a. Validation of the Danckwerts-plot technique by simultaneous chemical absorption of CO_2 and physical desorption of O_2. Chem. Eng. Sci. 60(21) 5809–5818.

Cents, A.H.G., Brilman, D.W.F., Versteeg, G.F., 2005b. CO_2 absorption in carbonate/bicarbonate solutions: The Danckwerts-criterion revisited. Chem. Eng. Sci. 60, 5830–5835, 21.

Cerri, M.O., Baldacin, J.C., Cruz, A.J.G., Hokka, C.O., Badino, A.C., 2009. Prediction of mean bubble size in pneumatic reactors. Biochem. Eng. J. 53, 12–17.

Chaumat, H., A.M. Billet-Duquenne, F. Augier, C. Mathieu, H. Delmas, 2005. Mass transfer in bubble column for industrial conditions—effects of organic medium, gas and liquid flow rates and column design Chem. Eng. Sci. 60(22), 5930–5936.

Chen, C.M., Leu, L.P., 2001. Hydrodynamics and mass transfer in three-phase magnetic fluidized beds. Powder Technol. 117, 198–206.

Chen, P.-C., 2012. Greenhouse Gases—Capturing, Utilization and Reduction, Chapter 5. In: Guoxiang, L. (Eds.) InTech, Shanghai (China).

Clift, R., Grace, J.R., Weber, M.E., 1978. Bubbles, Drops and Particles. Academic Press, New York.

Cramers, P.H.M.R., Beenackers, A.A.C.M., Van Dierendonck, L.L., 1992. Hydrodynamics and mass transfer characteristics of a loop–venturi reactor. Chem. Eng. Sci. 47, 3557–3564.

Cussler, E.L., 1984. Diffusion—Mass Transfer in Fluid Systems. Cambridge University Press, Oxford, UK.

Danckwerts, P.V., 1970. Gas–Liquid Reactions. McGraw-Hill, New York.

Dargar, P., Macchi, A., 2006. Effect of surface-active agents on the phase holdups of three-phase fluidized beds. Chem. Eng. Proc. 45 (9), 764–772.

De Billerbeck, G.M., Condoret, J.S., Fonade, C., 1999. Study of mass transfer in a novel gas-liquid contactor: the aero-ejector. Chem. Eng. J. 72, 185–193.

Deckwer, W.D., 1992. Bubble Column Reactors. Wiley, New York.

Deckwer, W.-D., Burckhart, R., Zoll, G., 1974. Mixing and mass transfer in tall bubble columns. Chem. Eng. Sci. 29, 2177–2188.

Deckwer, W.-D., Nguyen-Tien, K., Kelkar, B.G., Shah, Y.T., 1983. Applicability of axial dispersion model to analyze mass transfer measurements in bubble columns. AIChE J. 29 (6), 915.

Dewes, I., Schumpe, A., 1997. Gas density effect on mass transfer in the slurry bubble column. Chem. Eng. Sci. 52, 4105–4109.

Dutta, N.N., Raghavan, K.V., 1987. Mass transfer and hydrodynamics of loop reactors with downflow liquid jet ejector. Chem. Eng. J. 36, 111–121.

Eben, C.D., Pigford, R.L., 1965. Gas absorption with chemical reaction on a sieve tray. Chem. Eng. Sci. 20, 803–811.

Evans, G.M., Machniewski, P.M., 1999. Mass transfer in a confined plunging jet bubble column. Chem. Eng. Sci. 54, 4981–4990.

Evans, G.M., Bin, A.K., Machniewski, P.M., 2001. Performance of confined plunging liquid jet bubble column as a gas–liquid reactor. Chem. Eng. Sci. 56, 1151–1157.

Fair, J.R., Lambright, A.J., Anderson, J.W., 1962. Heat transfer and gas holdup in a sparged contactor. Ind. Eng. Chem. Process Des. Dev. 1, 33–36.

Funatsu, K., Hsu, Y.Ch., Noda, M., Sugawa, S., 1988. Oxygen transfer in the water-jet vessel. Chem. Eng. Commun. 73, 121–139.

Gerster, J.A., A.B. Hill, N.H. Hochgraf, D.G. Robinson, 1958. Tray efficiencies in distillation columns, Final report from the University of Delaware, A.I. Ch. E., New York.

Godbole, S.P., Schumpe, A., Shah, Y.T., Carr, N.L., 1984. Hydrodynamics and mass transfer in non-newtonian solutions in a bubble column. AIChE J. 30, 213–220.

Gómez-Díaz, D., Nelma Gomes, José A. Teixeira, Isabel Belo, Oxygen mass transfer to emulsions in a bubble column contactor. Chem. Eng. J. 152(2–3), 354–360.

Gupta, P., Ong, B., Al-Dahhan, M.H., Dudukovic, M.P., Toseland, B.A., 2001. Hydrodynamics of churn turbulent bubble columns: gas–liquid recirculation and mechanistic modeling. Catal. Today 64 (3–4), 253–269.

Han, L., Muthanna, H., Al-Dahhan, 2007. Gas–liquid mass transfer in a high pressure bubble column reactor with different sparger designs. Chem. Eng. Sci. 62, 131–139.

Harned, H.S., Owen, B.B., 1958. The Physical Chemistry of Electrolyte Solutions. Reinhold Publishing Co, New York.

Haut, B., Cartage, T., 2005. Mathematical modeling of gas–liquid mass transfer rate in bubble columns operated in the heterogeneous regime. Chem. Eng. Sci. 60 (22), 5937–5944.

Haut, B., Halloin, V., Cartage, T., Cockx, A., 2004. Precipitation of sodium bicarbonate in industrial bubble columns. Chem. Eng. Sci. 59, 5687–5694.

Havelka, P., Linek, V., Sinkule, J., Zahradnik, J., Fialova, M., 2000. Hydrodynamic and mass transfer characteristics of ejector loop reactors. Chem. Eng. Sci. 55, 535–549.

Higbie, R., 1935. The rate of absorption of a pure gas into still liquid during short periods of exposure. Trans. Am. Inst. Chem. Eng. 31, 365–389.

Hikita, H., Asai, S., Tanigawa, K., Segawa, K., Kitao, M., 1981. The volumetric liquid-phase mass transfer coefficient in bubble columns. Chem. Eng. J. 22, 61–69.

Hsu, S., Lee, W., Yang, Y., Chang, C., Maa, J., 2000. Bubble formation at an orifice in surfactant solutions under constant-flow conditions. Ind. Eng. Chem. Res. 39 (5), 1473–1479.

Huynh, X., Briens, C.L., Catros, A., Bernard, J.R., Bergougnou, M.A., 1991. Hydrodynamics and mass transfer in an upward venturi/bubble column combination. Can. J. Chem. Eng. 69, 711–722.

Ide, M., Ucheiyama, H., Ishikura, T., 2001. Mass transfer characteristics in gas bubble dispersed phase generated by plunging jet containing small solute bubbles. Chem. Eng. Sci. 56, 6225–6231.

Inga, J.R., 1996. Scale-up and Scale-down of Slurry Bubble Column Reactor, a New Methodology (unpublished PhD Dissertation, University of Pittsburgh).

Jamialahmadi, M., Müller-Steinhagen, H., 1992. Effect of alcohol, organic acid and potassium chloride concentration on bubble size, bubble rise velocity and gas hold-up in bubble columns. Chem. Eng. J. 50, 47–56.

Jia, X., Hu, W., Yuan, X., Yu, K., 2015. Effect of surfactant type on interfacial area and liquid mass transfer for CO_2 absorption in a bubble column. Chinese J. Chem. Eng. 23 476–481.

Jin, H., Yang, S., He, G., Liu, D., Tong, Z., Zhu, J., 2014. Gas–liquid mass transfer characteristics in a gas–liquid–solid bubble column under elevated pressure and temperature. Chinese J. Chem. Eng. 22(9), 955–961.

Jordan, U., Schumpe, A., 2001. The gas density effect on mass transfer in bubble columns with organic liquids. Chem. Eng. Sci. 56, 6267–6272.

Jordan, U., Terasaka, K., Kundu, G., Schumpe, A., 2002. Mass transfer in high-pressure bubble columns with organic liquids. Chem. Eng. Technol. 25, 262.

Kabdasli, I., Arslan-Alaton, I., 2010. Chemical Oxidation Applications for Industrial Wastewaters. IWA Publishing, London.

Kang, Y., Cho, Y.J., Woo, K.J., Kim, S.D., 1999. Diagnosis of bubble distribution and mass transfer in pressurized bubble columns with viscous liquid medium. Chem. Eng. Sci. 54, 4887–4893.

Kasturi, G., Stephanek, J.B., 1974. Mass transfer in vertical tube. Chem. Eng. Sci. 29, 1849–1852.

Kawase, Y., Halard, B., Moo-Young, M., 1987. Theoretical prediction of volumetric mass transfer coefficients in bubble columns for Newtonian and Non-Newtonian fluids. Chem. Eng. Sci. 42, 1609–1617.

Kim, J.O., Kim, S.D., 1990. Gas–liquid mass transfer in a three phase fluidized bed floating bubble breakers. Can. J. Chem. Eng. 68, 368.

Kister, H.Z., 1992. Distillation Design. McGraw-Hill, New York.

Koide, K., Takazawa, A., Komura, M., Matsunaga, H., 1984. Gas holdup and volumetric liquid-phase mass transfer coefficient in solid-suspended bubble columns. J. Chem. Eng. Jpn 17, 459–466.

Kojima, H., Sawai, J., Suzuki, H., 1997. Effect of pressure on volumetric mass transfer coefficient and gas holdup in bubble column. Chem. Eng. Sci. 52, 4111–4116.

Kolthoff, I.M., Tomsicek, W.J., 1935. The oxidation potentials of the system potassium ferrocyanide-potassium ferricyanide at various ionic strengths. J. Phys. Chem. 39, 945–954.

Kottke, V., Blenke, H., 1970. Meßmethoden konvektiver Stoffubertragung. Chem. Ing. Tech. 50 (2), 81–90.

Kozinski, A.A., King, C.J., 1966. The influence of diffusivity on liquid phase mass transfer to the free interface in a stirred vessel. A. I. Ch. E. J. 12 (1), 109–116.

Krishna, R., Van Baten, J.M., 2003. Mass transfer in bubble columns. Catal. Today 79–80, 67–75.

Kuerten, H., Zehner, P., 1978. On interfacial area in aerated stirred vessels fitted with a flat-blade turbine. Ger. Chem. Eng. 1, 347–351.

Kulkarni, A., Shah, Y.T., 1984. Gas phase dispersion in a downflow bubble column. Chem Eng. Commun. 28, 311–326.

Lau, R., Peng, W., Velazquez-Vargas, L.G., Yang, G.Q., Fan, L.-S., 2004. Gas–liquid mass transfer in high-pressure bubble columns. Ind. Eng. Chem. Res. 43, 1302–1311.

Lemoine, R., Behkish, A., Morsi, B.I., 2004. Hydrodynamic and mass-transfer characteristics in organic liquid mixtures in a large-scale bubble column reactor for the toluene oxidation process. Ind. Eng. Chem. Res. 43 (19), 6195–6212.

Levenspiel, O., 1998. Chemical Reaction Engineering, third ed. John Wiley & Sons, New York, Chapters 23–24.

Levich, V.G., 1959. Fiziko-Khimicheskaya Gidrodinamika (Physico Chemical Hydro-dynamics). Fizmatgiz, Moscow.

Lewis, D.A., Davidson, J.F., 1985. Mass transfer in a recirculating bubble column. Chem. Eng. Sci. 40 (11), 2013–2017.

Lewis, W.K., Whitman, W.G., 1924. Principles of gas absorptions. Ind. Eng. Chem. 16, 1215–1220.

Lin, C.S., Denton, E.B., Gaskil, N.S., Putnam, C.L., 1951. Diffusion controlled electrode reactions. Ind. Eng. Chem. 43, 2136–2143.

Linek, V., Benes, P., Sinkule, J., 1989. Dynamic pressure method for kla measurement in large-scale bioreactors. Biotechnol. Bioeng. 33, 1406.

Maceiras, R., Nóvoa, X.R., Álvarez, E., Cancela, M.A., 2007. Local mass transfer measurements in a bubble column using an electrochemical technique. Chem. Eng. Process. Intensification 46 (10), 1006–1011.

Majumder, S.K., 2008a. Efficiency of non-reactive isothermal bubble column based on mass transfer. Asia-Pacific J. Chem. Eng. 3 (4), 440–451.

Majumder, S.K., 2008b. Analysis of dispersion coefficient of bubble motion and velocity characteristic factor in down and upflow bubble column reactor. Chem. Eng. Sci. 63 (12), 3160–3170.

Majumder, S.K., Kundu, G., Mukherjee, D., 2005. Mixing mechanism in a modified co- current downflow bubble column. Chem. Eng. J. 112 (1–3), 45–55.

Mandal, A., 2010. Characterization of gas–liquid parameters in a down-flow jet loop bubble column. Braz. J. Chem. Eng. 27 (02), 253–264.

Mandal, A., Kundu, G., Mukherjee, D., 2003. Interfacial area and liquid-side volumetric mass transfer coefficient in a downflow bubble column. Can. J. Chem. Eng. 81, 212–219.

Mandal, A., Kundu, G., Mukherjee, D., 2004. Gas-holdup distribution and energy dissipation in an ejector induced downflow bubble column: the case of non-newtonian liquid. Chem. Eng. Sci. 59, 2705–2713.

Manish, P., Majumder, S.K., 2009. Quality of mixing in downflow bubble column based on information entropy theory. Chem. Eng. Sci. 64 (8), 1798–1805.

McClure, D.D., Norris, H., Kavanagh, J.M., Fletcher, D.F., Barton, G.W., 2014a. Validation of a computationally efficient computational fluid dynamics (CFD) model for industrial bubble column bioreactors. Ind. Eng. Chem. Res. 53, 14526–14543.

McClure, D.D., Kavanagh, J.M., Fletcher, D.F., Barton, G.W., 2014b. Development of a CFD model of bubble column bioreactors: Part two—Comparison of experimental data and CFD predictions. Chem. Eng. Technol. 37, 131–140.

McClure, D.D., Norris, H., Kavanagh, J.M., Fletcher, D.F., Barton, G.W., 2015a. Towards a CFD model of bubble columns containing significant surfactant levels. Chem. Eng. Sci. 127, 189–201.

McClure, D.D., John M. Kavanagh, David F. Fletcher, Geoffrey W., 2015b. Barton Oxygen transfer in bubble columns at industrially relevant superficial velocities: experimental work and CFD modelling, Chem. Eng. J. 280, 138–146.

Miyahara, T., Hamanaka, H., Takino, T., Akagi, Y., 1997. Gas holdup gas–liquid interfacial area and mass transfer coefficient in external-loop airlift bubble column containing low density particles. J. Chem. Eng. Jpn 30, 958–961.

Muroyama, K., Imai, K., Oka, Y. Hayashi, J., 2013. Mass transfer properties in a bubble column associated with micro-bubble dispersions. Chem. Eng. Sci.100(30), 464–473.

Nagel, O., Kuerten, H., Sinn, R., 1970. Power jet reactors. I. Application of the ejector principle to the improvement of gas absorption in bubble columns. Chemie Ingenieur Technik 42, 774–780.

Nedeltchev, S., Ookawara, S., Ogawa, K., 1999. A fundamental approach to bubble column scale-up based on quality of mixedness. J. Chem. Eng. Jpn 32 (4), 431–439.

Neme, F., Coppola, L., Böhm, U., 1997. Gas holdup and mass transfer in solid suspended bubble columns in presence of structured packings. Chem. Eng. Technol. 20, 297–303.

Newman, J.S., 1973. Electrochemical Systems. Prentice-Hall, Englewood Cliffs, NJ.

Nijsing, R.A.T.O., Hendriksz, R.H., Kramers, H., 1959. Absorption of CO_2 in jets and falling films of electrolyte solutions, with and without chemical reaction. Chem. Eng. Sci. 10, 88–104.

Ohkawa, A., Kawai, Y., Ksabiraki, D., Saki, N., Endoh, K., 1987. Bubble size, interfacial area and volumetric liquid-phase mass transfer coefficient in downflow bubble columns with gas entrainment by a liquid jet. J. Chem. Eng. Jpn 20, 99–101.

Ozturk, S., Schumpe, S.A., Deckwer, W.D., 1987. Organic liquids in a bubble column: holdups and mass transfer coefficients. A. I. Ch. E. J. 33, 1473–1480.

Peeva, L., Ben-zvi Yona, S., Merchuk, J.C., 2001. Mass transfer coefficients of decane to emulsions in a bubble column reactor. Chem. Eng. Sci. 56, 5201.

Perry, R.H., Green, D., 1984. Perry's Chemical Engineers Handbook, McGraw-Hill, New York.

Porter, K.E., King, M.B., Varshney, K.C., 1966. Interfacial areas and liquid-film mass transfer coefficients. Trans. Instn. Chem. Eng. 44, T274–T283.

Raghuram, P.T., Mukherjee, A.K., Das, T.R., 1992. Studies on gas-liquid ejector-contactor. Indian Chem. Eng. 34, 40–49.

Ravoo, E., 1971. Mass Transfer in Electrochemical Systems. Enschede: Technische Hogeschool Twente, The Netherlands.

Ravoo, E., Rotte, J.W., Sevenstern, F.W., 1970. Theoretical and electrochemical investigation of free convection mass transfer at vertical cylinders. Chem. Eng. Sci. 25, 1637–1652.

Rosén, C., Trägårdh, C., 1995. Prediction of turbulent high Schmidt number mass transfer using a low Reynolds number k-ε turbulent model. Chem. Eng. J. 59, 153–159.

Rubia, M.D.L., García-Abuín, A., Gómez-Díaz, D., Navaza, J.M., 2010. Interfacial area andmass transfer in carbon dioxide absorption in TEA aqueous solutions in a bubble column reactor. Chem. Eng. Process. 49 (8), 852–858.

Sada, E., Kumazawa, H., Lee, C.H., 1983. Chemical absorption in a bubble column loading concentrated slurry. Chem. Eng. Sci. 38, 2047–2051.

Salvacion, J.L., Murayama, M., Ohtaguchi, K., Koide, K., 1995. Effects of alcohols on gas holdup and volumetric liquid-phase mass transfer coefficient in gel-particle-suspended bubble column. J. Chem. Eng. Jpn 28, 434–441.

Sardeing, R., Painmanakul, P., Hebrard, G., 2006. Effect of surfactants on liquid-side mass transfer coefficients in gas–liquid systems: a first step to modeling. Chem. Eng. Sci. 61 (19), 6249–6260.

Schumpe, A., Deckwer, W.D., 1980. Analysis of chemical methods for determination of interfacial areas in gas-in-liquid dispersions with non-uniform bubble sizes. Chem. Eng. Sci. 35, 2221–2233.

Schumpe, A., Saxena, A.K., Fang, L.K., 1987. Gas–liquid mass transfer in a slurry bubble column. Chem. Eng. Sci. 42, 1787–1796.

Schumpe, A., Deckwer, W.-D., Nigam, K.D.P., 1989. Gas–liquid mass transfer in three-phase fluidized beds with viscous pseudoplastic liquids. Can. J. Chem. Eng. 67, 873.

Shah, Y.T., Kelkar, B.G., Godbole, S.P., Deckwer, W.D., 1982. Design parameter estimation for bubble column reactors journal review. AIChE J. 28, 353–379.

Sharma, M.M., Danckwerts, P.V., 1970. Chemical methods of measuring interfacial area and mass transfer coefficients in two-fluid systems. Brit. Chem. Eng. 15, 522–528.

Sherwood, T.K., Pigford, R.L., 1952. Absorption and Extrction, second ed. McGraw Hill, New York.

Sherwood, T.K., Pigford, R.L., Wilke, C.R., 1975. Mass Transfer. McGraw-Hill Inc, New York.

Simonnet, M., Gentric, C., Olmos, E., Midoux, N., 2007. Experimental determination of the drag coefficient in a swarm of bubbles. Chem. Eng. Sci. 62, 858–866.

Singh, M.K., Majumder, S.K., 2011. Co- and counter-current mass transfer in bubble column. Int. J. Heat Mass Transfer 54, 2283–2293.

Sivaiah, M., Majumder, S.K., 2013. Mass transfer and mixing in an ejector-induced downflow slurry bubble column. Ind. Eng. Chem. Res. 52, 12661–12671.

Stein, W.A., Schafer, H., 1984. Aeration of viscous liquids with a downward directed two-fluid nozzle. Ger. Chem. Eng. 7, 115–125.

Sun, Z.M., Liu, B.T., Yuan, X.G., Yu, K.T., 2005. New turbulent model for computational mass transfer and its application to a commercial scale distillation column. Ind. Eng. Chem. Res. 44 (12), 4427–4434.

Tang, W.-T., Fan, L.-S., 1990. Gas–liquid mass transfer in a threephase fluidized bed containing low-density particles. Ind. Eng. Chem. Res. 29, 128.

Terasaka, K., Hieda, Y., Tsuge, H., 1999. SO_2 bubble formation at an orifice submerged in water. J. Chem. Eng. Jpn. 32 (4), 472–479.

Terasaka, K., Hullmann, D., Schumpe, A., 1998. Mass transfer in bubble columns studied with an oxygen optode. Chem. Eng. Sci. 53 (17), 3181.

Therning, P., Rasmuson, A., 2006. Mass transfer measurements in a non-isothermal bubble column using the uncatalyzed oxidation of sulphite to sulphate. Chem. Eng. J. 116(2), 97–103.

Tojo, K., Miyanami, K., 1982. Oxygen transfer in jet mixer. Chem. Eng. J. 24, 89–97.

Treybal, R.E., 1968. Mass-Transfer Operations. McGraw-Hill, New York.

Vafopulos, I., Sztatescny, K., Moser, F., 1975. Der einflub des partial-und gesamtdruckes auf den stoffaustausch. Chem. Eng. Technol. 47, 681–786.

Van de Sande, E., Smith, J.M., 1975. Mass transfer from plunging water jets. Chem. Eng. J. 10, 225–233.

Vandu, C.O., Krishna, R., 2004. Volumetric mass transfer coefficients in slurry bubble columns operating in churn-turbulent flow regime. Chem. Eng. Process. 43, 987–995.

Vázquez, G., Cancela, M.A., Riverol, C., Alvarez, E., Navaza, J.M., 2000. Determination of interfacial areas in a bubble column by different chemical methods. Ind. Eng. Chem. Res. 39, 2541–2547.

Verma, A.K., Rai, S., 2003. Studies on surface to bulk ionic mass transfer in bubble column. Chem. Eng. J. 94, 67–72.

Voyer, R.D., Miller, A.I., 1968. Improved gas–liquid contacting in co-current flow. Can. J. Chem. Eng. 46, 335–341.

Wachsmann, U., Rabiger, N., Vogelpohl, 1985. Effect of geometry on hydrodynamics and mass transfer in the compact reactor. Ger. Chem. Eng. 8, 411–418.

Wang, T., Wang, J., 2007. Numerical simulations of gas–liquid mass transfer in bubble columns with a CFD–PBM coupled model. Chem. Eng. Sci. 62 (24), 7107–7118.

Wilkinson, P., Haringa, H., 1994. Mass transfer and bubble size in a bubble column under pressure. Chem. Eng. Sci. 49 (9), 1417–1427.

Xigang, Y., Guocong, Y., 2008. Computational mass transfer method for chemical process simulation. Chin. J. Chem. Eng. 16 (4), 497–502.

Xue, J., Al-Dahhan, M., Dudukovic, M.P., Mudde, R.F., 2003. Bubble dynamics measurements using four-point optical probe. Can. J. Chem. Eng. 81 (3–4), 375–381.

Yang, W., Jinfu Wang, Bin Zhao, Yong Jin, 2003. Gas–liquid mass transfer in slurry bubble systems: II. Verification and simulation of the model based on the single bubble mechanism. Chem. Eng. J. 96(1–3), 29–35.

Zheng, C., Chen, Z., Feng, Y., Hofmann, H., 1995. Mass transfer in different flow regimes of three-phase fluidized beds. Chem. Eng. Sci. 50 (10), 1571.

9

Heat Transfer Characteristics

Introduction

It is well accepted that the heat, mass transfer, and hydrodynamics are crucial to the performance of gas–liquid reactors. Many hydrodynamic studies investigate the heat transfer between the heating objectives and the system flow to understand the effects of hydrodynamic structures on the heat transfer for improving the design and operation of bubble column reactors (Deckwer, 1992). Many chemical and biochemical reactions are carried out by contacting a gas component with a liquid phase reactant under the favorable heat transfer properties of gas–liquid dispersions. The most widely used are stirred tank and bubble column reactors because of their cheap and simple construction. Bubble columns are frequently superior over stirred tanks, particularly at higher temperatures and pressures (Deckwer, 1992). In bubble columns, the heat transfer coefficients are larger by a factor of 100 than in single-phase flow (Deckwer, 1992). Thermal control in bubble columns is of importance because the chemical reactions are usually accompanied by heat supply (endothermic) or removal (exothermic) operations. Therefore, maintaining desirable bulk media temperature is necessary because it plays an important role in the performance of the reactor. Hence, the knowledge and understanding of heat transfer phenomena and quantity of the heat transfer coefficients are essential and required for proper safe, efficient design and operation of these reactors. Some typical examples of exothermic gas–liquid reactions which takes place under the environment of the heat energy in bubble columns are shown in Table 9.1.

The selectivity of reactions in bubbly flow condition is generally governed by the residence time and temperature distributions. Hence, applications of gas–liquid reactors are particularly recommended if temperature-sensitive processes are to be performed (e.g., Flscher-Tropsch synthesis). An important and growing area of application of bubble column reactor systems is production of clean and renewable fuels such as production of sulfur-free diesel by the Fischer-Tropsch process, dimethyl ether (DME), and bioethanol processes at a certain temperature. For a large number of these processes, there is a need for proper design of heat removal in these reactor systems to allow optimal temperature control for the desired product quality and yield (Duduković et al., 2002).

Heat Transfer Mechanism in Bubbly Flow

The mechanism of heat transfer from wall to the gas–liquid dispersion was studied by several investigators. Kolbel et al. (1958) were the first who supposed that the enhancing effect of heat transfer produced by the gas bubble on the heat transfer rate in bubble columns is related to the removal of stagnant liquid portions from the transfer surface through the

Hydrodynamics and Transport Processes of Inverse Bubbly Flow. http://dx.doi.org/10.1016/B978-0-12-803287-9.00009-6

Table 9.1   Some Typical Examples of Gas–Liquid Reactions Carried Out in Bubble Columns (Deckwer, 1992)

Gas Phase	Liquid Phase	Product	Heat of Reaction ($-\Delta HR$, kJ/mol)
Propene	Benzene	Cumene	113
Chlorine	Benzene	Chlorobenzene	132
Chlorine	Ethylene	Ethylene dichloride	180
Chlorine	Paraffin	Chloro paraffins	110
Hydrogen	Benzene	Cyclohexane	216
Hydrogen, Carbon monoxide	Water	Hydrocarbons	205
Hydrogen	Nitrobenzene	Aniline	493
Oxygen	Ethylene	Acetaldehyde	244
Oxygen	Acetaldehyde	Acetic acid	293
Oxygen	o-Xylene	Phthalic acid	1110
Oxygen	Butane	Acetic acid	1270

boundary layer. A summary of the published work on heat transfer conducted in bubble columns is shown in Table 9.2. The findings of these authors clearly showed that the main impact on the heat transfer coefficients results from gas velocity and liquid phase properties. Generally, the heat transfer coefficient increases with an increase in gas velocity, liquid thermal conductivity, and heat capacity and decreases with increasing liquid viscosity (Deckwer, 1980; Magiliotou et al., 1988). The flow structure or liquid velocity induced by bubble movement has a dominant influence on the heat transfer coefficient. This means that the bubble movement can enhance the heat transfer in liquid because of bubble wake–induced turbulence (Kumar et al., 1992). This bubble movement in the reactors is dictated by bubble characteristics. Kast (1962) analyzed the motion of fluid elements around a gas bubble moving with a velocity. It was pointed out that the radial component of the liquid velocity, which is induced by the moving bubble, is responsible for the high heat transfer coefficients in bubble columns. Liquid circulation is also one of the most important characteristics of bubble columns, which represents the liquid flow induced by the moving bubbles and governs the rate of heat and mass transfer (Joshi et al., 1980). The main features of Kast's model is shown in Figure 9.1. According to him, a fluid element in front of the moving bubble receives a radial momentum and thus moves toward the heat-exchanging surface. They reported that the lateral mass transport forced by the axial bubble motion weakens and breaks up the boundary layer at the wall surface, and the fluid elements are sucked into the region at the rear of the bubble (Deckwer, 1980). The fast radial exchange flows of mass are extended to the wall. Therefore, a capacitive heat transport results. According to Kast, the pertinent heat transfer coefficient is proportional to the radial velocity component of the fluid element. Kast deduced the Stanton number ($St = h/(u\rho C_p)$) as a characteristic group for the heat transfer between gas–liquid dispersions and walls. Kast proposed the relation ($St = f(Re, Fr, Pr^2)$), which led to a good description of the heat transfer coefficient with the different operating variables in bubbly flow. Since that time, several

Table 9.2 Important Correlations in the Literatures on Heat Transfer Coefficient in Bubble Column

Reference	System	Range of Liquid Properties: ρ, kg/m³/μ, Pas/σ, N/m	Range of Column Diameter (m)/liquid height (m)	Range of u_{sg}/Range of u_{sl} (m/s)	Range of Temperature, K/Pressure, kPa	Gas Distributor Type/Hole Diameter, m/Number of Holes	Correlation
Fair et al. (1962)	Air–water	1000/ 0.001/ 0.072	0.45-1.07/3.2	0.006-0.045/-; 1.07-3.04/-	295/101.3	Sparger/-; Ring/-	$h_w = 8850 u_{sg}^{0.22}$
Kast (1962)	Air–water, isopropanol (45 wt.%)	1000/ 0.001/ 0.072	0.29/4.0	0.0025-0.06/-	295/101.3	Sparger	$\dfrac{h}{\rho_l C_{pl} u_{sg}} = 0.1\left[\left(\dfrac{\rho_l d_c u_{sg}}{\mu_g}\right)\left(\dfrac{u_{sg}^2}{g d_c}\right)\left(\dfrac{C_{pl}\mu_l}{k_l}\right)^2\right]^{-0.22}$
Konsetov (1966)	—	—	—	—	—	—	$\dfrac{h_w d_c}{k_l} = 0.25\varepsilon_g^{1/3}\left(\dfrac{g\rho_l^2 d_c^3}{\mu_l^2}\right)^{1/3}\left(\dfrac{\mu_l C_{pl}}{k_l}\right)^{1/3}\left(\dfrac{\mu_l}{\mu_w}\right)^{0.14}$ $\dfrac{h d_{tube}}{k_l} = 0.18\varepsilon_g^{1/3}\left(\dfrac{g\rho_l^2 d_{tube}^3}{\mu_l^2}\right)^{1/3}\left(\dfrac{d_c}{d_{tube}}\right)^{1/3}\left(\dfrac{\mu_l C_{pl}}{k_l}\right)^{1/3}\left(\dfrac{\mu_l}{\mu_w}\right)^{0.14}$ Theoretical treatment only
Burkel (1972)	Air–water, methanol, mercury	1000/ 0.001/ 0.072	0.19/-	0-0.5/-	295/101.3	Sparger/-	$\dfrac{h}{\rho_l C_{pl} u_{sg}} = 0.11\left[\left(\dfrac{\rho_l d_c u_{sg}}{\mu_g}\right)\left(\dfrac{u_g^2}{g d_c}\right)\left(\dfrac{C_{pl}\mu_l}{k_l}\right)^{2.48}\right]^{-0.22}$
Hart (1976)	Air-water, ethylene	993/ 0.00038/ 0.0637	0.106/0.812	0.000058–0.0158/ –	344/ 101.3	Single nozzle/ 0.0063/1	$\dfrac{h_w}{\rho_l C_p u_{sg}}\left(\dfrac{C_p\mu_l}{k_l}\right)^{0.6} = 0.125\left(\dfrac{u_{sg}^3\rho_l}{\mu_l g}\right)^{-0.25}$
Baker et al. (1978)	Air–water	—	—	0-0.25/ 0-0.125	295/101.3		$h = 1977 u_{sl}^{0.07} u_{sg}^{0.059} d_p^{0.106}$
Deckwer et al. (1980)	Nitrogen–xylene	—	0.1/0.6	0.003-0.04/.	-/400-1100	Sintered	$St = 0.1\left[\left(\text{Re}_m\, Fr\, \text{Pr}_m^2\right)^{-0.25}\right]$
Joshi et al. (1980)	-	—	—	—	—	—	$\dfrac{h_w d_c}{k_l} = 0.031\left[\dfrac{d_c^{1.33} g^{1/3}(u_{sg} - \varepsilon_g u_{bt})^{1/3}\rho_l}{\mu_l}\right]^{1/3}$ $\times\left(\dfrac{C_{pl}\mu_l}{k_l}\right)^{1/3}\left(\dfrac{\mu_l}{\mu_w}\right)^{0.14}$ Based on analogy with mechanically agitated contactor and liquid flow through the pipes

(Continued)

Table 9.2 Important Correlations in the Literatures on Heat Transfer Coefficient in Bubble Column (*cont.*)

Reference	System	Range of Liquid Properties: ρ, kg/m³/μ, Pas/σ, N/m	Range of Column Diameter (m)/liquid height (m)	Range of u_{sg}/Range of u_{sl} (m/s)	Range of Temperature, K/Pressure, kPa	Gas Distributor Type/Hole Diameter, m/Number of Holes	Correlation
Hikita et al. (1981)	Air–water, sucrose, methanol, butanol solution	976–1230/ 0.0008– 0.013/ 0.024–0.074	0.10–0.19/ 1.45–2.30	0.053–0.35/ 0–0.0034	295–318/ 101.3	Single hole 0.009,0.013, 0.020/1	$$\frac{h_w}{\rho_l C_p u_{sg}}\left(\frac{C_p \mu_l}{k_l}\right)^{2/3} = 0.411\left(\frac{u_{sg}\mu_l}{\sigma}\right)^{-0.851}\left(\frac{\mu_l^4 g}{\sigma \rho_l}\right)^{0.308};$$ $5.4\times10^{-4} < u_{sg}\mu_l/\sigma < 7.65.4\times10^{-2}$; $4.9 < C_p\mu_l/k_l < 93$; $7.7\times10^{-12} < (\mu_l^4 g)/(\sigma\rho_l) < 1.6\times10^{-6}$
Kato et al. (1981)	Air–liquid	-/ 0.001- 0.036/-	0.052, 0.12	0-0.15/ 0.004-0.06	295/101.3	—	$Nu' = 0.044(\text{Re}'\text{Pr})^{0.78} + 2.0Fr_g'^{0.17}$; $Nu' = hd_p\varepsilon_l/k_l(1-\varepsilon_l)$; $\text{Pr} = \mu_l C_{Pl}/k_l$; $\text{Re}' = \rho_l d_p u_{sl}\varepsilon_l/[\mu_l(1-\varepsilon_l)]$; $Fr_g' = u_{sg}^2/gd_p$
Muroyama et al. (1984)	Air–liquid	330 – 3500/-/-	0.095/-	0.03-0.259/ 0.012-0.26	295/101.3	Perforated plate/ 0.0015/ 120	$$\frac{h_3}{h_2} = 1 + 0.0413\left(\frac{u_{sg}}{u_{sl}}\right)^{0.30}\left(\frac{d_p(\rho_s - \rho_l)}{d_c\rho_l}\right)^{0.61}$$ h_2 = heat transfer coefficient in liquid–solid fluidized bed (kW/m °C) and h_3 = heat transfer coefficient in three phase fluidized bed (kW/m °C)
Zehner (1986)	Air–water	1000/ 0.001/ 0.072	0.139/ 0.298	0-0.1/-	295/101.3	Perforated plate/ 0.0015/ 120	$h_w = 0.18(1-\varepsilon_g)\left((k_l^2\rho_l^2 C_{pl}u_{ed}^2)/(l\mu_l)\right)^{1/3}$ $l = d_b(\pi/6\varepsilon_g)^{1/3}$; $u_{ed} = \left[\frac{1}{2.5}\left(\frac{\rho_l - \rho_g}{\rho_l}\right)gd_c u_{sg}\right]$ l is mean distance between bubbles and u_{ed} is the eddy velocity.
Kawase and Moo-Young (1987)	Non-Newtonian fluids	Theoretical treatment only	—	—	—	—	$$\frac{hd_c}{k_l} = 0.134\left(\frac{\mu_l C_{pl}}{k_l}\right)^{1/3}\left(\frac{u_g^2}{gd_c}\right)^{-1/4}\left(\frac{\rho d_c u_g}{K}\right)^{3/4}; K = \text{power-law consistency index (Pa.s}^n)$$
Verma (1989)	Air–water	1000/ 0.001/ 0.072	0.11/1.7	0.1-0.4/-	295/101.3	Perforated plate/ 0.0015/ 120	$$\frac{h_w}{\rho_l C_p u_{sg}} = 0.121(1-\varepsilon_g)\left(\frac{C_p\mu_l}{k_l}\right)^{-0.5}\left(\frac{\mu_l^3\rho_l}{\mu_l g}\right)^{-0.851}$$

Saxena (1989)	Air–water, air-water-magnetite	1000/ 0.001/ 0.072	0.11/2.25	0.015-0.333/-	295/101.3	Porous plate/-/-	$h_{w,max} = 0.12\left(\dfrac{g^2\rho_m}{\mu_m}\right)^{1/6}\left(\dfrac{\rho_m - \rho_g}{\rho_m}\right)^{1/3}\left(k_m\rho_m C_{pm}\right)^{1/2}$
Suh and Deckwer (1989)	Theoretical treatment only	—	—	—	—	—	$h = 0.1\left[K_l\rho_l C_{pl}\left(\dfrac{P_v}{\eta_{sl}\mu_l}\right)^{\frac{1}{2}}\right]^{\frac{1}{2}}$, where $P_v = [(u_l + u_g)(\varepsilon_l\rho_l + \varepsilon_s\rho_s + \varepsilon_g\rho_g) - u_{l\rho l}]g/\varepsilon_l;$ $\eta_{sl} = \dfrac{2.5\beta}{1 - 39\beta/64}$; and $\beta = \varepsilon_g/(1 - \varepsilon_g)$
Yang et al. (2000)	Nitrogen–paratherm	864-870/ 0.0053-0.024/ 0.026–0.029	0.106/ 1.09	0.06-0.21/ 0	308-354/ 101.3-4200	Perforated Plate/ 0.0015/ 120	$St = 0.037\left[\left(\text{Re}_m\,Fr\,\text{Pr}_m^{1.87}\right)\left(\dfrac{\varepsilon_g}{1-\varepsilon_g}\right)\right]^{-0.22}$
Jamialahmadi and Müller-Steinhagen (2000)	Air–water, isopropanol and sodium sulfate	—	0.15, 0.30/-	0.002-0.035/-	295/101.3	Perforated plate	$h = h_c + 4\varepsilon_g(1-\varepsilon_g)^{2.39}(h_b - h_c)$ h_c = heat transfer coefficient produced by forced convection h_b = heat transfer coefficient in the bulk
Kantarci et al. (2005)	Air–water-yeast,	1000/ 0.001/ 0.072	0.17/0.6	0.03-0.20/-	295/101.3	Six-arm	$St = A(\text{Re}\,Fr)^\beta \text{Pr}^C\left(\dfrac{x}{H}\right)^D\left(\dfrac{r}{R}\right)^E$
Rozenblit et al. (2006)	air–water flow	1000/ 0.001/ 0.072	0.025/-	0- 1.0/0.1-1.0	295/101.3	Perforated plate	$\dfrac{hd_c}{k_l} = 125\left(\dfrac{u_{sg}}{u_{sl}}\right)^{0.125}\left(\dfrac{\mu_g}{\mu_l}\right)^{0.6}\text{Re}_l^{0.25}\text{Pr}_l^{0.333}\left(\dfrac{\mu_{l,b}}{\mu_{l,w}}\right)$
Jhawar and Prakash (2007)	Air–water	1000/ 0.001/ 0.072	0.15/ 1.05	0.028–0.32/ 0	295/ 101.3	Ring sparger and porous plate/ 0.000015–0.0019/ 20–87,000	$h = 8.65(u_{sg}/\varepsilon_g) + 1.32;\ for\ u_{sg}/\varepsilon_g \leq 0.3\ m/s$ $h = 2.0(u_{sg}/\varepsilon_g) + 3.3;\ for\ u_{sg}/\varepsilon_g \geq 0.3\ m/s$
Fazeli et al. (2008)	Air–paraffin, slurry	866.9-1190.1/ 0.00899-0.012841	0.3/3.0	0.026–0.230	295/101.3	Perforated plate/0.001 /89	$St = 1.26985(\text{Re}\,Fr\,\text{Pr}^{3.119286})^{-0.26365}(H_m/R)^{0.06336}$ $\times (1-(r/R)^2)^{0.04046}$

(Continued)

Table 9.2 Important Correlations in the Literatures on Heat Transfer Coefficient in Bubble Column (*cont.*)

Reference	System	Range of Liquid Properties: ρ, kg/m^3/μ, Pas/σ, N/m	Range of Column Diameter (m)/liquid height (m)	Range of u_{sg}/Range of u_{sl} (m/s)	Range of Temperature, K/Pressure, kPa	Gas Distributor Type/Hole Diameter, m/Number of Holes	Correlation
Our work; Sivaiah and Majumder (unpublished 2013); inverse bubbly flow	Air–water	997.5/ 0.0008/ 0.071	0.05/1.65	0.0085- 0.034/ 0.028 -0.0169	295/101.3- 303.9	Ejector, nozzle/ 0.003- 0.005/1	$Nu = \lambda \operatorname{Pr}^{1/3}$ where $$\lambda = 1.110 \frac{Su_n^{0.23} Mo^{0.085} A_R^{0.408} \operatorname{Re}_{\ln}^{0.634}}{H_m^{0.0625}} \left(\frac{H_m \varepsilon_g}{u_{rel} t_m} \right)^{0.08}$$ $$\times \left(\frac{d_c \bar{t}^{0.08}}{u_j^{0.5} d_n^{0.69} H_m^{0.063}} \right); \bar{t} = \frac{H_m \varepsilon_g}{u_{rel}}$$
Kim et al. (2014)	Air– paraffin oil, squalane	Paraffin: 880, squalane:800/ paraffin oil: 16.9 mPa s and squalane; 25.9 mPa s/-	0.0508/1.5	0.002- 0.164/0	295/101.3	Perforated plate distributor/0005/30	Homogeneous $$Nu = \frac{hd_p(1-\varepsilon_{sl})}{k_l \varepsilon_{sl}} = 1.86 \times 10^{-3} \left(\frac{d_p \rho_l u_{sg}}{\mu_l \varepsilon_{sl}} \right)^{1.024} \left(\frac{C_{pl} \mu_l}{k_l} \right)^{2.487}$$ Heterogeneous $$Nu = \frac{hd_p(1-\varepsilon_{sl})}{k_l \varepsilon_{sl}} = 3.12 \times 10^{-3} \left(\frac{d_p \rho_l u_{sg}}{\mu_l \varepsilon_{sl}} \right)^{0.522} \left(\frac{C_{pl} \mu_l}{k_l} \right)^{2.260}$$
Tow and Lienhard (2014)	Air–water		0.157 $\times$ 0.110/0.284; hydraulic diameter (0.202 m),	0-0.06/-	295/101.3	Sparger plate/0.003 /16	$Nu = C_H \operatorname{Pr}^{1/2}$; $$C_H = 0.02 + 0.18 \left[1 - \exp\frac{-H_c}{300\eta} \right]; \ \eta = (v^3 u_{sg} g)^{1/4}$$

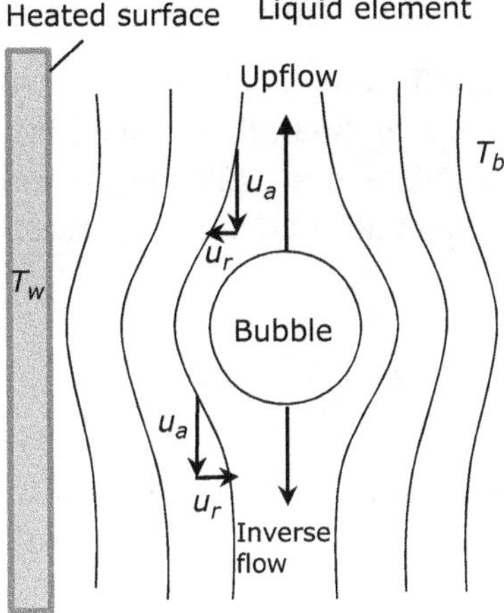

FIGURE 9.1 Flow around a moving bubble (as per Kast, 1962).

other authors, as shown in Table 9.2, investigated heat transfer coefficient at various operating conditions. It is observed that the correlations proposed by these investigators almost completely correspond to the correlation originally proposed by Kast (1962).

Zehner (1986) explained that a thinned thermal boundary layer on the heat transfer surface exists when bubbles are present. The length of the boundary layer is taken to be the same distance l between successive bubbles where the distance l depends on the bubble diameter and gas hold-up as ($l = d_b(\pi/6\varepsilon_g)^{1/3}$). The heat transfer through the boundary layer is described as that over a flat heated plate. Jamialahmadi and Müller-Steinhagen (2000) suggested a heat transfer mechanism in which the heat transfer surface is divided into two zones. First the area, which is affected by bubbles through which heat is transferred into the fluid by transient heat conduction from the heat transfer surface to the attached liquid. The hot liquid film is transported into the liquid bulk and replaced by cooler liquid. In the remaining heat transfer surface area, heat is transferred to the fluid by forced convection. Both mechanisms are assumed to occur in parallel in separate zones of the heat transfer surface.

Measurement of the Heat Transfer Coefficient

The main heat transfer studies in bubble columns carried out into two ways concerning inserted objects-to-bed and the wall-to-bed methods. Most of the previous studies on heat transfer rate in bubble columns are concerned with the steady-state time-averaged heat transfer coefficient (Wu et al., 2007). These measurements of heat transfer coefficients normally require a heat source and measurements of surface and bed temperatures.

Wall-to-Bed Method (by Measurement of the Energy Input)

In this method, heat transfer coefficient is calculated based on measurement of the energy input from the heated source to the bulk media and the temperature difference between them (Baker et al., 1978, Hikita et al., 1981, Deckwer et al., 1980, Kang et al., 1985, Saxena, 1989; Cho et al., 2002; Wu et al., 2007). In this case, the following equation is used (Deckwer et al., 1980):

$$h = \frac{q}{A(T_s - T_b)} = \frac{IV}{A\Delta T} \tag{9.1}$$

where h is the heat transfer coefficient ($kW/(m^2 K)$), q is the heat supplied by electric power or by steam which is equal to the product of electric current, I, in Ampere and voltage, V in unit volt, A is the heat transfer area (m^2), and ΔT is the temperature difference between the heat source (T_s) and the bulk media (T_b). In this method, there may be experimental error due to the heat loss in heating up all the surrounding materials, including the connecting fittings or column wall. So this error should be counted into the heat transfer from the heat source to the bulk flow. A schematic of the experimental setup is shown in Figure 9.2. Oil-free air is fed to the column through a pressure regulator, a filter, and a calibrated rotameter or by any other mechanical device such as an ejector system. As a heating source, a cylindrical heater with an outside diameter (suitably, 15% of column diameter) and a height of 80% of the liquid column to be installed vertically at the center of the column. The heat source can also be mounted on the wall of the column with its inside diameter the same as the column diameter. Temperatures at the heater surface and the column proper can be measured by means of temperature sensing device. A suitable number of temperature-detecting probes can be installed at the center of the annulus between the heater and the column wall with a certain interval of distance. A suitable number of temperature-detecting probes also need to be attached at the heater surface to measure the heater surface temperatures. These temperature-detecting probes have to be connected to a temperature-indicating system, data acquisition system, and a personal computer to record all points' temperatures simultaneously and continuously. The output voltage is then to be sampled at a certain rate of fraction of second in such a way that the total sampling time can generate at least 5000 data points. The output voltage is proportional to the temperature fluctuations at each point. This combination of sampling rate and time detect the full spectrum of temperature fluctuations. The processing systems should have enough fast response time to measure the dynamic temperature fluctuations at the heater surface as well as in the column proper simultaneously.

Inserted Object-To-Bed Method (by Measurement of Heat Flux)

In this method, the instantaneous and the local time-averaged heat transfer coefficient can be determined by direct measurement of heat flux and the difference between surface and bulk temperatures at a given time (Kumar et al., 1992, Luo et al., 1997, Yang et al., 2000; Li

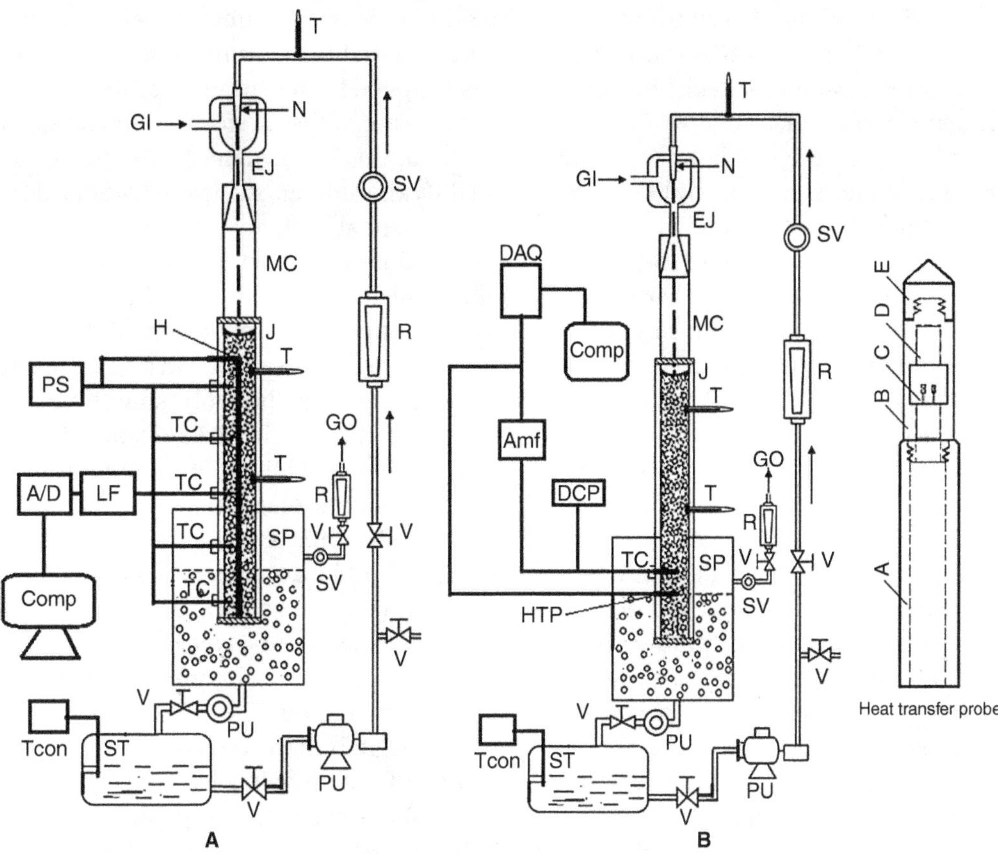

FIGURE 9.2 Experimental apparatus for heat transfer coefficient. **A,** Wall bed. **B,** Inserted object bed. *A/D,* A/D voltage converter; *A,* Teflon tube; *Amf,* amplifier; *B,* brass shell; *C,* heat flux sensor; *Comp,* computer; *D,* heater; *DAQ,* Data acquisition system; *DCP,* DC power supply; *E,* Teflon cap; *EJ,* ejector; *GI,* gas inlet; *GO,* gas outlet; *GO,* gas outlet; *H,* an inserted object acting as heater; *HTP,* heat transfer probe; *J,* insulating Jacket; *LF,* low-pass filter; *MC,* main column; *N,* nozzle; *PS,* power supply; *PU,* pump; *R,* rotameter/flowmeter; *SP,* separator; *ST,* liquid reservoir; *SV,* solenoid valve; *T,* thermometer; *TC,* thermocouple probe; *Tcon,* temperature controller; *V,* valve.

and Prakash, 2002, and Kantarci et al., 2005). For this method, the following equation is used:

$$h_i = \frac{q_i}{\Delta T_i} \tag{9.2}$$

$$h_{ave} = \frac{1}{n} \sum_{i=1}^{n} \frac{q_i}{\Delta T_i} \tag{9.3}$$

where h_i is the instantaneous local heat transfer coefficient $(kW/(m^2K))$, q_i is the heat flux across the sensor $(kW/(m^2))$, ΔT_i is the instantaneous temperature difference between the heat source and the bulk (K), h_{ave} is the time averaged heat transfer coefficient $(kW/(m^2K))$, and n is the total number of the samples. The schematic of experimental setup in an inverse bubbly flow system is shown in Figure 9.2. The bubble column of inverse bubbly flow is made of stainless steel. Compressed gas is supplied from compressor after it passed through a dryer and several air filter units. The flow rate of the filtered dry air can be adjusted by a pressure regulator and rotameter system. Gas is introduced into the column through a suction chamber by liquid jet. During the experiments, the fixed dynamic liquid height should be maintained by varying the static height at each operating condition. It is to be noted that the range of static height variation does not affect the column hydrodynamics at the conditions. A thermocouple probe, which contained no of thermocouples (as per experimental design) can be used to measure the bulk temperature of the media in the column adjacent to the heat transfer probe. The heat flux sensor can also be used to measure both the local heat flux and the surface temperature of the probe simultaneously. During the experiments, the heat transfer probe should be horizontally installed in the fully developed region of the bubble column at a suitable axial height (measured from the jet plunging point to the heat flux sensor). A typical schematic diagram of the heat transfer measurement probe (HTP) developed by Li and Prakash (1997) is shown in Figure 9.2.

One can also follow the development of heat transfer probe developed by Saxena (1989). The heat flux sensor is attached to the brass shell on the outer surface, and the direction of the sensor indicates the probe orientation. At most experimental conditions, the probe is set to be lateral because the surfaces of heat transfer elements are oriented laterally to the flow field, as opposed to upward and downward orientation. The orientation effect of the probe can be investigated as well. The measured signals of the heat flux may be in the range of microvolts. Therefore, it should be amplified before being received by the data acquisition system. After being amplified, the heat flux signals, together with the signals from the thermocouples, are sampled at a suitable frequency for a certain time. From the local inside wall temperature, the local peripheral inside wall heat flux, and the local bulk fluid temperature, the local peripheral heat transfer coefficient can be calculated as follows:

$$h_i = \frac{q_i}{T_{si} - T_{bi}} \tag{9.4}$$

Based on the assumption that the bulk temperature increases linearly in the pipe from the inlet to the outlet according to the following equation:

$$T_b = T_{in} + (T_{out} - T_{in})z / L \tag{9.5}$$

The local average heat transfer coefficient at each station can be calculated by

$$\bar{h}_i = \frac{\bar{q}_i}{\bar{T}_{si} - \bar{T}_{bi}} \tag{9.6}$$

where i is the index of a thermocouple station. The local average peripheral values for inside wall temperature, inside wall heat flux, and heat transfer coefficient can be obtained by averaging all the appropriate individual local peripheral values at each axial location.

The large variation in the circumferential wall temperature distribution, which is typical for two-phase gas–liquid flow, leads to different heat transfer coefficients depending on which circumferential wall temperature is selected for calculations. Especially in a bubbly flow system, to overcome the unbalanced circumferential heat transfer coefficient, equation (9.7) is recommended for calculation of the overall mean two-phase heat transfer coefficient, $h_{tp,ave}$.

$$h_{tp,ave} = \frac{1}{n}\sum_{i=1}^{n}\frac{q_i}{T_{si}-T_{bi}} \tag{9.7}$$

where h_i is the instantaneous local heat transfer coefficient ($kW/(m^2K)$), q_i is the instantaneous heat flux across the sensor ($kW/(m^2)$), T_{bi} is the instantaneous bulk temperature of the media (K), T_{si} is the instantaneous surface temperature of the probe (K), h_{ave} is the time averaged heat transfer coefficient ($kW/(m^2K)$), and n is the total number of the samples station. In this regard, it is worth mentioning that the use of average heat transfer may cause the loss of information regarding the effect of instantaneous bubble dynamics on heat transfer (Chen et al., 2003). For this reason, Chen et al. (2003) concluded that most correlations on the heat transfer coefficient do not remain valid over a wide range of gas flow rates. The local maximum instantaneous heat transfer coefficient is associated with the passage of the bubble wake. They examined the data using a rescaled range analysis and chaos analysis (including evaluation of the correlation dimension and Hurst exponent), which indicated the behavior in the bubble column is highly nonlinear and differed with scale.

Radial Average Heat Transfer Coefficient

The radial average heat transfer coefficient over cross-section area of the column can be obtained from the radial heat transfer coefficient at different locations as follows:

$$h_r = \frac{1}{R^2}\int_0^R h(r)r\,dr \tag{9.8}$$

Centerline Average Heat Transfer Coefficient

Generally, the centerline heat transfer coefficient values are higher than at the wall. Based on the radial profiles of heat transfer coefficients Kagumba (2013) reported that it is approximately 16% higher the wall region heat transfer coefficient and is approximately 7% higher than radial average heat transfer coefficient. Hence, it can provide useful indication and close approximation from an engineering point of view. He proposed a correlation to predict the centerline heat transfer coefficient in bubble columns, which can be expressed as

$$h_c = 3.413\times10^{-5}u_{sg}^{-0.52}\rho_l C_{pl}d_c^{0.14}\left[\left(\frac{d_c u_{sg}\rho_l}{\mu_l}\right)\left(\frac{u_{sg}^2}{d_c g}\right)\left(\frac{C_{pl}\mu_l}{k_l}\right)^{2.54}\right]^{0.254} \tag{9.9}$$

Li and Prakash (2001) reported that the differences between heat transfer coefficients in the wall and central regions are relatively constant. Therefore, reasonable estimates of heat transfer coefficients in the ceneal region can be obtained from the values in the wall region by multiplying them with a suitable enhancement factor. Therefore, for the central region, the correlation of Deckwer et al. (1980) was modified as

$$h_c = 0.1217 \left[k_l \rho_l C_{p,l} \left(\frac{P_v}{\mu_l} \right)^{0.5} \right]^{0.5} \tag{9.10}$$

They observed that the coefficient in the central region is about 22% higher than the original Deckwer et al. (1980) correlation. In the central region, the bubble wake behind larger bubbles may also contribute to surface renewal as described by Deckwer et al. (1980). The contribution due to bubble wake can be defined as a bubble wake enhancement factor (E_w), as a function of which, the centerline heat transfer coefficient can be expressed as

$$h_c = 0.1(1 + E_{wk}) \left[k_l \rho_l C_{p,l} \left(\frac{P_v}{\mu_l} \right)^{0.5} \right]^{0.5} \tag{9.11}$$

They reported that the bubble wake enhencement factor can be considered as 0.217 in case of slurry bubble column.

Overall and Volumetric Heat Transfer Coefficient

There are several problems that exist for determination of area-based heat transfer coefficient such as (1) the area-based heat transfer coefficient on the fundamental bubble level may be difficult to estimate due to difficulty in determination of the area of interface, (2) the interfacial area may vary considerably throughout the reactor, (3) the design of the probe surface by which the transfer of heat to be measured is difficult, and so on. A more easily evaluated approach that has been used in many studies reported in the literature is to replace the area-based heat transfer characteristics with volume-based values. The overall heat transfer coefficient can be calculated by using the log mean temperature difference (LMTD) approach. If the area between the two separate fluids (i.e., gas and liquid) is known, the heat energy exchanged can be calculated as

$$\dot{q} = h_{ov} A_i \Delta t_{lm} \tag{9.12}$$

where h_{ov} (W/m^2 °C) is the overall heat transfer coefficient based on the fluid interface area A_i (m^2), and ΔT_{lm} (°C) is the LMTD, which can be expressed as

$$\Delta t_{lm} = \frac{T_{g,in} - T_{g,out}}{\ln\left(\dfrac{T_{g,in} - \bar{T}_l}{T_{g,out} - \bar{\bar{T}}_l} \right)} \tag{9.13}$$

$T_{g,in}$ and $T_{g,out}$ are the inlet and outlet gas temperatures, respectively, and $\bar{T}_l$ is the average temperature of the liquid. The exchange of heat energy dissipation rate ($\dot{q}$) can be calculated as (Robert, 2003)

$$\dot{q} = h_v V_R \Delta T_{lm} \tag{9.14}$$

where V_R is the total internal volume of the reactor and h_v is the volumetric heat transfer coefficient based on that volume. The exchanged heat energy dissipation rate ($\dot{q}$) can be calculated from the heat energy dissipation rate by the gas phase, which can be expressed as

$$\dot{q} = \dot{m}_g C_{p,g} (T_{g,in} - T_{g,out}) \tag{9.15}$$

where $\dot{m}_g$ (kg/s) is the gas mass flow rate and C_{pg} (J/kg °C) is the gas-specific heat at constant pressure. Substitute equations (9.13) and (9.15) into equation (9.14) and solving for h_v, one can get

$$h_v = \frac{\dot{m}_g C_{p,g}}{V_R} \ln\left(\frac{T_{g,in} - \bar{T}_l}{T_{g,out} - \bar{T}_l} \right) \tag{9.16}$$

The gas properties can be calculated at the average gas temperature ($\bar{T}_g = (T_{g,in} - T_{g,out})/2$).

Effect of Different Variables on Heat Transfer Coefficient in Bubbly Flow

Although bubble columns are simple in construction, the design, scale-up, and modeling of bubble columns is a difficult task because of lack of information on the effect of different parameters such as operating conditions, column geometry, distributor type, physicochemical properties of the phases, and slurry concentrations (in case of slurry bubble column) on hydrodynamics, heat, and mass transfer. The crucial design considerations include phase hold-up structure, operating flow regimes, mixing and circulation pattern, axial and radial dispersion of liquid (slurry) and gas phase, heat and mass transfer coefficient, and interfacial area. These parameters are affected by various variables such as superficial gas and liquid/slurry velocities; operating conditions (temperature and pressure); physical characteristics of gas phase (density) and liquid/slurry phase properties (surface tension, viscosity, density and thermal conductivity, particle concentration and size distribution); and column diameter; internals, and their design (e.g., heat transfer surface). An important relation between column hydrodynamics and the heat transfer process has also been well established by literature studies (Lin and Hung-Tzu., 2003; Jhawar and Prakash, 2011). Therefore, the design and operating variables that affect hydrodynamics may also influence the heat transfer rate. These parameters can be divided into two main categories: operational parameters and geometric parameters. The first category includes variables such as superficial gas and liquid velocity, operating pressure and temperature, liquid thermophysical properties, particles properties, and concentration. The second category includes column dimensions, internal design and location, and axial and radial position of heat transfer surface. The important effects of these variables on the heat transfer in bubble columns are discussed as follows based on the studies on literature:

Effect of Gas and Liquid Velocities

Generally, the introduction of gas into a liquid or slurry bed augments the respective heat transfer coefficients. It has been reported that the heat transfer coefficient increases with the superficial gas velocity irrespective of the liquid flow rate, bubble properties (diameter, shape, and densities), liquid viscosity, and slurry concentrations. The rate of increase of heat transfer coefficients with gas velocity is rapid at low gas velocity, and then the rate decreases with a further increase of gas velocity (Fair et al., 1962; Baker et al. 1978; Deckwer et al., 1980; Kang et al. 1985; Saxena et al., 1989). The introduction of gas into the column enhances the turbulence in the medium and thus increases the heat transfer coefficients. Fair et al. (1962) studied the heat transfer coefficients in a commercial-scale bubble column (0.45 and 1.07 m in diameter) by using an air–water system. The authors concluded that the heat transfer coefficients between the gas and the heating surface are high and vary directly with the superficial gas velocity. Baker et al. (1978) observed an increase in the heat transfer coefficient with gas flow rate caused by an increase in the bed turbulence by gas introduction and consequently, the heat transfer coefficient. Kang et al. (1985) also reported that the introduction of a gas stream generates mixing of the liquid phase, which results in an increase of the heat transfer coefficient. A higher gas velocity increased the overall liquid circulation but not micro-scale eddies in the bed. Thus, the heat transfer coefficient does not increase appreciably at a higher gas velocity. Deckwer (1980) explained that at high gas velocities when heterogeneous flow prevails, the major part of the gas flow penetrates the column in the form of large bubbles and slugs. These large bubbles and slugs induce an overall liquid circulation in the column instead of transferring energy into micro-scale eddies. Therefore, the contact time no longer increases with increasing gas velocity, and hence no significant changes in the heat transfer coefficient result at high gas velocities. Jamialahmadi and Müller-Steinhagen (2000) investigated the effects of heat addition on the performance of bubble column reactors using distilled water, isopropanol, and sodium sulfate solutions as the liquid phase and air as gas phase. They studied the effect of operating parameters on the heat transfer coefficients from a submerged heat transfer surface to the fluid and of fluid temperature on the gas hold-up. They observed significant improvement of heat transfer because of the presence of air bubbles in the column. They reported that in the bubbly flow regime, the heat transfer coefficient increases with gas velocity, and when the flow regime changes to the churn-turbulent flow regime, no further significant improvement in the heat transfer coefficient were observed. According to Wu et al. (2007), at low gas velocity, the bubble population increases under high pressure, but the liquid circulation does not change much because bubble coalescence and break-up are insignificant in this bubbly regime. They reported that at a high superficial gas velocity, more large bubbles are formed in the transition flow regime and churn-turbulent flow regime. Even though the average size at a high pressure is smaller than that under atmospheric pressure, the enhanced bubble frequency and increased bubble number under high pressure cause intense interaction between bubble–bubble, bubble–liquid, liquid–wall, and liquid–probe surfaces. Such enhanced interactions due to pressure decrease the thickness of the contact film between the probe and the bulk and then increase the heat transfer coefficient to some extent. However, Yang et al. (2000) stated that these enhancements on heat transfer coefficient due to increasing bubble number and bubble frequency with increasing pressure are not as strong as the effect of the bubble size reduction. Hence, at high superficial gas velocities (churn-turbulent flow regime), the differences in the heat transfer coefficient between low pressure and high pressure become smaller compared with low-range superficial gas velocities (bubbly flow regime).

The liquid or slurry velocity increases the heat transfer coefficient because bubble wake–induced circulation velocity in bubble or slurry bubble columns reactors has a dominant effect (Yu and Kim, 1991; Michele and Hempel, 2002). Muroyama et al. (1984) reported that if bed is operated with large particle size and the bed operation moves toward a three-phase sparged reactor or fluidized bed, liquid velocity effects become more significant. In three-phase bubbling beds, particle movements and collision affect the heat transfer surface. This effect increases with liquid velocity up to a critical value and then decreases as the particles agitation effect becomes smaller with increasing gas hold-up. In an ejector-induced liquid jet–driven inverse bubbly flow system, the heat transfer coefficient increases for the same reason. Most of the studies reported that the heat transfer coefficient initially increases, passes through a maximum, and then decreases as a function of progressiveiy increasing liquid veiocity. The maximum value is mainly dependent on the liquid and bubble properties (Kang et al., 1985). It increases with increasing density and bubble size but decreases at increasing liquid viscosity. The initial increase of the heat transfer coefficient with liquid velocity due to an increase of turbulence and oscillatory motion of the bubbles results in an erosion of the thermal boundary layer (Baker et al., 1978). At higher liquid velocities, the heterogeneity and the slug pattern may reduce the heat transfer coefficient.

Effect of Fluid Properties

The heat transfer coefficient decreases with an increase in the viscosity of the liquid in gas–liquid or gas–liquid–solid systems (Deckwer et al., 1980; Kang et al., 1985; Saxena et al., 1992; Li and Prakash, 1997; Yang et al., 2000). Kang et al. (1985) explained that the decrease in heat transfer coefficient may be due to two factors. First, in the region adjacent to the heater surface, the thickness of the thermal boundary layer around the heating surface increases with increasing viscosity, thus increasing resistance to heat trasfer. Second, bubble movement is retarded because of increasing viscosity, diminishing their attack on the thermal boundary layer around the heating surface. Kantrachi et al. (2005) reported that the decrease in heat transfer coefficient with viscosity is actually attributed to lower turbulence attained in viscous media. The heat transfer coefficient increases with the heat capacity and thermal conductivity of the liquid. Surface tension also plays a crucial role in heat transfer in bubbly flow based on the size of the bubble and by governing the rate at which bubbles coalesce when they come into contact with the heated surface.

Effect of Operating Pressure and Temperature

Pressure has a significant effect on the hydrodynamics and bubble dynamics (especially the bubble size) in bubble columns (Luo et al., 1999). There are only a few studies about heat transfer coefficient in high-pressure reactors (Deckwer et al., 1980, Luo et al., 1997; Yang et al., 2000) that need further evaluation. Deckwer (1980) conducted the heat transfer study under coal liquefaction conditions with pressure range of 0.4 to 1.1 MPa. They found that the heat transfer coefficient does not have apparent variation with pressure because there is no notable effect on bubble characteristics. However, if pressure has a significant effect on bubble characteristics for some systems, it should cause some levels of changes on heat transfer. Luo et al. (1997) reported that the heat transfer coefficient between an immersed surface and the bed at a given gas and liquid velocity in a three-phase fluidized bed increases to a maximum at a pressure of 6 to 8 MPa and then decreases with a further increase in pressure. Yang et al. (2000) investigated the effect of pressure on the local heat transfer coefficient in the center of a slurry bubble column. They found that the heat transfer coefficient decreases with an increase in pressure. However, all of these reported heat transfer coefficient measurements under high pressure were carried out in small-diameter columns (<11 cm). According to Wilkinson et al. (1992), the inner diameter of the bubble column should be equal or larger than 15 cm to avoid wall effect. Wu et al. (2007) reported that with the change of pressure from 1 to 10 bar, the heat transfer coefficients decreased in both the center and the wall regions of the column. They explained that pressure can affect the liquid properties such as density, viscosity, and surface tension. With the increase in the pressure, the gas density increases, which causes the decrease of the initial bubble size from the sparger and the increase of bubble breakup rate. Because of these effects, they pointed out that the pressure decreases the bubble size and narrows the range of bubble size distribution, hence the decrease in the heat transfer coefficient at an elevated pressure. They also pointed out that the overall decrease trend of the heat transfer coefficient with increasing pressure is mainly due to the decrease of bubble size. The heat transfer coefficient will increase if the bed temperature is increased (Saxena et al., 1992). The liquid viscosity decreases as the bed temperature increased and enhanced turbulence at higher temperatures. Yang et al. (2000) reported increase in heat transfer coefficient in a slurry bubble column with an increase in temperature. This is mainly attributed to a decrease in the liquid viscosity of the liquid with an increase in temperature compared with other effects.

Effect of Particle Properties and Concentration

In a bubble column, the heat transfer coefficient increases with particle size at low gas velocities (<5 cm/s) in a bubble column (Kantarci et al., 2005). At higher gas velocities, it passes through a minimum value at a particle size of about 1.5 mm. In general, the effect of particle size on heat transfer coefficients is negligible at particle sizes larger than 3.0 mm, particularly at high gas velocities. Li et al. (2003) observed that the effect of particle size is not significant in the wall region, but the heat transfer coefficient decreases with an increase in particle size in the core region of the column. The larger effect of the particle size at the column core is attributed to its effects on wake formation process that is dominant in the central region of the column. Deckwer (1980) reported that the heat transfer coefficients increased with increasing slurry concentrations because of alteration of thermophysical properties of the slurry with the introduction of solids and also enhanced exchange rate of fluid elements on the heated surface of the probe because of motion of solid particles. Kantarci et al. (2005) reported that the alteration of the bubble properties such as bubble coalescence with solid addition results higher heat transfer rates. Li and Prakash (1997) reported that as solid concentration increased, the heat transfer coefficient decreased, which is a contradiction with the results of the Deckwer et al. (1980). They explained that the promotion of viscosity of the medium with an increase in the solid concentration results in a decrease of turbulence in the system, which may decrease the heat transfer coefficient. Addition of fine solid particles into a liquid changes the average thermophysical properties of the suspension and increases its apparent viscosity (Deckwer et al., 1980). Whereas an increase in thermal conductivity and heat capacity of suspension would have a positive effect on the heat transfer rate, increase in suspension viscosity would have a negative effect because of an increase in the hydrodynamic boundary layer thickness. Because the suspension viscosity will increase with the particle concentration, any increase in the heat transfer coefficient for a given gas velocity can be attributed to the dominant effect of thermophysical properties (Jhawar and Prakash, 2011). There is still a need to systematically study the effects of slurry rheology on heat transfer coefficient in bubble column reactors.

Effect of Probe Orientation

The position of the heat transfer probe in the column may also change the values of the heat transfer coefficient. It depends on the measurement of the distance from the gas distributor to the probe position and radial differences from the bubble position. Several studies were performed by locating the heat transfer probe at various axial and radial locations in the column and determining the corresponding values of the heat transfer coefficients at those locations. Saxena et al. (1990a) reported that the heat transfer coefficient is higher if the probe is located nearer the distributor. This may be because of the design of the distributor. For an ejector-induced liquid jet gas distributor, the heat transfer coefficient will be higher at the vicinity of the plunging point of the liquid jet because of the high turbulence of the fluid mixture. It may also depend on the flow pattern developing by the different types of distributor. If the distance of the probe from the distributor is less than two times the column diameter (in case of jet-induced inverse bubbly flow, less than jet penetration length), the bubble sizes are definitely smaller than the ones in the bulk region. This is because the external pressure around the bubble decreases as the bubble moves in the column. Thus, large bubbles would be more dominant away from the distributor. Faster moving large bubbles would be more effective on heat transfer than smaller bubble. Li and Prakash (2002) reported that the column center heat transfer coefficients were higher than the near-wall heat transfer coefficients because large bubbles collect more dominantly at the center. In addition, there exist more turbulence in the center as compared with near the wall due to possible wall effects. Wu et al. (2007) reported that flow direction is important in understanding the flow structure in the bubble columns. Using the measurement of the heat transfer coefficients at different orientations of the sensor on the probe, the flow direction can be qualitatively identified (upward or inverse). They observed that in the center of the column, the heat transfer coefficient in the inverse direction was the largest, which indicates an enhanced upward gas–liquid flow in the center of the column. On the other hand, they reported that an enhanced inverse flow in the wall region of the column can be obtained.

Effect of the Column Diameter

The heat transfer coefficient was reported to be increased with increasing column diameter. Saxena and Patel (1991) studied the heat transfer coefficients in small column diameters (0.09 m) and in high column diameters (1.07 m) using the three different heat transfer probe diameters. They reported that the increased heat transfer coefficient is due to the change in intensity of the fluid mixing. However, Kim and Laurent (1991) analyzed the published results to sumrnarize the effect of column diameter on the heat transfer coefficient. They reported that the effect was found to be negligible for column diameters ranging from 0.052 to 0.152 m. In the case of inverse bubbly flow, the heat transfer coefficient may increase with an increase in the column diameter. This can be attributed to the fact that at higher column diameters, more bubbles may rise up because of buoyancy by lateral movement of the bubble. Hence, mixing increases the heat transfer. Jhawar and Prakash (2011) reported that column diameter effects seem to diminish above a diameter of 0.3 m. The increase in heat transfer coefficient is related to increasing liquid recirculation velocity with column diameter. Their observations are, however, based on a column diameter up to 0.3 m and need to be verified for larger diameter columns.

Effect of Bubble Size

Bubbles dispersed into liquid or liquid–solid systems enhance the heat transfer rate through the bubble wake based on their size (Kumar et al., 1992; Kumar and Fan, 1994). The efffect of bubble sizes on the instantaneous heat transfer coefficient is due to the passage of bubble in a liquid for probe located at the column center. The wake is proportional to the bubble size. The maximum heat transfer occurs in the wake region at a short distance behind the bubble. The heat transfer enhancement is thus attributed to the bubble wake created by the bubble(s) passing over the heat transfer surface. Kumar and Fan (1994) reported that larger bubbles would have larger wakes and stronger vortices associated with the wake, thereby enhancing the rate of heat transfer. The strong vortices and turbulence in the bubble wake region increase the heat transfer surface renewal rate. They demonstrated that the heat transfer rate is proportional to the bubble size. These studies did not make clear the effect of larger bubble population because they were limited to single bubbles or a chain of bubbles. They also did not cover the range of gas velocities suitable for most commercial applications. Therefore, further study is suggested.

Effect of Internals

The number of heat transfer tubes inside a column may have an effect on heat transfer coefficients. Saxena et al. (1990b) reported that the effect of internals on heat transfer coefficient in large-diameter columns (0.3 m) is not significant, but the effect is significant in smaller diameter columns (0.108 m). They compared the experimental results obtained with single and seven-tube bundles immersed inside the column. They reported that the heat transfer coefficients with seven-tube bundles were systematically higher than those with single tube. The authors attributed the difference to improved liquid mixing in the presence of the seven-tube bundle compared with the single tube case. They pointed out that the presence of internals can help in promoting better mixing in small-diameter columns by limiting the maximum stable size of the bubbles. It is, however, likely that the small-diameter column used by Saxena et al. (1990b) exhibited more sluglike flow behavior without the internals. The presence of internals, however, would have changed the hydrodynamic conditions, leading to an increase in turbulence and heat transfer enhancement they reported. Schlüter et al. (1995) observed that the tube pitch has no significant effect on heat transfer coefficient in low viscosity, and it has only a small effect in case of highly viscous liquid. Some important correlations related to the study based on the geometric parameters is shown in Table 9.3.

Table 9.3 Important Correlations for Heat Transfer Coefficient Studied with Different Geometric Parameters

Geometric parameters	Correlations	Reference
Internals	$h = 0.18\dfrac{k_g}{v^2}\left(\varepsilon_g \operatorname{Pr}\dfrac{d_c}{d_p}\right)^{1/3}\left(\dfrac{\mu_l}{\mu_p}\right)^{0.14}$	Konsetov (1966)
Wall	$St = 0.1(\operatorname{Re}\mathit{Fr}\operatorname{Pr}^2)^{-1/4}$	Deckwer (1980)
Tube bundle	$St = 0.139(\operatorname{Re}\mathit{Fr}\operatorname{Pr}^{2.26})^{-0.28}A_f^{-0.2}\left(\dfrac{d_c}{d_p}\right)^{0.14}\left(\dfrac{\mu_l}{\mu_p}\right)^{0.3}$	Korte (1987)
Internals	$h = 14.83\left(\dfrac{d_c}{d_p}\right)u_{sg}^{0.21}$	Saxena and Patel (1991)
Internals	$Nu_{d_p} = 0.133\operatorname{Pr}^{1/3}\left(\varepsilon^{1/3}d_p^{4/3}/v\right)^{0.709}$	Muroyama et al. (2001)
Tube bundle	$St = 0.139(\operatorname{Re}\mathit{Fr}\operatorname{Pr}^{2.26})^{-0.28}A_f^{-0.2}\left(\dfrac{d_c}{d_p}\right)^{0.14}\left(\dfrac{\mu_l}{\mu_p}\right)^{0.3}$	Kagumba (2013)

The included geometric parameters show the relationship between heat transfer coefficient and superficial velocity and the magnitude of the predicted heat transfer coefficient. Systematic studies are required to further investigate the effect of internais and their configurations on mixing and heat transfer coefficients in up and inverse bubble column.

Effect of Axial or Radial Location

Many authors (Prakash et al., 2001; Li and Prakash, 2002; Li et al., 2003; Saxena et al., 1992; Wu et al., 2007; Jhawar and Prakash, 2011) reported that heat transfer coefficient depend on axial and radial location in the column. In the radial direction, the highest heat transfer coefficient is obtained at the center and the lowest near the column wall (Li and Prakash, 2001; Wu et al., 2007). It is also pointed out that the wall region heat transfer coefficient is not affected by the particle size and column diameter (Jhawar and Prakash, 2011). Li and Prakash (2002) pointed out that the radial profile of the heat transfer coefficient is relatively flat in the distributor, intermediate, and disengagement zones compared with the bulk zone. The radial profile of heat transfer coefficient is reported to be affected by slurry concentration up to 20 volume %, and the effect became insignificant for higher slurry concentrations (Li and Prakash, 2001). Saxena et al. (1992) reported that the heat transfer coefficient increases as the axial distance from the distributor increases. They explained that this may be due to the variation of liquid mixing along the column height. Li and Prakash (2002) also reported that the heat transfer coefficient increases with increase in distance from the gas distributor, but in the fully developed bulk zone, the effect of axial position becomes insignificant. We agree with the statement for the inverse bubbly flow system. In gas introduction by the jet system for inverse bubbly flow, the introduction zone is highly turbulent, but beyond the jet penetration length, the bulk zone does not have an intense mixing zone (Majumder et al., 2005). These observations need to be taken into consideration for proper design and arrangement of heat transfer surface. To determine values at any radial location, the following equation (Jhawar and Prakash, 2007) is recommended.

$$\frac{h_c - h(r)}{h_c} = \left(\frac{n-1}{n}\right)\left(\frac{r}{R}\right)^n \tag{9.17}$$

where h_c is the heat transfer coefficient in core region of the column. The recommended value of n in the equation (9.17) is 1.4. This correlation is valid in central region ($r/R = 0.0$ to 0.75). For the wall region, correlation of Deckwer et al. (1980) is recommended. There is a need to test this procedure in columns of diameter larger than 0.3 m and inverse bubbly flow columns. A summary of the general consensus as to the effect of key process parameters on the heat transfer coefficient is shown in Table 9.4. Where the consensus is contradictory, the varying conclusions are presented along with a reference to the relevant sources.

Table 9.4 Summary of the Relative Effects of Process Variables on Heat Transfer Coefficient

Independent Variable	Relative Magnitude of Independent Variable		Remarks with Reference
	Low	**High**	
u_g	Increase	No change	Kang et al. (1985): initially decreases; approaches asymptotic value
	Increase	Decrease	Fair et al. (1962), Baker et al. (1978), Deckwer et al. (1980), Kang et al. (1985), Saxena et al. (1989)
u_l	Increase	Decrease	Yu and Kim (1991), Michele and Hempel (2002), Muroyama et al. (1984): local maximum
μ_l	Decrease		Deckwer et al. (1980), Kang et al. (1985), Saxena et al. (1992), Li and Prakash (1997), Yang et al. (2000)
σ_l	Decrease	No change	Wu et al. (2007): initially decreases; approaches asymptotic value because of decrease in bubble size with decrease in surface tension
d_b	Increase		Jhawar and Prakash (2011): "unbaffled" column (i.e., single tube)
	Decrease		Jhawar and Prakash (2011): "baffled" column,
d_p	Decrease		Quiroz et al. (2003)
	No change		Saxena and Chen (1994)
	Increase	No change	Kang et al. (1985)
	Decrease		Small particles in relatively low-viscosity liquids
	No change		Saxena et al. (1992)
d_t	Decrease		Jhawar and Prakash (2011): studies with internals
d_c	Increase	—	Kim and Laurent (1991), Jhawar and Prakash (2011): no studies for $d_c > 0.45$ m
H	Increase		Li and Prakash (2002)
T	Increase	No change	Saxena et al. (1992), Yang et al. (2000)
P	Decrease		Yang et al. (2000), Wu et al. (2007): for relatively large d_p ($d_p = 53$ μm)
	Increase-no change-decrease		Luo et al. (1997), Deckwer (1980), Yang et al. (2000): relatively large d_p ($d_p = 2.1$-3 mm) or g-l only

Model to Assess Heat Transfer in Bubbly Flow Condition

General Correlation Model

The heat transfer coefficient depends on the different operating variables as discussed in the previous section. The prediction of the heat transfer coefficient based on mechanistic model is still challenging because the phenomenon such as momentum transfer between the phases, the wall friction, and the shear at the phase interface affecting the heat transfer coefficient cannot be quantified directly from the experiments which are required for theoretical modelling. Therefore, prediction of transport coefficient by theoretical analysis alone does not depict well with significant accuracy. Several workers as mentioned in Table 9.2 have developed correlations for the two-phase heat transfer coefficient by dimensional analysis as a function of the directly measurable parameters. To improve the prediction of heat transfer rate in turbulent two-phase flow, regardless of fluid combination and flow pattern, correlation can be developed based on the concept of Kim et al. (2000) and Ghajar (2005). The correlation takes into account the appropriate contributions of both the actual liquid and and gas velocities. The total gas–liquid two-phase heat transfer is assumed to be the sum of the individual single-phase heat transfers of the gas and liquid, weighted by the volume of each phase present, which can be expressed as

$$h_{tp} = (1-\varepsilon_g)h_l + \varepsilon_g h_g = (1-\varepsilon_g)h_l\left[1+\left(\frac{\varepsilon_g}{1-\varepsilon_g}\right)\left(\frac{h_g}{h_l}\right)\right] \tag{9.18}$$

The Sieder and Tate (1936) equation can be used to calculate the single-phase heat transfer coefficient (h_l or h_g) because of its practical simplicity and wide use. Any other well-known single-phase turbulent heat transfer correlation can also be used in place of the Sieder and Tate (1936) equation. Based on this equation, the single-phase heat transfer coefficients in equation (9.19), h_l and h_g can be expressed as a function of dimensionless groups: Reynolds number, the Prandtl number, and the ratio of bulk to wall viscosities. Thus, equation (9.19) can be expressed as

$$h_{tp} = (1-\varepsilon_g)h_l\left[1+\left(\frac{\varepsilon_g}{1-\varepsilon_g}\right)\left(\frac{f(\mathrm{Re}_g,\mathrm{Pr}_g,\mu_{g,b}/\mu_{g,w})}{f(\mathrm{Re}_l,\mathrm{Pr}_l,\mu_{l,b}/\mu_{l,w})}\right)\right] \tag{9.19}$$

$$h_{tp} = (1-\varepsilon_g)h_l\left[1+\left(\frac{\varepsilon_g}{1-\varepsilon_g}\right)f\left\{\left(\frac{\mathrm{Re}_g}{\mathrm{Re}_l}\right)\left(\frac{\mathrm{Pr}_g}{\mathrm{Pr}_l}\right)\left(\frac{\mu_{g,b}/\mu_{g,w}}{\mu_{l,b}/\mu_{l,w}}\right)\right\}\right] \tag{9.20}$$

which on rearranging can be expressed as

$$\psi_h = h_{tp}/h_l = (1-\varepsilon_g)\left[1+\left(\frac{\varepsilon_g}{1-\varepsilon_g}\right)f\left\{\left(\frac{\rho_g}{\rho_l}\right)\left(\frac{u_{sg}/\varepsilon_g}{u_{sl}/(1-\varepsilon_g)}\right)\left(\frac{\mathrm{Pr}_g}{\mathrm{Pr}_l}\right)\left(\frac{\mu_{g,b}/\mu_{g,w}}{\mu_{l,b}/\mu_{l,w}}\right)\right\}\right] \tag{9.21}$$

Based on practical considerations, the viscosity ratios can be significant only for the wall temperature condition (Ghajar, 2005), so

$$\frac{\mu_{g,b}/\mu_{g,w}}{\mu_{l,b}/\mu_{l,w}} \approx \frac{\mu_l}{\mu_g} \tag{9.22}$$

Therefore, equation (9.21) can be expressed as

$$\psi_h = h_{tp}/h_l = (1-\varepsilon_g)\left[1+\left(\frac{\varepsilon_g}{1-\varepsilon_g}\right)f\left\{\left(\frac{x_g}{1-x_g}\right)\left(\frac{1-\varepsilon_g}{\varepsilon_g}\right)\left(\frac{\mathrm{Pr}_g}{\mathrm{Pr}_l}\right)\left(\frac{\mu_l}{\mu_g}\right)\right\}\right] \tag{9.23}$$

where x_g is the mass quality of gas which is defined as

$$x_g = \frac{m_g}{m_m} = \frac{\rho_g u_{sg} A_c}{\rho_g u_{sg} A_c + \rho_l u_{sl} A_c} \tag{9.24}$$

The functionality of the individual parameters that appear in equation (9.23) is assumed to be a power-law relationship based on which equation (9.23) can be expressed as

$$\psi_h = h_{tp}/h_l = (1-\varepsilon_g)\left[1+a\left(\frac{x_g}{1-x_g}\right)^b\left(\frac{\varepsilon_g}{1-\varepsilon_g}\right)^c\left(\frac{\mathrm{Pr}_g}{\mathrm{Pr}_l}\right)^d\left(\frac{\mu_g}{\mu_l}\right)^e\right] \tag{9.25}$$

The coefficients of equation (9.25) can be obttained by multiple linear regreassion with the experimental data. For multiple linear regression, Microsoft Excel's the regression analysis tool can be used. By considering the Sieder and Tate (1936) equation for calculation of single-phase heat transfer coefficient, Ghajar (2005) reported that the coefficients *a* to *e* are to be for two-phase flow heat transfer coefficient in bubbly flow regime in vertical pipe as: 0.27, –0.04, 1.21, 0.66, and –0.72 respectively. For horizontal bubbly, slug, or annular flow, these are 2.86, 0.42, 0.35, 0.66, and –0.72, respectively. Some important correlations for h_{tp}/h_l developed by different investigators are given in Table 9.5. Based on the experimental results of the wall to bed heat transfer coefficient in an ejector-induced inverse bubbly flow (the experimental setup is shown in Figure 9.2), a correlation is developed based on the different operating variables, which can be expressed as

$$Nu \propto \left(\frac{\rho_l\varepsilon^{1/3}d_c^{4/3}\overline{t}^{1/3}}{\mu_l}\right)^{3/4}\mathrm{Pr}^{1/3} \tag{9.26}$$

$$\varepsilon = K_m\left(\frac{1}{2}u_j^2\right)\Big/\overline{t} \tag{9.27}$$

$$\overline{t} = \frac{H_c\varepsilon_g}{u_{rel}} \tag{9.28}$$

and

$$K_m = 3.04\,\mathrm{Re}_{\mathrm{ln}}^{-0.465}\,A_R^{0.408}\,H_m^{-0.25}\,Su_n^{0.92}\,Mo^{0.339} \tag{9.29}$$

u_{rel} is the relative velocity between gas and liquid phases in the column, which can be calculated by equation $u_{rel} = u_{sl}/(1-\varepsilon_g) - u_{bt}$. By considering the energy dissipation per unit mass of fluid for an ejector-induced inverse bubble flow, the following correlation are developed.

$$Nu = \lambda\,\mathrm{Pr}^{1/3} \tag{9.30}$$

where

$$\lambda = 1.110\frac{Su_n^{0.23}\,Mo^{0.085}\,A_R^{0.408}\,\mathrm{Re}_{\mathrm{ln}}^{0.634}}{H_m^{0.0625}}\left(\frac{H\varepsilon_g}{u_{rel}\overline{t}}\right)^{0.08}\left(\frac{d_c\overline{t}^{\,0.08}}{u_j^{0.5}d_n^{0.69}H^{0.063}}\right) \tag{9.31}$$

The correlation for individual effect of variables on the Nusselt number can also be expressed for the inverse bubbly flow system as

Table 9.5 Correlations for Two-Phase Heat Transfer Coefficient Relative to Single-Phase Heat Transfer Coefficient

Reference	Ratio of Two-Phase to Single-Phase Heat Transfer Coefficient ($\psi_h = h_{tp}/h_l$)	Correlation for Single-Phase Correlation Used
King (1952)	$\psi_h = \dfrac{(1-\varepsilon_g)^{-0.52}}{1+0.025\text{Re}_{sg}^{0.5}}\left[\dfrac{(\Delta P/\Delta L)_{tp}}{(\Delta P/\Delta L)_l}\right]^{0.32}$	$h_l = 0.023\text{Re}_{sl}^{0.8}\,\text{Pr}_l^{0.4}$
Knott et al. (1959)	$\psi_h = \left(1+u_{sg}/u_{sl}\right)^{1/3}$	Sieder and Tate (1936) equations: $Nu_l = 1.86(\text{Re}_{sl}\,\text{Pr}_l\,d_c/L)^{1/3}(\mu_b/\mu_w)^{0.14}$ for $\text{Re}_{sl} < 4000$ $Nu_l = 0.027\text{Re}_{sl}^{0.8}\,\text{Pr}_l^{0.33}(\mu_b/\mu_w)^{0.14}$ for $\text{Re}_{sl} > 4000$
Dorresteijn (1970)	$\psi_h = (1-\varepsilon_g)^{-1/3}$ for $\text{Re}_{sl} < 4000$ $\psi_h = (1-\varepsilon_g)^{-0.80}$ for $\text{Re}_{sl} > 4000$	$Nu_l = 0.0123\text{Re}_{sl}^{0.9}\,\text{Pr}_l^{0.33}(\mu_b/\mu_w)^{0.14}$
Martin and Sims (1971)	$\psi_h = 1+0.64(u_{sg}/u_{sl})^{0.5}$	Sieder and Tate (1936) equations
Serizawa et al. (1975)	$\psi_h = 1+462X_{tt}^{-1.27}$ $X_{tt} = ((1-x)/x)^{0.9}(\rho_g/\rho_l)^{0.5}(\mu_g/\mu_l)^{-0.1}$ $X_{tt} = \left((dp/dz)_{sl}/(dp/dz)_{sg}\right)_{f,tt}$	Sieder and Tate (1936) equations
Aggour (1978)	$\psi_h = (1-\varepsilon_g)^{-1/3}$ for $\text{Re}_{sl} < 4000$ $h_{tp}/h_l = (1-\varepsilon_g)^{-0.83}$ for $\text{Re}_{sl} > 4000$	$Nu_l = 1.615(\text{Re}_{sl}\,\text{Pr}_l\,d_c/L)^{1/3}(\mu_b/\mu_w)^{0.14}$ for $\text{Re}_{sl} < 4000$ $Nu_l = 0.0155\text{Re}_{sl}^{0.83}\,\text{Pr}_l^{0.5}(\mu_b/\mu_w)^{0.30}$ for $\text{Re}_{sl} > 4000$
Shah (1981)	$\psi_h = \left(1+u_{sg}/u_{sl}\right)^{1/4}$	$Nu_l = 1.86(\text{Re}_{sl}\,\text{Pr}_l\,d_c/L)^{1/3}(\mu_b/\mu_w)^{0.14}$ for $\text{Re}_{sl} < 4000$ $Nu_l = 0.023\text{Re}_{sl}^{0.8}\,\text{Pr}_l^{0.4}(\mu_b/\mu_w)^{0.14}$ for $\text{Re}_{sl} > 4000$
Vijay et al. (1982)	$\psi_h = \left(\Delta P_{tpf}/\Delta P_l\right)^{0.451}$	$Nu_l = 1.615(\text{Re}_{sl}\,\text{Pr}_l\,d_c/L)^{1/3}(\mu_b/\mu_w)^{0.14}$ for $\text{Re}_{sl} < 4000$ $Nu_l = 0.0155\text{Re}_{sl}^{0.83}\,\text{Pr}_l^{0.5}(\mu_b/\mu_w)^{0.33}$ for $\text{Re}_{sl} > 4000$
Rezkallah (1987)	$\psi_h = (1-\varepsilon_g)^{-0.90}$	Sieder and Tate (1936) equations
Others	One can find the ψ_h from other correlations given in Table 9.2 multiplying by the factor $k_{tp}/(k_l Nu_l)$ where Nu_l can be calculated from Sieder and Tate's (1936) equations.	

$$Nu = \frac{hd_c}{k_l} \propto \rho_l^{0.78} \mu_l^{-0.833} \sigma_l^{0.145} u_j^{0.134} u_{rel}^{-0.08} d_c^{0.184} d_n^{0.992} H^{-0.062} \varepsilon_g^{0.08} \qquad (9.32)$$

The proportionality factor can be obtained from the experimental data as per design of experiment.

It was observed that the heat transfer coefficient in the inverse bubbly flow compared with the single-phase heat transfer coefficient calculated by Sieder and Tate (1936) equations at a condition of $Re = 10562.5$ is 4.95 times higher at its maximum operational condition at superficial liquid velocity of 0.0169 m/s. The correlation developed within a range of: $0.028 \leq u_{sl} \leq 0.0169$ m/s and controlled gas flow rate as $0.0085 \leq u_{sg} \leq 0.034$ m/s.

Mechanistic Models

To predict the heat transfer rates and coefficients in the bubbly flow condition, several correlations have been proposed as shown in Table 9.2. These correlations have been developed based on either experimental studies on heat transfer in bubble columns or based on bubble dynamics. Turbulence and mixing that are induced by gas bubbles play important role in heat and mass transfer in gas–liquid and gas–liquid-solid systems. The high heat transfer rate in multiphase flow systems, particularly for bubbly flow, is mainly due to bubble-induced turbulence (Yang et al., 2000, Kumar et al., 1992). Detailed critical review of the studies on heat transfer coefficient and bubble dynamics and hydrodynamics is available in literature (Fan, 1989). There is a close relation between the bubble dynamics and the heat transfer rate in two- and three-phase systems. Thus, there is a need to mechanistically asses how the heat transfer phenomenon in bubble columns is affected by different bubble properties that govern the flow behavior, including local gas hold-up, bubble passage frequency, bubble velocity, and bubble sizes, as well as their directions. Both experimental and theoretical results reported in the literature suggest that there is a series of film and surface renewal that governs the heat exchange between a heat transfer surface and flowing fluid adjacent to the surface (Karst, 1962; Wasan and Ahluwalia, 1969; Kumar and Fan, 1994; Yang et al., 2000).

Model Based on Film Theory

The film theory describes that the steady-state mass transfer and hence heat transfer occurs by molecular diffusion across a stagnant or laminar film at the interface between phases where the fluid is turbulent (Nernst, 1904). The mass transferred across a unit area of the interface per unit time is proportional to the concentration difference between the bulk fluid and the interface such that

$$J_m = -\frac{D}{\delta_e}(C_b - C_i) = -k(C_b - C_i) \qquad (9.33)$$

where D is the molecular diffusivity, δ_e is the effective film thickness, C_b is the average concentration in the bulk fluid and C_i is the average concentration in the interface, J_m is the rate of mass transfer across a unit area of the interface, and k is the the mass transfer coefficient. The coefficient of mass transfer is proportional to the diffusion coefficient and inversely proportional to the film thickness. This phenomenon is commonly known as the film theory. The form of heat transfer can be derived in a similar manner analogous to the mass transfer, which can be expressed as

$$J_h = -\frac{\alpha}{\delta_e}(T_b - T_i) = -h(T_b - T_i) \tag{9.34}$$

where α is the thermal diffusivity and T_b and T_i are the average temperature of the bulk fluid and of the interface, respectively. J_h is the rate of heat transfer per unit area, and h is the heat transfer coefficient. The mass transfer coefficient and heat transfer coefficient are then expressed as

$$k = \frac{D}{\delta_e} \tag{9.35}$$

$$h = \frac{\alpha}{\delta_e} \tag{9.36}$$

According to this model, there is a linear relationship between mass flux and the molecular diffusivity and hence the heat flux and the thermal diffusivity. The model also exaggerates the actual conditions near a phase boundary. Moreover, the theory is based on the assumption that mass or heat transfer occurs through a fluid layer (stagnant film) of a definite but unknown thickness around the phase boundary, which contains all the resistance to mass or heat transfer. The resistance depends on the velocity of the diffusive transport in each contacting phases, and it is localized near the interface between two stagnant liquid and gas films with finite thickness. There is a thermodynamic equilibrium between interfacial phase concentrations; thus, the interface itself does not represent a mass or heat transfer resistance. Beyond this film, concentration is homogeneous, and transfer through the film occurs at steady state. The difficulty with film theory models is that investigators are always faced with somehow predicting the fictitious film thickness. The concentration gradient of film theory will not develop if the exposure time of the phase is too short.

Model Based on Penetration Theory

Higbie (1935) proposed the penetration theory based on Fick's second law of diffusion. In this case, the transfer process occurs at an unsteady-state condition, which is not accounted for by the film theory. The transfer processes are assumed to occur during the repeated contacts of fluid element (gas/solid) with the liquid interface. Fresh liquid elements continually replace those interacting with the interface. During each contact period between the liquid element and the interface, mass/heat is transferred to the element. In this regard, the contact time of the small eddies with the interface is so short that it canot be interpreted by the steady-state characteristics. Therefore, the transfer of heat or mass is interpreted by unsteady-state process according to this model. Moreover, all the eddies are assumed to stay in contact with the interface for same length of time (θ) during which diffusion of heat or mass occurs into the eddy. During the time, θ, the liquid element is subjected to unsteady-state diffusion of penetration, which can be expressed for heat transfer as

$$\frac{\partial T}{\partial \theta} = \alpha \frac{\partial^2 T}{\partial x^2} \tag{9.37}$$

The proper boundary conditions are

$$T = T_i; \quad x = 0; \quad t \geq 0 \tag{9.38}$$

$$T = T_b; \quad x > 0; \quad t = 0 \tag{9.39}$$

$$T = T_b; \quad x = \alpha; \quad t > 0 \tag{9.40}$$

Integration of equations (9.37) to (9.40) can be done by means of Laplace transforms. The average heat transfer coefficient during the contact time between the fluid eddy and the heat transfer surface according to Higbie (1935) becomes

$$h_{avg} = \sqrt{\left(\frac{\alpha}{\pi\theta}\right)\rho C_p} \tag{9.41}$$

The limitation of Higbie's penetration theory is that the time of exposure for all eddies of particles of liquid is considered as constant, but in a real situation, it may not.

Model Based on Surface Renewal theory

The surface renewal theory is also called the modified penetration theory. It was developed by Dankwarts (1970). He pointed out that the time of contact or exposure is not constant as explained in penetration theory but varying lengths of time. Surface renewal theory has been applied to explain the heat transfer phenomena in bubbly flow system such as follows.

WASAN AND AHLUWALIA (1969) MODEL

Wasan and Ahluwalia (1969) proposed a model based on surface renewal mechanism. In this model, a uniform film was supposed to lie adjacent to the wall, and a mass of fluid was assumed to exchange heat by unsteady-state conduction at the outer edge of this film. In addition, the film is assumed thinner than that predicted by film theory because of the fluid convection and the motion of particles near the surface. This is explained in Figure 9.3. To apply this theory to predict heat transfer coefficient between a bubbling bed and the heat transfer surface, the following assumption can be made (Wasan and Ahluwalia, 1969):

- The temperature of the intefiace between the film and the fluid element changes with time.
- The temperature of the mass of fresh fluid as it swept the outer surface of the film is assumed uniform and equal to the bulk fluid temperature.
- For the instantaneous heat transfer rate to this fluid mass, there is no heat storage in the film.
- The bulk temperature of the bed is assumed to be constant.

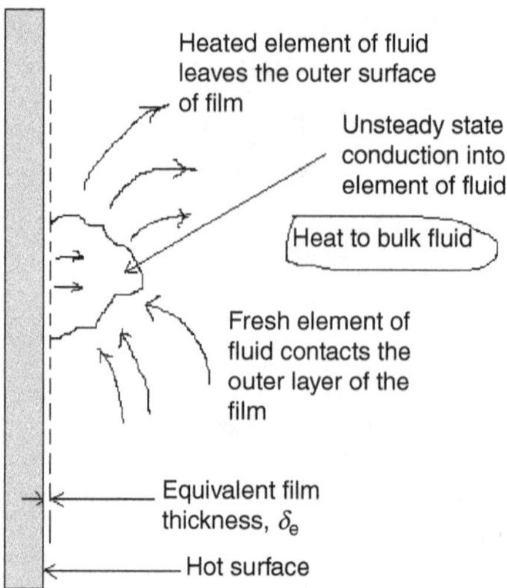

FIGURE 9.3 Main feature of surface renewal model proposed by Wasan and Ahluwalia (1969).

- Resistance to heat transfer is due to the film followed by penetration and unsteady-state heating of an element resting at the outer edge of the film.

From the second assumption, the heat flux on the surface of the heat transfer wall can be described as follows:

$$q = -k_l \left(\frac{dT}{dx} \right)_{x=0} = h(T_w - T_{x=0}) \tag{9.42}$$

where q is instantaneous heat flux, k_l is thermal conductivity, and x is the distance measured fiom the edge of the film. T_w and $T_{x=0}$ are wall and bulk temperature. h is the heat transfer coeficient, defined as $h = k_l/\delta_e$ where δ_e is the thickness of the thermal boundary layer. To describe instantaneous temperature profile, the unsteady balance equation (equation [9.37]) cm be applied. By integrating equation (9.37), Wasan and Ahluwalia (1969) obtained an expression of total heat transfer rate as

$$q_t = \int_0^\theta q\, d\theta = \frac{2k_l \theta^{1/2}}{\sqrt{\pi\alpha}}(T_w - T_b) + \frac{k_l \delta_e}{\alpha}(T_w - T_b)\left[\exp\left(\frac{\alpha\theta}{\delta_e^2}\right) \times \left(1 - erf\sqrt{\frac{\alpha\theta}{\delta_e}}\right) - 1\right] \tag{9.43}$$

The time-averaged heat transfer rate can then be obtained as

$$q_{ave} = \frac{1}{\theta}\int_0^\theta q\, d\theta = h_{ave}(T_w - T_b) \tag{9.44}$$

Combining equations (9.43) and (9.44), the average heat transfer coefficient can be obtained as

$$h_{ave} = \frac{2k_l}{\sqrt{\pi\alpha\theta}} + \frac{k_l \delta_e}{\alpha\theta}\left[\exp\left(\frac{\alpha\theta}{\delta_e^2}\right) \times \left(1 - erf\sqrt{\frac{\alpha\theta}{\delta_e}}\right) - 1\right] \tag{9.45}$$

The term $(\alpha\theta / \delta_e)$ accounts for the contribution of film resistance to heat transfer (Luo, 1997). If $\delta_e \to 0$, equation (9.45) can be written as

$$h_{ave} = \frac{2k_l}{\sqrt{\pi\alpha\theta}} = 2\left(\frac{\rho_l C_{pl} k_l}{\pi\theta}\right)^{1/2} \tag{9.46}$$

The contact time, θ, can be estimated by applying Kolmogoroff concept of isotropic turbulence (Deckwer, 1980), which can be given by

$$\theta = \left(\frac{\mu_l}{\rho_l \varepsilon}\right)^{1/2} \tag{9.47}$$

where ε is the energy dissipation rate per unit mass of fluid. The film thickness of thermal conductions, δ_e, is equivalent to the thickness of the diffusion sublayer and is related to the laminar viscous sublayer, δ_o, as

$$\delta_e = \frac{\delta_0}{\text{Pr}_l^{1/3}} \tag{9.48}$$

The thickness of laminar viscous sublayer δ_o depends on the geography of the surface of heat transfer source. The thickness of the laminar sublayer is defined as the distance from solid surface while velocity inside laminar layer is larger than 0.99 of bulk velocity. For the cylindrical probe, the thickness of laminar viscous sublayer at different angular locations with respect to the point of incidence can be expressed as (Schlichting, 1960)

$$\delta_0 = \frac{\varphi R_{probe}}{\text{Re}_{probe}^{1/2}} \tag{9.49}$$

where R_{probe} is the radius of the cylindrical probe, parameter φ depends on the angular location on the probe (Schlichting, 1979), and Re_{probe} is Reynolds number based on probe radius. If the heat flux sensor is located at an angle of 90 degrees in front of liquid flow. The parameter φ is to be 2.5 (Schlichting, 1979).

DECKWER (1980) MODEL

According to Deckwer (1980), the movement of bubble bring on an irregular flow of fluid eddies from the bulk fluid to the wall and vice versa, which results in the transport of heat. The occurrence of fast lateral exchange flow rates can be regarded as a lateral eddy diffusivity. Hence, it is impractical to assume the existence of a boundary layer at the wall. The fluid elements reside for a certain contact time at the surface and then leave it and enter the bulk fluid again. Thus, applying Higbie's penetration ideas to heat exchange with bubble-aerated flow in the fluid elements adjacent to the surface at unsteady-state condition with varying length of contact time will be more reasonable. Deckwer (1980) described the instanteneous instationary heat flux profile during the contact time θ of the fluid eddy at the interfacial area, which can be expressed by solving equations (9.37) to (9.40) as

$$q = \left(2\sqrt{\left(\frac{\alpha}{\pi\theta}\right)}\right)\rho C_p (T_i - T_b) \tag{9.50}$$

Comparison with the common mathematical expression of the heat transfer coefficient, it yields

$$h = 1.128 \left(\frac{k_l \rho_l C_{pl}}{\theta} \right)^{1/2} \tag{9.51}$$

Danckwerts (1970) showed a more realistic theory and introduced a normal distribution of contact times for the fluid elements. From this theory, it can be expressed as

$$h \sim (k_l \rho_l C_{pl} s)^{1/2} \tag{9.52}$$

where s is the Danckwerts' surface renewal parameter. The exposure time of a fluid element at the bubble surface can be written for inverse flowing bubble as

$$\theta = \frac{d_b}{u_{b,rel}} = \frac{d_b}{u_{sl}/(1-\varepsilon_g) - u_{bt}} \tag{9.53}$$

The bubble terminal rise velocity (u_{bt}) can be calculated from the correlation given by Clift et al. (1978). More information on bubble terminal rise velocity can be obtained from review article of Kulkarni and Joshi (2005). For upward bubbly flow, the contact time can be calculated as $d_b/(u_{sg}/\varepsilon_g)$ (Deckwer, 1980). Deckwer (1980) also reported that the contact time, θ, can be estimated by applying Kolmogoroff's concept of isotropic turbulence. The contact time of turbulent eddies at the heat exchange surface depends on the size of the micro-scale eddy governed by the intensity of the turbulence. The size of the micro-scale eddy is given by (Deckwer, 1980)

$$l_{ed} = \left(\frac{(\mu/\rho)^3}{\varepsilon} \right)^{1/4} \tag{9.54}$$

where ε is the energy dissipation rate per unit mass of fluid. He reported that in bubbly flow system, the moving bubbles or the wake behind them, respectively, can be considered as the large-scale eddies, which break down into micro-scale eddies because of turbulence, which extend their action toward the wall because of radial flow as proposed by Kast (1962). He pointed out that it is reasonable to assume that contact time introduced by application of the surface renewal theory is connected with the length and velocity scale of the micro eddies. According to him, the contact time can be expressed as (Deckwer, 1980)

$$\theta = \frac{l_{ed}}{u_{ed}} \tag{9.55}$$

where

$$u_{ed} = \left(\frac{\mu \varepsilon}{\rho} \right)^{1/4} \tag{9.56}$$

Substituting equations (9.55) and (9.56) in equation (9.51), the heat transfer coefficient can then be expressed as

$$h = 1.128 \left(\frac{k_l^2 \rho^3 C_p^2 \varepsilon}{\mu} \right)^{1/4}$$

(9.57)

JOSHI ET AL. (1980) MODEL

Joshi et al. (1980) developed the model based on the axial liquid circulation velocity in a bubble column. This approach does not require all the energy to be dissipated into micro-scale eddies. It does not predict the heat transfer coefficient at superficial gas velocity greater than 0.1 m/s. The average liquid circulation velocity is given by

$$u_{c,ave} = 1.31 \left\{ g d_c (u_{sg} - \varepsilon_g u_b) \right\}^{1/3}$$

(9.58)

The axial component of the liquid circulation velocity is

$$u_{c,ave} = 1.18 \left\{ g d_c (u_{sg} - \varepsilon_g u_b) \right\}^{1/3}$$

(9.59)

They used this velocity in the expression for heat transfer in pipe flow to estimate the value of heat transfer coefficient in a bubble column, giving the following equation

$$\frac{h d_c}{k_l} = 0.087 \left\{ \frac{d_c^{1.33} g^{1/3} (u_{sg} - \varepsilon_g u_{sg})^{1/3} \rho_l}{\mu_l} \right\}^{0.8} \left(\frac{C_{pl} \mu_l}{k_l} \right)^{1/3}$$

(9.60)

The shortcoming of the model is that it requires the knowledge of the exact amount of energy to be transferred to micro eddies as the gas velocity increases.

ZEHNER (1986) MODEL

Zehner (1986) proposed a model based on the concept of heat transfer through a boundary layer. The boundary layer exists at the places where the bubbles penetrate the thermal boundary layer. This is determined from the known mean bubble diameter and gas hold-up. He assumed the heat transfer for flow over a flat plate is analogous to the case of bubbly flow in a bubble column. At the boundary layer where the bubble penetrates, the heat transfer is inhibited because of the thermal conductivity of the gas being much lower than that of the liquid. He has considered the fact by assuming heat transfer coefficient proportional to liquid hold-up as

$$h = 0.18 \varepsilon_l \left(\frac{k_l^2 \rho_l^2 C_{pl} u_{ed}^2}{l_{b-b} \mu_l} \right)^{1/3}$$

(9.61)

where l_{b-b} is the mean distance between bubbles. The flow over the surface was assumed to have an eddy velocity u_{ed} given by

$$u_{ed} = \left\{ \frac{1}{2.5} \left(\frac{\rho_l - \rho_g}{\rho_l} \right) g d_c u_{sg} \right\}$$

(9.62)

He reported that the values of h by equation (9.61) become constant above a certain value of gas velocity because as the gas velocity increases, the liquid hold-up decreases, and the term within the bracket in equation (9.61) increases. As a result, the value of the heat transfer coefficient remains almost constant. Hence, the Zehner's boundary layer concept gives better prediction of the heat transfer coefficient compared with the surface renewal concept of Deckwer (1980). However, because of robust mixing in bubble columns, the surface renewal mechanism appears to be the more coherent.

VERMA (1989) MODEL

Verma (1989) explained the heat transfer in the bubble column by the mechanism of heat conduction through a thin layer of liquid between the surface and the bubble. Verma considered that at the places where the bubble is very close to the heat transfer surface, there is a very thin layer of liquid between the surface and the bubble as shown in Figure 9.4.

Verma developed the model based on the following assumptions:

- The heat transfer through this layer takes place only by conduction and not by the surface renewal mechanism.
- Heat is transferred through the fraction of the area of interface of the bubble and the heat transfer surface because the bubble change its shape according to the geometry of the surface.
- The fraction of the area not covered by the bubble is proportional to the liquid hold-up as $(1-\varepsilon_g)^n$.
- The heat transfer by conduction through the area covered by the bubble is negligible in comparison with heat transfer at other places.

The heat transfer at places not covered by the bubble can be expressed as

$$\frac{h}{\rho_l C_{pl} u_{sg}} = 0.1 \left(\frac{u_{sg}^3 \rho_l}{\mu_l g} \right)^{-1/4} \left(\frac{C_{pl} \mu_l}{k_l} \right)^{-1/2} \tag{9.63}$$

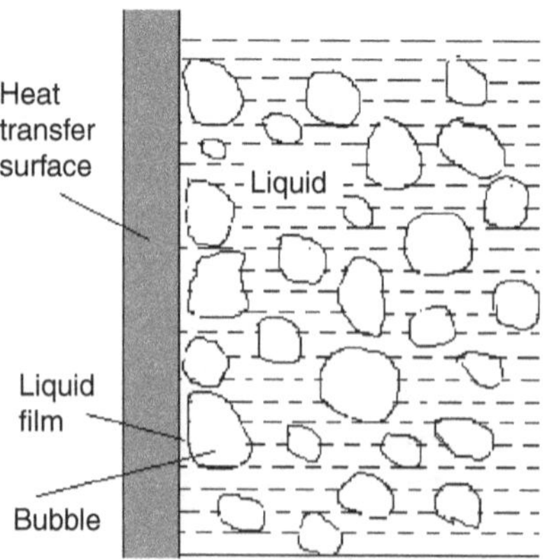

FIGURE 9.4 Heat conduction mechanism beyween bubble and heat transfer surface through thin liquid layer (as per Verma, 1989).

The heat transfer coefficient over the entire surface is therefore given by the following equation:

$$\frac{h}{\rho_l C_{pl} u_{sg}} \propto \varepsilon_l \left(\frac{u_{sg}^3 \rho_l}{\mu_l g} \right)^{-1/4} \left(\frac{C_{pl}\mu_l}{k_l} \right)^{-1/2}$$

(9.64)

YANG ET AL. (2000) MODEL

Yang et al. (2000) also used a consecutive film and surface renewal model to analyze their heat transfer results. On the basis of the model, they claimed that the main resistance to heat transfer in a bubble column lies within a fluid film surrounding the heating surface. However, they assumed that the liquid elements move at the same velocity as the bubbles around the heat transfer resistance film, and thus the contact time between the liquid elements and the film is equal to the contact time between the bubbles and the film when the bubble motion is considered as the driving force of the liquid elements.

Momentum, Heat, and Mass Transfer Analogy

To predict momentum and transport process of bubble flow, it is important to explain turbulent structure of the continuous liquid phase, which may result in how to describe the contribution of bubble existence to the flow characteristics (Sato and Sekoguchi, 1975). According to Sato and Sekoguchi (1975), the momentum in bubble flow is subdivided into the two components, one caused by the inherent liquid turbulence independent of relative motion of bubbles and the other caused by the additional turbulence caused by bubble agitation. Based on this concept of momentum transfer, it can be applied to heat and mass transfer problems. The turbulent heat flux is assumed to be subdivided into the two components, one caused by the inherent liquid turbulence and the other caused by bubble agitation. Let us consider a two-dimensional flow in a bubble column as shown in Figure 9.5 represented with a coordinate system.

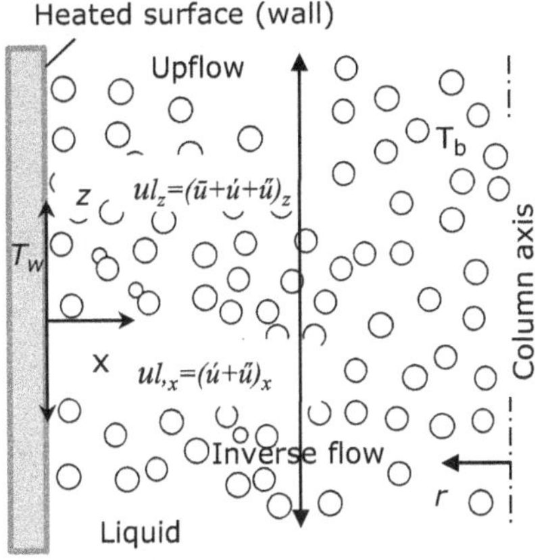

FIGURE 9.5 Two-dimensional flow system in a bubble column.

The z-axis is parallel to the main flow direction, and x- and r-axes are normal to it, considered from the wall and the column center, respectively. At any position (z, x) of the column, the liquid flow has velocity components (u_{lz}, u_{lx}) and a density ρ_l. Velocity fluctuation from the mean may be split further into the two parts on the assumption that there are two kind of turbulence in the liquid phase independent of and dependent on bubble agitation, designated as (u'_z, u'_x) and (u''_z, u''_x), respectively. According to Sato and Sekoguchi (1975), if it is assumed that gas bubbles can be treated as mere voidages, no momentum transfer takes place in the gas phase, and uses of liquid phase are sufficient to describe the flow in the bubble column. They derived the momentum equation in the z-direction in terms of mean liquid velocity and fluctuations and averaging with respect to time, which can be expressed as

$$
\tau = \underbrace{\mu_l \frac{d\bar{u}_{l,z}}{dx}}_{\text{effect of viscosity}} - \underbrace{\rho_l \overline{u'_{l,z} u'_{l,x}}}_{\substack{\text{additional stress} \\ \text{independent of bubble motion}}} - \underbrace{\rho_l \overline{u''_{l,z} u''_{l,x}}}_{\substack{\text{additional stress} \\ \text{dependent on bubble motion}}}
\tag{9.65}
$$

Taking the existence of gas hold-up into consideration, the total shear stress at any point can be written as

$$
\tau = (1 - \varepsilon_g)\left(\mu_l \frac{d\bar{u}_{l,z}}{dx} - \rho_l \overline{u'_{l,z} u'_{l,x}} - \rho_l \overline{u''_{l,z} u''_{l,x}} \right)
\tag{9.66}
$$

where the factor $(1 - \varepsilon_g)$ is regarded as the time-averaged volume fraction occupied by the liquid at the point. It can also be expressed by introducing the eddy diffusivities, e' and e'', for the two additional stresses by substituting

$$
\rho_l \overline{u'_{l,z} u'_{l,x}} = \rho_l e' \frac{d\bar{u}_{l,z}}{dx}
\tag{9.67}
$$

$$
\rho_l \overline{u''_{l,z} u''_{l,x}} = \rho_l e'' \frac{d\bar{u}_{l,z}}{dx}
\tag{9.68}
$$

Equation (9.66) yields after substitution of equations (9.67) and (9.68) as

$$
\tau = \rho_l (1 - \varepsilon_g)(v_l + e' + e'') \frac{d\bar{u}_{l,z}}{dx}
\tag{9.69}
$$

Or the velocity distribution (not an average) from equation (9.69) can be represented as

$$
\frac{du_{l,z}}{dx} = \frac{\tau}{\rho_l (1 - \varepsilon_g)(v_l + e' + e'')}
\tag{9.70}
$$

Equation (9.70) can be represented by the dimensionless form as

$$
\frac{du^*_{l,z}}{dx^*} = \frac{\tau^*}{(1 - \varepsilon_g)(v_l + e' + e'')/(Ru^*_l)}
\tag{9.71}
$$

where

$$u^*_{l,z} = u_{l,z}/u^f_l \tag{9.72}$$

$$x^* = x/R \tag{9.73}$$

$$\tau^* = \tau/\tau_w \tag{9.74}$$

$$u^f_l = \sqrt{(\tau_w/\rho_l)} \tag{9.75}$$

where τ_w is the wall shear stress, u^f_l is the friction velocity, and R is the radius of the bubble column. Sato et al. (1981) reported that it may be reasonable to postulate that even in bubbly flow condition, the wall turbulence is suppressed (i.e., $e' \sim 0$) when the liquid Reynolds numbers (based on bulk velocity) is lower than 2300, that is,

$$Re_l = \frac{\rho_l d_c u_{sl}}{(1-\varepsilon_g)\mu_l} < 2300 \tag{9.76}$$

Under such a condition, the velocity distribution (not an average) from equation (9.71) can be represented as

$$\frac{du^*_{l,z}}{dx^*} = \frac{\tau^*}{(1-\varepsilon_g)(v_l + e'')/(Ru^*_l)} \tag{9.77}$$

The fully developed turbulent motion in the bubble column occurs at a distance from the wall while a viscous sublayer exists in the immediate neighborhood of the wall. Therefore, the expressions for both the eddy diffusivities e' and e'' has to be examined properly. In Table 9.6, some recommended formulas are given to estimate the eddy diffusivities. The eddy diffusivities may be influenced by deformation of bubble near the wall. So in regard to the bubble size, it should be considered because the scale of eddies generated by bubbles must depend on the dimension of bubble. According to Sato et al. (1981), there are two kinds of bubbles, sliding bubble that slide on the wall and coring bubbles that flow away from the wall. The shape of a sliding bubble is different from an ellipsoid or a coring bubble. There may be a reduction of bubble dimension toward the wall. There is a thin gap between the wall surface and the end of bubble. The following empirical expression for the bubble size is recommended as an approximation of d_b by Sato et al. (1981).

$$d_b = \begin{cases} 0 & \text{if } 0 < x < 20 \,\mu m \\ 4x(\hat{d}_b - y)/\hat{d}_b & \text{if } 20 \,\mu m < x < \hat{d}_b/2 \\ \hat{d}_b & \text{if } \hat{d}_b/2 < x < R \end{cases} \tag{9.78}$$

where $\hat{d}_b$ is the cross-sectional mean diameter of the bubbles.

The description of the momentum transfer can be extended to the heat transfer as described follows. If temperature fluctuation of the liquid phase is assumed to be split into

Table 9.6 The Formula for Eddy Diffusivities

Reference	Formula for Eddy Diffusivities	Remarks
Reichardt (1951)	$$e' = \frac{\kappa R u_l^f}{6}\left(1 - r^{*2}\right)\left(1 + 2r^{*2}\right)$$ $$e'' = k_1 \varepsilon_g \left(\frac{d_b}{2}\right) u_{b,rel}$$ $r^* = r/R.\ k_1 = 1.2\text{-}1.4$ as reported by Sato et al. (1981) (depending on experimental condition)	Equation e' is valid for the region of $r^* < 0.9$. Equation for e'' taking account of the "drift" phenomena of liquid particles due to relative motion of gas bubbles. The theory is based on the assumption that the turbulence structure of the liquid phase consists of superposition of the two independent mechanisms, the wall turbulence and the free turbulence due to wakes of bubbles. Gives high value of e'' near the wall.
Prandtl (1942)	$e'' = \kappa_1 b u_{lmax}$; b is the width of the mixing zone and u_{lmax} is the maximum deficit velocity	Similar form to the well-known virtual kinematic viscosity of a free turbulent flow such as a wake behind a solid body
Schlichting (1979)	$$e' = \kappa^2 x^2 \frac{du}{dx} = \kappa x u_l^f$$	Not applicable close to the wall
van Driest (1956)	$$e' = \left\{1 - \exp\left(-\frac{x}{A}\right)\right\}^2 \kappa^2 x^2 \frac{du}{dx}$$ $$= \frac{v}{2}\left[-1 + \sqrt{\left(1 + 4\kappa^2 x^{+2}\left\{1 - \exp(-x^+/A^+)\right\}^2\right)}\right]$$ $x^+ = x u_l^f / v$; $A^+ = A u_l^f / v$ $A = \text{constant}$	Less accurate than the Reichardt (1951) formula in the determination of the velocity in the core region. It is unfit for the wall region; it tends to close with the Prandtl's (1942) expression when the wall is approached.
Sato et al. (1981)	$$e' = \left\{1 - \exp\left(-\frac{x^+}{A^+}\right)\right\}^2 \times$$ $$\left\{1 - \frac{11}{6}\left(\frac{x^+}{R^+}\right) + \frac{4}{3}\left(\frac{x^+}{R^+}\right)^2 - \frac{1}{3}\left(\frac{x^+}{R^+}\right)^3\right\} v_l \kappa x^+$$ $R^+ = R u_l^f / v$; $A^+ = 16$; $\kappa = 0.4$; $Re = 5 \times 10^4\text{-}5 \times 10^5$ $$e'' = \left\{1 - \exp\left(-\frac{x^+}{A^+}\right)\right\}^2 k_1 \varepsilon_g \left(\frac{d_b}{2}\right) u_{b,rel}$$ $k_1 = 1.2$, $A^+ = 16$	Combination of Reichardt (1951) and van Driest (1956). The combination of both formulas is expected to be suitable for the flow prediction of pipe flows over the entire cross-section (Sato et al., 1981).

the two independent components in a manner similar to the velocity fluctuations, the velocities and temperature can be expressed as

$$u_z = \bar{u}_z + u_z' + u_z'' \tag{9.79}$$

$$u_x = \bar{u}_x + u_x' + u_x'' \tag{9.80}$$

$$T = \bar{T} + T' + T'' \tag{9.81}$$

Sato et al. (1981) reported that introducing equations (9.79) to (9.81) into the equation of energy for a two-dimensional system and then forming averages with respect to time, the two additional heat fluxes are to $\rho_l C_p \overline{u'_{lx} T'}$ and $\rho_l C_p \overline{u''_{lx} T''}$. The eddy diffusivities for heat, e'_h and e''_h, are defined by

$$\rho_l C_p \overline{u'_{lx} T'} = -\rho_l C_p e'_h \frac{d\overline{T}}{dx} \tag{9.82}$$

$$\rho_l C_p \overline{u''_{lx} T''} = -\rho_l C_p e''_h \frac{d\overline{T}}{dx} \tag{9.83}$$

The total heat flux involving the heat conduction term can be written by

$$q = -\rho_l C_p (\alpha_l + e'_h + e''_h) \frac{d\overline{T}}{dx} \tag{9.84}$$

in which α_l is the thermal diffusivity of liquid. Applying this to the bubble flow problem, the following equation can be obtained, provided that gas bubbles can he treated as true voids.

$$q = -\rho_l C_p \varepsilon_l (\alpha_l + e'_h + e''_h) \frac{dT}{dx} \tag{9.85}$$

Both eddy diffusivities for heat transfer (e'_h, e''_h) are regarded as equal to those for momentum transfer. Based on the wall heat flux q_w and the temperature difference between the wall and the bulk $(T_w - T_{lb})$ as a reference, the following equations can be derived for the liquid temperature gradient in a dimensionless form:

$$\frac{dT_l^*}{dx^*} = -\frac{q^*}{\varepsilon_l \left\{ 1 + \mathrm{Pr}_l \left(\dfrac{e'_h + e''_h}{\nu_l} \right) \right\} \Big/ \dfrac{1}{2} Nu} \tag{9.86}$$

where

$$q^* = q/q_w \tag{9.87}$$

$$T_l^* = (T_l - T_{lb})/(T_w - T_{lb}) \tag{9.88}$$

$$h_{tp} = q_w/(T_w - T_{lb}) \tag{9.89}$$

$$Nu = h_{tp} d_c / k_l \tag{9.90}$$

Equation (9.86) is the basic equation in determining the liquid temperature distribution. The description of mass transfer in a bubble flow can be obtained simply by replacing the temperature and the heat flux in the above equations with the concentration of mass and the mass flux, respectively.

Correlation-Based Analogy of Heat and Mass Transfer

The effects of the operating variables on the heat and mass transfer are found to be a similar trend in bubbly flow system. The heat transfer coefficient in bubbly flow system is an increasing function of gas and liquid velocities but a decreasing function of the dynamic viscosity of the liquid. The gas–liquid mass transfer coefficient also increases with increasing gas and liquid velocities but decreases with increasing liquid viscosity. Some correlations are expressed in a similar fashion for heat and mass transfer. Lin et al. (1953) first proposed the heat- and mass-transfer analogy based on heat, mass, and momentum analogy in pipe flow. Based on Higbie's surface renewal model and Kolmogoroff's theory of isotropic turbulence, Deckwer (1980) proposed the equation in terms of dimensionless number for heat transfer as

$$St = 0.1 \left(\mathrm{Re}\, Fr\, \mathrm{Pr}^2 \right)^{-0.25} \tag{9.91}$$

Joshi et al. (1980) have used an analogy between bubble column and flow through circular pipes to predict the heat transfer coefficient in bubble columns based on the liquid circulation velocity in place of the fluid velocity. Zehner (1986) used the expression for heat transfer over a flat plate in a model based on the boundary layer concept. Analogous to this heat transfer correlation, Patil and Sharma (1983) reported the mass transfer correlation by replacing the coefficient 0.1 by 0.052 and 0.047, respectively, and Prandtl number by Schmidt number. Kim and Kang (1997) suggested similar correlations for heat and mass transfer, respectively, as

$$Nu = 0.436\, \mathrm{Pr}^{0.333}\, \mathrm{Re}_b^{0.196} \quad \text{(for heat transfer)} \tag{9.92}$$

$$Sh = 2.694\, \mathrm{Pr}^{0.333}\, \mathrm{Re}_b^{0.196} \quad \text{(for mass transfer)} \tag{9.93}$$

within the ranges of conditions u_{sg} up to 0.25 m/s, u_{sl} from 0.004 to 0.25 m/s, d_b from 1 to 8 mm, and μ_l from 0.001 to 0.0985 Pas. The Reynolds number based on bubble diameter (Re_b) is defined as $\mathrm{Re}_b = \rho_l d_b u_{sg}/\mu_l$. The Chilton-Colburn analogy between heat and mass transfer can also be used express the correlation for mass transfer coefficient by adjusting the coefficient using experimental data. Yasunishi et al. (1988) developed the correlations of heat and mas transfer coefficients based on the energy dissipation rate per unit mass of liquid according to Colburn J-factor in wall-to-bed heat and mass transfer process. The correlation for mass transfer coefficient in the wall-to-bed mass transfer coefficient can be expressed as

$$Sh = 0.13 \left(\frac{\rho_l \varepsilon^{1/3} d_c^{4/3}}{\mu_l} \right)^{3/4} Sc^{1/3} \tag{9.94}$$

Based on the analogy between the heat and mass transfer, equation (9.94) can be converted into the corresponding dimensionless groups for heat transfer as

$$Nu = 0.13 \left(\frac{\rho_l \varepsilon^{1/3} d_c^{4/3}}{\mu_l} \right)^{3/4} \mathrm{Pr}^{1/3} \tag{9.95}$$

Yasunishi et al. (1988) reported that the wall-to-bed heat transfer coefficients in slurry bubble column satisfied well with the correlation given by equation (9.95). Thus, it is evident that an analogy between heat and mass transfers exists in the wall-to-bed system. According to Lin et al. (1953), the mass-momentum analogy in pipe flow can be represented by (Verma, 2002)

$$Sh = \frac{1}{\phi_d}\left(\frac{f}{2}\right)\text{Re} \times Sc \tag{9.96}$$

where

$$\phi_d = 1 + \left(\frac{f}{2}\right)^{1/2}\left[a_1 Sc^{2/3} F_d + a_2 \ln\left(\frac{1 + a_3 Sc}{a_4(1 + a_5 Sc)}\right) - a_6\right] \tag{9.97}$$

$$F_d = \frac{1}{2}\ln\left(\frac{(1 + a_7 Sc^{1/3})^2}{1 - a_7 Sc^{1/3} + a_7^2 Sc^{2/3}}\right) + \sqrt{3}\tan^{-1}\frac{2a_7 Sc^{1/3} - 1}{\sqrt{3}} + \frac{\pi\sqrt{3}}{6} \tag{9.98}$$

The coefficients a_1 to a_7 are experimentally obtained by fitting the experimental data. The values for upflow bubbly system are 14.5/3, 5, 5.64, 6.64, 0.041, 4.77, and 5/14.5, respectively. One can obtain the value in the same fashion for an inverse bubbly flow system. In the case of heat transfer, analogous equations may be written according to Verma (2002) as

$$Nu = \frac{1}{\phi_h}\left(\frac{f}{2}\right)\text{Re} \times \text{Pr} \tag{9.99}$$

in which ϕ_h is obtained by substituting Pr in place of Sc in equations (9.97) and (9.98). In the case of bubble columns, fluid velocity can be replaced by circulation velocity for upward bubbly flow or by simply inverse liquid velocity for inverse bubbly flow systems. The Stanton number for heat transfer in a bubble column may be expressed as ($Nu/(Re.Pr)$).

Nomenclature

A Heat transfer area (m²)
A Cross-sectional area
A_f Free area fraction (-)
A_R Area ratio of nozzle to column diameter (-)
C Concentration (mol/m³)
C_i Concentration in the interface (mol/m³)
C_p Specific heat capacity (J/kg K)
D Molecular diffusivity (m²/s)
d Diameter (m)
d_b Cross-sectional mean diameter of the bubbles (m)
e Eddy diffusivity (m²/s)
E_{wk} Bubble wake enhancement factor (-)
f Friction factor (-)

Fr Froude number (u^2/gd) (-)
h Heat transfer coefficient (W/m^2K)
H Height (m)
H_{rn} Ratio of height to nozzle diameter (-)
J_h Rate of heat transfer across a unit area of the interface $(J/m^2.s)$
J_m Rate of mass transfer across a unit area of the interface $(kg/m^2.s)$
k Mass transfer coefficient (m/s)
K Thermal conductivity (W/mK)
l Mean distance between bubbles (m)
l_{b-b} Mean distance between bubbles (m)
l_{ed} Length scale of the eddy (m)
m Mass (kg)
Mo Morton number $(g\mu_l^4/\rho_l\sigma_l^3)$ (-)
n Number of sample station, coefficient
Nu Nusselt number $(hd_c/k_l,$ unless not defined) (-)
P Pressure (N/m^2)
Pr Prandtl number $(C_{pl}\mu_l/k_l,$ unless not defined) (-)
P_v Power per unit volume $(N/m^2.s)$
q Heat energy (J), heat flux $(J/m^2/s)$
r Radial distance (m)
R Radius (m)
Re Reynolds number (-)
Re_{ln} Reynolds number $(\rho_l d_n u_{sl}/\mu_l)$ (-)
Re_{probe} Reynolds number based on probe radius (-)
s Danckwerts' surface renewal parameter (-)
Sc Schmidt number $(\mu_l/(\rho_l D))$ (-)
St Stanton number $(h/(u\rho C_p$ or $Nu/(Re.Pr))$ (-)
Su_n Suratmann number based on the nozzle diameter, $(\rho_l\sigma_l d_n)/\mu_l^2$ (-)
T Temperature (K)
t Time (s)
$\bar{t}$ Mean residence time of bubble (s)
u Velocity (m/s)
u_l^f Friction velocity (m/s)
u_{ed} Velocity scale of the micro eddy (m/s)
V_R Total internal volume of the reactor (m_3)
x Horizontal axial distance (m)
z Vertical axial distance (m)

Greek Letters

φ Parameter; depends on angular location on probe (-)
τ Shear stress (N/m^2)
α Thermal diffusivity (m^2/s)
ρ Density (kg/m^3)
ε Energy dissipation rate per unit mass (N-m/kg)
ν Kinematic viscosity (m^2/s)
θ Length of time (s)
ϕ Parameter (-)
σ Surface tension (N/m)

μ Viscosity (Pa-s)
δ_e Effective film thickness (m)
ε_g Gas hold-up (-)
ψ_h Ratio of two-phase to single-phase heat transfer coefficient (= h_{tp}/h_l) (-)
ε_l Liquid hold-up (-)
δ_o Thickness of laminar viscous sublayer (m)

Subscripts

ave average
b bubble, bulk
c column, centerline, circulation
d for mass transfer
ed eddy
g gas
h heat transfer
j jet
l liquid
m mixture
max maximum
n nozzle
ov overall
p particle
rel relative
s surface, source
sg superficial gas
sl superficial liquid, slurry
t turbulent, terminal
tp two-phase
tt turbulent-turbulent
v volumetric
w wall

Superscripts

$'$ independent of bubble agitation
$''$ dependent on bubble agitation
f friction

References

Aggour, M.A., 1978. Hydrodynamics and heat transfer in two-phase two-component flow, PhD Thesis. University of Manitoba, Canada (cited by Ghajar, 2005).

Baker, C.G.J., Armstrong, E.R., Bergougnou, M.A., 1978. Heat transfer in three-phase fluidized beds. Powder Technol. 21, 195–204.

Burkel, W., 1972. Der, warmeubergang an heiz- und kuhlflachen in begasten flussigkeiten. (Heat transfer at heating and cooling surfaces in gassed liquids). Chem. Ing. Tech. 44, 265–268.

Chen, W., Hasegawa, T., Tsutsumi, A., Otawara, K., Shigaki, Y., 2003. Generalized dynamic modeling of local heat transfer in bubble columns. Chem. Eng. J. 96, 37–44.

Cho, Y.J., Woo, K.J., Kang, Y., Kim, S.D., 2002. Dynamic characteristics of heat transfer coefficient in pressurized bubble columns with viscous liquid medium. Chem. Eng. Process 41, 699–706.

Clift, R., Grace, J.R., Weber, M.E., 1978. Bubbles, Drops and Particles. Academic Press, New York.

Danckwerts, P.V., 1970. Gas–Liquid Reactions. McGraw-Hill, New York.

Deckwer, W.-D., 1980. On the mechanism of heat transfer in bubble column reactors. Chem. Eng. Sci. 35 (6), 1341–1346.

Deckwer, W.D., 1992. Bubble Column Reactors. Wiley, New York.

Deckwer, W.-D., Louisi, Y., Zaidi, A., Ralek, M., 1980. Hydrodynamic properties of the Fischer–Tropsch slurry process. Ind. Eng. Chem. Process Des. Dev. 19 (4), 699–708.

Dorrestejin, W.R., 1970. Experimental study of heat trasnfer in upward and downward two-phase flow of air and oil through 70 mm tubes. Proceedings of Fourth International Heat Transfer Conference, B., 5., 9, vol. 5, pp. 1–10 (cited by Ghajar, 2005).

Duduković, M.P., Larachi, F., Mills, P.L., 2002. Multiphase catalytic reactors: a perspective on current knowledge and future trends. Catal. Revi. 44, 123–246.

Fair, J.R., Lambright, A.J., Anderson, J.W., 1962. Heat transfer and gas holdup in a sparged contactor. Ind. Eng. Chem. Process De. Dev. 1, 33–36.

Fan, L.S., 1989. Gas–Liquid–Solid Fluidization Engineering. Butterworths, Boston.

Fazeli, A., Fatemi, S., Ganji, E., Khakdaman, H.R., 2008. A statistical approach of heat transfer coefficient analysis in the slurry bubble column. Chem. Eng. Res. Des. 86 (5), 508–516.

Ghajar, A.J., 2005. Non-boiling heat transfer in gas-liquid flow in pipes – a tutorial. J. Braz. Soc. Mech. Sci. Eng XXVII (1), 46–73.

Hart, W.F., 1976. Heat transfer in bubble-agitated systems. A general correlation. Ind. Eng. Chem. Process De. Dev. 15, 109–114.

Higbie, R., 1935. The rate of absorption of a pure gas into still liquid during short periods of exposure. Trans. Am. Inst. Chem. Eng. 31, 365–389.

Hikita, H., Asai, S., Kikukawa, H., Zaike, T., Ohue, M., 1981. Heat transfer coefficient in bubble columns. Ind. Eng. Chem. Process Des. Dev. 20, 540–545.

Jamialahmadi, M., Müller-Steinhagen, H., 2000. Hydrodynamics and heat transfer of liquid fluidized bed systems. Chem. Eng. Commun. 179, 35–79.

Jhawar, A.K., Prakash, A., 2007. Analysis of local heat transfer coefficient in bubble column using fast response probes. Chem. Eng. Sci. 62, 7274–7281.

Jhawar, A.K., Prakash, A., 2011. Influence of bubble column diameter on local heat transfer and related hydrodynamics. Chem. Eng. Res. Des. 89, 1996–2002.

Joshi, J.B., Sharma, M.M., Shah, Y.T., Singh, C.P.P., Ally, M., Klinzing, G.E., 1980. Heat transfer in multiphase contactors. Chem. Eng. Commun. 6, 257–271.

Kagumba, M. O. O., 2013. Heat transfer and bubble dynamics in bubble and slurry bubble columns with internals for Fischer–Tropsch synthesis of clean alternative fuels and chemicals, Doctoral Dissertations. Paper 2032.

Kang, Y., Suh, I.S., Kim, S.D., 1985. Heat transfer characteristics of three phase fluidized beds. Chem. Eng. Commun. 34, 1–13.

Kantarci, N., Ulgen, K.O., Borak, F., 2005. A study on hydrodynamics and heat transfer in a bubble column reactor with yeast and bacterial cell suspensions. Can. J. Chem. Eng. 83, 764–773.

Kast, W., 1962. Analyse Des Warmeubergang in Blasensaulen. Int. J. Heat Mass Transfer 5, 329–336.

Kato, Y., Uchida, K., Kago, T., Morooka, S., 1981. Liquid holdup and heat transfer coeficient between bed and wall in liquid–solid and gas–liquid–solid fluidized beds. Powder Technol. 28, 173–179.

Kawase, Y., Moo-Young, M., 1987. Heat transfer in bubble column reactors with Newtonian and non-Newtonian fluids. Chem. Eng. Res. Des. 65, 121–126.

Kim, D., Ghajar, A.J., Dougherty, R.L., 2000. Robust heat transfer correlation for turbulent gas-liquid flow in vertical pipes. J. Thermophys. Heat Transfer 14 (4), 574–578.

Kim, H.S., Kim, J.H., Lee, C.G., Kang, S.H., Woo, K.J., Jung, H.J., Kim, D.W., 2014. Bubble and heat transfer phenomena in viscous slurry bubble column. Adv. Chem. Eng. Sci. 4, 417–429.

Kim, S.D., Laurent, A., 1991. The state of knowledge on heat transfer in three-phase fluidized beds. Int. Chem. Eng. 31, 284–302.

Kim, S.D., Kang, Y., 1997. Heat and mass transfer in three-phase fluidized-bed reactors an overview. Chem. Eng. Sci. 52, 3639–3660.

King, C.D.G., 1952. Heat transfer and pressure drop for an air–water mixture flowing in a 0.737 inch I.Dh horizontal tube, MS Thesis. University of California, Berkeley, CA (cited by Ghajar, 2005).

Knott, R.F., Anderson, R.N., Acrivos, A., Petersen, E.E., 1959. An experimental study of heat transfer to nitrogen-oil mixtures. Ind. Eng. Chem. 51 (11), 1369–1372.

Kolbel, H., Siemes, W., Mass, R., Muller, K., 1958. Warmeubergang an Blasensaulen. Chem. Eng. Tech. 30, 400–404.

Konsetov, V., 1966. Heat transfer during bubbling of gas through liquid. Int. J. Heat Mass Transfer 9 (10), 1103–1108.

Korte, H., 1987. Heat transfer in bubble columns with and without internals, PhD Thesis. University of Dortmund.

Kulkarni, A.A., Joshi, J.B., 2005. Bubble formation and bubble rise velocity in gas–liquid systems: a review. Ind. Eng. Chem. Res. 44, 5873–5931.

Kumar, S., Fan, L.S., 1994. Heat-transfer characteristics in viscous gas–gas–liquid–solid systems. AIChE J. 40, 745–755.

Kumar, S., Kusakabe, K., Raghunathan, K., Fan, L.S., 1992. Mechanism of heat transfer in bubbly liquid and liquid-solid systems: single bubble injection. AIChE J. 38, 733–741.

Li, H., Prakash, A., 1997. Heat transfer and hydrodynamics in a three-phase slurry bubble column. Ind. Eng. Chem. Res. 36, 4688–4694.

Li, H., Prakash, A., 2001. Survey of heat transfer mechanisms in a slurry bubble column. Can. J. Chem. Eng. 79, 717–725.

Li, H., Prakash, A., 2002. Analysis of flow patterns in bubble and slurry bubble columns based on local heat transfer measurements. Chem. Eng. J. 86, 269–276.

Li, H., Prakash, A., Margaritis, A., Bergougnou, M.A., 2003. Effects of micron-sized particles on hydrodynamics and local heat transfer in a slurry bubble column. Powder Technol. 133, 171–184.

Lin, C.S., Moulton, R.W., Putnam, G.L., 1953. Mass transfer between solid wall and fluid stream. Ind. Eng. Chem. 45, 636.

Lin, T.J., Hung-Tzu, C., 2003. Effects of macroscopic hydrodynamics on heat transfer in a three-phase fluidized bed. Catal. Today, 159–167, 79-80.

Luo, X., Jiang, P., Fan, L.-S., 1997. High-pressure three-phase fluidization: hydrodynamics and heat transfer. A. I. Ch. E. J. 43, 2432–2444.

Luo, X., Lee, D.J., Lau, R., Yang, G.Q., Fan, L.S., 1999. Maximum stable bubble size and gas holdup in high-pressure slurry bubble columns. AIChE J. 45, 665–680.

Magiliotou, M., Chen, Y.-M., Fan, L.-S., 1988. Bed-immersed object heat transfer in a three-phase fluidized bed. A. I. Ch. E. J. 34, 1043–1047.

Majumder, S.K., Kundu, G., Mukherjee, D., 2005. Mixing mechanism in a modified co-current downflow bubble column. Chem. Eng. J. 112, 45–55.

Martin, B.W., Sims, G.E., 1971. Forced convection heat transfer to water with air injection in a rectangular duct. I&EC Fund. 14, 1115–1134.

Michele, V., Hempel, D.C., 2002. Liquid flow and phase holdup—measurement and CFD modeling for two- and three-phase bubble columns. Chem. Eng. Sci. 57, 1899–1908.

Muroyama, K., Fukuma, M., Yasunishi, A., 1984. Wall-to-bed heat transfer coefficient in gas–liquid–solid fluidized beds. Can. J. Chem. Eng. 62, 199–208.

Muroyama, K., Okumichi, S., Goto, Y., Yamamoto, Y., Saito, S., 2001. Heat transfer from immersed vertical cylinders in gas–liquid and gas–liquid–solid fluidized beds. Chem. Eng. Technol. 24 (8), 835–842.

Nernst, W.Z., 1904. Theorie der reaktions geschwindigkeit in heterogenen systemen (Theory of reaction speed in heterogeneous systems). Phys. Chem. 47, 52–55.

Patil, V.K., Sharma, M.M., 1983. Solid–liquid mass transfer coefficient in bubble columns up to one metre diameter. Chem. Eng. Res. Des. 61, 21.

Prakash, A., Margaritis, A., Li, H., Bergougnou, M.A., 2001. Hydrodynamics and local heat transfer measurements in a bubble column with suspension of yeast. Biochem. Eng. 9, 155–163.

Prandtl, L., 1942. Bemerkungen zur Theorie der freien Turbulenz. ZAMM 22, 241–243, (cited by Sato et al., 1981).

Quiroz, I., Herrera, I., Gonzales-Mendizabal, 2003. Experimental study on convective coefficients in a slurry bubble column. Int. Comm. Heat Mass Transfer 30 (6), 775–786.

Reichardt, H., 1951. Vollst/indige Darstellung der turbulenten Geschwindigkeitsverteilung in glatten Leitungen. ZAMM 31, 208–219 (cited by Sato et al., 1981).

Rezkallah, K.S., 1987. Heat transfer and hydrodynamics in two- phase two-component flow in a vertical tube, PhD Thesis. University of Manitoba, Canada (cited by Ghajar, 2005).

Robert, F.B., 2003. Direct contact heat transfer. In: Bejan, A., Kraus, A.D. (Eds.), Heat Transfer Handbook. John Wiley & Sons, Inc, New York, p. 1374.

Rozenblit, R., Gurevich, M., Lengel, Y., Hetsroni, G., 2006. Flow patterns and heat transfer in vertical upward air–water flow with surfactant. Int. J. Multiphase Flow 32 (8), 889–901.

Sato, Y., Sekoguchi, K., 1975. Liquid velocity distribution in two-phase bubble flow. Int. J. Multiphase Flow 2, 79–95.

Sato, Y., Sadatomi, M., Sekoguchi, K., 1981. Momentum and heat transfer in two-phase bubble flow-I: theory. Int. J. Multiphase Flow 7, 167–177.

Saxena, S., Patel, B., 1991. Heat transfer investigations in a bubble column with immersed probes of different diameters. Int. Commun. Heat Mass Transfer 18 (4), 467–478.

Saxena, S.C., 1989. Heat transfer from a cylindrical probe immersed in a bubble column. Chem. Eng. J. 41, 25–39.

Saxena, S.C., Chen, Z.D., 1994. Hydrodynamics and heat transfer of baffled and unbaffled slurry bubble columns. Rev. Chem. Eng. 10 (3-4), 193–400.

Saxena, S.C., Rao, N.S., Saxena, A.C., 1990a. Heat transfer from a cylindrical probe immersed in a three-phase slurry bubble column. Chem. Eng. J. 44, 141–156.

Saxena, S.C., Patel, B.B., 1990b. Heat transfer and hydrodynamic investigations in a baffled bubble column: air-water-glass bead system. Chem. Eng. Commun. 98, 65–88.

Saxena, S.C., Rao, N.S., Saxena, A.C., 1992. Heat transfer and gas holdup studies in a bubble column: air-water-sand system. Can. J. Chem. Eng. 70, 33–41.

Saxena, S.C., Vandivel, R., Saxena, A.C., 1989. Gas holdup and heat transfer from immersed surfaces in two- and three- phase systems in bubble columns. Chem. Eng. Commun. 85, 63–83.

Schlichting, H., 1960. Boundary layer theory. McGraw-Hill Publication House, New York.

Schlichting, H., 1979. Boundary layer theory, seventh ed. McGraw-Hill, New York, p. 588.

Schlüter, S., Steiff, A., Weinspach, P.-M., 1995. Heat transfer in two- and three-phase bubble column reactors with internals. Chem. Eng. Process. 34, 157–172.

Serizawa, A., Kataoka, I., Michiyoshi, I., 1975. Turbulence structure of air–water bubbly flow—iii. transport properties. Int. J. Multiphase Flow 2, 247–259.

Shah, M.M., 1981. Generalized Prediction of heat transfer during two component gas–liquid flowin tubes and other channels. AIChE Symp. Ser. 77, 140–151.

Sieder, E.N., Tate, G.E., 1936. Heat transfer and pressure drop of liquids in tubes. Ind. Eng. Chem. 28 (12), 1429–1435.

Suh, I.S., Deckwer, W.D., 1989. Unified correlation of heat transfer coefficient in three-phase fluidized beds. Chem. Eng. Sci. 44, 1455–1 458.

Tow, E.W., Lienhard, V.J.H., 2014. Heat transfer to a horizontal cylinder in a shallow bubble column. Int. J. Heat Mass Transfer 79, 353–361.

van Driest, E.R., 1956. On turbulent flow near a wall. J. Aeron. Sci. 1036, 1007–1011 (cited by Sato et al., 1981).

Verma, A.K., 1989. Heat transfer mechanism in bubble columns. Chem. Eng. J. 42, 205–208.

Verma, A.K., 2002. Heat- and mass-transfer analogy in a bubble column. Ind. Eng. Chem. Res. 41, 882–884.

Vijay, M.M., Aggour, M.A., Sims, G.E., 1982. A correlation of mean heat transfer coefficients for two-phase two-component flow in a vertical tube. Proceedings of Seventh International Heat Transfer Conference, vol. 5, pp. 367–372 (cited by Ghajar, 2005).

Wasan, D.T., Ahluwalia, M.-S., 1969. Consecutive film and surface renewal mechanisrn for heat or mass transfer fiom the wall. Chem Eng. Sci. 24, 1535–1542.

Wilkinson, P.M., Spek, A.P., van Dieredonck, L.L., 1992. Design parameters estimation for scale-up of high-pressure bubble columns. AIChE J. 38, 544–554.

Wu, C., Al-Dahhan, M., Prakash, A., 2007. Heat transfer coefficients in a high pressure bubble columns. Chem. Eng. Sci. 62, 140–147.

Yang, G.Q., Luo, X., Lau, R., Fan, L.-S., 2000. Heat transfer characteristics in slurry bubble columns at elevated pressures and temperatures. Ind. Eng. Chem. Res. 39, 2568–2577.

Yasunishi, A., Fukwna, M., Muroyama, K., 1988. Wall-to-liquid mass transfer in packed and fluidized beds wi th gas–liquid concurrent upflow. J. Chem. Eng. Jpn. 21, 522–528.

Yu, Y.H., Kim, S.D., 1991. Bubble properties and local liquid velocity in the radial direction of cocurrent gas-liquid flow. Chem. Eng. Sci. 46, 313–320.

Zehner, P. Momentum, 1986. Mass and heat transfer in bubble columns. Part 2. Axial blending and heat transfer. Int. Chem. Eng. 26, 29–35.

Suggestions for Further Study

Today many industries are facing interfacial challenges because of complex phenomena of gas–liquid or gas–liquid-solid flow in multiphase reactors for different chemical and biochemical processes. A strong need exists for new and innovative concepts to achieve hydrodynamics and transport phenomena in multiphase process. The inverse bubbly flow system is one of the ways to intensify the transport process to get more interfacial area and increased residence time of dispersed phase in a bubbly flow device. This book focuses on some hydrodynamics and transport process in the inverse bubbly flow system relative to the conventional bubbly flow system. However, more studies on the inverse bubbly flow system are required for further understanding and the industrial application of the system. Some recommendations for further study are as follows:

- During the past three decades, fair progress in understanding the flow pattern in conventional bubbly flow system has been made. There is a lack of studies on the details of flow patterns in the inverse bubbly flow condition. Additional work is needed on the flow pattern for inverse bubbly flow conditions at different operating conditions. Because most of the industrial process is carried out with different non-Newtonian liquids, a study of the flow pattern mechanism with the non-Newtonian liquid is required. Detailed study of the flow pattern map in the inverse bubbly flow system is required to establish to infer optimized process yield. The transition of the flow pattern is a crucial factor to design the bubbly flow reactor at its optimized operating conditions for better yield of the transport processes. More studies on the stability of the inverse bubbly flow system are required. The computational fluid dynamics (CFD) analysis of the flow patterns are also suggested for further study.
- The gas entrainment for inverse flow of bubble is a primary requirement to the bubble flow in the inverse direction. The different mechanisms of the entrainment of gas bubble are yet not fully understood. The physics of the entrainment by different techniques should be studied in details with different Newotonian and non-Newtonian liquids. The optimization of gas entrainment is recommended for further study. Numerical analysis of the entrainment process of the gas is also required. The effect of distributor design on the flow pattern and the entrainment efficiency needs to be investigated. Further studies should include the modeling of the distributor region and the break-up region at the column top where gas starts to entrain in the liquid column.
- More studies of gas hold-up characteristics with different types of non-Newtonian liquid and its analysis with different models are required. Studies on the radial distribution of the gas hold-up with different sizes of column are scarce regarding the inverse bubbly flow system. Futher study is required for the distribution of the gas

Hydrodynamics and Transport Processes of Inverse Bubbly Flow. http://dx.doi.org/10.1016/B978-0-12-803287-9.00010-2

hold-up because it is important to optimize the distribution of gas in the gas–liquid reactive process.

- Different empirical models are available to analyze the frictional pressure drop in the bubbly flow system. The models, however, can be modified to analyze the frictional pressure drop in inverse bubbly flow system. It is required to develop a comprehensive mechanistic model to analyze the frictional pressure drop. The estimation of drag coefficient, even for a single bubble, is difficult in the case of inverse flow of the bubble. This is because of the uneven energy distribution of the phases around the bubble.

- The experimental techniques for measurement of gas hold-up, pressure drop, bubble size, mixing characteristics including the use of Pitot tubes, hot-wire anemometry, laser-Doppler anemometry, photography, particle image velocimetry, laser sheeting, computer-automated radioactive particle trackin,g and pressure cross-correlation methods are described in the respective chapters. There is a strong need to undertake measurements throughout the column, right from the gas entranment region upto the bottom zone and over a wide range of column diameters. A spectrum of gas–liquid systems should be covered within a wide range of slip and relative velocity. Furthermore, Reynolds stress, energy estimation of spectra, and various space and time correlations should be studied, which will provide a very valuable base for understanding the relations between the flow pattern and the design parameters.

- Phase mixing in inverse bubbly flow is analyzed with axial dispersion models. There is a lack of studies on the radial dispersion coefficient in the invesre bubbly flow column with large diameter. The intensity of the mixing in the inverse bubbly flow column is a significant order higher than that the conventional bubble column. However, more studies are required with differen types of Newtonian and non-Newtonian liquids. The effects of the physical properties and the flow pattern on the mixing mechanism are yet to be studied in details. There is no numerical analysis of the mixing process in the inverse bubbly flow system. A CFD approach is required to analyze the mixing process in the inverse bubbly flow column. The mixing time, residence time distribution, pressure drop, bubble size distribution, interfacial area transport, and wall heat and mass transfer coefficient to be analyzed by the CFD models.

- More detailed study is required when estimating the nonadjustible operating conditions on the characteristics determining dispersion coefficient and the transport coefficient and the rate of heat and mass transfer between the phase during the inverse flow of bubble against its buyancy. These quantities all depend in a complex manner on geomatric size, gas entrainment, and phase velocity as well as on the physical properties, especially viscosity and surface tension of liquid. Theoretical correlations are based solely on the experimental results and may not be suitable for a specific applications beyond a range within which the correlation is made. The question of most interest for practical purpose is which whether the correlation is suitable over the other range of operating conditions. The correlation sometimes cannot be recommended unless it is based on the measurements and material systems investigated from the same material system using aparatus of the same geometric size and in particular with the same gas distributor. Therefore, physical models are required to develop to predict the accurate physics of the transport phenomena of the inverse bubbly flow system. The correlation that is made should be checked by a sensitivity analysis, which involves simulating the sensitivity of the target reactor features such as conversion, space time, yield to variations, and transport processes.

- Studies on the heat transfer phenomena in the inverse bubbly flow system with different liquids and gases are very scanty. More study is required with various liquids and with wide range of temeparture and pressure.

- Various process involve gas, liquid, and solid. The work with suspended solid or slurry in the inverse bubbly flow system is yet to be studied in detail. The study available is only about some hydrodynamics and mass transfer processes. More studies are required with different types of fluid system. The different hydrodynamics and transport processes with a slurry system are yet to studied.

- There is a lack of development in the formulation of basic equations for the two-phase gas–liquid turbulent flows in inverse bubbly flow systems. Research is needed for the modeling of interface forces, higher order correlations between phase hold-ups, phase velocities, and turbulent dispersion in the phases in case of inverse flow of bubble.

- The CFD simulation of the inverse bubbly flow is not yet reported. The correspondence between the CFD simulation and the real gas–liquid system is not available in the literature. The simulation experiments are recommended to performe in a very wide range of column diameter, distributor design, and other various operating conditions. Parametric sensitivity of turbulence parameters is an important aspect in the numerical investigation of the bubbly flow pattern.

- The quantitative effects of superficial phase velocity, column diameter, and the slip/ relative velocity on hydrodynamics of inverse flow bubble columns is recommended to be thoroughly investigated by the CFD simulation. It is very important to establish complete correspondence between the conditions of numerical simulation and the experimental measurements.

- In the case of inverse bubbly flow, the levels of shear rate and turbulence in the vicinity of any bubble is much different than a single isolated bubble. Under these conditions, the mechanism of momentum transfer is not well understood. One of the problems is the boundary condition at the interface. The quantitative effect depends on boundary conditions. Moreover, in the inverse bubbly flow regime, there is a possibility of a wide bubble size distribution because of distribution of the bubble based on its balance of buyancy aganist liquid momentum. The bubble also experiences radial forces. For ellipsoidal or spherical cap bubbles, the drags experienced in the vertical and horizontal directions are markedly different. This difference has not been quantified either experimentally or theoretically. All of the above issues need further investigation. When a bubble moves in the field of transverse velocity and pressure gradients, it experiences a radial force. A systematic analysis of this problem is needed, particularly when the flow is turbulent and the bubble is nonspherical or is in a wide distribution of bubble.

- The gas bubbles repeatedly undergo coalescence, break-up, and dispersion in the vicinity of liquid jet plunging region. In addition, bubble–bubble interaction may change the hydrodynamic characteristics. Mass transfer may result in bubble shrinkage or inflation caused by absorption or stripping, respectively. Furthermore, in the case of tall bubble columns, the hydrodynamics and the transport processes may change with height because of a change in the hydrostatic head. All of these issues need to be studied by CFD simulation.
- Study should be further continued for the prediction of bubble size distribution, effective interfacial area, gas–liquid mass transfer, transport phenomena for solid suspension, solid–liquid mass transfer, and so on in the case of the inverse bubbly fllow system aided by a plunging liquid jet by CFD.
- Moreover, studies are required by considering the problem in real industrial processes in the inverse bubbly flow system.

O

One-dimensional axial dispersion model, 316
One-dimensional integral analysis, 173
Open-open boundary conditions, 194, 205
Operating variables, 157
Overall volumetric mass transfer coefficient, correlations for, 335
Oxygen-enriched air, 312

P